Mitochondria and Microsomes

in honor of **LARS ERNSTER**

Lars Ernster

Mitochondria and Microsomes

in honor of **LARS ERNSTER**

Edited by

C. P. Lee
Wayne State University
School of Medicine

G. Schatz
Biocenter
University of Basel

G. Dallner
Department of Biochemistry
University of Stockholm

1981

ADDISON-WESLEY PUBLISHING COMPANY
Advanced Book Program/World Science Division
Reading, Massachusetts

London · Amsterdam · Don Mills, Ontario · Sydney · Tokyo

MITOCHONDRIA and MICROSOMES

The electron micrograph on the cover shows the cytoplasm of the rat liver cell, in particular the close relationship of the mitochondria with membranes of the rough endoplasmic reticulum, a special relationship apparently not involving direct physical connections. (Illustration courtesy of Dr. Ulf Brunk, University of Uppsala, Uppsala, Sweden.)

Library of Congress Cataloging in Publication Data

Main entry under title:

Mitochondria and microsomes.

 Bibliography: p.
 Includes index.
 1. Mitochondria—Addresses, essays, lectures.
2. Microsomes—Addresses, essays, lectures.
3. Ernster, L.   I. Lee, C. P. (Chuan Pu), 1931–
II. Ernster, L.   III. Schatz, G. (Gottfried)   IV. Dallner,
G. (Gustav), 1933–
QH603.M5M54   574.87'342   81-8029
ISBN 0-201-04576-1   AACR2

Reproduced by Addison-Wesley Publishing Company, Inc., Advanced Book Program/World Science Division, Reading, Massachusetts, from camera-ready copy prepared by the authors and the offices of the editors.

Manufactured in the United States of America.

ABCDEFGHIJ-HA-8987654321

CONTENTS

PREFACE

This book attempts to summarize the current status of research on mitochondria and the endoplasmic reticulum - the major membrane systems of eukaryotic cells. These membranes carry out so many different functions that the scope of this book necessarily transcends that of monograph written for the specialist. Indeed, there is no scarcity of extensive recent reviews on virtually each of the topics covered here. However, it is much more difficult to find an integrated collection of brief and up-to-date summaries on topics which are related, but not identical, to one's own research. To fill this gap is one of the major aims of this book.

An equally important aim, however, is to honor Lars Ernster who has made many fundamental contributions to our understanding of how mitochondria and the endoplasmic reticulum function. When Lars celebrated his sixtieth birthday last year, his friends and collaborators could think of no more fitting tribute than a comprehensive review of the fields he has so decisively helped to shape. His legendary talent for spotting, summarizing and evaluating new developments is a formidable standard by which each of the chapters of this book will have to be judged.

But to those of us who have been privileged to know Lars personally, he is not only a scientific challenge; his warmth, his intellectual honesty and his innate tact during difficult arguments have often helped to remind us that science is, after all, a supremely human endeavor that can bring joy and satisfaction to those who practice it. We present this book to Lars as a small sign of our admiration and affection.

During the preparation of this book the editors received valuable assistance from Ms. Doina Maria Chesches.

The Editors

LIST OF CONTRIBUTORS

Milton Adesnik, Department of Cell Biology, New York University School of Medicine, 550 First Avenue, New York, New York 10016, USA.   (pp. 563–583)

A. Alexandre, Department of Physiological Chemistry, Johns Hopkins University School of Medicine, Baltimore, Maryland 21205, USA.   (pp. 459–479)

Alain Amar-Costesec, Laboratoire de Chimie Physiologique, Université Catholique de Louvain and International Institute of Cellular and Molecular Pathology, B-1200 Brussels, Belgium.   (pp. 629–653)

Anders Åström, Department of Biochemistry, Arrhenius Laboratory, University of Stockholm, S-106 91 Stockholm, Sweden.   (pp. 585–610)

G. F. Azzone, NCR Unit for the Study of Physiology of Mitochondria and Institute of General Pathology, University of Padova, Padova, Italy.   (pp. 481–518)

Lennart Balk, Department of Biochemistry, Arrhenius Laboratory, University of Stockholm, S-106 91 Stockholm, Sweden.   (pp. 585–610)

Herrick Baltscheffsky, Department of Biochemistry, Arrhenius Laboratory, University of Stockholm, S-106 91 Stockholm, Sweden.   (pp. 519–540)

Margareta Baltscheffsky, Department of Biochemistry, Arrhenius Laboratory, University of Stockholm, S-106 91 Stockholm, Sweden.   (pp. 519–540)

Henri Beaufay, Laboratoire de Chimie Physiologique, Université Catholique de Louvain and International Institute of Cellular and Molecular Pathology, B-1200 Brussels, Belgium.   (pp. 629–653)

A. Beavis, Department of Physiological Chemistry, Johns Hopkins University School of Medicine, Baltimore, Maryland 21205, USA.   (pp. 459–479)

S. D. Black, Department of Biological Chemistry, Medical School, The University of Michigan, Ann Arbor, Michigan 48109, USA.   (pp. 707–727)

Paul D. Boyer, Department of Chemistry and Molecular
    Biology Institute, University of California, Los
    Angeles, California 90024, USA.   (pp. 407-426)

Ronald A. Butow, Department of Biochemistry, The University
    of Texas, Health Science Center at Dallas, Dallas,
    Texas, 75235, USA.   (pp. 67-91)

Barbara Cannon, The Wenner-Gren Institute, University of
    Stockholm, Norrtullsgatan 16, S-113 45 Stockholm,
    Sweden.   (pp. 93-119)

J. Capdevila, Department of Biochemistry, Southwestern
    Medical School, The University of Texas Health Science
    Center at Dallas, Dallas, Texas 75235, USA.
                                             (pp. 683-705)

Ernesto Carafoli, Laboratory of Biochemistry, Swiss Federal
    Institute of Technology (ETH), 8092 Zurich,
    Switzerland.   (pp. 357-374)

B. Chance, Johnson Research Foundation, University of
    Pennsylvania, Philadelphia, Pennsylvania 19104, USA.
                                             (pp. 271-292)

Y. Ching, Bell Laboratories, 600 Mountain Avenue, Murray
    Hill, New Jersey 07974, USA.   (pp. 271-292)

T. E. Conover, Department of Biological Chemistry, Hahnemann
    Medical College, Philadelphia, Pennsylvania 19102, USA.
                                             (pp. 481-518)

M. J. Coon, Department of Biological Chemistry, Medical
    School, The University of Michigan, Ann Arbor,
    Michigan 48109, USA.   (pp. 707-727)

Gustav Dallner, Department of Biochemistry, University of
    Stockholm, Department of Pathology, Karolinska
    Institutet, S-106 91 Stockholm, Sweden. (pp. 655-681)

P. Davies, Department of Physiological Chemistry, Johns
    Hopkins University School of Medicine, Baltimore,
    Maryland 21205, USA.   (pp. 459-479)

Christian de Duve, Laboratoire de Chimie Physiologique,
    Université Catholique de Louvain and International
    Institute of Cellular and Molecular Pathology, B-1200
    Brussels, Belgium.   (pp. 629-653)

Joseph W. DePierre, Department of Biochemistry, Arrhenius
   Laboratory, University of Stockholm, S-106 91 Stockholm,
   Sweden.   (pp. 585-610)

R. W. Estabrook, Department of Biochemistry, Southwestern
   Medical School, The University of Texas Health Science
   Center at Dallas, Dallas, Texas 75235, USA.(pp. 683-705)

Osamu Hayaishi, Department of Medical Chemistry, Kyoto
   University Faculty of Medicine, Kyoto 606, Japan.
                                                      (pp. 611-628)

Frank W. Hemming, Department of Biochemistry, Medical School,
   University of Nottingham, Nottingham, England.
                                                      (pp. 655-681)

Dean P. Jones, Department of Biochemistry, Emory University,
   Atlanta, Georgia 30322, USA.   (pp. 749-772)

M. Klingenberg, Institute for Physical Biochemistry,
   University of Munich, Goethestrasse 33, 8000 Munich 2,
   Federal Republic of Germany.   (pp. 293-316)

D. R. Koop, Department of Biological Chemistry, Medical
   School, The University of Michigan, Ann Arbor,
   Michigan 48109, USA.   (pp. 707-727)

Gert Kreibich, Department of Cell Biology, New York
   University School of Medicine, 550 First Avenue,
   New York, New York 10016, USA.   (pp. 563-583)

C. P. Lee, Department of Biochemistry, Wayne State
   University School of Medicine, Detroit, Michigan
   48201, USA.   (pp. 121-153)

A. L. Lehninger, Department of Physiological Chemistry,
   Johns Hopkins University School of Medicine, Baltimore,
   Maryland 21205, USA.   (pp. 459-479)

Olov Lindberg, The Wenner-Gren Institute, University of
   Stockholm, Norrtullsgatan 16, S-113 45 Stockholm,
   Sweden.   (pp. 93-119)

Bengt Mannervik, Department of Biochemistry, Arrhenius
   Laboratory, University of Stockholm, S-106 91
   Stockholm, Sweden.   (pp. 729-748)

B. S. S. Masters, Department of Biochemistry, Southwestern Medical School, The University of Texas Health Science Center at Dallas, Dallas, Texas 75235, USA.
(pp. 683–705)

Johan Meijer, Department of Biochemistry, Arrhenius Laboratory, University of Stockholm, S-106 91 Stockholm, Sweden.   (pp. 585–610)

Peter Mitchell, Glynn Research Institute, Bodmin, Cornwall, PL30 4AU, England.   (pp. 427–457)

E. T. Morgan, Department of Biological Chemistry, Medical School, The University of Michigan, Ann Arbor, Michigan 48109, USA.   (pp. 707–727)

Ralf Morgenstern, Department of Biochemistry, Arrhenius Laboratory, University of Stockholm, S-106 91 Stockholm, Sweden.   (pp. 585–610)

Takashi Morimoto, Department of Cell Biology, New York University School of Medicine, 550 First Avenue, New York, New York 10016, USA.   (pp. 563–583)

Jan Nedergaard, The Wenner-Gren Institute, University of Stockholm, Norrtullsgatan 16, S-113 45 Stockholm, Sweden.   (pp. 93–119)

B. Dean Nelson, Department of Biochemistry, Arrhenius Laboratory, University of Stockholm, S-106 91 Stockholm, Sweden.   (pp. 217–247)

Tomoko Ohnishi, Department of Biochemistry and Biophysics, School of Medicine, University of Pennsylvania, Philadelphia, Pennsylvania 19104, USA.   (pp. 191–216)

Sten Orrenius, Department of Forensic Medicine, Karolinska Institutet, S-104 01 Stockholm, Sweden.   (pp. 749–772)

Christian Paech, Molecular Biology Division, Veterans Administration Medical Center, San Francisco, California 94121, and Department of Biochemistry and Biophysics, University of California, San Francisco, California 94143, USA.   (pp. 155–190)

A. V. Persson, Department of Biological Chemistry, Medical
    School, The University of Michigan, Ann Arbor, Michi-
    gan 48109, USA.   (pp. 707-727)

J. A. Peterson, Department of Biochemistry, Southwestern
    Medical School, The University of Texas Health Science
    Center at Dallas, Dallas, Texas 75235, USA. (pp. 683-705)

L. Powers, Bell Laboratories, 600 Mountain Avenue, Murray
    Hill, New Jersey 07974, USA.   (pp. 271-292)

R. A. Prough, Department of Biochemistry, Southwestern
    Medical School, The University of Texas Health Science
    Center at Dallas, Dallas, Texas 75235, USA.(pp. 683-705)

Efraim Racker, Section of Biochemistry, Molecular and Cell
    Biology, Division of Biological Sciences, Cornell
    University, Ithaca, New York 14853, USA.   (pp. 337-355)

Rona R. Ramsay, Molecular Biology Division, Veterans Admin-
    istration Medical Center, San Francisco, California
    94121 and Department of Biochemistry and Biophysics,
    University of California, San Francisco, California
    94143, USA.   (pp. 155-190)

B. Reynafarje, Department of Physiological Chemistry, Johns
    Hopkins University School of Medicine, Baltimore,
    Maryland 21205, USA.   (pp. 459-479)

Jan Rydström, Department of Biochemistry, Arrhenius Labora-
    tory, University of Stockholm, S-106 91 Stockholm,
    Sweden.   (pp. 317-335)

David D. Sabatini, Department of Cell Biology, New York
    University School of Medicine, 550 First Avenue,
    New York, New York 10016, USA.   (pp. 563-583)

Gottfried Schatz, Biocenter, University of Basel, CH-4056
    Basel, Switzerland.   (pp. 45-66)

Janeric Seidegård, Department of Biochemistry, Arrhenius
    Laboratory, University of Stockholm, S-106 91
    Stockholm, Sweden.   (pp. 585-610)

Philip Siekevitz, Rockefeller University, New York, New York
    10021, USA.   (pp. 541-562)

Thomas P. Singer, Molecular Biology Division, Veterans Administration Medical Center, San Francisco, California 94121, and Department of Biochemistry and Biophysics, University of California, San Francisco, California 94143, USA.   (pp. 155-190)

Vladimir P. Skulachev, Department of Bioenergetics, Laboratory of Molecular Biology and Bioorganic Chemistry, Moscow State University, Moscow 117234, USSR.
(pp. 375-405)

E. C. Slater, Laboratory of Biochemistry, B.C.P. Jansen Institute, University of Amsterdam, TV 1018 Amsterdam, The Netherlands.   (pp. 1-3, 15-43)

B. T. Storey, Henry Watts Neuromuscular Disease Research Center and Department of Physiology, Obstetrics and Gynecology, University of Pennsylvania School of Medicine, Philadelphia, Pennsylvania 19104, USA.
(pp. 121-153)

Robert L. Strausberg, Department of Biology, Southern Methodist University, Dallas, Texas 75275, USA.
(pp. 67-91)

A. Villalobo, Department of Physiological Chemistry, Johns Hopkins University School of Medicine, Baltimore, Maryland 21205, USA.   (pp. 459-479)

J. Werringloer, Department of Biochemistry, Southwestern Medical School, The University of Texas Health Science Center at Dallas, Dallas, Texas 75235, USA.
(pp. 683-705)

Mårten Wikström, Department of Medical Chemistry, University of Helsinki, SF-00170 Helsinki 17, Finland.
(pp. 249-269)

Ryotaro Yoshida, Department of Medical Chemistry, Kyoto University Faculty of Medicine, Kyoto 606, Japan.
(pp. 611-628)

LARS ERNSTER

As seen by a friend, colleague, fellow-European and competitor

E.C. Slater

Laboratory of Biochemistry, B.C.P. Jansen Institute,
University of Amsterdam, The Netherlands

Lars was born in Budapest on May 4, 1920 and moved to Sweden with his wife Edit shortly after the war. He obtained his University education at the University of Stockholm, receiving his Licentiat of Philosophy (in animal physiology) in 1954 and his Doctor of Philosophy in 1956. In the same year he was appointed "Docent" in experimental zoology and cell research and in 1962 in physiological chemistry. He also received a Licentiat of Philosophy in chemistry (biochemistry) from the University of Lund in 1962.

From 1953 to 1967 Lars worked at the Wenner-Gren Institute for Experimental Biology of the University of Stockholm, first as Research Associate, then as Assistant Professor and from 1963 as Associate Professor of the Swedish Natural-Science Research Council. He was head of the Department of Physiological Chemistry of the Wenner-Gren Institute from 1957 to 1967. In 1967 he was appointed Professor and Chairman of the Department of Biochemistry of the University of Stockholm, a position that he still holds.

C. P. Lee, G. Schatz, G. Dallner (eds.), Mitochondria and Microsomes
in honor of Lars Ernster   ISBN 0-201-04576-1

1

Lars tells me that we first met on the lawn outside the Molteno Institute, University of Cambridge, during the First International Congress of Biochemistry in the summer of 1949. I do not remember this, but Lars is 3 years younger than I, so has the better memory. I do remember meeting him and his teacher Olov Lindberg in a café at the corner of Boulevard St. Michel and Rue Soufflot during the Second International Congress of Biochemistry in Paris three years later, and in his laboratory in Stockholm in the following year. I have a vivid recollection of a very nervous young man whose mastery of the English language can best be described as much better than my Hungarian. Nowadays, when I am listening to Lars giving a beautifully clear, logical and elegant talk in the most impeccable English, I sometimes ask myself if this is really that terribly nervous, stuttering young man I talked with in Paris and Stockholm in the early fifties. But of course, it is. The burning interest in science, the desire to get to the truth of the matter, the intense but courteous questioning and, above all, his charming and warm smile were all present in Lars model 1952-3.

Hans Krebs once wrote a genealogy of Nobel Prize winners in which the genetic parents are replaced by teachers. One could do the same for the fields in which a research scientist works. This is usually related to, or has logically evolved from, the research interests of his teacher. So it is with Lars. Olov Lindberg was a student of J. Runnström, an authority on the biochemistry of the sea-urchin egg, a favourite study object of the animal physiologist, and one of the early workers on oxidative phosphorylation. Lars' first paper (1) published in 1948, in, I am happy to relate, Volume 2 of *Biochimica et Biophysica Acta* "On carbohydrate metabolism

in homogenised sea urchin eggs" followed this tradition. In
1950, four papers appeared in as many different journals,
three dealing with phosphate turnover. The one published in
Acta Chemica Scandinavica (2) on "A method for the determination of tracer phosphate in biological materials", of which
Lars is the first author, is a widely quoted article.

The interaction between Lars' work and my own, which
has always been very strong, started in 1952 with the appearance in *Experimental Cell Research* of a long paper (3) by
Olov and Lars, entitled "On the mechanism of phosphorylative
energy transfer in mitochondria" - not the last time that Lars
has published a paper with a similar title - and with my work
with Ken Cleland on the swelling of mitochondria brought about
either by hypotonicity or by uptake of calcium. Olov and Lars
were then writing a book entitled 'Chemistry and Physiology
of Mitochondria and Microsomes' (4) that eventually appeared
in 1954 as part of *Protoplasmatologia, Handbuch der Protoplas-
maforschung*, in which our work is given some prominence.
Indeed, it includes a Table entitled *Chemical constitution of
cat heart sarcosomes prepared in sucrose medium containing
versene*, reported unpublished analytical values by Cleland.
Since I am still waiting for Ken Cleland to get around to
publishing this work, and I believe that the data are still
valid, this is the opportunity of pointing out two errors.
First, that the work was carried out on rat not cat sarcosomes
(mitochondria) - this error I ascribe to Ken's minute hand-
writing. Secondly, that the value given for calcium (0.099
μmol/mg protein) refers to mitochondria isolated in the absence of versene(EDTA) which prevents the uptake of calcium
from extramitochondrial and extracellular fluid during isolation of the mitochondria. I might also draw attention to a

little-known fact that Mg^{2+} (0.29 μmol/mg protein) must contribute very largely to the osmotic activity of the matrix.

Lars and I also had our first scientific disagreements
around about this time. Using mitochondria that had very little
adenylate kinase (from cat - yes cat - heart), Francis Holton and
I believed that we had shown quite conclusively that ADP and
not AMP is the sole substrate of oxidative phosphorylation and
that phosphorylation of AMP proceeds via adenylate kinase.
Lars and Olov, on the other hand, believed that AMP can be
directly phosphorylated although at only 5% of the rate of ADP.
This question of the adenine nucleotide specificity of oxidative phosphorylation has turned out to be a hardy perennial,
breaking into bloom every five years or so. Although the later
discovery by Martin Klingenberg of the adenine nucleotide
translocator and of its specificity for ADP re-opened the
question, there is, in my opinion, still no reason to doubt
that neither the translocator nor the phosphorylation machinery
reacts with AMP.

The paper "Manganese, a co-factor of oxidative phosphorylation" (5) published in Nature in 1954 worried me a good
deal since Olov and Lars suggested that our EDTA treatment removed manganese present in mitochondria. Although I confirmed
their main (and to me still unexplained) observation on which
this conclusion was based, namely that Mn^{2+} prevents Ca^{2+}-
induced inactivation of respiration, I could find no evidence
that EDTA removed an essential component of oxidative phosphorylation, and put this down to the first of the "red herrings"
that were to strew both our paths in the coming years.

From early days, specific inhibitors have been an
essential tool in studying the mechanism of biological oxidations and oxidative phosphorylation. In 1955, cyanide was used

routinely to inhibit the terminal oxidation step, antimycin to inhibit between cytochromes b and c (c_1), malonate to inhibit succinate dehydrogenase and 2,4-dinitrophenol to uncouple oxidative phosphorylation. There was, however, no specific inhibitor of NADH oxidation that left succinate oxidation intact. This was particularly necessary in order to study succinate oxidation by mitochondria, uncomplicated by oxidation of the malate formed. In 1955, Lars showed that the barbiturate, Amytal, has this property (6). Amytal immediately joined the arsenal of inhibitors in Amsterdam where we paid little attention to Ef Racker's stricture "only the uninhibited use inhibitors". The study of the Amytal- (and antimycin-)resistant pathway for NADH oxidation (7) led Lars towards his interest in the outer-mitochondrial membrane and the microsomes where this pathway resides. Later the much more active and specific rotenone (8) replaced Amytal, and barbiturates became the object of another line of research.

In the middle fifties the first of some well-known names of 'post-doc' visitors to the Wenner-Gren Institute start to appear as Lars' co-authors - R.E. Beyer, P.D. Boyer and P. Siekevitz. Indeed, a stream of papers, on quite different subjects, was now emerging from Lars' group. Like most of us at that time, he believed that the study of the mechanism of uncoupling could give an important clue to the coupling mechanism. This research led into profitable and often unexpected side-ways, such as the discovery, together with R. Luft, of an inborn error of metabolism, in which the oxidative phosphorylation in the skeletal mitochondria is uncoupled (9) and, with T.E. Conover, of a dicoumarol-sensitive NAD(P)H diaphorase (10,11).

Related to the latter topic was the study of the nico-

tinamide nucleotide specificity of mitochondrial isocitrate
dehydrogenase, which was the subject of a controversy with
Jack Purvis in our laboratory. This research led naturally to
the examination of the nicotinamide nucleotide transhydrogenase and to the discovery of the energy-linked transhydrogenase (12), which I consider one of Lars' most important discoveries.

We are now dealing with the heroic pre-Mitchellian era
in the study of bioenergetics during which great upheavals and
quick subsidences followed each other in rapid succession. The
arrival of G.F. Azzone in Stockholm heralded one of these upheavals, when it was suggested that NAD^+ is on the normal
pathway of oxidation of succinate to fumarate (13) - a revival,
but in the reverse, of the Szent-György proposal from the
thirties that the succinate-fumarate system is involved in the
oxidation of NADH. Fortunately for my blood pressure, this up-
heaval subsided fairly quickly after a hectic informal discussion in the Wenner-Gren Institute in the summer of 1960, immediately following the Symposium on Biological Structure and
Function, organized by Lars. I still think that Lars, like
many before and after him, was led astray by another hardy
perennial - inhibition of succinate dehydrogenase by oxaloacetate.

This misstep led Lars, however, into an intensive
study of the mechanism of a reaction earlier studied by
Britton Chance, namely the *energy-requiring* reduction of NAD^+
by succinate. In 1960-1, there developed an intense competition
between Lars, who used added acetoacetate or 2-oxoglutarate
(+ NH_3) as the acceptor and our own group who measured directly
the reduction of mitochondrial nicotinamide nucleotide as well
as reduction of added 2-oxoglutarate + NH_3. The competition

was essentially a question of who did what first rather than
scientific differences between two groups, since they reached
essentially the same conclusions independently. The most important conclusion was reported and discussed extensively at
the Fifth International Congress of Biochemistry in Moscow in
1961, namely that oligomycin does not inhibit the utilization
of the energy liberated in the oxidation of succinate by oxygen
for the energy-requiring reduction of NAD^+ by succinate. Also
it finally became generally accepted that phosphate is not
required for that reaction. This was the final proof that
energy from respiration may be conserved and utilized in mitochondria without the synthesis of ATP or even the intervention
of inorganic phosphate. That both Lars and I considered that
the energy is conserved in the form of a high-energy intermediate of oxidative phosphorylation, my $A{\sim}C$ or as Lars believed
$BH_2{\sim}C$ where BH_2 is reduced flavoprotein (14), in the year of
birth of the chemiosmotic hypothesis is one of the ironies of
scientific history.

The Congress in Moscow in 1961 marked Lars' emergence as
one of the leading actors on the bioenergetics stage (or
should one say in the travelling bioenergetics circus). It is
hard to think of a symposium in the next 20 years in which
Lars has not given in a major talk a clear and objective
survey of the field.

The appearance of a number of papers in 1964 dealing
with oxidative reactions catalysed by microsomes heralded the
opening of another major field of research in Lars' laboratory
which were later to be developed extensively by G. Dallner and
S. Orrenius. The enormous increase of oxidative demethylating
enzymes in the microsomes following administration of pheno-
barbitol is especially noteworthy (15).

The year 1964 also marks an especially productive period of co-operation with C.P. Lee. From this work emerged
what I would consider to be the fourth major contribution by
Lars to mitochondrial research (the others being the use of
Amytal and rotenone, the discovery of the energy-linked nicotinamide transhydrogenase and bypassing of ATP in the utilization of respiratory energy) - namely the introduction of
carefully characterized submitochondrial particles that still
retain the ability to carry out oxidative phosphorylation but
are free from the barriers to substrates that cause difficulties in many types of experiments with intact mitochondria.
In an important lecture at the first Bari symposium (16) in
1965, C.P. and Lars proposed that the inner mitochondrial membrane is turned "inside-out" during the preparation of the
submitochondrial particles, so that the inaccessible "inside"
of the membrane becomes exposed to the suspension medium. A
very important finding also reported in this lecture was the
stimulation of oxidative phosphorylation by low concentrations
of oligomycin. Later work showed that oligomycin can replace
F_1 in restoring oxidative phosphorylation to particles from
which much F_1 had been stripped. These findings were first
interpreted in terms of chemical hypothesis, but later were
seen to provide important support for Mitchell's chemiosmotic
theory. It is now generally believed that oligomycin blocks
the proton channel in F_o. In intact systems, it prevents the
passage of protons through the channel to F_1, where phosphorylation of ADP takes place. In F_1-deficient preparations it
blocks the back leak of protons through proton channels in F_o
that are exposed by removal of F_1, thereby facilitating the
full utilization of the protonmotive force for the synthesis
of ATP on the still intact $F_o \cdot F_1$ assemblies.

During this period, too, there was much interaction between the work in the Stockholm and Amsterdam laboratories. To my lasting regret, visa difficulties prevented a planned stay of C.P. Lee in the Amsterdam laboratory. We were more fortunate with a name-sake of C.P., namely In-Young, who after two years in Stockholm and with an intervening year in Martin Klingenberg's laboratory joined us for three years. In-Young brought to Amsterdam the highly coupled preparations of mito-chondria and submitochondrial particles then used in Stockholm. The paper by Lars, In-Young and others on "Interactions of ubiquinone and cytochrome b in the respiratory chain" and published in one of the Johnson Foundation Symposia (17) is still very close to current activities in our laboratory. The demonstration, in collaboration with Lars Ehrenberg (18), of the ubisemiquinone radical in submitochondrial particles has also turned out to be very important.

We confirmed and extended work done in Stockholm and sometimes disagreed but usually agreed. Maybe we were sometimes first. One finding from this period that I have always consi-dered important was the demonstration by Lars (19) and by Rob van de Stadt and Karel van Dam in Amsterdam of the dissociation, induced by coupled electron transfer in the respiratory chain, of Maynard Pullman's ATPase inhibitor from the ATPase F_1. Another investigation in which both the Stockholm and Amsterdam groups put much effort was trying to obtain a conclusive answer to the question whether there is a diffusible intermediate between electron transfer and phosphorylation (20).

As exchange for In-Young we sent Jan Hoek. In an im-portant paper (21), published jointly from the two laboratories, Jan and Lars, together with Ed de Haan and Joseph Tager, pro-vided the long-sought explanation for a phenomenon that had given headaches to Piet Borst, Joseph Tager, Ed de Haan, Rob Charles and Jan - not to mention myself - namely why is so little glutamate oxidized by glutamate dehydrogenase in isolated mitochondria. The answer is that it is, but the products are immediately reduced again to glutamate via the $NADPH/NADP^+$ couple which is kept highly reduced by the operation of Lars' energy-linked nicotinamide transhydrogenase system. The net result is a futile cycle driven by energy supplied by the oxidation of glutamate to aspartate.

Mention has already been made to Lars' performances in the travelling bioenergetics circuses. He is also a prolific writer of reviews, starting with the book with Olov Lindberg written while he was still a student (4). In this connection, I would like to mention his contributions to the various "Bari symposia". Indeed, he was the first to join and to succeed the Amsterdam laboratory in collaborating with our Italian colleagues in organizing these meetings. His first contribution to Annual Reviews in Biochemistry, written with C.P. Lee, appeared in 1964 (22). There are two more (23,24). Is this a record? Most of us who have once managed a review for Annual Reviews take a pledge never to do it again.

The joint review written together with Paul Boyer, Britton Chance, Peter Mitchell, Ef Racker and myself (24) is, I think, worthy of particular mention, since I believe that this review, written largely on the initiative of Lars, did much to signal to those outside the field that a broad consensus, based on the Mitchell hypothesis, had been reached concerning the main features of the mechanism of oxidative phosphorylation.

Lars has deservedly received many honours and distinctions, including an Honorary Doctor of Medicine of the Karo-

linska Institute, membership of the Royal Swedish Academy of
Sciences and Honorary Membership of the American Society of
Biological Chemists. Naturally, he is a much sought-after mem-
ber of Editorial Boards and for Visiting Professorships. Among
his functions might be mentioned that of Secretary of the
Swedish National Committee on Biochemistry, Chairman of the
Organizing Committee of the Ninth International Congress of
Biochemistry and Member of the Nobel Committee for Chemistry.

The writer has in front of him Lars' publication list
which was up-to-date a few months ago. It has 385 entries, and
no doubt the 400 mark is now approaching. It is impossible in
this chapter to give adequate attention to more than a few of
these. The selection that has been made is to some extent arbitrary being based as it is on personal interests and a by no
means infallible memory. A balanced account would require re-
reading and in some cases reading for the first time all 385
publications, which is beyond the writer's capacities or possibilities, if this tribute to Lars on the occasion of his 60th
birthday is to be ready before his 65th.

Looking back over the last 30-odd years, I am once
again struck by how closely Lars' approach and the conclusions
that he reached at a given period resemble those of our own.
It is not just because we belong to the same generation. The
leaders of the other groups that come to mind when I make this
comparison are only a little older or, in some cases, younger
than the writer, and Lars is only 3 years younger - an important difference perhaps in 1949 and 1952 - but no longer significant.

Lars' upbringing and mine are quite different - being
separated geographically by almost half the world's circumfe-

rence. There is the resemblance that we both left our homeland at about the same time, but for quite different reasons. Perhaps a more significant resemblance is that, although both of us became heads of large Biochemical Institutes, belonging (at least in part) to Faculties of Chemistry, we received our pre- and post-doctoral training in laboratories deeply entrenched in the European biological tradition.

REFERENCES

1. Lindberg, O. and Ernster, L. (1948) Biochim. Biophys. Acta 2, 471-477.
2. Ernster, L., Zetterström, R. and Lindberg, O. (1950) Acta Chem. Scan. 4, 942-947.
3. Lindberg, O. and Ernster, L. (1952) Exptl. Cell Research 3, 209-239.
4. Lindberg, O. and Ernster, L. (1954) Chemistry and Physiology of Mitochondria, Vol. III A4, Springer Verlag, Vienna.
5. Lindberg, O. and Ernster, L. (1954) Nature 173, 1038.
6. Ernster, L., Jalling, O., Löw, H. and Lindberg, O. (1955) Exptl. Cell Research, Suppl. 3, 124-132.
7. Ernster, L. (1956) Exptl. Cell Research 10, 721-732.
8. Ernster, L., Dallner, G. and Azzone, G.F. (1963) Biochem. Biophys. Res. Commun. 10, 23-27.
9. Luft, R., Ikkos, D., Palmieri, G., Ernster, L. and Afzelius, B. (1962) J. Clin. Invest. 41, 1776-1804.
10. Ernster, L., Danielson, L. and Ljunggren, M. (1962) Biochim. Biophys. Acta 58, 171-188.
11. Conover, T.E. and Ernster, L. (1962) Biochim. Biophys. Acta 58, 189-200.
12. Danielson, L. and Ernster, L. (1963) Biochem. Z. 338, 188-205.
13. Azzone, G.F. and Ernster, L. (1960) Nature 187, 65-67.
14. Ernster, L. (1963) Proc. Vth Intern. Congr. of Biochemistry, Moscow, 1961, pp. 115-145, Pergamon Press, Oxford.
15. Orrenius, S. and Ernster, L. (1964) Biochem. Biophys. Res. Commun. 16, 60-65.
16. Lee, C.P. and Ernster, L. (1966) Symp. on Regulation of Metabolic Processes in Mitochondria, Bari, 1965. BBA Library 7, 218-236.
17. Ernster, L., Lee, I.-Y., Norling, B., Persson, B., Juntti, K. and Torndal, U.-B. (1971) in Probes of Structure and Function of Macromolecules and Membranes, Vol. 1, Probes

and Membrane Function, p. 377, Academic Press, New York.

18. Bäckström, D., Norling, B. Ehrenberg, A. and Ernster, L. (1970) Biochim. Biophys. Acta 197, 108-111.

19. Juntti, K., Asami, K. and Ernster, L. (1971) Abstracts 7th FEBS Meeting, Varna, p. 243.

20. Ernster, L. (1975) Proc. 10th FEBS Meeting 40, 253-276.

21. Hoek, J.B., Ernster, L., De Haan, E.J. and Tager, J.M. (1974) Biochim. Biophys. Acta 333, 546-559.

22. Ernster, L. and Lee, C.P. (1964) Ann. Rev. Biochem. 33, 729-788.

23. DePierre, J.W. and Ernster, L. (1977) Ann. Rev. Biochem. 46, 201-262.

24. Ernster, L. (1977) Ann. Rev. Biochem. 46, 981-995.

Part I
Mitochondria

A SHORT HISTORY OF THE BIOCHEMISTRY OF MITOCHONDRIA

E.C. Slater

Laboratory of Biochemistry, B.C.P. Jansen Institute,
University of Amsterdam, The Netherlands

INTRODUCTION

The word history means different things to different
people. To Henry Ford it was "bunk". In my school, "history"
without any qualifying adjective, meant the History of England
and it ended in 1901 with the death of Queen Victoria, despite
the fact that this was also the year of birth of my country.

In the Introduction to his book 'The history of cell
respiration and cytochrome' (1) my teacher David Keilin pointed out how arbitrary and difficult it is to fix the date of
the beginning of the history of a subject. I find it equally
difficult to fix its end. I have no difficulty in considering
as belonging to history events that took place before I became
actively involved in science. I have slightly more difficulty
in putting events in the 1940's and 1950's also into this category, but daily contact with people beginning their research
career who were born in the late fifties brings the realization

C. P. Lee, G. Schatz, G. Dallner (eds.), Mitochondria and Microsomes
in honor of Lars Ernster   ISBN 0-201-04576-1

15

that they have other ideas about what (and who!) belong to
history. The gentle comment by my co-author concerning the
opening sentence of a recently submitted paper - 'In 1948, one
of us published a paper .....' - that he did not then exist,
speaks for itself.

Most will agree that papers published in 1980, even if
the author considers them to be of historical importance, are
not yet history. History as a discipline is more than drawing
up a catalogue. The historian seeks to trace trends in the
development of his subject and to give the correct significance
to a historical event. Even in such a rapidly developing field
as biochemistry, some time must elapse before this is possible.

When, then, should I bring my short story to a close?
I can hardly stop before 1961, a 'historical year' in bioenergetics. Surely, at least 10 years is necessary to allow even
a preliminary assessment of the significance of historical
events. Thus, I have to choose between 1961 and 1970. In principle, I have included the entire decade, but I may well have
given insufficient attention to important discoveries made
towards its end. This can be corrected in a subsequent history
and by other historians.

A biochemist is not capable of writing a professional
history any more than a historian is capable of doing biochemical research, although the latter fact is more generally
recognized than the former. There are particular and clear
dangers when the amateur historian-professional biochemist has
himself participated in the events described. All he can do is
to recognize these dangers and to minimize bias, one way or
the other, by being as objective as possible, insofar that this
is possible when the subject is himself. It is for others to
assess whether he has been successful.

OBSERVATIONS OF MITOCHONDRIA IN THE CELL

What are now known as mitochondria were first described in striated muscle by Henle in his book "Allgemeine Anatomie" (2) published in 1841. The unusually large and abundant granules in insect muscle were first described by Aubert in 1853 (3). Kölliker (4,5) made a thorough study of these "interstitial granules" of muscle and in the second of these papers, published in 1888, he described the first experiment in what we may call the experimental cytology of mitochondria. He was able to separate the granules by teasing the fibres of insect muscle and to study the effect of various treatments on their appearance under the microscope. He observed that when these granules are immersed in water, they swell greatly and are transformed to vesicular-shaped bodies with a clearly defined membrane. The contents of these granules take the form of a half-moon lying at one side of the vesicle.

Shortly afterwards, Retzius (6) described these granules in the following words:

> They exist in many animals in different positions and arrangements – frequently in longitudinal rows, as has been shown by Kölliker – and they constitute, as the latter has pointed out, 'Körperchen sui generis'; they are not fat granules, as several authors have maintained, even recently. Because of their pronounced and characteristic properties (resistance to reagents, specific staining properties, etc.) and also in order to distinguish them clearly from pathologically formed fat granules ..... I have named the normal interstitial or more correctly, intercolumnar, Körperchen, as 'sarcosomes'.

Between 1909 and 1916, Regaud (7), Holmgren (8) and Bullard (9,10) showed that the granules are more abundant in muscles required for sustained activity, such as the flight

muscles of birds and insects and the heart muscle of vertebrates.

In the meantime granules had also been observed in
tissues other than muscle. In 1890 Altmann published his book
'Die Elementarorganismen und ihre Beziehungen zu den Zellen'
(11). Benda (12) introduced the name mitochondrion (from mitos,
a thread, and chondros, a grain) to describe these rod-like
granules, and in 1909 Regaud (7) concluded on the basis of
staining reactions that sarcosomes are the mitochondria of
muscle.

IDENTIFICATION OF MITOCHONDRIA AS SITE OF INTRACELLULAR
RESPIRATION

A suspension of small particles isolated by grinding
heart muscle, precipitating the extract with acid and resus-
pending the precipitate was used as early as 1912 by Battelli
and Stern (13) for their studies of succinate oxidase and indophenol oxidase. Battelli and Stern proposed that the 'Haupt-
atmung' of the cell takes place on these particles. A similar
heart-muscle preparation was used by Keilin and his school
since about 1925 for the study of the cytochrome system (1).
Although Keilin and Hartree (14) realized that these particles
contain a delicately organized complex enzyme system, which
must be derived from some pre-existing structure in the intact
muscle, no attempt was made to determine its origin in the
cell .

It was much easier to separate the respiratory granules
from ground muscle than from other tissue, since the large mass
of fibres in which the nuclei are embedded is sedimented on

standing, leaving the suspension of what we would now call
submitochondrial particles in the supernatant. Differential
centrifugation is necessary to separate components of the
cytoplasm in other cells, such as the liver, and centrifuges
of sufficient speed did not become generally available until
the 1940's. However, already in 1913, Warburg (15) succeeded
by centrifugation in concentrating the large granules to a
considerable extent and demonstrated that most of the respiration of the cell was in the fraction containing these
granules.

It was not until 1946, however, that, thanks to the
work of Claude and his associates, isolated mitochondria became widely used in biochemistry. Hogeboom, Claude and Hotchkiss (16) showed that isolated mitochondria contain all the
cytochrome oxidase and succinate oxidase. Shortly afterwards,
Lehninger and Kennedy showed that all the reactions involved
in the oxidation of pyruvate, fatty acids and oxidative phosphorylation also take place in isolated liver mitochondria
(17,18).

Slater (19) showed in 1950 that granules isolated from
heart muscle by following a procedure based on that of Claude
and colleagues oxidize Krebs-cycle intermediates with concomitant phosphorylation and these granules were identified as
sarcosomes in 1953 (20,21). At about the same time, Harman and
Feigelson (22) showed that similar granules separated from
heart "cyclophorase preparations" possess respiratory activity.
Watanabe and Williams (23) obtained similar results with blow-
fly thoracic muscle and Chappell and Perry (24) and Harman and
Osborne (25) with pigeon-breast muscle. Thus, by 1953 it had
been established on the basis of biochemical criteria that
sarcosomes have the function of mitochondria in muscle.

At about this time Palade (26) and Sjöstrand (27) showed by thin-section electron microscopy that the substructures of mitochondria from liver and muscle mitochondria are very similar, and that the greater respiratory activity of muscle mitochondria is correlated with a larger amount of internal membranes (26).

Cleland and Slater (20,21) forged the final link with the earlier work of Battelli and Stern (13) and of the Keilin school by showing that the Keilin and Hartree heart-muscle preparation is formed by the disintegration of the membrane in swollen mitochondria. The respiratory chain (systems oxidizing succinate and NADH) were shown to be present in this membrane, whereas many dehydrogenases required for the operation of the Krebs cycle are dissolved and remain in the supernatant after sedimentation of the membranes.

RECOGNITION OF MITOCHONDRIA AS AN OSMOTIC SYSTEM

As mentioned above, already in 1888 Kölliker (5) observed the swelling and other structural changes that take place when mitochondria of insect thoracic muscle are placed in water. He clearly described a boundary membrane in swollen mitochondria. Sixty-five years later, Kölliker's findings were confirmed and extended to heart-muscle mitochondria (20,21).

Swelling of isolated liver mitochondria (28) or of mitochondria in situ (29) gave indications of the presence of an osmotic barrier but the boundary membrane was not clearly seen in these mitochondria. Indeed, in 1950, Harman (30) denied the existence of a membrane. In 1952-3, however, the question was clearly settled by thin-section electron microscopy which showed the presence of two membrane systems - an outer membrane and a highly convoluted inner membrane system (the cristae).

In 1952, Cleland (31) and Raaflaub (32) introduced the light-scattering technique to measure swelling of the mitochondria and the permeability of the membrane to different solutes. Cleland showed that mitochondria suspended in isotonic potassium succinate do not swell, despite the fact that succinate is oxidized. Phosphate, on the other hand, rapidly penetrates the mitochondria. Ca^{2+} was also shown to be rapidly taken up into the mitochondrion (33).

Thus, by the early 1950's it was established that the respiratory chain is located in a membrane with selective permeability to certain anions and cations. The importance of this was generally recognized. It was, for example, given prominence in Lindberg and Ernster's book 'Chemistry and Physiology of Mitochondria and Microsomes' (34) that was published at this time. For the next decade, this membrane was seen mostly as a barrier and also sometimes as a suitable hydrophobic milieu in which the reactions of oxidative phosphorylation could take place without subjecting hypothetical water-sensitive intermediates to hydrolysis.

In 1961, Mitchell (35) suggested a physiological function for this barrier, namely to separate protons liberated to one side of the membrane by the operation of the respiratory chain and consumed on the other side. Specific translocators for substrate anions, exchanging across the membrane with OH^- or with other substrate anions, were discovered in 1964 by Chappell (36), using the light-scattering method to follow mitochondrial swelling. Another type of porter, now established to be an electrogenic ATP^{4-}-ADP^{3-} translocator, was discovered by Klingenberg (37).

THE DISCOVERY OF THE COMPONENTS OF THE RESPIRATORY

CHAIN

As already mentioned, in their early studies Battelli
and Stern (13) and Keilin (1) made use of enzyme preparations
derived by fragmentation of the membrane of heart mitochondria,
although they could not at the time establish the origin of
the particles in these prepartions. The controversy between
Wieland and Thunberg, who believed that biological oxidations
occur by virtue of activation by specific *dehydrogenases* of
hydrogen atoms in oxidizable substrates and Warburg, who recognized a rôle only for an oxygen-activating enzyme, *Atmungs-
ferment*, was resolved in 1925 by Keilin (38) who showed that
both dehydrogenases and an O_2-reducing enzyme are required,
the former to reduce the cytochrome system and the latter to
oxidize it. At the same time, the concept of a respiratory
chain was born, but the term was not then used.

Although Keilin's discovery resolved the controversy
between Wieland and Warburg, it opened up a new one, this time
between Keilin and Warburg, concerning the nature of the O_2-
reducing enzyme and, perhaps even more, the nomenclature. In
1925 (38) Keilin observed that the oxidation of cytochrome is
sensitive to cyanide, but did not relate this to Warburgs's
theory of the central importance of cyanide-sensitive, iron-
containing Atmungsferment. In 1926, Warburg (39) showed that
cellular respiration is sensitive, light reversibly, to CO.
In 1927, Keilin (40) showed that indophenol oxidase is also
inhibited by CO and identified Warburg's Atmungsferment with
indophenol oxidase. In 1930, Keilin (41) showed that the oxidation of isolated cytochrome *c*, catalysed by heart-muscle
preparations, is sensitive to CO as well as to cyanide and,

when in 1938 Keilin and Hartree (42) found that cytochrome c is also involved in the indophenol oxidase reaction they re-named the cyanide- and CO-sensitive oxygen-reducing enzyme as cytochrome oxidase. Finally, in 1939 Keilin and Hartree (43) showed that what had hitherto been considered a single cyto-chrome (a) consisted of two components, one of which (a_3) has all the properties expected of cytochrome oxidase or Atmungs-ferment. In the meantime, Ball (44) had established from measurements of the redox potentials that the three cytochromes of Keilin (38) react in the order of $b \rightarrow c \rightarrow a$ so that it was now possible to describe the respiratory chain for the oxida-tion of succinate by the scheme

$$\text{succinate} \rightarrow b \rightarrow c \rightarrow a \rightarrow a_3 \rightarrow O_2$$

succinate dehydrogenase

It had also been demonstrated, largely by the work of Warburg, that the oxidation of many substrates such as malate requires the additional participation of nicotinamide-adenine nucleo-tides (NAD^+ and NADP) and flavoproteins.

Since 1939, many components have been added to the respiratory chain. In chronological order:

1. In 1940, Yakushiji and Okunuki (45) showed that an addi-tional c-type cytochrome is present acting on the substrate side of the hitherto studied cytochrome c

$$\rightarrow c_1 \rightarrow c \rightarrow$$

This work was not generally accepted until it was confirmed by Keilin and Hartree (46) in 1955.

2. In 1948, Slater (47) found that treatment of heart-muscle particles with BAL (2,3-dimercaptopropanol) in the presence

of O_2 irreversibly destroyed a factor supposed to be necessary for electron transfer between cytochromes b and c.

$\rightarrow b \rightarrow$ factor $\rightarrow c \rightarrow$

3. In 1951, Tsou (48) established that succinate dehydrogenase itself functions as an electron carrier.

succinate $\rightarrow$ succinate dehydrogenase $\rightarrow$

4. In 1951, Gunsalus (49) discovered that the low-potential substrates pyruvate and 2-oxoglutarate require lipoic acid for the reduction of NAD^+.

dehydrogenase

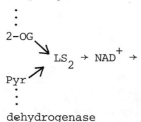

dehydrogenase

5. In 1956, Wang and Tsou (50) and Singer (51) established that succinate dehydrogenase is a flavoprotein.

6. In 1957, Crane et al. (52), working in Green's laboratory, discovered that a quinone, ubiquinone, is involved. Soon afterwards, it was proposed that it functions between flavo-protein and cytochrome b.

$\rightarrow$ fp $\rightarrow$ Q $\rightarrow b$

7. In 1958, Massey (53) discovered that what had previously been thought to be the enzyme responsible for oxidizing NADH in the respiratory chain (54) has, in fact, the function of transferring electrons from reduced lipoic acid to NAD^+.

$\rightarrow LS_2 \rightarrow$ fp $\rightarrow NAD^+$

8. In 1958, Chance (55) published the first observations on the effect of antimycin on cytochrome b which, after many years of confusion and controversy (see ref. 56), finally led to the acceptance that there are at least two b cytochromes acting in the order

$$b\text{-566} \rightarrow b\text{-562}$$

9. In 1960-2, Takemori (57) and Beinert (58) finally established previous suggestions (in particular by Wainio (59)) that cytochrome oxidase contains copper as an integral component.

10. In 1960, Beinert (60) discovered a new class of iron-containing components in the respiratory chain, most of which are associated with flavoprotein. The chemical structure of the active centres of these proteins, now called Fe-S proteins, was largely established before the end of the decade.

11. In 1964, Rieske (61) discovered a Fe-S protein which differs from the others present in the respiratory chain in that it is not associated with a flavoprotein, but with electron transfer in the bc_1 region.

The integration of all this knowledge into an acceptable model of the respiratory chain belongs to the present and the future rather than the past. If the writer may be permitted to transgress his self-imposed limitations, it might be added that the BAL-labile factor of 1948 has now been identified with the Rieske Fe-S protein (62).

THE NATURE OF ENERGY CONSERVATION

The first demonstration of the synthesis of ATP coupled to respiration was made in 1930 by Engelhardt (63) in erythro-

cytes. Rabbit erythrocytes which do not respire were found to
form inorganic phosphate when the glycolysis was inhibited by
fluoride, whereas the respiratory inhibitor cyanide had no
effect. Pigeon erythrocytes, on the other hand, rapidly formed
phosphate when respiratory inhibitors (cyanide, urethane or CO)
were added. The amount of P_i formed agreed closely with the
disappearance of "7-min P" (ATP): after removal of the cyanide
by washing, a resynthesis of acid-labile P could be demonstrated. Engelhardt concluded that the splitting of ATP occurs
both in the absence and in the presence of respiratory inhibitors, but that in its absence the splitting is compensated
by a resynthesis of ATP.

Engelhardt's work seems to have had relatively little impact when it was published, and interest in oxidative phosphorylation, as the process came to be called, stems from the
work of Kalckar in 1937 (64), who showed that phosphorylation
of glucose, glycerol or AMP takes place when cell-free homogenates of kidney and other tissues catalyse the oxidation of
citrate, glutamate, fumarate or malate. Since no reaction takes
place under anaerobic conditions and no phosphoglycerate is
formed, it was concluded that glycolysis is not responsible
for the phosphorylation. In 1939, Lipmann (65) showed that the
oxidation of pyruvate by *Bacterium delbruckii* is coupled with
phosphorylation.

In 1939-40, two groups of workers - Belitzer and Tsibakowa (66) in Engelhardt's laboratory in Leningrad and Ochoa
(67) in Oxford - made the important observation that more than
one atom of P is esterified for every atom of oxygen consumed
by respiring preparations under conditions in which there is
no glycolysis. Both groups drew the conclusion that phosphorylation must occur not only when the substrate is dehydrogenated

- as in the oxidation reaction of glycolysis - but during the
further passage of the hydrogen atoms (or the electrons) along
the respiratory chain to oxygen. Thus, these workers were the
first to postulate what became known as respiratory-chain
phosphorylation. This postulate was directly confirmed by
Lehninger nearly 10 years later, using NADH as substrate (68).

      During the 1950's and up to the middle 1960's the
question of the stoicheiometry of oxidative phosphorylation -
the P:O ratio - was controversial, and indeed the chapter is
not yet closed. The reader is referred elsewhere to a review
(69) published in 1966 dealing with this question and a related one, namely the location of the phosphorylation "sites"
in the respiratory chain. By this time, there was a general
consensus that these lie between NADH and Q, between QH_2 and
cytochrome c, and between cytochrome c and oxygen. The concept
of specific "sites" belongs, however, more to the chemical or
conformational hypothesis than to the chemiosmotic description.

      In the early 1950's the discovery by Racker (70) and
Lynen (71) of the rôle of thiol esters as "high-energy" intermediates in the oxidation of glyceraldehyde 3-phosphate and
pyruvate, respectively, opened up the way for the elucidation
of the mechanism of the substrate-linked phosphorylation step
in mitochondria associated with the oxidation of 2-oxogluta-
rate. It was also the first demonstration that energy of oxidation reactions can be conserved without the formation of
"high-energy" phosphate compounds.

      This, and the observation that heart particles oxidize
succinate and NADH maximally in the absence of inorganic phosphate, provided the stimulus to the proposal in 1953 (72) of
what became known as the chemical hypothesis of oxidative
phosphorylation. This hypothesis had two tenets: (1) that

energy is conserved before the intervention of inorganic phosphate; (2) that energy is conserved in a high-energy compound formed by liganding (with a compound C) of the oxidized product of the reaction. The high-energy intermediate then reacts with phosphate and ADP.

$$AH_2 + B + C \rightleftharpoons A{\sim}C + BH_2$$
$$A{\sim}C + ADP + P_i \rightleftharpoons A + C + ATP$$

It was also predicted that A~C might be utilized directly for certain energy-requiring processes in the cell, without prior conversion of its energy to ATP.

Chance and Williams (73), in adopting this hypothesis, replaced "C" by "I" in order to emphasize that the liganded component of the respiratory chain (they preferred $BH_2{\sim}I$) is an inhibited form since it cannot participate further in electron-transfer reactions before the ligand (inhibitor) is removed. Later it became customary to speak of A~C or $BH_2{\sim}I$ as a low or high-potential form of A or BH_2 , respectively.

The first of the two tenets of the hypothesis was found to be correct when it was found possible by adding oligomycin to allow various energy-requiring processes to proceed although synthesis of ATP in the respiratory chain was inhibited (74, 75). The energy-requiring process studied in that time was the so-called reversal of the respiratory chain (e.g. reduction of NAD^+ by succinate). Lars Ernster played an important rôle in establishing the concept of the direct utilization of the hypothetical high-energy intermediate, which formed one of the highlights of the International Congress of Biochemistry in Moscow in 1961 (76). That the uptake of Ca^{2+} by mitochondria, already known since 1953 (77), is energy-linked was foreshadowed by Chance in 1955 (78) and demonstrated by de Luca

and Engström (79) in 1961. Even before 1961, however, Saris
(80) had made an observation the significance of which became
apparent only later. He found that when ATP and Ca^{2+} were
added to mitochondria, an acidification of the medium far in
excess of that expected from hydrolysis of ATP ensued. This was
interpreted as being due to an active ion transport with extrusion of H^+ and uptake of Ca^{2+} leading to gradients of the two
ions across the membrane.

The demonstration of active uptake of Ca^{2+} opened up a
large chapter in mitochondrial research which is not yet
closed. Somewhat later, Pressman (81) discovered that, in the
presence of the ionophore, valinomycin, K^+ behaves in the same
way as Ca^{2+} in the absence of exogenous ionophore.

The energy-linked nicotinamide transhydrogenase, discovered by Danielson and Ernster (82), was also found to be
driven by the so-called high-energy intermediate.

After the demonstration of the direct utilization of
the energy of biological oxidations the search for the hypothetical intermediate A∿C was intensified. The rest of the
1960's is not one of the happiest periods in the history of
mitochondrial research. Apparently spectacular successes proved
unfounded, so that by 1966 the ground was ripe for a new theory
proposed in order, among other reasons,

> *to acknowledge the elusive character of the*
> *C∿I intermediates by admitting that they may*
> *not exist (83).*

Paradoxically, this new theory, the "chemiosmotic
theory" of Mitchell, had been proposed in 1961 (84), the peak
year in the fortunes of the chemical hypothesis. It had, however, received little attention except in one extensive review
(69) until a new version of the hypothesis, modified to take
account of the observed stoicheiometry of proton extrusion from

mitochondria as the result of a pulse of oxygen in the presence
of oxidizable substrate (85), was published in August 1966 in
what came to be known as "the little grey book" (83).

Although this was certainly not emphasized by Mitchell,
the new hypothesis retained the first tenet of the chemical
hypothesis, namely that energy is conserved (and may be utilized) without formation of ATP. It differed fundamentally,
however, concerning the nature of the conserved energy. According to the chemiosmotic hypothesis, this is the electrochemical
activity of protons (protonmotive force) across the membrane,
created by the translocation of protons from inside to outside
the membrane, driven by a proton pump linked with electron
transfer along the respiratory chain. An ATP-driven proton pump
drives protons in the same direction, and the protonmotive
force built up by the operation of the proton pump linked with
electron transfer drives the ATP-driven pump in the opposite
direction with the concomitant synthesis of ATP.

The gradual acceptance of the essential features of the
chemiosmotic theory belongs mainly to the period that has been
excluded from this history. The reader is referred to a review
article published in 1977(86). The so-called "acid bath" experiment of Jagendorf and Uribe (87) which was published in
1966, the significance of which slowly became apparent to
opponents of the chemiosmotic synthesis, should, however, be
mentioned, even if it was not carried out with mitochondria.
These workers created a proton gradient across the thylakoid
membrane by placing chloroplasts in a solution containing a
permeant acid, successively at low and high pH. Protons accumulating inside the chloroplast during the acid bath move out
when the pH is raised, with concomitant synthesis of ATP.

Between 1961 and 1966, an important modification of

the chemical hypothesis was brought forward by Boyer (88), who proposed that energy is conserved by a change of conformation in the respiratory protein. In other words, he replaced the A$\sim$C of the chemical hypothesis by A*, which is a non-reducible conformation of A. The so-called conformation hypothesis is still very much alive as a possible mechanism of ATP synthesis.

ENZYMIC MACHINERY FOR ATP SYNTHESIS

The study of the enzyme system responsible for the synthesis of ATP was initially carried out without consideration for the nature of the initially conserved energy. In the 1950's the hope existed that the secrets of oxidative phosphorylation would be laid bare by fractionation procedures that had been so successful in the study of other biochemical systems. Extraction of heart mitochondria by salt yielded a particle with decreased oxidative and phosphorylative ability, both (but particularly the phosphorylative) being restored by the soluble extract. The active component of the soluble extract turned out to be cytochrome c (79). A more useful fractionation is into a particle that catalyses maximal oxidation without phosphorylation and a supernatant fraction specifically required for phosphorylation.

The first successful fractionations of this type were achieved in 1958 by Linnane (89) with beef-heart mitochondria, in 1959 by Hovenkamp (90) with *Azobacter vinelandii*, and in 1960 by Racker and coworkers (91) with beef-heart mitochondria. The latter publication was particularly significant, since the soluble factor was found to possess ATPase activity and to possess a binding site for ADP. These workers suggested that this preparation, called F_1 (Coupling Factor no. 1), contains

the terminal enzyme of oxidative phosphorylation, a view that
is now generally accepted.

The designation Coupling Factor 1 in their first publication indicated that this group had more Coupling Factors
to report. Coupling Factor 2 was identified with Linnane's
Factor. So far as the writer is aware, its rôle is not yet
established. For a time, the relationship between the different
factors was not immediately apparent, nor was the nomenclature
always clear to others, but gradually it became clear that the
ATP-synthesizing machinery is made up of three types of components - the F_1 ATPase, a large globular protein visible in
electron micrographs of negatively stained material as pro-
jecting on the outside of submitochondrial particles (and inside the matrix in intact mitochondria), F_o (o for oligomycin,
not zero) embedded in the membrane (92) and some factors, one
called Fc (93), or OSCP (94) and the other Fc_2 (95), necessary
for attachment of F_1 to the membrane. The isolation of F_o was
particularly important, since it was found that the complex
$F_1 \cdot F_o$ (Fc and Fc_2 were present in the F_o preparation used)
formed vesicles that retain a rudimentary system for oxidative
phosphorylation in the sense that they catalyse a P_i-ATP
exchange. It is now generally accepted that F_o contains the
proton channel leading to the F_1 and that this is blocked by
oligomycin or DCCD. The identification by Beechey (96) of the
DCCD-binding protein as a very hydrophobic subunit of F_o turned
out to be important in the decade not covered in this history.

THE BIOGENESIS OF MITOCHONDRIA

The idea that mitochondria might be self-reproducing
bodies is almost implicit in the early description of them as

organelles. Indeed, many workers in the nineteenth century and as late as the 1920's considered them to be symbiotic micro-organisms. This is not surprising in view of the striking resemblances between mitochondria and bacteria, and it was only after the development of more refined morphological and biochemical criteria that a clear distinction could be drawn between them. Indeed, the resemblances were the cause of a considerable controversy in the 1960's.

The independence, at least in part, of the biogenesis of mitochondria from genetic control by the nucleus was established by Ephrussi (97) who isolated 'petite' mutants of *Saccharomyces cerevisiae* that are respiratory deficient. Crosses between normal and deficient strains do not give results expected of a Mendelian inheritance. Ephrussi concluded that an inheritable cytoplasmic factor coupled to the mitochondria is responsible for the extrachromosal or cytoplasmic inheritance.

The biochemical study of the biogenesis of mitochondria was opened by Simpson and coworkers who showed that mitochondria isolated from skeletal muscle are able to incorporate amino acids into protein, both in vivo and in vitro (98). This was soon confirmed by others, and the properties of the system responsible differed sufficiently from those of the well-known system present in endoplasmic reticulum, e.g. in resistance to ribonuclease (99) and inhibition by chloramphenicol (100) to warrant the conclusion that the incorporation was not due to contaminating microsomes that are always present in mitochondrial preparations. It was more difficult, however, to counter the criticism that the protein synthesis was, in fact, due to contaminating bacteria, since as already mentioned mitochondria resemble bacteria very closely and protein synthesis measured with bacterial suspensions is also insensitive to ribonuclease and chloramphenicol. It was only in 1967 when mitochondria isolated under sterile and bacteriologically controlled conditions were found to synthesize proteins with undiminished vigour (101) that this ghost was set to rest.

In the meantime, those who believed in protein synthesis in mitochondria were, on the basis of inhibition by actinomycin D (102) and chloramphenicol, convinced that the mitochondria must also contain DNA, and there was also cytological evidence in support of this view (103,104). Contamination of the mitochondrial fraction with even a small amount of nuclei presented a problem, however, since as the name suggests the nucleic acid content in nuclei is much greater than that in mitochondria. The first demonstration of an organelle DNA with a unique buoyant density in CsCl came from work with chloroplasts (105). In 1965, Luck and Reich (106) reported similar results with a DNA concentrated in the mitochondrial fraction and Borst and coworkers showed in 1966 that mitochondrial DNA has a unique renaturation behaviour (107) and is circular (108).

The study of mitochondrial protein synthesis and nucleic acids became suddenly respectable: the first symposium on the biochemical aspects of the biogenesis of mitochondria was held in 1967 (109). In recent years this has become one of the most actively worked fields in molecular biology, culminating in the determination of the complete nucleotide sequence in human and beef mitochondrial DNA. Most of this story is, however, post-history as we have defined it.

Mention should be made, however, of Linnane's introduction (110) of antibiotics such as chloramphenicol and erythromycin to inhibit synthesis of most of the components of the repiratory chain without affecting the synthesis of many other mitochondrial enzymes. Later it was found, in fact, that the synthesis of only a few of the subunits of the respiratory enzymes are under the control of mitochondrial DNA but that when these are not synthesized cytoplasmically-synthesized components are not properly incorporated into the mitochondrial membrane. Already in his first review on purified DNA, Borst (111) pointed out that the information content is far too low to code for all the proteins that make up the mitochondria. It is now established that indeed this DNA codes only for the three heavy subunits of cytochrome oxidase, cytochrome b, 2-3 subunits in F_o , a ribosomal protein, 2 rRNAs and all t-RNAs. Proteins that are not synthesized on the mitochondrial ribosome are imported from the cytoplasm.

Thus, the mitochondrion exists in a symbiotic relationship with the cytoplasm. The cytoplasm provides the oxidizable substrates and most of the material required for the growth or regeneration of the mitochondrion. The latter exports most of the energy it generates to the rest of the cell, but it retains some for its own purposes including the synthesis of some of its own membrane proteins and for the incorporation of extramitochondrially synthesized proteins.

The old idea that mitochondria are free living bacteria within the cell is certainly wrong. It is not possible to cultivate mitochondria outside the cell. However, there is good reason to believe that evolutionarily speaking there is a germ of truth in the idea, in the sense that mitochondria have evolved from bacteria captured by the early eucaryote cells.

INTERMEDIARY METABOLISM

Most of the early work establishing the main path-

ways of oxidation of carbohydrates, fatty acids and proteins
was carried out with homogenates of cells in which the mitochondria were suspended in a solution of the cytoplasm. Only
later were these preparations replaced by suspensions of mitochondria washed free from cytoplasm. However, misleading results can be obtained with such preparations, even when all
the enzymes concerned are present in the mitochondria. For
example, isolated rat-liver mitochondria oxidize glutamate
almost quantitatively to aspartate, without any appreciable
formation of ammonia, despite the presence of an active glutamate dehydrogenase (112). In the intact cell, however, this
enzyme is largely responsible for the formation of the NH_3 that
is converted to urea. In order to study the factors controlling
the functioning of glutamate dehydrogenase and the urea cycle
in vivo it is necessary to turn to the intact cell as study
object. In the last decade, the biochemist has returned to the
perfused liver and more recently to the isolated liver cell in
order to study these problems. The introduction by Bücher (113)
of methods based on nitotinamide nucleotide-linked dehydrogenases in equilibrium to study the effective redox potential of
the intra- and extramitochondrial compartiments in the intact
cell was particularly important. The reader is referred to a
recent review dealing with the use of isolated liver cells and
kidney tubules in metabolic studies (114).

METHODOLOGY

It would be incomplete even in a short history of mitochondria to neglect the important rôle played by methodology.
Since most of the proteins in the mitochondria are firmly
attached to membranes, methods that played such an important

rôle in the development of classical biochemistry in the 1930's
- for example, extraction, purification and crystallization of
the glycolytic enzymes - have little application to mitochondria. Biophysical methods have played an important rôle since
the early studies on cytochromes by Keilin (1) using the visual
spectroscope and of Warburg (39) to determine the photochemical
action spectrum of what we now recognize as cytochrome c oxidase. Advances in the field have often had to await the introduction of suitable instruments. As already mentioned, the
isolation of mitochondria became a routine procedure only after
the introduction of high-speed centrifuges into the average
biochemical laboratory.

In the 1950's Britton Chance was responsible more than
anyone else for introducing sophisticated instrumental methods
into mitochondrial research (72). The polarographic method
largely replaced manometry for measuring oxygen consumption
and the dual-wavelength spectrophotometer was introduced
in order to measure light absorption at specific wavelengths
with minimum disturbance from light scattering. He also introduced fluorimetry into this field. Helmut Beinert (60)
introduced EPR spectrometry as a powerful tool for detecting
paramagnetic species of the respiratory chain and Mildred Cohn
introduced NMR. Further developments in the latter technique have even enabled a discrimination between ATP and ADP.

Preparative methods have also played an important rôle.
Early on the large-scale preparation of heart-muscle particles
(42), making use of a mechanical mortar and pestle used in the
paint industry but without a high-speed centrifuge or even a
cold room (certainly not necessary in Cambridge in the winter
of 1946-7) provided Keilin and his collaborators with ample
material for their studies. Later, the large-scale preparation

of beef-heart mitochondria (115) introduced by David Green in
Madison and the fractionation (116) into the four complexes of
the respiratory chain - succinate:Q oxidoreductase; NADH:Q
oxidoreductase; QH_2:ferricytochrome c oxidoreductase and ferrocytochrome c:O_2 oxidoreductase - provided the basis for the
further elucidation of the mechanism of electron transfer in
the respiratory chain. The provision of highly coupled submitochondrial particles, mainly thanks to the efforts of
Lars Ernster and his collaborators in Stockholm (117,118), was
later also of great value.

More recently, and outside the period covered by this
history, might be mentioned electron diffraction of two-dimensional crystals of cytochrome c oxidase and QH_2:ferricyto-
chrome c oxidoreductase which has revealed the overall shape
of these proteins in the membrane. Edge-on synchotron radiation is giving information on the orientation of the haem
molecules in the membrane. Finally, recent developments of the
recombinant-DNA technique and in methods of determining the
nucleotide sequence in DNA has provided us, more quickly than
could have been obtained by sequencing of the amino acids,
the primary structure of the three large subunits of cytochrome
oxidase and of cytochrome b.

REFERENCES

1. Keilin, D. (1966) The History of Cell Respiration and
   Cytochrome, pp. 2-3, University Press, Cambridge.
2. Henle, J. (1841) Allgemeine Anatomie, Leipzig.
3. Aubert, H. (1853) Z. wiss. Zool. 4, 388
4. Kölliker, A. (1856) Z. wiss. Zool. 8, 311
5. Kölliker, A. (1888) Z. wiss. Zool. 47, 689
6. Retzius, G. (1890) Biol. Untersuch. N.F. 1, 51
7. Regaud, C. (1909) C.R. Acad. Sci. Paris 149, 426
8. Holmgren, E. (1910) Arch. mikr. Anat. 75, 240
9. Bullard, H.H. (1913) Amer. J. Anat. 14, 1
10. Bullard, H.H. (1916) Amer. J. Anat. 19, 1
11. Altmann, R. (1890) Die Elementarorganismen und ihre
    Beziehungen zu den Zellen, von Veit, Leipzig.
12. Benda, C. (1898) Arch. Anat. Physiol.,393-398
13. Battelli, F. and Stern, L. (1912) Ergebn. Physiol. 15,
    96-268
14. Keilin, D. and Hartree, E.F. (1940) Proc. Roy. Soc. B
    129, 277-306
15. Warburg, O. (1913) Arch. Ges. Physiol. 154, 599
16. Hogeboom, G.H., Claude, A. and Hotchkiss, R.D. (1946)
    J. Biol. Chem. 165, 615-629
17. Kennedy, E.P. and Lehninger, A.L. (1948) J. Biol. Chem.
    172, 847-848
18. Lehninger, A.L. and Kennedy, E.P. (1948) J. Biol. Chem.
    173, 753-771
19. Slater, E.C. (1950) Nature 166, 982-983
20. Cleland, K.W. and Slater, E.C. (1953) Quart. J. Micr. Sci.
    94, 329-346
21. Cleland, K.W. and Slater, E.C. (1953) Biochem. J. 53,
    547-566
22. Harman, J.W. and Feigelson, M. (1952) Exp. Cell Res. 3,
    47-58
23. Watanabe, M.I. and Williams, C.M. (1953) J. gen. Physiol.
    37, 71
24. Chappell, J.B. and Perry, S.V. (1953) Biochem. J. 55,
    586-595
25. Harman, J.W. and Osborne, U.H. (1953) J. exp. Med. 98,
    81-98
26. Palade, G.E. (1952) Anat. Rec. 114, 427
27. Sjöstrand, F.S. (1953) Nature 171, 30-31
28. Claude, A. (1944) J. exp. Med. 80, 19-29
29. Zöllinger, H.U. (1948) Amer. J. Path. 24, 569
30. Harman, J.W. (1950) Exp. Cell Res. 1, 394-402
31. Cleland, K.W. (1952) Nature, 170, 497-499
32. Raaflaub, J. (1953) Helv. physiol. Pharmacol. Acta 11, 142

33. Slater, E.C. and Cleland, K.W. (1953) Biochem. J. 55, 566-580
34. Lindberg, O. and Ernster, L. (1954) Chemistry and Physiology of Mitochondria and Microsomes, Springer, Wien.
35. Mitchell, P. (1961) Nature, 191, 144-148
36. Chappell, J.B. (1964) Biochem. J. 90, 225-237
37. Pfaff, E., Klingenberg, M. and Heldt, H.W. (1965) Biochim. Biophys. Acta 104, 312-315
38. Keilin, D. (1925) Proc. Roy. Soc. B 98, 312-339
39. Warburg, O. (1926) Biochem. Z. 177, 471-486
40. Keilin, D. (1927) Nature 119, 670-671
41. Keilin, D. (1930) Proc. Roy. Soc. B 106, 418-444
42. Keilin, D. and Hartree, E.F. (1938) Proc. Roy. Soc. B 125, 171-186
43. Keilin, D. and Hartree, E.F. (1939) Proc. Roy. Soc. B 127, 167-191
44. Ball, E.G. (1938) Biochem. Z. 295, 262-264
45. Yakushiji, E. and Okunuki, K. (1940) Proc. imp. Acad. Japan 16, 299-302
46. Keilin, D. and Hartree, E.F. (1955) Nature 176, 200-206
47. Slater, E.C. (1948) Nature, 161, 405-406
48. Tsou, C.L. (1949) Biochem. J. 49, 512-520
49. Dolin, M.I. and Gunsalus, I.C. (1951) J. Bacteriol. 62, 199
50. Wang, T.Y., Tsou, C.L. and Wang, Y.L. (1956) Scientia Sinica 5, 73-90
51. Singer, T.P., Kearney, E.B. and Bernath P. (1956) J. Biol. Chem. 223, 599-613
52. Crane, F.L., Hatefi, Y., Lester, R.L. and Widmer, C. (1957) Biochim. Biophys. Acta 25, 220-221
53. Massey, V. (1957) Biochim. Biophys. Acta 30, 205-206
54. Straub, F.B. (1939) Biochem. J. 33, 787-792
55. Chance, B. (1958) J. Biol. Chem. 233, 1223-1229
56. Wikström, M.K.F. (1973) Biochim. Biophys. Acta 301, 155-193
57. Takemori, S., Sekuzu, I. and Okunuki, K. (1960) Biochim. Biophys. Acta 38, 158-160
58. Beinert, H., Griffiths, D.E., Wharton, D.C. and Sands, R.H. (1962) J. Biol. Chem. 237, 2337-2346
59. Eichel, B., Wainio, W.W., Person, P. and Cooperstein, S.J. (1950) J. Biol. Chem. 183, 89-103
60. Beinert, H. and Sands, R.H. (1960) Biochem. Biophys. Res. Commun. 3, 41-46
61. Rieske, J.S., Hansen, R.E. and Zaugg, W.S. (1964) J. Biol. Chem. 239, 3017-3022
62. Slater, E.C. and de Vries, S. (1980) Nature, in the Press.
63. Engelhardt, W.A. (1930) Biochem. Z. 227, 16-38
64. Kalckar, H. (1937) Enzymologia 2, 47-52

65. Lipmann, F. (1939) Nature 143, 281
66. Belitzer, V.A. and Tsibakowa, E.T. (1939) Biokhimiya 4, 516
67. Ochoa, S. (1940) Nature, 146, 267
68. Friedkin, M. and Lehninger, A.L. (1948) J. Biol. Chem. 174, 757-758
69. Slater, E.C. (1966) in Comprehensive Biochemistry (M. Florkin and E.H. Stotz, ed.), Vol. 14, pp. 327-396, Elsevier.
70. Racker, E.F. (1951) J. Biol. Chem. 190, 685-696
71. Lynen, F. and Reichert, E. (1951) Angew. Chem. 63, 47
72. Slater, E.C. (1953) Nature 172, 975-978
73. Chance, B. and Williams, G.R. (1956) Adv. Enzymology, 17, 65-134
74. Ernster, L. (1961) in IUB/IUBS Symposium on Biological Structure and Function, Stockholm, 1960, Vol. 2, p. 139, Academic Press, New York.
75. Snoswell, A.M. (1961) Biochim. Biophys. Acta 52, 216-218
76. Slater, E.C., ed. (1963) Symposium on Intracellular Respiration: Phosphorylating and Non-Phosphorylating Oxidation Reactions. Proc. 5th Intern. Congr. Biochem., Moscow, 1961, Vol. 5, Pergamon, Oxford.
77. Slater, E.C. and Cleland, K.W. (1953) Biochem.J. 55, 566-580
78. Chance, B. (1956) Proc. 3rd Intern. Congr. Biochem. Brussels, 1955, p. 300, Academic Press, New York.
79. De Luca, H.F. and Engstrom, G.W. (1961) Proc. Natl. Acad. Sci. U.S.A. 47, 1744-1750
80. Saris, N.-E.L. (1959) Suom. Kemistiseuran Tied. 68, 98-107
81. Pressman, B.C. (1965) Proc. Natl. Acad. Sci. U.S.A. 53, 1076-1082
82. Danielson, L. and Ernster, L. (1963) Biochem. Z. 338, 188-205
83. Mitchell, P. (1966) Chemiosmotic Coupling in Oxidative and Photosynthetic Phosphorylation, Glynn Research Ltd., Bodmin.
84. Mitchell, P. (1961) Nature 191, 144-148
85. Mitchell, P. and Moyle, J. (1965) Nature 208, 147-151
86. Boyer, P.D., Chance, B., Ernster, L., Mitchell, P., Racker, E. and Slater, E.C. (1977) Ann. Rev. Biochem. 46, 955-1026
87. Jagendorf, A.T. and Uribe, E. (1966) Proc. Natl. Acad. Sci. U.S.A. 55, 170-177
88. Boyer, P.D. (1965) in Oxidases and Related Redox Systems (T. King, H.S. Mason and M. Morrison, ed.), pp. 994-1008, Wiley, New York.
89. Linnane, A.W. (1958) Biochim. Biophys. Acta 30, 221-222
90. Hovenkamp, H.G. (1959) Nature 184, 471
91. Pullman, M.E., Penefsky, H.S., Datta, A. and Racker, E. (1960) J. Biol. Chem. 235, 3322-3329

92. Kagawa, Y. and Racker, E. (1966) J. Biol. Chem. 241, 2461-2466
93. Bulos, B. and Racker, E. (1966) J. Biol. Chem. 243, 3891-3900
94. Tzagaloff, A., MacLennan, D.H. and Byington, K.H. (1968) Biochemistry 7, 1596-1602
95. Knowles, A.F., Guillory, R.J. and Racker, E. (1971) J. Biol. Chem. 246, 2672-2679
96. Beechey, R.B., Roberton, A.M., Holloway, C.T. and Knight, I.C. (1967) Biochemistry 12, 3867-3879
97. Ephrussi, B. (1953) Nucleocytoplasmatic Relations in Micro-organisms, Clarendon Press, Oxford.
98. Simpson, M.V. and McLean, J.R. (1955) Biochim. Biophys. Acta 18, 573-575
99. McLean, J.R., Cohn, G.L., Brandt, I.K. and Simpson, M.V. (1958) J. Biol. Chem. 233, 657-663
100. Rendi, R. (1959) Exptl. Cell Res. 18, 187-189
101. Kroon, A.M., Saccone, C. and Botman, M.J. (1967) Biochim. Biophys. Acta 142, 552-554
102. Kroon, A.M. (1963) Biochim. Biophys. Acta 76, 165-167
103. Chevremont, M. (1963) in Cell Growth and Cell Division (R.J. Harris, ed.), p. 323, Academic Press, New York.
104. Nass, M.M.K. and Nass, S. (1962) Exptl. Cell Res. 26, 424-427
105. Chun, E.H.L., Vaughan, Jr., N.H. and Rich, A. (1963) J. Mol. Biol. 7, 130-141
106. Luck, D.J.L. and Reich, E. (1964) Proc. Natl. Acad. Sci. U.S.A. 52, 931-938
107. Borst, P. and Ruttenberg, G.J.C.M. (1966) Biochim. Biophys. Acta 114, 645-647
108. Van Bruggen, E.F.J., Borst, P., Ruttenberg, G.J.C.M., Gruber, M. and Kroon, A.M. (1966) Biochim. Biophys. Acta 119, 437-439
109. Slater, E.C., Tager, J.M., Papa, S. and Quagliariello, E. (ed.) (1968) Biochemical Aspects of the Biogenesis of Mitochondria, Adriatica Editrice, Bari.
110. Huang, M., Biggs, D.R., Clark-Walker, G.D. and Linnane, A.W. (1966) Biochim. Biophys. Acta 114, 434-436
111. Borst, P., Kroon, A.M. and Ruttenberg, G.J.C.M. (1967) in Genetic Elements, Properties and Function (D. Shagar, ed.), pp. 81-116, Academic Press, London and New York, and PWN, Warsaw.
112. Borst, P. (1962) Biochim. Biophys. Acta 57, 256-269
113. Hohorst, H.J., Kreuz, F.H. and Bücher, T. (1959) Biochem. Z. 332, 18-46
114. Tager, J.M., Söling, H.D. and Williamson, J.R. (ed.) (1976)

Use of Isolated Liver Cells and Kidney Tubules in Metabolic Studies, North Holland, Amsterdam.
115. Crane, F.L., Glenn, J.L. and Green, D.E. (1956) Biochim. Biophys. Acta 22, 475-487
116. Hatefi, Y. (1966) in Comprehensive Biochemistry (M. Florkin and E.H. Stotz, ed.), Vol. 14, pp. 199-231, Elsevier, Amsterdam.
117. Löw, H. and Vallin, I. (1963) Biochim. Biophys. Acta 69, 361-374
118. Lee, C.P. and Ernster, L. (1968) Eur. J. Biochem. 3, 385-390

IAL GENETIC SYSTEM

entity and the arrangement of genes on mitochon-
om several species (in particular Saccharomyces
and man) are now known and DNA sequencing studies
mitochondrial DNAs are progressing rapidly. Incomplete nucleotide sequence of human mitochondrial
ailable although it has not yet appeared in print
nce these developments are discussed in greater de-
the next chapter of this book, I will only mention a
findings. First, mitochondrial DNAs from yeast,
pora, Aspergillus and humans appear to contain nearly
ame type of genes but the arrangement of these gene on
ircular genome is widely different (14, 15). Second, the
iderably greater size of mitochondrial DNAs from yeast
Neurospora ($\sim 5 \times 10^7$ daltons) as compared to those from
ergillus ($\sim 3 \times 10^7$ daltons) or humans ($\sim 1 \times 10^7$ daltons)
es not reflect the presence of more genes, but a higher
ontent of intervening sequences ("introns") as well as of
T-rich sequences between genes. The introns present in the
yeast mitochondrial cytochrome b gene are particularly interesting since mutations within these introns can impair the
expression not only of the cytochrome b gene itself but also
that of the gene for cytochrome c oxidase subunit I. This
fact, together with the observation that some of these introns have an "open" reading frame has led to the suggestion
that some introns may code for polypeptide sequences involved
in RNA splicing (16). Fourth, the genetic code used in mitochondria differs from that which has up to now been consider-

THE BIOGENESIS OF MITOCHONDRIA

Gottfried Schatz

Biocenter, University of Basel
CH-4056 Basel, Switzerland

INTRODUCTION

When I first met Lars Ernster almost twenty years ago,
research on mitochondrial formation was still a somewhat exo-
tic enterprise. Indeed, the first international symposium exclusively devoted to this topic was not held until 1967 (1).
Today, mitochondrial biogenesis is one of the most rapidly
developing areas of "mitochondriology" and the task of re-
viewing its recent achievements in a brief review appears almost hopeless. I will therefore only point out some of the
most exciting new developments and speculate where they may
lead us.

SOME KEY FACTS ABOUT MITOCHONDRIAL BIOGENESIS

Before discussing recent findings, it might be helpful
to restate some important facts that have emerged during the
last two decades. Some of these facts are better documented
than others but all of them are now widely accepted.

(1) Mitochondria are formed by growth and division of pre-
existing mitochondria (2).

C. P. Lee, G. Schatz, G. Dallner (eds.), Mitochondria and Microsomes
in honor of Lars Ernster   ISBN 0-201-04576-1

(2) Mitochondria contain their own genetic system that consists of a relatively small (usually $1-5 \times 10^6$ daltons) DNA and a complete machinery for replicating, transcribing and translating the genetic information residing in this DNA (3).

(3) The genetic code utilized by the mitochondrial genetic system differs slightly from that utilized by most other biological systems (4-8).

(4) Each component of the mitochondrial genetic system differs from its counterpart in the nucleo-cytoplasmic genetic system (9).

(5) Mitochondrial DNA codes for all of the mitochondrial RNA components and for 8-20 (depending on the species) polypeptides (3,9).

(6) All mitochondrially-translated polypeptides are coded by mitochondrial DNA; there is no evidence for the exchange of mRNAs between mitochondria and the nucleus (3,9).

(7) Mitochondrially-made polypeptides identified so far are not complete enzymes, but subunits of multimeric complexes. In yeast-or Neurospora mitochondria, all but one of them are hydrophobic subunits of oxidative phosphorylation enzymes (Table I).

(8) Mitochondrially-made polypeptides account for only about one-tenth of the protein mass of the organelle. The remainder of the proteins (including those of the mitochondrial genetic system itself) are coded by nuclear genes, synthesized on extramitochondrial ribosomes and imported into the mitochondria (9).

(9) Importation of polypeptide int[...] dependent, usually accompanied [...] cessing of larger precursor polyp[...] tely coupled to simultaneous prote[...]

(10) There is no evidence for protein exp[...] chondria (12). Communication between [...] and the nucleo-cytoplasmic genetic sys[...] only mediated via small molecules or via [...] imported into mitochondria.

Table I. Polypeptides synthesized by mitochondr[...]

Polypeptide	Function
Cytochrome c oxidase subunit I	Binds heme a and/or [...]
Cytochrome c oxidase subunit II	Binds heme a and/or C[...] as well as cytochrome [...]
Cytochrome c oxidase subunit III	H^+-channel (?)
ATPase complex subunit VI	?
ATPase complex subunit IX*	H^+-channel; binds dicyclo-hexyl-carbodiimide
Apocytochrome b	Binds protoheme
var 1 (protein of the small mitochondrial ribosomal subunit)	Protein synthesis (?)

* Made outside the mitochondria in Neurospora, Aspergillus and humans.
See refs. 3 and 9 for details.

Since several of these statements are based on negative evidence, they may have to be modified as new information becomes available.

ed to be "universal". There are even slight differences in
codon usage between mitochondria from different species
(Table II).

TABLE II. Unusual codon usage in mitochondria

Codon	"normal" assignment	Assignment in mitochondria from		
		humans	yeast	Neurospora
UGA	stop	trp	trp	trp
CUN	leu	leu	thr	leu
AUA	ile	met	?	?
AGA or AGG	arg	stop	arg	?

See ref. 14 for details

The fact that the triplet UGA codes for tryptophan in mitochondria but for "stop" in bacteria and the eukaryotic cytoplasm explains the many earlier failures to obtain faithful
translation of mitochondrial mRNAs in cell-free translation
systems from E.coli, wheat germ or rabbit reticulocytes (17).

    Future efforts to characterize other mitochondrial
genomes will undoubtedly be helped by the observation that
purified pieces of mitochondrial genes from yeast hybridize
to specific regions on mitochondrial DNA from Neurospora
(18) or Aspergillus (15). Because defined probes of virtually any region of yeast mitochondrial DNA are readily available in considerable quantity through cloning in "petite" mutants, an uncharacterized mitochondrial DNA can now be digested with restriction enzymes and fragments adjacent on the

DNA genome as well as the probable genetic function of these
fragments can be identified by hybridization tests. This approach has already led to considerable information on
Aspergillus mitochondrial DNA (15, 20) and is also applicable
to plant mitochondrial DNA (19). The correlated genetic and
physical maps of three mitochondrial genomes are given in
Figure 1.

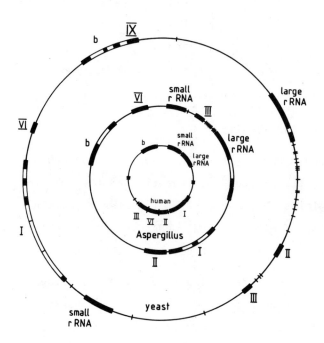

Fig. 1, legend (see next page)

Fig. 1   Physical and genetic maps of mitochondrial genomes
         from humans, Aspergillus and Saccharomyces

         The blocks superimposed on the circles represent gen-
         es for ribosomal RNAs or proteins. Filled blocks sig-
         nify EXONS, open blocks INTRONS. The short thin lines
         pointing towards the center identify tRNA genes. The
         different sizes of the circles are roughly proportio-
         nal to the sizes of these mitochondrial DNAs. The
         evidence for some of the introns of tRNAs genes is
         still tentative.
         b: cytochrome b; I, II and III: subunits I, II and
         III of cytochrome c oxidase; VI, IX: subunits VI and
         IX associated with the F_O part of the ATPase complex.
         Subunit IX is the "DCCD-binding protein". Mitochon-
         drial DNA of Saccharomyces and probably also of
         Neurospora contains an additional genetic locus con-
         trolling the expression of a polypeptide associated
         with the small subunit of mitochondrial ribosomes.
         Since it is still unclear whether this gene codes for
         any proteins in yeast (see following article by Butow
         and Strausberg in this volume) this gene is not shown
         here.

The next few years will certainly bring us a wealth
of additional structural information on mitochondrial DNAs
from different species. It will be particularly interesting
to learn more about mitochondrial DNAs from higher plants
since these DNAs are much larger (up to 2×10^8 daltons) and
more complex than other mitochondrial DNAs (21). A greater informational content of plant mitochondrial DNA is also suggested by the observation that isolated maize mitochondria appear to synthesize at least 18 major polypeptides, more than
twice as many as yeast mitochondria (21). Does plant mitochondrial DNA carry genes which are present in the nucleus of
other species? If so, what is the function of these "extra"
polypeptides and why are they coded for in mitochondria of

plants but not in those of many other species? These questions are given additional significance by the observation
that mitochondrial mutations in maize can prevent normal
pollen development and thereby render the plants "male-
sterile" (21). Use of such male-sterile maize strains has
had a major impact on maize production in the United States
and other countries. Three distinct mitochondrial mutations
causing male sterility have so far been analyzed in some detail. All of them result in the production of at least one
abnormal mitochondrially-made polypeptide. Even more interestingly, one of these mitochondrial mutations is accompanied by the presence of two linear plasmid-like DNA molecules (3.5×10^6 and 4.1×10^6 d) inside the mitochondria
(22). These DNA species may code for the abnormal polypeptides produced and may perhaps open up the experimental possibility of transferring genes between different mitochondrial genomes or between mitochondria and the nucleus. Transfer of a mitochondrial gene between two yeast strains by
transformation of spheroplasts has recently been reported (23).
If the validity and general applicability of this technique
can be documented by additional evidence, it would represent
a major advance in mitochondrial genetics.

IMPORT OF POLYPEPTIDES INTO MITOCHONDRIA

The question of how polypeptides made on cytoplasmic
ribosomes are transported into mitochondria has long been a
baffling problem. How does a polypeptide "know" that it is destined to be imported into mitochondria and not into lysosomes or

peroxisomes? How does it find its correct intramitochondrial location? How can a polypeptide move through the hydrophobic core of one or even two mitochondrial membranes? And, finally, what drives these specific and unidirectional transport processes?

In the early seventies, the study of protein movement through the endoplasmic reticulum of secretory cells (24) suggested some answers to these questions. It was found that secretory proteins (such as digestive enzymes, immunoglobulins or protein hormones) are synthesized as precursors with N-terminal extensions of around 20 amino acids. Polysomes carrying the nascent precursors then become bound to the endoplasmic reticulum through a specific interaction between N-terminal extension (the "signal" sequence) and some corresponding receptor on the cytoplasmic face of the endoplasmic reticulum. The growing polypeptide is moved across the membrane barrier and the N-terminal extension is cleaved off by a specific membrane protease before, or shortly after synthesis of the polypeptide chain has been completed. In this process, polypeptide chain elongation and transmembrane movement of the polypeptide are obligately coupled: it is strictly "cotranslational". This process has been termed "vectorial translation". Later it was realized that a similar mechanism could also explain the insertion of polypeptides into membranes if it was assumed that a suitable (probably hydrophobic) region close to the C-terminus prevented escape of the polypeptide from the membrane.

These results led to the suggestion that cytoplasmi-cally-made mitochondrial proteins are vectorially translated into mitochondria by a special subclass of cytoplasmic ribo-somes which are bound to the cytoplasmic face of the mito-chondrial outer membrane. Those proteins which have to be transported across both mitochondrial membranes were suggest-ed to travel through contact sites between the inner and outer membrane (25). Electron micrographs of yeast cells did indeed suggest the presence of mitochondria-bound cytoplas-mic polysomes (26). However, the polypeptides made by these polysomes could not be identified and similar polysomes were not reported to be present in other organisms, so the pro-blem seemed as intractable as ever.

In the late seventies, pulse-chase studies with Neurospora (27) and, later, yeast (28) strongly argued against a major role of "vectorial translation" in mitochondrial pro-tein import. It was found that import of a finished polypep-tide into mitochondria occurred only after a lag and that it was insensitive to an inhibitor of cytoplasmic protein syn-thesis. Also, newly-made mitochondrial proteins appeared to accumulate first outside the mitochondria. However, since the newly-synthesized mitochondrial polypeptides found outside the mitochondria were chemically indistinguishable from those inside the organelles it remained open whether these results reflected redistribution artefacts which had accompanied sub-cellular fractionation.

This changed, however, when it was observed that se-veral cytoplasmically-made mitochondrial proteins of yeast

are initially synthesized as precursors which are between
2000 and 6000 daltons larger than the corresponding "mature"
proteins found inside the mitochondria (29-31). These prote-
ins include the three largest subunits (α, β and γ) of the
F_1-ATPase (located in the matrix), two subunits of the cytochrome $\underline{bc_1}$ complex (located in the mitochondrial inner membrane) and cytochrome $\underline{c}$ peroxidase (located in the· intermembrane space). The precursors were identified as follows:
first, yeast polypeptides were synthesized in the presence
of ^{35}S-methionine in a nuclease-pretreated reticulocyte ly-
sate programmed with total yeast RNA. Second, in vitro synthesized individual mitochondrial polypeptides were isolated
by immunoprecipitation with monospecific antisera raised
against the mature polypeptides. Finally, the apparent molecular weights of the in vitro made polypeptides were compared with those of the corresponding mature polypeptides by
SDS-polyacrylamide gel electrophoresis followed by fluoro-
graphy. When the radiolabeled in vitro-made precursors were
incubated with isolated yeast mitochondria under conditions
preventing protein synthesis, they were converted to their
mature size and became resistant to externally added proteases (29, 31). Radiolabeled mature mitochondrial polypeptides or an in vitro-synthesized cytosolic enzyme such as
glyceraldehyde-3-P-dehydrogenase were neither detectably
changed by isolated mitochondria nor rendered protease-resi-
stant. These results suggested that proteolytic maturation
of the precursors is accompanied by their uptake into the
mitochondria. This post-translational uptake process was
termed "vectorial processing" (32) in order to emphasize the
apparent coupling between transmembrane movement and covalent

polypeptide modification. Soon thereafter, larger precursor forms were also detected for mitochondrial proteins from Neurospora and rat liver (Table III).

What drives this uptake process? In contrast to vectorial translation, vectorial processing appears to proceed without an obligate ribosome-membrane junction; it is, thus, not possible to invoke the ribosomal chain elongation machinery as a driving force. At least in yeast, import of proteins into or across the mitochondrial inner membrane needs high-energy phosphate bonds in the matrix: if the ATP-level in the matrix of intact yeast cells is lowered by blocking both oxidative phosphorylation and ATP import via the adenine nucleotide carrier, polypeptides destined for the inner membrane or the matrix are no longer imported and accumulate as their larger precursors in vivo (33, 40). Control experiments suggest that high-energy phosphate bonds, not a membrane potential, drive import (33, 40). It is not clear how phosphate bond energy can be used to translocate polypeptides across a membrane; one testable possibility would be that translocation requires the phosphorylation of a component on the matrix-side of the mitochondrial inner membrane. Whether energy is also needed for import of polypeptides into the intermembrane space remains open since that space cannot be experimentally depleted of ATP without lowering the ATP level in the cytosol.

Experiments with yeast cells have identified a soluble protease in the mitochondrial matrix that appears to process those precursors that are transported into the matrix space

Table III. Precursors to cytoplasmically-made mitochondrial proteins

Polypeptide	Intramitochon-drial location	Organism	Reference
F_1-ATPase	Matrix		
- α-subunit		yeast	29, 33
- β-subunit		yeast	39, 33
- γ-subunit		yeast	29, 33
carbamyl phosphate synthetase		rat	34, 35
Mn^{++}-superoxide dismutase		yeast	36
citrate synthase		Neurospora	37
cytochrome c_1	Inner membrane	yeast and Neurospora	33, 38
subunit V of cytochrome bc_1-complex		yeast	30
ADP/ATP carrier*		Neurospora and yeast	39, 40
cytochrome c oxidase			
- subunit IV		yeast and rat	41, 42
- subunit V		yeast	40, 41, 43
- subunit VI		yeast	40, 41, 43
- subunit VII		yeast	41
Proteolipid of ATPase complex		Neurospora	44
cytochrome c*	Intermembrane space	Neurospora and yeast	45 46
cytochrome c peroxidase		yeast	31

* Precursor polypeptide has the same molecular weight as the
mature form but appears to differ from it in conformation

or into the mitochondrial inner membrane (47). The protease
is sensitive to EDTA and o-phenanthroline and fails to affect all non-mitochondrial polypeptides tested so far. Even
more significantly, it also fails to process the precursor
of cytochrome c peroxidase (an intermembrane space enzyme)
to its mature size. It has the properties of a "limited protease" since, even if added in excess, it only cuts the precursors to their mature size and no further. However, "correct" processing by the protease has so far only been infer-
red from the mobility of the products on SDS-polyacrylamide
gels. N-terminal analysis will be required to corroborate
this point. A similar matrix-located protease has been identified in mitochondria from rat-liver, rat heart and maize
(47).

The identification of the matrix-localized processing
protease has furnished the basis for the following experiment (46) which strongly suggests that the in vitro synthesized precursor to the F_1-ATPase β-subunit is transported into the matrix (i.e. its correct intramitochondrial location).
The strongly polar bathophenanthroline disulfonate completely
inhibits the solubilized protease but not the protease inside
the matrix since only the outer, but not the inner mitochondrial membrane is permeable for the polar inhibitor. Since in
vitro synthesized precursor of the F_1-ATPase β-subunit is
processed to its mature form by intact yeast mitochondria in
the presence of bathophenanthroline disulfonate (46), the
precursor must reach the matrix. Processing of precursors in
the presence of this polar inhibitor is thus a simple and
convenient assay for protein import.

The presence of a processing protease in the matrix space predicts the existence of at least one additional protease which acts on precursors imported into the intermembrane space. Yet another protease might be involved in the maturation of precursors inserted into the outer membrane.

These results would imply that import of proteins into mitochondria is post-translational and obligately coupled to the processing of larger precursor polypeptides. However, the situation is not as simple as that. At least two mitochondrial proteins (cytochrome c and the adenine nucleotide carrier) are not made as larger precursors (see Table III), yet are imported posttranslationally (11). Cytochrome c is imported as the apoprotein which is covalently attached to the heme inside the mitochondria (probably in the intermembrane space). Even with cytochrome c, therefore, import is coupled to covalent modification of the polypeptide. However, this is not true for the adenine nucleotide carrier which lacks any prosthetic group. Clearly, different polypeptides may be imported by (at least partly) different mechanisms.

Another unsolved problem is the function of mitochondria-bound cytoplasmic polysomes. Their existence has been confirmed and they could be shown to be highly enriched in mRNAs for mitochondrial polypeptides (48, 49). Still, a major part of these mRNAs is associated with "free" (i.e. unattached) polysomes, at least after cell breakage. The most plausible explanation for these observations would be that some "import receptor" on the mitochondrial surface is recognized not only by finished precursors but also by nascent ones.

Polysomes synthesizing these precursors would thus be attrac-
ted to the mitochondrial surface with the result that at
least part of the import process might be co- rather than
post-translational. The relative importance of these two import routes might well depend on the physiological status of
the cell. This model is unorthodox, but consistent with
the conclusion of previous studies which argue against an
obligate coupling between protein synthesis and import. An
alternate model would be that precursors destined for import
into or across the mitochondrial inner membrane are first
vectorially translated across the mitochondrial outer membrane without being processed. In a second step, they are
then moved into or across the inner membrane post-transla-
tionally and processed to their mature form. This second model appears less plausible since it is not easily reconciled
with the experiments demonstrating post-translational import
in vitro or with the evidence suggesting the existence of extramitochondrial pools of newly-made mitochondrial proteins
(cf. above). Clearly, the role of mitochondria-bound cytoplasmic polysomes deserves further study.

    In the next few years it should be possible to iso-
late mitochondrial precursors in significant amounts and to
define the conformational features that undoubtedly set them
apart from the corresponding mature polypeptides. Is the N-
terminal extension only involved in recognizing a mitochondrial import receptor or is its major role to lock the precursor into a special conformation? How do mitochondrial processing proteases recognize the correct processing site on the
precursors? Can one identify components of the mitochondrial

outer membrane that function as "import receptors"? There might well be separate receptors for import into the various mitochondrial compartments. How are proteins inserted into the mitochondrial outer membrane? Can one reconstruct the import process with liposomes containing purified outer membrane components and an internally trapped processing protease? Many of these questions can be approached with presently available techniques.

CONCLUSION

What is the future of research on mitochondrial biogenesis? In the next few years most or all of the nucleotide sequence of a few representative mitochondrial genomes will be known. The logical next steps would then be to learn more about the expression of mitochondrial genes. Present knowledge in this area is surprisingly scant. Can different genes on a mitochondrial DNA be transcribed at different rates and, if yes, how is this controlled? Do RNA splicing events have a regulatory function in those mitochondria which have "split" genes? Another major unsolved problem is the interaction between mitochondrial and nuclear genes. The isolation of nuclear genes for mitochondrial proteins (50) and the study of nuclear suppressors of specific mitochondrial mutations (51, 52) will undoubtedly feature prominently in efforts to tackle this problem. At a still higher level of complexity, possible interactions between mitochondria and the cytoskeleton should definitely be investigated. There already exists indirect evidence for interactions between mitochondria and microtubuli (e.g. 53, 54) but the significance of these interactions is not known. Do such interactions play a role in mitochondrial fusion or division? These questions should serve as a reminder that some of the most fascinating problems of mitochondrial biogenesis are only now coming into view.

REFERENCES

1. Slater, E.C., Tager, J.M., Papa, S. and Quagliariello, E.
    (eds.) (1968), Biochemical Aspects of the Biogenesis of
    Mitochondria (Proceedings of a Round-table Discussion
    held May 15-18, 1967), Adriatica Editrice, Bari.

2. Review: Schatz, G. (1968) in: Membranes of Mitochondria
    and Chloroplasts (E. Racker, ed.), pp. 251-314,
    Van Nostrand Reinhold Co., New York.

3. Review: Borst, P. and Grivell, L.A. (1978) Cell 15,
    705-723.

4. Barrell, B.G., Bankier, A.T. and Druin, J. (1979)
    Nature 282, 189-194.

5. Macino, G., Coruzzi, G., Nobrega, F.G., Li, M. and
    Tzagoloff, A. (1979) Proc. Natl. Acad. Sci. USA 76,
    3784-3785.

6. Fox, T.D. (1979) Proc. Natl. Acad. Sci. USA 76,
    6534-6538.

7. Li, M. and Tzagoloff, A. (1979) Cell 18, 47-53.

8. Heckman, J.E., Sarnoff, J., Alzner-de Weerd, B., Yin, S.
    and RajBhandary, U.L. (1980) Proc. Natl. Acad. Sci. USA
    77, 3159-3163.

9. Review: Schatz, G. and Mason, T.L. (1974) Annu. Rev.
    Biochem. 43, 51-87.

10. Review: Schatz, G. (1979) FEBS Letters 103, 201-211.

11. Review: Neupert, W. and Schatz, G. (1980)
    Trends Biochem. Sci. 6, 1-4.

12. Van't Sant, P., Mak, F.C. and Kroon, A.M. (1980) in:
    The Organization and Expression of the Mitochondrial
    Genome (Kroon, A.M. and Saccone, C., eds.), pp. 387-390,
    North Holland, Amsterdam.

13. Walker, J. (1980) reported at the Symposium on the Orga-
    nization and Expression of the Mitochondrial Genome,
    Martina Franca, Italy, June 23-28.

14. Kroon, A.M. and Saccone, C. (eds.). (1980), The Organi-
    zation and Expression of the Mitochondrial Genome,
    North Holland, Amsterdam.

15. Macino, G., Scazzocchio, C., Waring, R.B., Berks, M.M. and Davies, R.W. (1980) Nature 288, 404-406.

16. Lazowska, J., Jacq, C. and Slonimski, P.P. (1980) Cell 22, 333-348.

17. Moorman, A.F.M., Verkley, F.N., Asselbergs, F.A.M. and Grivell, L.A. (1977) in: Mitochondria 1977 (Bandlow, W., Schweyen, R.J., Wolf, K. and Kaudewitz, F., eds.), pp. 385-399, De Gruyter, Berlin

18. Agsteribbe, E., Samallo, J., De Vries, H., Hensgens, L.A.M. and Grivell, L.A. (1980) see ref. 14, pp. 51-60.

19. Fox, T.D. and Leaver, C., personal communication.

20. Köchel, H.G., Lazarus, C.M., Basak, N. and Küntzel, H. (1980) Cell, in press.

21. Review: Leaver, C. (1980) Trends Biochem. Sci. 5, 248-252.

22. Pring, D.R., Levings, C.S. and Conde, M.F. (1980) in: The Plant Genome (Davies, D.R. and Hopwood, D.A., eds.), The John Innes Charity.

23. Nagley, P., Atchison, B.A., Devenish, R.J., Vaughan, P.R. and Linnane, A.W. (1980), see ref. 14, pp. 75-78.

24. Review: Palade, G.W. (1975) Science 189, 347-358.

25. Review: Butow, R.A., Bennett, W.F., Finkelstein, D.B. and Kellems, R.E. (1975) in: Membrane Biogenesis (Tzagoloff, A., ed.), pp. 155-199. Plenum, New York.

26. Kellems, R.E., Allison, V.F. and Butow, R.A. (1975) J. Cell. Biol. 65, 1-14.

27. Hallermayer, G., Zimmermann, R. and Neupert, W. (1977) Eur. J. Biochem. 81, 523-532.

28. Dockter, M.E., Frey, T.G., Chan, S.H.P., Woodrow, G., Schär, M., Grüninger, H. and Schatz, G. (1978) in: Biochemistry and Genetics of Yeasts (Bacila, M., Horecker, B.L. and Stoppani, A.O.M., eds.), pp. 549-562, Academic Press, New York.

29. Maccechini, M.-L., Rudin, Y., Blobel, G. and Schatz, G. (1979) Proc. Natl. Acad. Sci. USA 76, 343-347.

30. Coté, C., Solioz, M. and Schatz, G. (1979) J. Biol. Chem. 254, 1437-1439.

31. Maccechini, M.-L., Rudin, Y. and Schatz, G. (1979) J. Biol. Chem. 254, 7468-7471.

32. Review: Schatz, G. (1979) FEBS Letters 103, 201-211.

33. Nelson, N. and Schatz, G. (1979) Proc. Natl. Acad. Sci. USA 76, 4365-4369.

34. Shore, G.C., Carignon, P. and Raymond, Y. (1979) J. Biol. Chem. 254, 3141-3143.

35. Mori, M., Miura, S., Tatibana, M. and Cohen, P. (1979) Proc. Natl. Acad. Sci. USA 76, 5071-5075.

36. Autor, A. and Schatz, G., unpublished.

37. Harmey, M.A. and Neupert, W. (1979) FEBS Letters 108, 385-389.

38. Zimmermann, R., Weiss, H. and Neupert, W., unpublished.

39. Zimmermann, R., Paluch, U., Sprinzl, M. and Neupert, W. (1979) Eur. J. Biochem. 99, 247-252.

40. Nelson, N. and Schatz, G. (1979) in: Membrane Bioenergetics (Lee, C.P., Schatz, G. and Ernster, L., eds.), pp. 133-152. Addison Wesley, Reading.

41. Mihara, K. and Blobel, G. (1980) Proc. Natl. Acad. Sci. USA 77, 4160-4164.

42. Schmelzer, E. and Heinrich, P.C. (1980) J. Biol. Chem. 255, 7503-7506.

43. Lewin, A., Gregor, I., Mason, T.L., Nelson, N. and Schatz, G. (1980) Proc. Natl. Acad. Sci. USA 77, 3998-4002.

44. Michel, R., Wachter, E. and Sebald, W. (1979) FEBS Letters 101, 373-376.

45. Zimmermann, R., Paluch, U. and Neupert, W. (1979), FEBS Letters 108, 141-151.

46. Gasser, S. and Schatz, G., unpublished.

47. Böhni, P., Gasser, S., Leaver, C. and Schatz, G. (1980), see ref. 14, pp. 423-433.

48. Suissa, M. and Schatz, G., unpublished.

49. Ades, I.Z. and Butow, R.A. (1980) J. Biol. Chem. 255, 9918-9924.

50. Montgomery, D.L., Hall, B.D., Gillam, S. and Smith, M.
     (1978) Cell 14, 673-680.

51. Fox, T.D. (1980) in: Cell Compartmentation and Metabolic
     Channeling (Nover, L., Lynen, F. and Mothes, K., eds.),
     pp. 303-311, Elsevier / North Holland, Amsterdam.

52. Dujardin, G., Pajot, P., Groundinsky, O. and Slonimski,
     P.P. (1980) Mol. Gen. Genet. 179, 469-482.

53. Heggeness, M.H., Simon, M. and Singer, S.J. (1978)
     Proc. Natl. Acad. Sci. USA 75, 3863-3866.

54. Howard, R.J. and Aist, J.R. (1980) J. Cell Biol. 87,
     55-64

THE MITOCHONDRIAL GENOME

Ronald A. Butow[*] and Robert L. Strausberg[+]

[*] Department of Biochemistry, The University of Texas
Health Science Center at Dallas, Dallas, Texas   75235
and [+] Department of Biology, Southern Methodist University, Dallas, Texas 75275.

The study of mitochondrial structure, function, and
development spans a very broad range of conceptual and
technical disciplines.   Over the years, the mitochondrion has been subjected to relentless physical and chemical
scrutiny to reveal a high degree of structural and functional complexity.   More recently, the organelle has been the
object of study by geneticists and molecular biologists.
Some of their findings have been at the least, surprising,
and in some cases almost revolutionary in biological impact.

In the past few years, considerable progress has been
made in delineating the more formal aspects of mitochondrial
genetics and the cellular and molecular mechanisms governing the transmission of mitochondrial genes (1-3).   In addition, in a number of organisms most mitochondrial genes have
now been identified and mapped and details of their fine

---

[1] Supported by NIH Grant GM-19090, GM-22525, GM-26546 and
Grant I-642 from The Robert A. Welch Foundation.

C. P. Lee, G. Schatz, G. Dallner (eds.), Mitochondria and Microsomes
in honor of Lars Ernster   ISBN 0-201-04576-1

structure are becoming clear. With these successes, efforts are now focused on elaborating the regulation of expression of mitochondrial genes. In this article we shall briefly review some of the more recent developments which have contributed to our understanding of the organization, function, and genetic behavior of the mitochondrial genome.

MITOCHONDRIAL GENETICS. Saccharomyces cerevisae has served as the pre-eminent model system for studies in mitochondrial genetics for several reasons. First, the ease of mutant isolation in this system is particularly attractive; second, the nuclear genome is well characterized, facilitating genetic manipulations; and third, mitochondrial genomes are generally transmitted biparentally in this organism, unlike the uniparental transmission of mitochondrial genes in so many other systems (2,4-7).

Before considering the principle features of mitochondrial transmission genetics, we will briefly survey the major classes of mitochondrial mutations and indicate how these have been especially useful in the analysis of mitochondrial genes.

i) Cytoplasmic petites. Discovered over 30 years ago by Ephrussi (8), these mutations result in a respiratory incompetent phenotype and are represented by strains which have either completely lost all their mtDNA, the so-called ρ° petites (9), or are massively deleted for wild-type mtDNA sequences. The interesting and useful property of the latter (ρ^{-} petites) is that in many such strains, the sequences retained are often faithful colinear copies of wild-type sequences and are present in the petite in an amplified configuration (10,11). The great utility of such

petites is that they can be used for genetic and physical
mapping studies since their DNAs are capable of recombining
with wild-type mtDNA (12).

ii) Mutations to drug resistance. Mutations in this
class are of two types - those which confer resistance to inhibitors of mitochondrial protein synthesis like chloramphenicol and erythromycin (13,14) and those which confer
resistance to inhibitors of the oxidative phosphorylation
machinery (ATPase or components of the electron transport
chain), like oligomycin (15) antimycin (16), diuron (17)
etc. Antimycin and diuron are representative of a number
of drug resistance mutations which fall within defined
coding sequences (exons) of the cytochrome b (cob/box)
gene (18). Mutations conferring resistance to inhibitors
of protein synthesis are of interest because most of them
appear to lie within sequences encoding rRNAs.

iii) Mit⁻ mutations. These represent a very important
class of mutations which result in loss of function or
absence of synthesis of respiratory chains components (19).
Analysis of mutants in this class have been extremely
useful not only in helping to define and locate mitochondrial
genes, but also in the development of important concepts in
gene organization and regulation. Perhaps most striking in
this regard, discussed in some greater detail below, is
the discovery of mutations which lie within intervening
sequences, particularly at the cob/box locus.

iv) Syn⁻ mutations. These are defined as mutations which
affect mitochondrial protein synthesis. Specific examples
now exist where mutants with low or undetectable levels of
mitochondrial protein synthesis can be traced to specific
base changes in mitochondrial tRNAs (20).

v) <u>Suppressors</u>. Recently, mitochondrial mutants have
been obtained which have the interesting property of
suppressing various mit⁻ mutations (21). Some of these
act as informational suppressors and have been localized to
the 21S rRNA gene. Other suppressors have been observed
which appear to be gene-specific and suppress mutations
within introns.

vi) <u>Polymorphisms</u>. At the DNA level, these show up as
stable, strain-dependent insertions and deletions and include intron sequences as well as sequences located outside
of genes (22,23). Of the mitochondrial translation products,
varl polypeptide represents the most striking example of a
protein polymorphism (24) (See below).

<u>Transmission genetics</u>. One of the key concepts in consider-
ing the transmission of mitochondrial genes is that each cell
is polyploid for the mitochondrial genome. The average yeast
cell contains about 50 mtDNA molecules, but this number can
vary from about 4 to 100 molecules per cell depending on
the strain and cellular growth conditions (25-27). Thus,
the concept of an intracellular population of mtDNA molecules is important to consider. In addition, subpopula-
tions may exist further complicating genetic analysis.
In some cases many mtDNA molecules are packaged within
an individual mitochondrion and some of these may be
clustered within specific organelle regions (nucleoids).
Certainly, DNA interactions between molecules in these
nucleoids, or chondriolites, may occur more frequently than
interactions between molecules packaged within different
organelles.

The fusion of two haploid cells containing alternate

forms of the mitochondrial genome produces a heteroplasmic
zygote. Genetic and biochemical evidence suggest that in
this zygote there is ample opportunity for genome interaction and recombination (2). The zygote reproduces by
generating diploid buds which contain a sample of the mtDNA
molecules of the zygote. A critical question, which has not
been satisfactorily answered, is how many organelle genomes
are transmitted to the diploid buds. Sena et al. (28) have
measured the mtDNA population of zygotes and first zygotic
buds immediately following bud formation and found that the
zygotes contained between 44-67 mtDNA molecules, whereas the
zygote itself contained 112 mtDNA molecules. These numbers
suggest that a fairly high proportion of the available mtDNA
molecules are transmitted from zygote to bud. Genetic analysis also reveals that zygotic buds containing greater than
99% alleles from one parent are fairly common, a result
also consistent with the transmission of many mtDNA molecules to the zygote (29). In addition, Strausberg and
Perlman (30) observed that in some zygotes producing a
first bud from a zygotic end, all available genomes of the
parent from which that end was generated are transmitted to
the buds, leaving only mitochondrial genomes of the other
parent in the zygote. In spite of these data supporting
the transmission of many mtDNA molecules to the zygote,
there are still some doubts regarding the actual situation
because of the observation that the frequency of homoplasmic
first buds is much higher than expected given the transmission of roughly 50 mtDNA molecules to the bud. Part
of the problem in resolving this issue is our lack of
knowledge of the physical basis for a mitochondrial
segregating (heritable) unit. It has been suggested that

the segregating unit could consist of individual mtDNA
molecules, organelles, or perhaps nucleoids within
organelles (25).

Dujon et al. (1) have proposed a mathematical model to
explain the distribution of mitochondrial genes in crosses.
The model relies on several assumptions. In the zygote,
a panmictic mating pool of mtDNA molecules is rapidly
established, allowing for random pairing and recombination.
Thus, if equal numbers of molecules are contributed to the
zygote by each haploid parent, half of the pairings will be
between homogenic mtDNA molecules therefore reducing the detectable mtDNA recombinations. It is also assumed that all
mtDNA molecules contributed to the zygote participate in
these processes and are available for distribution to the
buds and that the transmission is a random process with no
selective advantage to any mtDNA molecule. The essential
features of the model are based on an analogy with transmission of bacteriophage genes in crosses (31) and is
therefore referred to as the phage analogy model. A prediction of the model, suitable for experimental analysis,
is that the maximal percent recombination between two loci
will be observed when parental contributions to the zygote
are equal since half of the pairings will be heterogenic;
any variation in favor of one parent will result in a decrease in the freqeuncy of heterogenic pairings. This input
fraction can be manipulated experimentally either by increasing or decreasing the mtDNA content of one of the haploid
parents (32,33). A disparity in parental contributions to the
zygote is termed bias and, if the model is valid, a relationship should hold such that as bias is increased, the observed
recombination frequency should decrease. Experimentally,

these predictions have been confirmed (34), thereby giving credence to the phage analogy model. However, a number of the assumptions proposed by the architects of the model are apparently not completely valid, at least for some crosses. The first problem is that the establishment of the panmictic mating pool is apparently delayed for some time after cell fusion, so that bud position effects are observed in mitochondrial gene transmission (30,35). Such position effects are not consistent with the rapid establishment of a panmictic pool of mtDNA molecules.

A second problem arises in considering the actual proportion of mtDNA molecules in the zygote which participate in recombination and subsequent transmission to progeny. Birky et al. (36) have observed that in some crosses, a proportion of the zygotic clones contain only individual alleles or entire genomes from one parent. This phenomenon is termed uniparental inheritance, but should be distinguished from those cases in which all progeny of a cross inherit mitochondrial genomes from one parent, often maternally (2). The frequency of uniparental zygotes observed is directly related to the input fraction of the cross, and clearly indicates that at least in some zygotes, not all mitochondrial genomes survive for transmission to zygotic buds. This phenomenon is clearly reflected in cases of bias amplification in which the output frequency of a cross is more extreme than would have been predicted from the measured input fraction. The molecular mechanisms which could generate uniparental zygotes have been discussed by Birky et al. (36). In spite of these reservations, the phage analogy model does explain at least quantitatively mitochondrial genetic phenomena in many crosses.

Recombination. The mitochondrial genome offers some exciting
opportunities for dissecting mechanisms of DNA recombination.
Two cases of asymmetric, unidirectional gene conversion have
been detected and characterized in yeast. The first involves the transmission of gene segments of the 21S rRNA
locus. Genetic analysis revealed two allelic forms of this
gene which could be distinguished based on transmission of
antibiotic resistance markers within the locus (37). The
alleles are termed ω^+ and ω^- and are characterized in the
following manner: a cross involving the same ω allele
(homopolar) results in the appearance of reciprocal recombinants at equal frequency among the 21S rRNA markers.
Individual drug resistance alleles are transmitted to the
same proportion of homoplasmic progeny (that is, they are
transmitted coordinately). However, a cross involving alternative ω alleles (heteropolar) results in the preferential
transmission of markers associated with the ω^+ allele (a polarity of transmission starting at the ω site); reciprocal recombinants are now found at unequal frequency, always showing
a preference for the recombinant containing the ribl allele
of the ω^+ parent (11,37,38). Formally, this process of gene
conversion is similar to conversion observed for nuclear
genetic loci since there is an apparent fixed starting point
for the initiation of conversion, and markers proximal to
this initiation point more frequently participate in the
conversion reaction than do distal markers (that is, the conversion is polarized). Subsequently, Dujon et al. (39) iso-
lated a new allele of ω from an ω^- strain, termed ω^n. This
new allele is characterized by the lack of asymmetric gene
conversion when crossed to either an ω^- or ω^+ testers.

The molecular basis for the polarity of recombination

at the 21S rRNA locus lies, in part, with the fact that ω^+
alleles contain a large 1143 bp intron within the 21S rRNA
gene (40-43) along with mini-insert of some 66 bp (43).
In the conversion of ω^- to ω^+, these intervening sequences
are effectively copied into the ω^- allele. Dujon (43)
has recently sequenced parts of the ω region from ω^+, ω^- and
ω^n strains. In addition to uncovering the 66 bp mini-
insert, these studies demonstrated that the alteration in
genetic transmission associated with ω^n results from a
single bp substitution located four base pairs from the
position of the intron in ω^+ strains. While it is not
clear why the phenotypic effect is observed, the change
does occur within a 6 bp direct repeat which is also found
25 bp upstream. In ω^+ strains the repeat is also found
within the intron region.

Also of considerable interest is the finding that the
intron has a unique open reading frame containing coding information for 235 amino acids. Clearly, such a sequence
would not arise by chance, but until the polypeptide potentially encoded by this sequence can be identified, we cannot
rule out the possibility that the sequence represents a gene
expressed in ancestral mitochondria but no longer serving a
direct function. Since splicing of this transcript can occur
in the absence of mitochondrial protein synthesis (in petites)
this would not be the function of this theoretical polypeptide.
The other obvious possibility is that this species plays a
role in polarity at the ω locus, but again evidence exists
suggesting that recombination at this locus can proceed in
the absence of mitochondrial protein synthesis (44).

The second instance of asymmetric gene conversion involves the genetic locus, varl, specifying allelic forms of

the var1 polypeptide, a component of the small ribosomal subunit (45-49). Alleles at this locus are characterized by the
size of the var1 polypeptide synthesized, with forms ranging in apparent molecular weight from 40,000 to 44,000.
Crosses involving different forms of var1 led to the
observation that new forms of var1 polypeptide can be
generated in crosses (50). Furthermore, marker transmission
analysis revealed that the recombinant species were always
generated at the expense of the shorter parental var1 form
suggesting a unidirectional asymmetric gene conversion
event. Further support for this conclusion comes from a
detailed analysis of crosses involving all possible combinations of var1 alleles (51). The results could be co-
herently explained by postulating the existence of two or
more DNA segments (insertions) present in var1 alleles
specifying the longest form of var1 polypeptide and absent
in the alleles specifying the shortest form of the protein.
Moreover, these DNA segments could participate independently
in the asymmetric, unidirectional gene conversion resulting
in the generation of var1 recombinant species. Restriction
endonuclease analysis of different var1 alleles does indeed
reveal small DNA insertions (30-100 bp) which correlate with
the size of var1 polypeptide (52). However, the nature of the
relationship between these segments and the alteration in
var1 polypeptide size remains to be determined. Sequence
analysis of these alleles should provide clues to the
nature of the elements controlling asymmetric gene conversion.

Mannella and Lambowitz (53) have recently reported the
unidirectional, asymmetric gene conversion of two sequences
present in Neurospora crassa mtDNA, estimated to be 50 and

1200 bp in length. Although these sequences have not yet been associated with any mitochondrial gene product, the results indicate that the phenomenon of unidirectional, asymmetric gene conversion involving specific inserted sequences may be a general characteristic of mitochondrial transmission genetics.

Although these are the only cases of unidirectional gene conversion observed thus far in mitochondrial gene transmission, evidence exists suggesting that many, if not all, mitochondrial genetic events may result from gene conversions. Dujon et al. (1) proposed that reciprocal recombination would not account for the observed recombination frequencies. Recently, VanWinkel-Swift and Birky (54) analyzed the progeny of individual zygotes and observed that the frequency of reciprocal recombinants occurring together in a zygotic clone was no more frequent than expected by chance. These results are not consistent with reciprocal recombination being the sole mechanism for generating recombination of mtDNA molecules. These results suggest that bidirectional gene conversion may also play an important role in mitochondrial genetic phenomena (for example, in the generation of homoplasms from heteroplasms), particularly within relatively small mtDNA populations such as nucleoids (25).

The above discussion has been confined to the transmission of mitochondrial genes from wild-type (ρ^+) cells. However, additional information concerning mechanisms of organelle genome transmission may be obtained by observing the behavior of petite genomes in genetic crosses. Petites may be placed in two categories, neutral and suppressive, based on transmission of their genomes in crosses with wild-

type testers (55). Zygotic progeny of a cross involving
neutral petites always include some diploids containing the
wild-type genome. These cases can be explained by simple
vegetative segregation of the two genomes. However, a
cross involving a suppressive petite results in a proportion
of the zygotic clones which only contain petite genomes
(56). This is a second case of uniparental inheritance observed in mitochondrial gene inheritance in yeast. Blanc and
Dujon (57) have further categorized suppressive petites into
those which are highly suppressive (hyper-suppressive) and
those which are moderately suppressive. No evidence was found
for recombination between highly suppressive petites and
wild-type genomes (all of the progeny contained sequences
only found in the petite), thereby, arguing against a recom-
binational mechanism to account for suppressiveness. A
similar conclusion has been arrived at by Bernardi and
colleagues (56). Sequence analysis of highly suppressive
petites shows a conservation of a 300 bp sequence which is
present in multiple copies on wild-type mtDNA. These
sequences have been likened to replication origins and, as
such, could explain the replicative advantage of mtDNA from
highly suppressive petites. Clearly, these sequences are
not absolutely required for mtDNA replication since some
petites exist which lack them. However, it would be expected
that their presence in an amplified configuration on petite
mtDNA should confer some selective advantage to that DNA
over wild-type.

Complementation. The ability to perform complementation
analysis in yeast has been a powerful approach in the
analysis of mitochondrial gene organization and regulation.
Recently, two experimental approaches have been developed

which allow for the detection of <u>trans</u> interactions between
mtDNA molecules in a heteroplasmic cell.

   Zygotic gene rescue (58) is based on the notion that
genes retained on petite mtDNA could be expressed in a
heteroplasmic zygote utilizing the transcription-translation
apparatus of a ρ^+ tester. This approach depends on the
ability to detect products of the petite genomes which
are distinct from those encoded by the wild-type. The var1
polypeptide provided the needed phenotypic difference in
developing this approach since petite and wild-type testers
could be isolated with alternate alleles at this genetic
locus. It was then possible to show that <u>var1</u> determinant
alleles retained in petite mtDNA could be expressed in
zygotes within 3-4 hours of mating. Quantitative
estimates of the extent of var1 rescue and the lack of
appearance early in the mating of recombinant var1 forms,
led to the conclusion that the <u>var1</u> determinant region
specifies the apparent molecular weight of var1 polypeptide
by a <u>trans</u> acting element. Sequence analysis of the <u>var1</u>
determinant region (59) suggests that the entire var1
polypeptide cannot be encoded within the determinant or
flanking segments; thus it appears that the generation.of
var1 polypeptide requires the function of at least two gene
segments, coding or regulatory, capable of <u>trans</u> interaction.

   A second protocol for detection of <u>trans</u> interactions
between mitochondrial genomes involves an analysis of complementation between mit⁻ loci in heteroplasmic cells. Foury
and Tzagoloff (60) observed that in heteroplasmic zygotes containing mitochondrial genomes with different mit⁻ lesions
(generating a respiration deficient phenotype), oxygen
consumption approached wild-type levels in cases where the

mit- lesions were in loci encoding different polypeptides.
Since the rate of oxygen consumption showed no correlation
with the proportion of homoplasmic diploid progeny which are
respiration sufficient, genetic recombination does not
account for the result. Since the complementing mutations
were in different genetic loci, these results could satisfactorily be explained by invoking extensive interactions
between mitochondria in heteroplasmic cells with
reconstitution of active complexes in the inner mitochondrial
membrane.

A detailed complementation analysis of the cob/box gene
(61,62) provided an important framework for the finding that
the gene is mosaic, containing at least 4 introns and 5
exons (18,63-65). While the details of these studies are beyond the scope of this discussion, the complementation analysis of cob/box showed that exons constitute a single complementation group while introns may comprise different complementation groups. From this finding, two important facts
emerged: 1) sequences within introns play a crucial role in
RNA processing, and 2) introns can regulate RNA processing
steps through trans acting elements. Although the nature
of the trans acting elements has not yet been resolved, we
may anticipate exciting developments in mechanisms of gene
regulation made possible by the unique genetic and
biochemical approach offered by the yeast mitochondrial
genome, whereby mutations may be obtained and studied not
only in exon sequences but within introns as well.

ORGANIZATION OF MITOCHONDRIAL GENOMES. Because of the
availability of genetic analysis, together with rapid
methods for DNA and RNA sequencing, it has been possible to

learn more about the organization of the yeast mitochondrial
genome than any other mtDNA studied thus far.

Early chemical analysis of yeast mtDNA revealed a
significant asymmetry in its base composition (66). Wild-
type yeast mtDNA is only 18% G+C and contains stretches of
A+T (> 95%) which have been termed spacer sequences (67).
More recently, it has been shown by DNA sequencing that
a number of mitochondrial structural genes are indeed
flanked by stretches of almost pure A+T (68-69). Also
present on the yeast mitochondrial genome are clusters of
G+C rich sequences often arranged as palindromes (70).
Although the function of these G+C clusters is not known,
it has been speculated that they might be involved in RNA
processing (59). Some evidence has been advanced that the
G+C clusters are primary sites of intragenic recombination
events which could explain the spontaneous formation of
cytoplasmic petites (71).

Most of the genes on yeast mtDNA, some of which have
already been mentioned, have been identified and located to
defined segments of the wild-type genome (3). In addition to
some 24 tRNAs and the genes for the 21S and 15S rRNA,
yeast mtDNA encodes three subunits of cytochrome oxidase
(oxi1, oxi2, and oxi3 corresponding to subunits 2,3, and 1,
respectively), two subunits of the oligomycin sensitive
ATPase (oli1 and oli2, encoding most likely subunits 9 and
6, respectively), the apoprotein of cytochrome b (cob/box),
and the var1 determinant which specifies the apparent molecular weight of var1 polypeptide.

The gene encoding the proteolipid subunit 9 of the
oligomycin-sensitive ATPase presents an extremely interesting
example of a difference in gene allocation. In yeast, this

gene (oli1) is located on mtDNA and its complete DNA sequence
has recently been obtained (68,69). Remarkably, in Neurospora, the gene for the proteolipid is found on the nuclear
genome (72), translated on cytoplasmic ribosomes, and the
product imported into mitochondria. Given this case, we must
consider the possibility that other examples will be found
where a mitochondrial protein is encoded by mtDNA in one organism and by nuclear DNA in another.

The rRNA and tRNA genes provide additional examples suggesting that there has been some divigence among mtDNAs of
different species. In yeast (73) and Tetrahymena (74), the
rRNA genes are far apart from each other and located at distinctly separate regions of the mitochondrial genome. By contrast, in most other species that have been examined, the rRNA
cistrons are quite close together. Striking differences have
also been noted in the arrangement of tRNA genes. In yeast,
most of the tRNA genes are clustered in one quadrant of the
genome flanked by the 21S and 15S rRNA genes (3). In HeLa
cells, however, the tRNA genes are distributed rather uni-
formly along the genome (75). Whether such differences in
gene arrangement imply fundamentally different transcriptional
controls, remains to be established. Furthermore, until the
genes on many more mtDNAs have been identified and mapped, it
may not be possibile to present a coherent picture for the
evolutionary and cellular bias which underlie differences
in the arrangement and allocation of genes encoding mitochondrial constituents.

Animal mtDNAs are generally smaller than mtDNAs from
lower eukaryotes and are in the range of 10×10^6 daltons.
Thus far there is no evidence that mitochondrial genes in
higher eukaryotes have intervening sequences.

Human mtDNA has been studied most extensively and some
interesting features can be pointed out. Attardi and his
colleagues (76) have carried out extensive transcriptional
mapping of mtDNA from HeLa cells and found that transcription is symmetric and complete; both the H and L strands are
transcribed but the L strand transcript turns over much
more rapidly than the H strand transcript (77,78). The bulk
of the informational and structural RNAs are represented in
the H strand transcripts with only a few tRNAs and one
poly(A)-containing RNA encoded by the L strand (76).

A number of observations now lead to the conclusion
that human mtDNA is a very compact genome with an unusual
arrangement of structural and tRNA gene sequences. Discrete
stable transcripts have been identified and mapped to HeLa
mtDNA. The total length of these transcripts is significantly greater than the unit length of all available sequences
on the H strand, that is, sequences exclusive of the rRNA
and tRNA genes (78). This finding means that some of the
RNAs must be related to each other either as precursor-pro-
ducts or as overlapping transcripts.

Two specific examples of the compactness of HeLa mtDNA
are noteworthy. The gene for subunit II of cytochrome
oxidase has been identified and sequenced (79). It is
flanked directly with no intervening nucleotides on its 5'
end by a tRNA^{Asp} gene and is separated by only 25 nucleotides
from a tRNA^{Lys} gene at its 3' end. Similarly, the 5' end
of the 12S rRNA gene is directly flush with a tRNA^{Phe} gene
(80). These findings are particularly interesting in light
of the suggestion that there is only a single promotor for
both H and L strand transcription located near the origin of
replication (76). Since the 5' end of the cytochrome oxidase

subunit II gene has an ATG translational start signal (79),
either the tRNA must serve as a ribosome binding site and
is then processed out of the transcript, or the tRNA is
immediately processed and the mitochondrial translation
apparatus initiates protein synthesis directly at a 5'
AUG start with no upstream leader sequence.

THE MITOCHONDRIAL GENETIC CODE IS NOT UNIVERSAL. According
to wobble rules, a minimum of 32 tRNAs are required to
decode all of the codons of the universal genetic code.
One of the puzzling questions raised from the analysis of
mitochondrial genomes from different species was that the
number of mitochondrial tRNAs found is considerably less
than the number required for protein synthesis, assuming a
universal genetic code and compliance with standard wobble rules. For example, almost the entire sequence of human mtDNA is now known and only 23 tRNAs can be identified
(81). In lower eukaryotes like yeast, DNA sequencing and
mapping studies reveal only about 24 tRNAs encoded by the
mitochondrial genome (82). Since it is not likely that
there is any significant import of nuclear-encoded tRNAs
into mitochondria (83), the question arises how the mitochondrial translation system is able to utilize this apparent limited number of tRNAs? The answer comes from
DNA and RNA sequencing studies, and the startling conclusion is that the mitochondrial genetic code is not
quite the same as the universal code.

If one examines codon families of the unmixed type
where all four codons of the family specify a single amino
acid, wobble considerations in the universal genetic code
require two tRNAs to decode the family. Since most, if not

all, of the mitochondrial tRNAs have now been identified in
human (81) and fungal mitochondria (82), it seems that only
one tRNA is used to specify a given amino acid in an unmixed
four codon family (81,82,84). In all of these cases, with the
exception of a CGN arg family in yeast which does not appear
to be a codon series generally used in the yeast mitochondrial
genetic code (82), the wobble base of the anti-codon is U.
Thus a single tRNA for a four codon family would suggest
that the mitochondrial genetic code might use a two out of
three reading method (85), or perhaps take advantage of the
possible U-U or U-C pairs (86). In cases of mixed codon
families where the third base change in a four codon family
specifies more than one amino acid, misreading by a relaxation of wobble rules appears to be avoided by a
modification of the U in the wobble position of the anti-
codon, so that only codons with an A or G in the third
position are read (82).

A comparison of DNA to amino acid sequences has uncovered another remarkable variation in the mitochondrial genetic code. UGA, a termination codon in the universal code, is
read as tryptophan in the mitochondrial code (in addition
to the "usual" UGG) (81,87,88). The anti-codon of all known
trp tRNAs in prokaryotes and eukaryotes is CCA. Following
usual wobble rules, these tRNAs can decode only the single
codon UGG. Since both UGG and UGA are evidently used as
tryptophan codons in the mitochondrial genetic code, it
was expected that the mitochondrial tRNA^{Trp} should have
a wobble base change from C to U to decode the "new" tryptophan UGA codon.   That this is indeed the case has now
been established recently in yeast from DNA sequencing of
the tRNA^{Trp} gene by Martin et al. (89) and by RNA

sequencing of the mitochondrial tRNA^{Trp} by Sibler et al.
(90). In the latter, the anti-codon of the tRNA^{Trp} was
found to be *UCA, where *U indicates a modified uridine
residue.

There are three additional examples where the mitochondrial genetic code is not only different from the universal
code, but also between mtDNAs of different species. First,
unlike the universal code, and in Neurospora and yeast
mitochondria where AUA is a isoleucine codon, both purine
codons of the AU family in human mtDNA are used to specify
methionine (81). Second, in yeast mitochondria the CUN
unmixed codon family is used to specify threonine instead
of the usual leucine (82), and third, in human mtDNA, AGA and
AGG are probably termination codons and not translated as
arginine (81).

CONCLUDING REMARKS. We have examined some recent developments in the study of mitochondrial genomes of different
organisms. The results of these studies have provided a
large body of new data and led to the formulation of new
concepts. However, rather than presenting a more clarified
picture of the evolution and genetic function of mtDNA, it
seems that many new and provocative questions have been
raised. To mention just a few, we may ask what is the basis
for the remarkable divergence of the mitochondrial genetic
code? What are the peculiar functions of intervening
sequences which, in some cases, are evidently dispensable
for a given gene in the same organism or among different
organisms? And what are the precise genetic functions of
the mitochondrial genome given the observation that mitochondrial genes may not be invariant among species?

Hopefully, the rate at which we may find the solutions to these and other questions will not be too different from the rate of progress we have already made in our understanding of the mitochondrial genome.

REFERENCES

1. Dujon, B., Slonimski, P.P. and Weill, L. (1974) Genetics 78, 415-437.
2. Birky, C.W., Jr. (1978) Ann. Rev. Genet. 12, 471-512.
3. Borst, P. and Grivell, L.A. (1978) Cell 15, 705-723.
4. Coen, D., Deutsch, J., Netter, P., Petrochillo, E. and Slonimski, P.P. (1970) Symp. Soc. Exp. Biol. 24, 449-496.
5. Lukins, H.B., Tate, J.R., Saunders, G.W. and Linnane, A. W. (1973) Mol. Gen. Genet. 120, 17-25.
6. Birky, C.W., Jr. (1975) Mol. Gen. Genet. 141, 41-58.
7. Dawid, I. and Blackler, A.W. (1972) Dev. Biol. 29, 152-161.
8. Ephrussi, B., Hottinguer, H. and Chimenes, A. M. (1949) Ann. Inst. Pasteur 76, 351-357.
9. Goldring, E.S., Grossman, L.I. Krupnick, D., Cryer, C.R., and Marmur, J. (1970) J. Mol. Biol. 52, 322-335.
10. Morimoto, R., Levin, A., Hsu, H.J., Rabinowitz, M., and Fukuhara, H. (1975) Proc. Natl. Acad. Sci. U.S.A. 72, 3868-3872.
11. Lazowska, J. and Slonimski, P.P. (1977) Mol. Gen. Genet. 156, 163-175.
12. Molloy, P.L., Linnane, A.W. and Lukins, H.B. (1975) J. Bact. 122, 7-18.
13. Thomas, D.Y. and Wilkie, D. (1968) Biochem. Biophys. Res. Commun. 30, 368-372.
14. Linnane, A.W., Saunders, G.W., Gingold, E.B. and Lukins, H.B. (1968) Proc. Natl. Acad. Sci. U.S.A. 59, 903-910.
15. Avner, P., Coen, D., Dujon, B. and Slonimski, P.P. (1973) Mol. Gen. Genet. 125, 9-25.
16. Michaelis, G. and Pratje, E. (1977) Mol. Gen. Genet. 156, 79-85.
17. Colson, A.M. and Slonimski, P.P. (1979) Mol. Gen. Genet. 167, 287-298.

88   RONALD A. BUTOW, ROBERT L. STRAUSBERG

18. Slonimski, P.P., Claisse, M.L., Foucher, M., Jacq, C., Kochko, A., Lamouroux, A., Pajot, P., Perrodin, G., Spyridakis, A., and Wambier-Kluppel, M.L. (1978) in Biochemistry and Genetics of Yeast (M. Bacila, et al. eds.) pp. 391-401.
19. Slonimski, P.P. and Tzagoloff, A. (1976) Eur. J. Biochem. 61, 27-41.
20. Miller, P.L. and Martin, N.P., personal communication.
21. Dujardin, G., Groudinsky, O., Kruszewska, A., Pajot, P., and Slonimski, P.P. (1980) in The Organization and Expression of the Mitochondrial Genome (C. Saccone and A. M. Kroon, eds.) North Holland, Amsterdam, in press.
22. Prunell, A. and Bernardi, G. (1977) J. Mol. Biol. 110, 53-74.
23. Sanders, J.P.M., Heyting, C., Verbeet, M. Ph.. Meijlink, F.C.P.W. and Borst, P. (1977) Mol. Gen. Genet. 157, 239-261.
24. Douglas, M.G. and Butow, R.A. (1976) Proc. Natl. Acad. Sci. U.S.A. 73, 1083-1086.
25. Williamson, D.H., Johnston, L.H., Richmond, K.U.M. and Gasse, J.C. (1977) in Mitochondria 1977 Genetics and Biogenesis of Mitochondria (Bandlow, W., ed.) de Gryter, Berlin, pp. 1-24.
26. Grimes, G.W., Mahler, H.R. and Perlman,P.S. (1974) J. Cell Biol. 61, 565-574.
27. Goldthwaite, C.D., Cryer, D.R. and Marmur, J. (1974) Mol. Gen. Genet. 133, 87-104.
28. Sena, E., Welch, J. and Fogel, S. (1976) Science 194, 433-435.
29. Birky, C.W., Jr., Strausberg, R.L., Perlman, P.S. and Forester, J.L. (1978) Mol. Gen. Genet. 158, 251-261.
30. Strausberg, R.L. and Perlman, P.S. (1978) Mol. Gen. Genet. 163, 131-144.
31. Visconti, N. and Delbrück, M. (1953) Genetics 38, 5-33.
32. Gunge, N. (1976) Mol. Gen. Genet. 148, 251-261.
33. Perlman, P.S., Birky, C.W., Jr., Demko, C.A. and Strausberg, R.L. (1976) in Genetics and Biogenesis of Chloroplasts and Mitochondria (Bucher, T., et al. eds.) North Holland, Amsterdam pp. 405-414.
34. Dujon, B. and Slonimski, P.P. (1976) in Genetics and Biogenesis of Chloroplasts and Mitochondria (Bucher, T., et al. eds.) North Holland, Amsterdam pp. 393-403.
35. Callen, D.F. (1974) Mol. Gen. Genet. 134, 65-76.
36. Birky, C.W., Jr., Demko, C.A., Perlman, P.S. and Strausberg, R.L. (1978) Genetics 89, 615-651.

37. Bolotin, M., Coen, D., Deutsch, J., Dujon, B., Netter, P., Petrochilo, E. and Slonimski, P.P. (1971) Bull. Inst. Pasteur 69, 215-239.

38. Perlman, P. and Birky, C.W.,Jr. (1974) Proc. Natl. Acad. Sci. U.S.A. 71, 4612-4616.

39. Dujon, B., Bolotin-Fukuhara, M., Coen, D., Deutsch, J.. Netter, P., Slonimski, P.P. and Weill, L. (1976) Mol. Gen. Genet. 143, 131-165.

40. Jacq. C., Kujawa, C., Grandchamp, C. and Netter, P. (1977) in Mitochondria 1977 (W. Bandlow et al. eds.) de Gruyter, Berlin, pp. 255-270.

41. Bos, J.L., Heyting, C., Borst, P., Arnberg, A.C. and Van Bruggen, E.F.J. (1978) Nature 275, 336-338.

42. Faye, G., Dennebouy, N., Kujawa, C. and Jacq, C. (1979) Mol. Gen. Genet. 1680, 101-109.

43. Dujon, B. (1980) Cell 20, 185-197.

44. Strausberg, S.L. and Birky, C.W., Jr. (1979) Current Genetics 1, 21-31.

45. Butow, R.A., Terpstra, P. and Strausberg, R.L. (1979) in Extrachromosomal DNA (D.J. Cummings et al. eds.) ICN-UCLA Symposium on Molecular and Cellular Biology Vol. XV. Academic Press pp. 269-286.

46. Groot, G.S.P., Mason, T.L. and Van Harten-Loosbroek (1979) Mol. Gen. Genet. 174, 339-342.

47. Terpstra, P., Zanders, E. and Butow, R.A. (1979) J. Biol. Chem. 254, 12653-12661.

48. Terpstra, P. and Butow, R.A. (1979) J. Biol Chem. 254, 12662-12669.

49. Butow, R.A., Lopez, I.C., Chang, H.P. and Farrelly, F. (1980) in The Organization and Expression of the Mitochondrial Genome (1980) (C.Saccone and A. M. Kroon, eds.). North Holland, Amsterdam, in press.

50. Strausberg,, R.L., Vincent, R.D., Perlman, P.S. and Butow, R.A. (1978) Nature 226, 577-583.

51. Strausberg, R.L. and Butow, R.A. Proc. Natl. Acad. Sci. U.S.A., in press.

52. Vincent, R.D., Perlman, P.S., Strausberg, R.L. and Butow, R.A. (1980) Current Genetics 2, 27-38.

53. Mannella, C.D. and Lambowitz, A.M. (1979) Genetics 93, 645-654.

54. Van Winkle-Swift, K.P. and Birky, C.W., Jr. (1978) Mol. Gen. Genet. 166, 193-209.

55. Ephrussi, B., Hottinguer, H. and Roman, H. (1955) Proc. Natl. Acad. Sci. U.S.A. 41, 1065-1071.

56. Goursot, R., deZamaroczy, M., Baldacci, G., and Bernardi, G. (1980) Current Genetics 1, 173-176.

57. Blanc, H. and Dujon, B. (1980) Proc. Natl. Acad. Sci. U.S.A. 77, 3942-3946.
58. Strausberg, R.L. and Butow, R.A. (1977) Proc. Natl. Acad. Sci. U.S.A. 74, 2715-2719.
59. Tzagoloff, A., Macino, G., Nobrega, M.P. and Li, M. (1979) in Extrachromosomal DNA (D.J. Cummings, et al. eds.) ICN-UCLA Symposium on Molecular and Cellular Biology Vol. XV, Academic Press, N.Y. pp. 339-355.
60. Foury, F. and Tzagoloff, A. (1978) J. Biol. Chem 253, 3792-3797.
61. Slonimski, P.P., Pajot, P., Jacq. C., Foucher, M., Perrodin, G., Kochko, A. and Lamouroux, A. (1978) in Biochemistry and Genetics of Yeast (M. Bacila et al. eds.) Academic Press, N.Y. pp. 339-368.
62. Lamouroux, A., Pajot, P.P., Kochko, A., Halbreich, A., Slonomski, P.P. (1980) in Organization and Expression of the Mitochondrial Genome (C. Saccone and A.M. Kroon eds.) North Holland, Amsterdam, in press.
63. Grivell, L.A., Arnberg, A.C., deBoer, P.H., Borst, P., Bos, J.L., Groot, G.S.P., Hecht, N.B., Hensgens, L.A., Van Ommen, G., and Tabak, H.F. (1979) in Extrachromosomal DNA (D. Cummings et al. eds.) Academic Press, New York, pp. 305-324.
64. Haid, A., Schweyan, R.J., Beckmann, H., Kaudewitz, F., Solioz, M. and Schatz, G. (1979) Eur. J. Biochem. 94, 451-464.
65. Alexander, N.J., Vincent, R.D., Perlman, P.S., Miller, D.H., Hanson, D.K. and Mahler, H.R. (1979) J. Biol. Chem. 254, 2471-2479.
66. Bernardi, G., Faures, M., Piperino, G., and Slonimski, P.P. (1970) J. Mol. Biol. 48, 23-42.
67. Ehrlich, S.D., Thiery, J.P. and Bernardi, G. (1972) J. Mol. Biol. 65, 207-212.
68. Lambertus, A., Hensgens, M., Grivell, L.A., Borst, P., and Bos, J.L. (1979) Proc. Natl. Acad. Sci. U.S.A. 76, 1663-1667.
69. Macino, G. and Tzagoloff, A. (1979) Proc. Natl. Acad. Sci. U.S.A. 76, 131-135.
70. Prunell, A. and Bernardi, G. (1977) J. Mol. Biol. 110. 53-74.
71. Baldacci, G., DeZamaroczy, M., and Bernardi, G. (1980) FEBS Lett. 114, 234-236.
72. Sebald, W. (1977) Biochim. Biophys. Acta 463, 1-27.
73. Sanders, J.P.M., Heyting, C. and Borst, P. (1975) Biochem. Biophys. Res. Commun. 65, 699-707.

74. Goldbach, R.W., Borst, P., Bollen-DeBoer, J.E. and Van Bruggen, E.F.J. (1978) Biochim. Biophys. Acta 521, 169-186.
75. Angerer, L., Davidson, N., Murphy, W., Lynch, D. and Attardi, G. (1976) Cell 9, 81-90.
76. Attardi, G., Palmiro, G., Ching, E.. Crews, S., Gilford, R., Merkel, C. and Ojala, D., in Extrachromo-somal DNA (D.J. Cummings et al. eds.) ICN-UCLA Symposium on Molecular and Cellular Biology Vol. XV Academic Press New York, pp. 443-469.
77. Aloni, Y. and Attardi, G.(1971) Proc. Natl. Acad. Sci. 68, 1757-1761.
78. Murphy, W., Attardi, B., Tu, C., and Attardi, G. (1975) J. Mol. Biol. 99, 809-814.
79. Barrell, B.G., Bankier, A.T. and Drouin, J. (1979) Nature 282, 189-194.
80. Crews, S. and Attardi, G. (1980) Cell 19, 775-784.
81. Barrell, B.G., Anderson, S., Bankier, A.T., DeBruijn, M.H.L., Chen, E., Coulson, A.R., Drouin, J., Eperon, I.C., Nierlich, D.P., Roe, B.A., Sanger, F., Schreier, P.H., Smith, A.J.H., Staden, R. and Young, I.G. (1980) Proc. Natl. Acad. Sci. U.S.A. 77, 3164-3166.
82. Bonitz, S.G., Berlain, R., Coruzzi, G., Li, M., Macino. G., Nobrega, F.G., Nobrega, M.P., Thalenfield, B.E. and Tzagoloff, A. (1980) Proc. Natl. Acad. Sci. U.S.A. 77, 3167-3170.
83. Aujame, L. and Freeman, K.B. (1979) Nucleic Acid Res. 6, 455-469.
84. Heckman, J.E., Sarnoff, J., Alzner-DeWeerd, B., Yin, S. and RajBhandari, U.L. (1980) Proc. Natl. Acad. Sci. U.S.A. 77, 3159-3163.
85. Lagerkvist, U. (1978) Proc. Natl. Acad. Sci. U.S.A. 75, 1759-1762.
86. Grosjean, H.J., DeHenan, S. and Crothers, D.M. (1978) Proc. Natl. Acad. Sci. U.S.A. 75, 610-614.
87. Macino, G., Coruzzi, G., Nobrega, F.G., Li, M. and Tzagoloff, A. (1979) Proc. Natl. Acad. Sci. U.S.A. 76, 3784-3785.
88. Fox, T.D. (1979) Proc. Natl. Acad. Sci. U.S.A. 76, 6534-6538.
89. Martin, N.C., Pham, H.D., Underbrink-Lyon, K., Miller, D. and Donelson,J. (1980) Nature 285-579-581.
90. Sibler, A.P., Berdonne, R. Dirheimer, G. and Martin, R. (1980) C.R. Acad. Sci. Paris 290, 695-698.

THERMOGENIC MITOCHONDRIA

Olov Lindberg, Jan Nedergaard, and Barbara Cannon

The Wenner-Gren Institute, University of Stockholm,
Norrtullsgatan 16, S-113 45 Stockholm, Sweden

INTRODUCTION

Fifteen years ago a review entitled "Thermogenic
Mitochondria" would have been either a truism or an attack
on all that mitochondriology at that time had taught us:
a truism because all working mitochondria release heat,
but this heat was considered only as a waste product and as
a reflection of the fact that biological processes are not
100% efficient; an attack on traditional mitochondriology,
because to suggest heat production as the main function for
certain mitochondria would have been to suggest a too simple
aim for a too complicated apparatus. For heat production
this organelle appeared overqualified, in the same manner
as an idling motor is overqualified for producing heat, compared to a stove. The thermodynamic aim - to produce heat -

C. P. Lee, G. Schatz, G. Dallner (eds.), Mitochondria and Microsomes
in honor of Lars Ernster   ISBN 0-201-04576-1

is so simple that the detour over the mitochondria would
have seemed totally unnecessary.

Today, we accept the existence of a type of mitochondria whose main function is not to be efficient in
conserving energy, but to be efficient in dissipating
energy. This introduction of a new type of mitochondria,
thermogenic mitochondria, is the result of biochemical
investigations of one specific type of mitochondria, those
from brown adipose tissue, and to date, these are the only
mitochondria which can be classified as thermogenic.

Our interest in the metabolism of brown fat mitochondria was initiated on one specific occasion:

> In February 1964 I [Lindberg] was invited to
> lunch by Robert E. Smith. The reason behind
> the invitation was that he had read some
> papers by Ernster and Lindberg on mitochon-
> drial oxidation, and now he wanted to test
> some ideas about the biochemical background
> to a physiological phenomenon he had dis-
> covered (1-3). As he talked, a most fasci-
> nating story emerged about how heat spreads
> within an arousing ground squirrel. He de-
> scribed how the point of ignition coincided
> with a little dark-brownish pad of fatty
> tissue called the hibernating gland. He told
> me also that the "gland" actually was an
> adipose tissue which could probably mobilize
> triglycerides by triggering lipolysis. Smith
> found it tempting to compare the process
> with known metabolic events in white adipose
> tissue, and he asked if I could accept a
> hypothesis involving a cyclic sequence,
> starting with lipolysis and finishing with
> reesterification of liberated fatty acids,
> as a nonsense reaction giving rise to heat.
> As Ernster and I had been advocators of the
> concept of mitochondrial respiratory control
> for more than 10 years, the discussion
> quickly came to involve mitochondria and the

possibility of a nonsense reaction consuming
ATP. This idea was obviously not farfetched,
because when I came back to Stockholm I was
met by an article of Dawkins and Hull (4)
who, in a substantial work on brown adipose
tissue, suggested reesterification as a basis
for nonshivering heat generation. Now I became
really interested and began discussing the
phenomenon with a previous collaborator, now
professor of pediatrics, Rolf Zetterström,
who convinced me that nonshivering heat
production was a very important concept in
pediatrics. These discussions initiated the
start of a new group in the Institute for
the study of brown adipose tissue. The start
coincided with the conclusion of 20 years of
close collaboration with Lars Ernster, who
at that time was appointed professor of biochemistry and consequently had to leave our
Institute, and this made a continued collaboration technically impossible.

Although electron microscopy of brown adipose tissue
revealed that the brown adipocyte consisted mainly of fat
droplets and mitochondria, our first attempts to study the
new field of thermogenesis focused on the possible role of
oxygenases. These could easily be conceived as releasers of
heat from substrates by oxidation without the inhibiting
generation of energy-rich intermediates. This short-cut was
unsuccessful but at least eliminated nonmitochondrial
reactions as heat generators. This forced us to concentrate
on the problem of how mitochondria, which since the dawn of
eucaryotic life have been specialized for efficient ATP
production, at the relatively recent emergence of homeo-
thermic mammals could entirely abandon their zeal of conserving energy.

THE DEVELOPMENT OF A NEW CONCEPT:

NOT A FUTILE CYCLE

The question we then faced was: How do brown fat
mitochondria participate in the heat production? We knew
the substrate and also the final result - but little else.
Our lack of knowledge could be visualized in a model like
this:

MODEL 1 :   fatty acids + O_2 ⟶ ? ⟶ ? ∿∿►heat

In the following we shall see how with time,
through a combination of efforts in our own laboratory and
the work of a series of other groups, it has been possible
to elaborate this model into our present understanding.
(Further reviews on studies of the metabolism of brown
adipose tissue (5,6), isolated brown fat cells (7) and
isolated brown fat mitochondria (8-10) can be found elsewhere.)

THE REESTERIFICATION HYPOTHESIS. This was the hypothesis
for heat production which, as said above, had first been
discussed. It had been suggested by Ball and Jungas (11)
as a reaction for heat generation in white adipose tissue,
and this hypothesis was extended to brown fat by Dawkins
and Hull (4). Thus the heat should result from the futile
cycling of fatty acids between the nonesterified and
esterified states; ATP produced in the mitochondria should
be consumed in the activation process for fatty acids i.e.
in the conversion of these into acyl-CoA esters:

MODEL 2: fatty acids + O_2 ⟶ ATP ⟶ triglycerides ⟵ NE / fatty acids

heat

Our experimental test of this model consisted of measurements of the norepinephrine-(NE)-stimulated oxygen consumption of tissue slices and of the norepinephrine--stimulated lipolysis. From the oxygen consumption we estimated ATP production simply by multiplication by 3; the resulting theoretical rate of ATP production was nearly two orders of magnitude higher than the measured rate of fatty acid release (12).

Thus the suggested ATP-consuming reaction was much too small to utilize the "theoretical" ATP production, and the theory had to be abandoned (12).

THE "CYTOPLASMIC ATPase" HYPOTHESIS. This was a mere generalization of the first hypothesis, in that it postulated the existence of some other cytoplasmic ATPase which could utilize the large amounts of ATP which should theoretically be produced:

MODEL 3: fatty acids + O_2 ⟶ ATP ⟶ "ATPase"

heat

The experimental test was to search for such an ATPase. We were not able to find any, and although it is dubious to argue through negative findings, we felt that such a putative potent ATPase ought to be measurable in some way. We therefore concluded that no potent cytoplasmic ATPase existed and Model 3 could not explain brown fat thermogenesis (12).

Model 3 has later been re-advanced. Horwitz in a series of contributions (e.g. (13,14)) has discussed the possibility that the cytoplasmic ATPase could be the plasma membrane Na^+/K^+-ATPase, which perhaps in preparations from brown fat is affected by catecholamines. However, based on the activities measured by Horwitz and Eaton (15), it seems to us (7) that this specific example of Model 3 has the same difficulties as Model 2: that the activity of the ATPase is much to low to account for the amount of ATP theoretically produced.

THE "NON-PHOSPHORYLATING MITOCHONDRIA" MODEL.   The ATP production discussed above was indeed very theoretical. We were at that time very frustrated because the methods we had used for more than 15 years in our laboratory to produce coupled mitochondria from a large variety of tissues failed to yield functional mitochondria from brown fat. Functional in this context meant phosphorylating mito-chondria, with acceptable P/O ratios and respiratory con-trol. We had earlier even studied whether the effects of e.g. thyroid hormones on metabolism could be explained on the basis of altered phosphorylation efficiency; we had had to conclude that this was not the case (16,17). The only known exception from the general rule that all mito-chondria could show respiratory control was a pathological case, the muscle mitochondria from a young girl suffering from what has become known as Luft's syndrome (18). However, we had been unable to demonstrate oxidative ATP production in brown fat mitochondria - even when the mitochondria were examined under a variety of different conditions (12), and also failing to find a potent ATP-consuming reaction, we somewhat sceptically formulated the following expression (19):

> "Particular attention is being directed to the possibility that heat production may constitute an additional energy-linked process that derives energy directly from intermediates of the respiratory chain-linked energy-transfer system."

In the contemporary nomenclature the model would look like this:

$$\textit{MODEL 4}: \quad \text{fatty acid} + O_2 \longrightarrow \sim X \rightsquigarrow heat$$

Although Model 4 explained all available data, and although similar thoughts were also formulated by R.E. Smith (in an article together with Karl Hittelman and Jane Roberts (20)) we felt that this model in some way was too simple, and time soon showed that this uneasiness was correct.

THE "UNCOUPLING BY FATTY ACIDS" MODEL. In two independent papers, published in 1967, Guillory and Racker (21) and Joel and coworkers (22) demonstrated that it was possible to bring brown fat mitochondria into a coupled state. The modification was in both cases to add albumin to the assay medium; we had earlier tried to wash the mitochondria with albumin (12), but had failed to find any beneficial effect

of this. (We later observed that stored mitochondria produced fatty acids with time (23); this probably explains the
difference between an albumin wash and albumin incubations.)
Thus the biochemical background for thermogenesis in brown
fat seemed clear: the free fatty acids (which were produced
by norepinephrine-stimulated lipolysis) were both the
substrate for thermogenesis and the direct uncouplers of
oxidative phosphorylation (as in vitro uncouplers, free
fatty acids were already well known (24)):

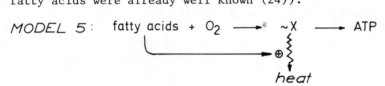

Model 5 was discussed at great length with Lars
Ernster. He warned us from the beginning against believing
in physiological uncoupling caused by free fatty acids.
One of his objections was that we had already seen too many
mistakes based on this inviting hypothesis, and with time he
was proved right.

THE UBIQUITOUS PURINES. The consequence of Model 5 would
be that once the uncoupling fatty acids had been removed,
brown fat mitochondria should be capable of phosphorylation
and show respiratory control, viz.:

MODEL 5 A: fatty acids + O_2 ⟶ ~X ⟶ ATP

albumin   heat

This was the implied action of albumin in the experiments of Guillory and Racker (21) and Joel et al. (22).
At that time Drahota et al. (25) demonstrated that brown

fat mitochondria would respire on endogenous substrate, if
they were provided with carnitine and ATP, and concluded that
this substrate was free fatty acids. When Karl Hittelman
arrived in our laboratory, he continued his earlier work on
brown fat mitochondria (20,26) by extending these studies of
Drahota et al.. According to the model, this new method of
removing fatty acids might also influence the coupling state:

MODEL 5 B: fatty acids + O_2 $\longrightarrow$ ~X $\longrightarrow$ ATP

$$\begin{array}{c} ATP \\ CoA \\ carn \end{array} \Bigg\rangle \quad\quad\quad\quad\quad\quad\quad\quad$$

$$CO_2 + H_2O \quad\quad\quad\quad\quad\quad heat$$

We were able to demonstrate that the addition of ATP,
carnitine and CoA not only led to a transient endogenous
respiration but it also transferred the mitochondria from
an uncoupled to a coupled state (27). It was not even
necessary to combust the fatty acids; it was sufficient to
remove them through the carnitine shuttle from the outside
of the mitochondrial membrane (27). Only a small fraction
of the endogenous fatty acids had to be removed in order
to couple the mitochondria (23,28); it thus seemed that a
specific endogenous fatty acid fraction played a role in the
coupling-uncoupling mechanism of the mitochondria (23).

It appeared then to be the unanimous opinion that
brown fat mitochondria were uncoupled due to the presence of
free fatty acids, and that the removal of some of these, by
either albumin or oxidation, would couple the mitochondria.

A major and at first sight contradictory development
occurred now, starting with a serendipidous observation of
Hohorst and Rafael (29). They were examining the degree of

respiratory control of brown fat mitochondria freely
respiring on added 2-oxoglutarate by studies with oxygen
electrodes. They consequently added ADP to the mitochondria
and observed that a small inhibition of respiration successively ensued. They concluded that perhaps the ATP produced
had a beneficial effect on the respiratory control of the
mitochondria, and by adding ATP they demonstrated that this
was the case. They discussed the specificity of this effect
of ATP and examined therefore the closest analogue: GTP.
GTP also showed this beneficial effect and not only with
oxoglutarate as substrate, but equally so with pyruvate,
glycerophosphate or succinate (29). These studies were
extended by Pedersen (30) who examined the coupling effect
of a series of nucleotides. He observed that ATP, GTP, GDP
and ITP were all about equally effective, whereas CTP and
UDP did not have any effect. However, the mere presence of
albumin, both as tested by the Norwegian group (31) and by
us, did not yield coupled mitochondria. Here was clearly a
contradiction. One series of experiments indicated that
removal of fatty acids was sufficient to couple the mitochondria; another series said that more was needed.

The clarification came when it was realized that
under all conditions where coupled mitochondria were
obtained, purine nucleotides were present in the incubation
medium (32). This apparently obvious point had been masked
by the fact that the purine nucleotides were present for
different reasons. Guillory and Racker (21) and Joel et al.
(22) had used manometric techniques and an ATP trap (ATP,
glucose and hexokinase) for their experiments. (Those of us
who could not confirm their results had used oxygen electrodes and added small amounts of ADP; actually successive

additions of ADP tended to increase the respiratory control
ratio (33)). Hohorst and Rafael (29) had added ATP (and
later GTP) because they believed that it was the ATP formed
which introduced respiratory control; and we ourselves had
added ATP (27) in order to activate the fatty acids. In all
these cases there was an extra effect, a coupling effect,
of added purine nucleotides, and we studied this by experiments with ADP (32).

As the coupling effect of added ADP was not influenced by atractylate or by oligomycin, we concluded that it
was an effect occurring on the outside of the mitochondrial
inner membrane (32) and quite different from the traditional
effects of ADP. This novel function of ADP was termed "the
purine nucleotide effect" because it could be demonstrated
with ATP, ADP, GTP or GDP. Thus the new model looked in
principle like this (with GDP as an illustrative nucleotide):

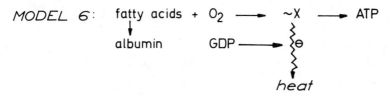

PERMEABILITY CHANGES. When Hans Grav and David Nicholls
joined our group in 1971/1972, a serendipidous finding
again led to an interesting development. For several years
we had used in our group the mitochondrial incubation
medium introduced by Karl Hittelman, and the mitochondrial
respiration proceeded vigorously in this medium (27).
However, when we tested respiration in a routine medium
respiration could not be evoked. The medium Karl Hittelman

had introduced was hypotonic (only 125 mOsm) - but was
nonetheless efficacious. It became clear (34) that the
mitochondria were condensed (and therefore inactive) when
isolated in sucrose, but they would swell quickly in a
hypotonic and slowly in an isotonic KCl solution. If also
the K^+ ionophore valinomycin was added, they would swell
very rapidly also in the isotonic KCl solution. This meant
that the mitochondria had a very high chloride permeability
(34).

The surprising finding was that this chloride permeability was also influenced by purine nucleotides (35).
We thus had to accept a complicated model:

MODEL 7: fatty acids + O_2 $\longrightarrow$ ~X $\longrightarrow$ ATP

albumin   GDP $\longrightarrow$ $\ominus$

$Cl^- \ominus$ perm   *heat*

The Mitchellian chemiosmotic theory had now become
generally accepted, and uncoupling was understood as a
phenomenon of increased H^+ permeability. One could therefore
hypothesize that the uncoupling was also a permeability
effect, viz.:

MODEL 8: fatty acids + O_2 $\longrightarrow$ $\Delta\mu_{H^+}$ $\longrightarrow$ ATP

H^+

albumin   GDP $\longrightarrow$ $\ominus$ $\sim\!\!\sim\!\!\sim$ *heat*

perm

$Cl^- \ominus$ perm

It was possible to confirm that GDP also inhibited
proton permeability and we demonstrated that the three
phenomena, uncoupling, H^+ permeability and Cl^- permeability,
were relatively equally sensitive to a long series of
different nucleotides (36). It therefore appeared that the

purine nucleotide effect on brown fat mitochondria consisted
in a parallel inhibition of two different permeabilities;
a somewhat cumbersome model.

ONLY ONE PERMEABILITY. Mitchell had stated (37) that H^+ and
OH^- movements in opposite directions are indistinguishable
in the chemiosmotic theory. A great simplification occurred
when Nicholls realized that this could explain the permeabilities of brown fat mitochondria (38). Thus Cl^- and OH^-
(and Br^-) permeabilities were probably manifestations of
only one (anion) permeability, and Cl^- and Br^- permeabilities could be disregarded as physiological phenomena
in themselves:

$$
MODEL\ 9: \quad \text{fatty acids} + O_2 \longrightarrow \Delta\mu_{H^+} \longrightarrow ATP
$$

A BINDING SITE. As purine nucleotides - of which GDP is
the preferred example for experimental studies - interact
with mitochondria in a very specific way, they ought also
to interact at a very specific site. An association of GDP
to brown fat mitochondria had already been measured by
Rafael et al. in 1971 (39). They found that the mitochondrial association amounted to about 0.4 nmoles of GDP per
mg protein, and that this association was saturated at

micromolar concentrations. As these experiments were performed at such an early date, it was still not realized that
the association was due to binding, and it was actually
interpreted as an uptake. From our earlier experiments (32)
it could be deduced that the binding site was on the outer
side of the inner mitochondrial membrane, and Nicholls (40)
investigated the nature of this site. A GDP-binding protein
could now be visualized:

MODEL 10 : fatty acids + O$_2$ ⟶ Δμ$_{H^+}$ ⟶ ATP

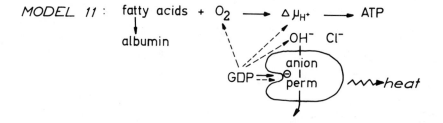

The indicated simple relationship between mitochondrial function and GDP-binding was confirmed by a series of
elegant studies. They were made possible through the method
Nicholls developed in our laboratory for the determination
of $\Delta\mu_H^+$ in mitochondria (41). The relationships between GDP-
-binding and the anion permeability (42), respiration (43),
both components of $\Delta\mu_H^+$ (43), the apparent proton conductance
C_H^+ (44) and the phosphorylation potential (45) were studied
as indicated in the model:

MODEL 11 : fatty acids + O$_2$ ⟶ Δμ$_{H^+}$ ⟶ ATP

THE NATURE OF THE BINDING SITE. In the models presented so
far, it has been implicated that brown fat mitochondria are
a homogeneous species. This is not the case. It was early
demonstrated by Rafael et al. (46,47) and by the Norwegian
group (48,49) that the degree of coupling of the isolated
mitochondria varied in accordance with the physiological
condition of the animal. According to the models this should
mean a change in the anion (OH$^-$) permeability - and perhaps
a change in the number of GDP-binding sites, viz.:

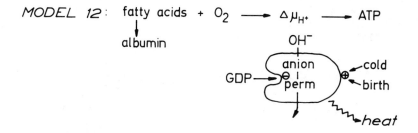

Such changes in GDP-association in the expected
directions had already been found at an early stage by
Rafael et al. in warm-acclimatizing guinea pigs after birth
(39), but these changes were at that time still interpreted
as changes in uptake. When the existence of the binding site
had been demonstrated, similar experiments were again undertaken by Rafael and Heldt in newborn guinea pigs (50), and
by Desautels et al. (51) and by us (52,53) in rats at
different environmental temperatures. As the simple
relationship between thermogenic capacity and the amount of
binding sites was evident, the next step would be to try to
identify the binding site. One possibility was to label the
site covalently with a radioactive azido-derivative of GDP.
Such experiments were also performed (54), but no site could

be identified as the specific activity of the azido-GDP was
too low.

Progress came from quite another series of investigations. Studies of cold-induced changes in the phospholipids of brown fat mitochondria had been performed by
Ricquier et al. (55,56), and as an extension of these,
Ricquier and Kader (57) studied the polypeptide composition
of brown fat mitochondria in cold-adapted and warm-adapted
rats. All peptides appeared unchanged, with one exception.
A polypeptide of molecular weight 32 000 significantly
increased in the mitochondria from cold-adapted rats. We
thus had this situation:

MODEL 13:

From the information available at that time, it now
seems that it should have been tempting to simplify the
model:

MODEL 14:

However, this simplification was not obvious until
Heaton et al. (58) successfully with a high-specific-activity
[32]P-labelled azido-ATP demonstrated that the 32 000 protein
and the GDP-binding protein are one and the same.

THE THERMOGENIN CONCEPT. As now realized, a series of
properties of brown fat mitochondria are all secondary to
the existence of one specific polypeptide in their inner
membrane. This polypeptide, however, has for technical
reasons been known under a series of different names: "the
32 000 protein", "the GDP-binding protein", "the uncoupling
protein", "the anion permease" etc.. To simplify terminology
as much as reality, we have reasoned that one name for this
polypeptide should be sufficient. We have suggested the name
THERMOGENIN (59):

MODEL 15:   fatty acids + O_2 $\longrightarrow$ $\Delta\mu_{H^+}$ $\longrightarrow$ ATP

albumin

OH^-

GDP $\longrightarrow$ thermo-genin $\leadsto$ heat

A PHYSIOLOGICAL ANTAGONIST. In the preceding models we have
illustrated the inhibitory effect of GDP, i.e. purine
nucleotides. However, as we early pointed out (32), the cytosolic purine nucleotides always surround the mitochondrion
in situ, and as the coupling effectiveness of adenosine and
guanosine di- and triphosphates is about the same (32,36),

we have the paradoxical situation of finding thermogenic
mitochondria in a milieu where they would apparently never
be uncoupled. Thus one has to postulate a physiological
antagonist (60). Already in 1973 we concluded that (32):

> "variations in the energy charge of the cyto-
> plasmic adenine nucleotide pool, and most
> probably changes in guanine nucleotides, do
> not provide a plausible regulatory mechanism
> but rather appear to be an omnipresent factor
> influencing respiration, whereas variations in
> the supply of fatty acids which are known to
> occur during sympathetic stimulation of the
> tissue provide a more likely regulation for
> the degree of respiratory control exhibited
> by the tissue."

We had thus returned to the fatty acids as the likely
physiological uncouplers of oxidative phosphorylation in
brown fat mitochondria. However, we considered them still
too unspecific. Later we realized that activated fatty acids,
acyl-CoAs, have much more of the specificity required of the
physiological uncoupler. We demonstrated that a long-chain
acyl-CoA, palmitoyl-CoA, could compete out bound GDP from
the thermogenin binding sites (60). This was not unexpected
because of the ADP-moiety which is part of CoA. In a more
functional test palmitoyl-CoA was able to reintroduce anion
permeability in mitochondria where this permeability had
been inhibited by GDP - i.e. acyl-CoA did function as a
GDP-antagonist (60). However, the experiments which
directly should demonstrate uncoupling of respiration by
palmitoyl-CoA did not succeed due to technical problems.
Although an important part of the evidence was thus missing,
we postulated the following model:

MODEL 16 : fatty acids + O_2 ⟶ $\Delta\mu_{H^+}$ ⟶ ATP

ATP, CoA ⟶ acyl – CoA ⟶ GDP⊖ (thermo-genin) OH⁻ ⟶⟶ heat

        In fact we already had available the evidence for a
direct uncoupling effect of acyl-CoAs. Fig. 1 is adapted
from a study published in 1971, concerning the fate of
added fatty acids (61). Note in Fig. 1A how the addition of
CoA, ATP and carnitine, as discussed in Model 5B, does
induce an endogenous respiration, and how the mitochondria
are then transferred into a coupled state. This is seen by
the fact that the addition of a substrate (acetyl-carnitine)
does not affect respiration. Only after addition of the
artificial uncoupler FCCP is the substrate combusted. If
palmitoyl-CoA is used as substrate, the effect is quite
different. Note how (Fig. 1B) added palmitoyl-CoA induces
its own respiration, i.e. exactly the expected, uncoupling
effect. Note also how in Fig. 1C addition of free fatty acid
is sufficient for similar uncoupling, provided that ATP and
CoA are present for activation of the fatty acid to acyl-
-CoA.

        After reinterpretation of these results, the only
part of the evidence lacking was to show that the uncoupling
by palmitoyl-CoA was specific, i.e. inhibitable by GDP. This
was found to be the case (59,62).

        Thus acyl-CoAs can function in a way that would be
expected of the physiological uncoupler. This naturally only

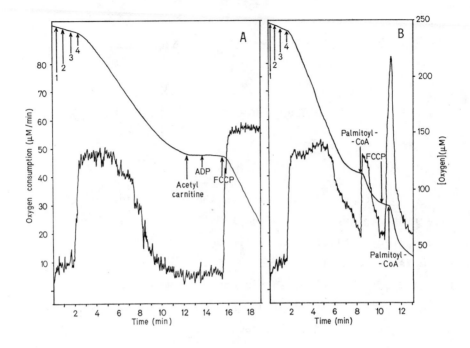

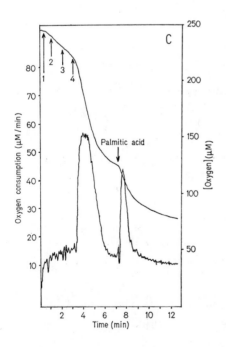

Fig. 1. Exogenous substrate respiration in brown fat mitochondria following a burst of respiration on endogenous substrate.

Additions:

1 = mitochondria
2 = CoA
3 = ATP
4 = carnitine

Modified from (61).

demonstrates that the acyl-CoAs <u>can</u> be the physiological
uncoupler, not that they do function as such within the
cell. However, to date no better candidate has been
suggested.

THE PRESENT MODEL

Fig. 2 is a scheme for the functioning of brown fat
mitochondria, as we see it today. The activity of the
respiratory chain results in the formation of a proton
electrochemical gradient over the mitochondrial membrane.
This gradient can be dissipated in two ways, both the traditional one and a brown-fat-specific one. In the traditional way protons can reenter the mitochondria through
the ATP synthetase; i.e. oxidative phosphorylation is

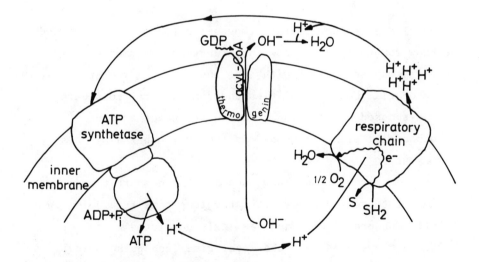

Fig. 2. <u>The present model for the function of brown fat
mitochondria.</u> (The figure is adapted from (63).)

possible, although the activity of the ATP synthetase is low compared to the activity of the respiratory chain (64). This is due to there being a low number of ATP synthetase complexes present, not to a structural alteration (64). The brown-fat-specific way is a dissipation of the gradient through the exit of hydroxide ions through thermogenin. The thermogenin molecule has recently been isolated (65,66) and may be a dimer (67). This channel for OH^- is open when acyl--CoAs occupy the purine nucleotide binding site. In the absence of acyl-CoAs the channel is closed by the cytosolic purine nucleotides (GDP, ADP, ATP, GTP). The passage of OH^- short-circuits the protonical circuit, i.e. the mitochondria lose respiratory control, and heat is produced. Thermogenin endows the mitochondria with thermogenic capacity.

OUTLOOK

Today we know that thermogenic mitochondria exist and we know something about how they function. Thermogenic mito-chondria may be a genus with just one species: the brown fat mitochondria; but similarities and similar reactions have been discussed in relation to at least one other type of mitochondria: plant mitochondria (68,69). Other aspirants may exist.

Whereas we today know much about the short-term regu-lation of thermogenin activity, the long-time regulation is totally obscure. A wide and seemingly unrelated number of different physiological, pharmacological and pathological states results in different amounts of thermogenin being present within the brown fat of animals (59) and - probably -

humans. What are the biochemical factors behind these changes?

Further, if we learn what is the messenger to brown fat for thermogenin synthesis, how then is this synthesis effected in molecular terms? We know that the concentration of thermogenin in the mitochondria may change, but what − teleologically speaking − is the reason for this? As the respiratory capacity of mitochondria is not changed by the same stimuli which alter thermogenin content (53,70), why does nature produce brown fat mitochondria with redundant respiratory capacity?

And how are mitochondria with different concentrations of thermogenin manufactured? Are thermogenin molecules inserted into preexisting mitochondria? Or − perhaps more probable − is a new generation of mitochondria initiated, all with a higher thermogenin concentration than their forerunners? And how are they then governed to become like this?

The development of these concepts will definitely not be a futile cycle.

Acknowledgements

The Swedish Natural Science Research Council has supported our studies on the mechanism of thermogenesis since 1968.

REFERENCES

1. Smith, R. E. (1961) The Physiologist 4, 113.
2. Smith, R. E. and Hock, R. J. (1963) Science 140,
    199-200.
3. Smith, R. E. (1964) Science 146, 1686-1689.
4. Dawkins, M. J. R. and Hull, D. (1964) J. Physiol.
    (Lond.) 172, 216-238.
5. Smith, R. E. and Horwitz, B. A. (1969) Physiol. Rev.
    49, 330-425.
6. Brown Adipose Tissue, (1970) (O. Lindberg, ed.)
    Elsevier, New York.
7. Nedergaard, J. and Lindberg, O. (1981) Int. Rev.
    Cytol., in press.
8. Flatmark, T. and Pedersen, J. I. (1975) Biochim.
    Biophys. Acta 416, 53-103.
9. Cannon, B. and Lindberg, O. (1979) Meth. Enzym. 55,
    65-78.
10. Nicholls, D. G. (1979) Biochim. Biophys. Acta 549,
    1-29.
11. Ball, E. G. and Jungas, R. L. (1961) Proc. Natl. Acad.
    Sci. U.S. 47, 932-941.
12. Lindberg, O., dePierre, J., Rylander, E. and Afzelius,
    B. A. (1967) J. Cell Biol. 34, 293-310.
13. Horwitz, B. A. (1973) Am. J. Physiol. 224, 352-355.
14. Horwitz, B. A. (1978) in Effectors of Thermogenesis,
    (L. Girardier and J. Seydoux, eds.) Exper. Suppl. 32,
    pp. 19-23, Birkhäuser Verlag, Basel and Stuttgart.
15. Horwitz, B. A. and Eaton, M. (1975) Eur. J. Pharmacol.
    34, 241-245.
16. Tata, J. R., Ernster, L., Lindberg, O., Arrhenius, E.,
    Pedersen, S. and Hedman, R. (1963) Biochem. J. 86,
    408-428.
17. Gustafsson, R., Tata, J. R., Lindberg, O. and Ernster,
    L. (1965) J. Cell Biol. 26, 555-578.
18. Ernster, L. and Luft, R. (1963) Exp. Cell Res. 32,
    26-35.
19. Lindberg, O., dePierre, J., Rylander, E. and Sydbom,
    R. (1966) Third FEBS Meeting, p. 139, M 48.
20. Smith, R. E., Roberts, J. C. and Hittelman, K. J.
    (1966) Science 154, 653-654.
21. Guillory, R. J. and Racker, E. (1967) Biochim. Biophys.
    Acta 153, 490-493.
22. Joel, C. D., Neaves, W. B. and Rabb, J. M. (1967)
    Biochem. Biophys. Res. Commun. 29, 490-495.

23. Bulychev, A., Kramar, R., Drahota, Z. and Lindberg, O.
     (1972) Exp. Cell Res. 72, 169-187.
24. Pressman, B. C. and Lardy, H. A. (1956) Biochim. Bio-
     phys. Acta 21, 458-466.
25. Drahota, Z., Honová, E. and Hahn, P. (1968)
     Experientia 24, 431-432.
26. Hittelman, K. J., Fairhurst, A. S. and Smith, R. E.
     (1967) Proc. Natl. Acad. Sci. U.S. 58, 697-702.
27. Hittelman, K. J., Lindberg, O. and Cannon, B. (1969)
     Eur. J. Biochem. 11, 183-192.
28. Cannon, B. (1971) Doctoral dissertation. University of
     Stockholm.
29. Hohorst, H.-J. and Rafael, J. (1968) Hoppe-Seyler's
     Z. Physiol. Chem. 349, 268-270.
30. Pedersen, J. I. (1970) Eur. J. Biochem. 16, 12-18.
31. Grav, H. J., Pedersen, J. I. and Christiansen, E. N.
     (1970) Eur. J. Biochem. 12, 11-23.
32. Cannon, B., Nicholls, D. G. and Lindberg, O. (1973)
     in Mechanisms in Bioenergetics, (G. F. Azzone, et al.,
     eds.) pp. 357-364, Academic Press, New York and
     London.
33. Pedersen, J. I. and Grav, H. J. (1972) Eur. J. Bio-
     chem. 25, 75-83.
34. Nicholls, D. G., Grav, H. J. and Lindberg, O. (1972)
     Eur. J. Biochem. 31, 526-533.
35. Nicholls, D. G. and Lindberg, O. (1973) Eur. J. Bio-
     chem. 37, 523-530.
36. Nicholls, D. G., Cannon, B., Grav, H. J. and Lindberg,
     O. (1974) in Dynamics of Energy-Transducing Membranes,
     (L. Ernster, et al., eds.) pp. 529-537, Elsevier,
     Amsterdam.
37. Mitchell, P. (1966) Chemiosmotic Coupling in Oxidative
     and Photosynthetic Phosphorylation, Glynn Res. Ltd.,
     Bodmin, Cornwall, England.
38. Nicholls, D. G. (1976) FEBS Lett. 61, 103-110.
39. Rafael, J., Heldt, H.-W. and Hohorst, H.-J. (1971)
     VII FEBS Meeting, Varna. Abstract No. 620.
40. Nicholls, D. G. (1976) Eur. J. Biochem. 62, 223-228.
41. Nicholls, D. G. (1974) Eur. J. Biochem. 50, 305-315.
42. Nicholls, D. G. (1974) Eur. J. Biochem. 49, 585-593.
43. Nicholls, D. G. (1974) Eur. J. Biochem. 49, 573-583.
44. Nicholls, D. G. (1977) Eur. J. Biochem. 77, 349-356.
45. Nicholls, D. G. and Bernson, V. S. M. (1977) Eur. J.
     Biochem. 75, 601-612.
46. Rafael, J., Klaas, D. and Hohorst, H.-J. (1968)
     Fifth FEBS Meeting. Abstract No. 446.

47. Rafael, J., Klaas, D. and Hohorst, H.-J. (1968) Hoppe-Seyler's Z. Physiol. Chem. 349, 1711-1724.
48. Pedersen, J. I., Christiansen, E. N. and Grav, H. J. (1968) Biochem. Biophys. Res. Commun. 32, 492-500.
49. Christiansen, E. N., Pedersen, J. I. and Grav, H. J. (1969) Nature 222, 857-860.
50. Rafael, J. and Heldt, H. W. (1976) FEBS Lett. 63, 304-308.
51. Desautels, M., Zaror-Behrens, G. and Himms-Hagen, J. (1978) Can. J. Biochem. 56, 378-383.
52. Cannon, B., Nedergaard, J., Romert, L., Sundin, U. and Svartengren, J. (1978) in Strategies in Cold: Natural Torpidity and Thermogenesis, (L. C. H. Wang and J. W. Hudson, eds.) pp. 567-594, Academic Press, New York.
53. Sundin, U. and Cannon, B. (1980) Comp. Biochem. Physiol. 65B, 463-471.
54. Cannon, B. (1976) Tenth Internat. Congress Biochem.. Abstract No. 06-7-112.
55. Ricquier, G., Mory, G. and Hemon, P. (1975) FEBS Lett. 53, 342-346.
56. Ricquier, D., Mory, G. and Hemon, P. (1976) Pflügers Arch. 362, 241-246.
57. Ricquier, D. and Kader, J.-C. (1976) Biochem. Biophys. Res. Commun. 73, 577-583.
58. Heaton, G. M., Wagenvoord, R. J., Kemp, A., Jr. and Nicholls, D. G. (1978) Eur. J. Biochem. 82, 515-521.
59. Cannon, B., Nedergaard, J. and Sundin, U. (1981) in Survival in Cold, (X. J. Musacchia and L. Janský, eds.) Elsevier/North Holland, Amsterdam, in press.
60. Cannon, B., Sundin, U. and Romert, L. (1977) FEBS Lett. 74, 43-46.
61. Cannon, B. (1971) Eur. J. Biochem. 23, 125-135.
62. Cannon, B., Nedergaard, J. and Sundin, U. (1981) Proc. Int. Symp. Thermal Physiol., Pécs, Hungary, in press.
63. Cannon, B. and Nedergaard, J. (1981) in Biochemical Development of the Fetus and the Neonate, (C. T. Jones, ed.) Elsevier, Amsterdam, in press.
64. Cannon, B. and Vogel, G. (1977) FEBS Lett. 76, 284-289.
65. Ricquier, D., Gervais, C., Kader, J.-C. and Hemon, P. (1979) FEBS Lett. 101, 35-38.
66. Lin, C. S. and Klingenberg, M. (1980) FEBS Lett. 113, 299-303.
67. Lin, C. S., Hackenberg, H. and Klingenberg, M. (1980) FEBS Lett. 113, 304-306.

68. Nedergaard, J. and Cannon, B. (1979) Meth. Enzym. <u>55</u>, 1-28.
69. Moore, A. L. and Rich, P. R. (1980) Trends Biochem. Sci. <u>5</u>, 284-288.
70. Himms-Hagen, J. and Desautels, M. (1978) Biochem. Biophys. Res. Commun. <u>83</u>, 628-634.

SUBMITOCHONDRIAL PARTICLES AND ENERGY TRANSDUCTION

C. P. Lee[a] and B. T. Storey[b]

[a]Department of Biochemistry
Wayne State University School of Medicine
Detroit, Michigan 48201

[b]Henry Watts Neuromuscular Disease Research Center
and Department of Physiology, Obstetrics and Gynecology
University of Pennsylvania School of Medicine
Philadelphia, Pennsylvania 19104

INTRODUCTION

It is well established that the inner mitochondrial membrane is the site of the enzymes of the respiratory chain and those of its associated energy transducing processes: in particular, the ATP synthetase and the energy-linked pyridine nucleotide transhydrogenase. The inner membrane also functions as a barrier of limited permeability between the cytosol and the mitochondrial matrix, and so contains translocation systems for Krebs Cycle intermediates, adenine nucleotides, and cations. Fragmentation of isolated mitochondrial preparations by mechanical disruption yields a preparation now universally known as submitochondrial particles. Since the outer membrane contributes less than 10% to the total mitochondrial

Abbreviations: 9AA, 9-aminoacridine; 9ACMA, 9-amino-3-chloro-7-methoxyacridine; DCCD, dicyclohexylcarbodiimide; PMS, phenazine methosulfate; QA, quinacrine; TMPD, tetramethyl phenylene diamine; S-13, 5-chloro-3-t-2'-chloro-4'-nitrosalicylanilide; RCI, Respiratory Control Index.

C. P. Lee, G. Schatz, G. Dallner (eds.), Mitochondria and Microsomes
   in honor of Lars Ernster   ISBN 0-201-04576-1

membrane mass (1), these fragments, as recovered by differential centrifugation, are effectively pieces of inner mitochondrial membrane. In the course of the fragmentation
and fractionation, the enzymes and cofactors in the mitochondrial matrix and intermembrane space, either free or
loosely bound to the membrane, are removed (2), while the
enzymes of the respiratory chain and energy transducing
processes are retained. The fragments usually take the
form of vesicles with inversion of membrane orientation
relative to intact mitochondria (3), so that the membrane-
bound enzymes originally interacting with components in
the matrix space are freely accessible to solutes in the
suspending medium. The dehydrogenases and the reversible
ATP synthetase, shielded in intact mitochondria by the
permeability barrier of the inner membrane, can interact
directly with their respective substrates. Submitochondrial particles constitute a simple and convenient experimental system for studying the respiratory chain-linked
electron transfer and energy transduction reactions, and,
so, have played an important role in the development of
the current concepts in mitochondrial bioenergetics.

One expected result of the utility of submitochondrial particles as an experimental system is an extensive
literature. A computer search revealed that more than
1300 publications related to submitochondrial particles
appeared in print during the last decade alone (1971 -
1980). In fact, much of the ground work was laid down
earlier, in the 1960's. For this review, even a brief
summary of all these studies would clearly be impossible
and impractical. We shall summarize first the well established energy transducing functions of submitochondrial
particles, then focus our attention on those issues which
remain controversial and unresolved.

PREPARATIONS OF SUBMITOCHONDRIAL PARTICLES

Before we discuss the energy transducing functions of
submitochondrial particles, let us first survey briefly
the most commonly employed submitochondrial particle preparations. Historically, the first may be considered to
be the Keilin-Hartree preparation (4) from heart muscle,
which was made directly from muscle mince before methods
for isolation of mitochondria had been developed. The
Keilin-Hartree heart muscle preparation has been used for
studying the reaction sequence of electron transfer along
the respiratory chain without the involvement of energy
coupling, and for the isolation of the electron transfer
complexes and individual components. The availability of
isolated mitochondria from liver and heart in quantity
made possible the preparation of submitochondrial particles from these sources by the mid-1950's. Both digitonin particles (5) (later identified as mainly mitoplasts
(6), which still retain at least part of the mitochondrial
matrix) and sonic particles (7) from liver mitochondria
were introduced at about the same time. Subsequently,
many kinds of submitochondrial particle preparations
derived from beef heart mitochondria by mechanical disruption were reported (cf. ref. 8), having been facilitated
by the introduction of methods for large scale preparations
of beef heart mitochondria in Professor Green's laboratory
(9) and by the discovery that frozen beef heart mitochondria can serve as well as fresh ones for preparing tightly
coupled submitochondrial particles. These submitochondrial
preparations are diverse and have borne much of the experimental load in studies of energy coupling. We have classified the types of preparations into 3 groups: conventional,
modified, and assorted other.

Conventional Preparations. These are submitochondrial particles prepared from isolated beef heart mitochondria, either fresh or frozen, by sonic disruption and recovered by differential centrifugation (10). The difference in functional activities among the preparations arises from the composition of the sonicating medium. Four of the most commonly used types are shown in Table I. Preparations denoted Mg-ATP (11, 12) and ETP_H (13) particles possess high phosphorylating efficiency and are often referred to as phosphorylating particles; they have been used extensively in studying various kinds of energy-linked reactions involving ATP, either as the product of energy conservation or as an energy-yielding substrate. Preparations denoted EDTA- (10, 14) and A-particles (15) possess very low phosphorylating efficiency unless they are supplemented with purified ATPase, F_1, (16) or an appropriate amount of either oligomycin (3, 17, 18) or DCCD (19). EDTA particles exhibit an oligomycin-induced respiratory control, whose magnitude shows that the energy yielding efficiency of these particles is close to that usually found with intact mitochondria (20). The oligomycin-recoupled EDTA particles can therefore serve as a useful tool in studying the respiratory chain-linked energy transducing processes without the involvement of the phosphorylation apparatus. A-particles exhibit properties qualitatively similar to EDTA particles, but the relatively high pH employed in making the A-particles results in very low succinate oxidase activity.

All these preparations are considered to be in the form of sealed vesicles with inverted membrane orientation; they are often called "inside-out" particles.

Table I

DEPENDENCE OF PHOSPHORYLATING EFFICIENCY OF SUBMITOCHONDRIAL PARTICLES

ON THE COMPOSITION OF THE SONICATING MEDIUM

Preparation	Essential reagent(s) in sonicating medium	ATP/O (succinate)	Reference
Mg-ATP particles	Mg^{++} , ATP	1.5 - 1.8	11, 12
ETP_H	Mg^{++} , Mn^{++} , ATP & succinate	0.8 - 1.2	13
EDTA particles	2 mM EDTA, pH 8.3 - 8.6	< 0.2	10, 14
A-particles	0.6 mM EDTA, NH_4^+ , pH 9.2	< 0.1	15

All conventional preparations of submitochondrial particles examined so far with the electron microscope have had the appearance of vesicles encrusted with the 90 Å subunits of the ATP synthetase, F_1 (21). The particles are poly-disperse in size. The EDTA particles fall in the size range of diameters between 500 to 3000 Å (22) with a mean diameter of 1850 Å as determined by light scattering (23).

Modified Preparations. These are conventional submito-chondrial particles subjected to further treatments which remove the readily dissociable part of the ATP synthetase, F_1 (21). Two preparations are derived from EDTA particles by treatment with either cardiolipin micelles + ATP (24) or urea (25), which are designated as D-ESP and ESU, respectively. The rate of NADH oxidation is unaltered in these preparations. On the other hand, the preparation derived from Mg-ATP particles by urea treatment exhibits a 4-fold stimulation in the rate of NADH oxidation (26). In all three preparations, energy coupling linked to NADH oxidation is retained virtually intact as shown by oligo-mycin-induced respiratory control. During the era of coupling factors in the 1960's, a large number of "depleted" or "nonphosphorylating" submitochondrial particle prepara-tions were reported in the literature. These were mainly used as tools to test the enzymic activities of coupling factors, since most of the isolated coupling factors, with the exception of those possessing the actual ATPase acti-vity, had no known intrinsic enzymic activity and could only be assayed by reconstitution with membrane prepara-tions deficient in the functional activity of interest. For an excellent review of interactions between coupling

factors and their reconstitution with depleted submitochondrial particles, the reader is referred to the one by Beechey
and Cattell (8).

Assorted Other Preparations. These are submitochondrial
particles prepared from other tissues, by methods other
than sonication, and/or of membrane orientation other than
"inside-out". We consider here a conventional preparation
from rat liver, preparations which are "right-side-out",
and preparations which appear to lack a defined membrane
orientation entirely.

Phosphorylating rat liver submitochondrial particles
are usually prepared by sonicating water-washed mitoplasts
in unbuffered water medium (27). Mg-ATP particles cannot
be prepared from liver mitochondria for technical reasons.
The combination of ATP + Mg^{++} induces a drastic change in
mitochondrial conformation from the orthodox to the condensed form (28), which, in turn, makes the fragmentation
very difficult. F_1-depleted SMP are obtained by treatment
of SMP with urea; both ATP synthesis and ATP utilization
can be restored upon reconstitution of these particles
with purified soluble F_1 (29).

"Right-side-out" submitochondrial particles can be
prepared from rat liver mitochondria by cholate treatment
of mitoplasts (30). A purer preparation of "right-side-
out" rat liver submitochondrial particles is obtained by
sonication of the mitochondria, followed by fractionation
of the sonic particles on a Sepharose-cytochrome c column
(31). This method of affinity chromatography (32) works
on the principle that the cytochrome c binding site is exposed to the medium in "right-side-out" particles, so that
these bind preferentially to the column. In practice,

particles of both orientations bind cytochrome $\underline{c}$ (10), so
that both bind to the column. The "inside-out" particles
from rat liver were readily eluted with 0.1 M KCl while the
"right-side-out" ones required 1 M KCl for elution, giving
a clean separation (31).

Two populations of particles, designated as X- and
Y-fragments, have been obtained by sucrose density gradient fractionation of EDTA particles (22). The X-fragments
are the predominant fraction. They consist of F_1-encrusted
particles with diameters ranging from 500 to 2000 $\overset{o}{A}$ and
possess high energy-linked functional activities comparable to those exhibited by conventional EDTA particles.
The Y-fragments consist of smooth particles with a diameter of 300 to 900 $\overset{o}{A}$. In contrast to X-fragments, the
endogenous cytochrome $\underline{c}$ of Y-fragments can be readily depleted by salt extraction. In addition, the Y-fragments
exhibit a high reactivity towards externally added cytochrome $\underline{c}$, as shown by a specific activity of NADH-cytochrome $\underline{c}$ reductase more than 4-fold, and a specific activity of cytochrome $\underline{c}$ oxidase more than 2.5-fold, that of
the X-fragments. On the basis of these properties, the
membrane orientation of the Y-fragments is opposite to
that of the X-fragments, namely "right-side-out". Yet the
rotenone-sensitive NADH dehydrogenase of the respiratory
chain, which is on the matrix side of the inner membrane,
is also fully accessible to its substrate. This would not
be expected in "right-side-out" particles with the same
membrane orientation as intact mitochondria, which are impermeable to NADH. The apparent conflict between these
properties disappears if one assumes that the Y-fragments
have no definite membrane orientation: they could be infi-
nitely leaky vesicles.

Submitochondrial particles obtained from swollen rabbit skeletal muscle mitochondria by disruption in a French pressure cell clearly have no unique membrane orientation. These particles exhibit respiratory control with NADH, succinate and L-3-glycerolphosphate (L-3-GP) as substrates (33). The maximal respiratory rate with each substrate catalyzed by these particles is comparable to that seen with intact mitochondria in the case of succinate and L-3-GP and to that seen with cholate-treated mitochondria in the case of NADH (33). Cholate does not increase the maximal rate of oxidation of any of the three substrates (34). It is now well established that L-3-GP dehydrogenase is on the cytosolic side of the inner mitochondrial membrane (35, 36), the side opposite to the other two dehydrogenases; yet all three dehydrogenases have full access to their substrates. With these particles, and presumably with the Y-fragments, there exists no functional "inside" or "outside", so that no transmembrane gradient can be generated or sustained. These types of membrane preparations are just beginning to be used in studies of the role of the transmembrane electrochemical potential in mitochondrial energy coupling.

ENERGY TRANSDUCING FUNCTIONS

Oxidative Phosphorylation. The reaction of oxidative phosphorylation of ADP into ATP is the primary raison d' etre of the mitochondrion in the intact cell. Submitochondrial particles carry out oxidative phosphorylation but lack the accessibility barrier encountered in the intact organelle. Consequently, one can study in detail the interaction of substrates and inhibitors with enzymes; and one can dissociate the enzyme subunits acting at the membrane aqueous

phase interface to acquire an entree into the resolution
of this complex system. The interaction between F_1 and F_1
inhibitor protein and the relationship of this interaction
with the energy level of the mitochondrial membrane have
been pursued (26). Gomez-Puyou et al. (37) recently showed
that the protein inhibitor inhibits both the hydrolytic and
synthetic activities of the ATPase complex by reversible
modification of the catalytic site of the F_1 component
modulated by the membrane potential as well as the concentration of ATP.

Respiratory Control. EDTA particles exhibit an oligomycin-
induced respiratory control which can be relieved by uncouplers. The respiratory activities of EDTA particles are
summarized in Table II. The oxidation rates of NADH, succinate, and ascorbate + PMS are low in oligomycin-treated
EDTA particles, but rapid in the presence of an uncoupler
of oxidative phosphorylation, indicating that all three
energy coupling sites of the respiratory chain are functional. The uncoupled rate of NADH oxidation is sufficiently rapid to demonstrate that the electron transfer components of the respiratory chain in these particles retain
kinetic competence characteristic of intact mitochondria.
Submitochondrial particles retain another aspect of energy
coupling seen in intact mitochondria: in the energized
state, cytochrome b is largely reduced, while in the de-
energized or uncoupled state, it is largely oxidized (20).
These criteria of respiratory control and uncoupler-
inducible oxidation of cytochrome b have proved to be useful
for demonstrating energy coupling capacity in particles
lacking an efficient phosphorylating system.

Table II

RESPIRATORY ACTIVITIES OF OLIGOMYCIN-COUPLED EDTA PARTICLES WITH
VARIOUS SUBSTRATES

Substrate	Respiratory rate (nAtom O/min·mg prot.)		RCI
	----	+ S-13	
NADH	104	980	9.4
Succinate	170	670	3.9
Ascorbate + PMS	670	980	1.5

Oxygen consumption was measured polarographically. The reaction mixture consisted of 150 mM sucrose, 30 mM Tris-sulfate, pH 7.5, and 0.6 mg oligomycin-treated EDTA particles (10). Other additions where indicated were: 0.8 mM NADH; 5 mM succinate; 8.3 mM ascorbate + 1.3 µM PMS; 0.8 µM S-13. Total volume: 3.0 ml; temperature: 25°C.

One aspect of oligomycin-induced respiratory control deserves further attention. This is the relationship between the extent of respiratory control and the rate of electron transport. Some years ago, it was found (20) that the respiratory control index (RCI) associated with NADH and succinate oxidation in EDTA particles is virtually unaffected when the respiration is partially inhibited by firmly-bound electron transport inhibitors: rotenone, antimycin A and KCN in the case of NADH oxidase, and antimycin A and KCN in the case of succinate oxidase. These findings were challenged (38) on technical grounds, but repetition of these experiments confirmed the earlier findings (20). Figure 1 shows the relationship between the RCI and respiratory rates of NADH oxidase (A) and succinate oxidase (B). The respiratory rates are regulated by partially inhibitory concentrations of antimycin A. A drop in the RCI is seen only when the rate of respiration decreases by more than 80%; however, in no case does the RCI fall below 2.5 and 1.6 for NADH and succinate oxidase, respectively, when the respiratory rate was inhibited by more than 95%. These findings would be very hard to interpret in terms of a transmembrane potential as the coupling device between energy transducing units. If that were the case, the RCI would be expected to decline as the respiratory rate decreases. On the other hand, these data can be accommodated readily with an intra-membrane hypothesis, namely, that respiratory chain-linked energy transduction involves a set of localized reactions of discrete molecular assemblies within the coupling membrane (39).

The phenomenon of respiratory control associated with the phosphorylation of ADP has been known for more than 2 decades. The respiratory control index, taken as the ratio

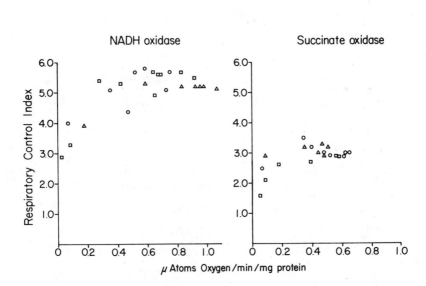

Figure 1.   Relationship between Respiratory Rate and Respiratory Control Index

NADH oxidase and succinate oxidase activities were determined by the spectrophotometric
and polarographic methods, respectively. The
reaction mixture consisted of 150 mM sucrose,
30 mM Tris-sulfate, pH 7.5, and oligomycin-
treated EDTA particles: (0.21 - 0.24) mg protein
in the case of NADH oxidase; (0.86 - 0.95) mg
protein in the case of succinate oxidase. Others
where indicated were: (25 - 250) ng antimycin A
per mg protein; 0.2 mM NADH; and 5 mM succinate.
Total volume: 3.0 ml; temperature: 25°C. (C. P.
Lee and S. J. Kopacz, unpublished results).

of the actively phosphorylating, or State 3, rate of O_2 uptake
to the resting, or State 4, rate (40), has proved invaluable
in intact mitochondria as a criterion for the integrity and
intactness of the energy conservation system, and as a
means for calculating the phosphorylating efficiency of
oxidative phosphorylation. On the other hand, with phosphorylating submitochondrial particles such as ETP_H or Mg-
ATP particles (cf. Table 1), ADP does not give significant
stimulation of the State 4 rate of O_2 uptake, and oligomycin has no appreciable inhibitory effect on the rate of O_2
uptake in these particles. Yet, ADP is phosphorylated
with great efficiency and this phosphorylation is inhibited
by oligomycin (41). Uncouplers of oxidative phosphorylation
give 2-3 fold stimulation of O_2 uptake, showing that respiratory chain-linked energy coupling is fully functional.
The lack of the ADP-induced respiratory control in these
particles is not understood at the present time. One of
the possibilities worth considering is the possible role
of the large amount of adenine nucleotides and Mg^{++} retained
in these particles (42, 43).

Energy-linked Reversed Electron Flow. The capacity for an
energy-linked reversal of electron transfer, first demonstrated in intact mitochondria, was shown to be retained
unaltered in tightly coupled submitochondrial particles.
In fact, it was the demonstration that "substrate" amounts
of NAD^+ were reduced by succinate with Mg-ATP particles
using ATP as energy source (44), which clarified the controversy surrounding reversed electron transport in the
earlier stages of its discovery, and confirmed the concept
that respiratory chain-linked oxidative phosphorylation is
a reversible process. ATP-induced reduction of cytochromes
c and a have also been demonstrated (45), showing that

the reversibility of electron transport occurs from substrate to cytochrome oxidase in the respiratory chain.
Ascorbate + PMS or TMPD can replace succinate as the reductant for ATP-supported NAD^+ reduction (11). This ATP
dependent reversal of electron transfer is inhibited by
both oligomycin and uncoupler, and also by antimycin A.
Alternatively, the energy of substrate oxidation provided by
ascorbate + PMS or TMPD can be used for the reduction of
NAD^+ by succinate in the presence of antimycin A. In this
case, oligomycin gives either no effect or a stimulation.
This reaction is completely inhibited by terminal electron
transfer inhibitors. As expected, all reactions leading
to energy-linked reduction of NAD^+ by succinate are inhibited by rotenone and malonate. Baum et al. (46) utilized
this reaction to address the question of global vs. localized
reactions in energy coupling (macroscopic vs. microscopic
coupling in their nomenclature). They reported that, when
the ATP-supported reaction was partially inhibited by
either rotenone or malonate, the titration profile with
oligomycin was virtually unchanged as compared with the
uninhibited system. Similar results were also found when
the reaction was titrated with either rotenone or malonate
with a system partially inhibited by oligomycin. These
results are analogous to those obtained in the respiratory
control experiments (cf. Figure 1), and are consistent with
localized reactions involving intramembrane molecular
assemblies. It should be pointed out that this interpretation has not been universally accepted.

Energy-Linked Pyridine Nucleotide Transhydrogenation. The
energy-linked reduction of $NADP^+$ by NADH was known to operate in intact mitochondria (47). However, since the mitochondrial inner membrane is impermeable to pyridine

nucleotides, this reaction is very difficult to study in the intact organelles. The ready accessibility of the transhydrogenase to pyridine nucleotides in submitochondrial particles has provided conclusive evidence for the operation of an energy-linked pyridine nucleotide transhydrogenase reaction associated with the inner mitochondrial membrane (48, 49). The energy can be supplied by the aerobic oxidation of succinate, ascorbate + PMS or TMPD, or by ATP hydrolysis. In the latter case it is inhibited by oligomycin, whereas in the former, oligomycin often gives a stimulation. The supply of energy which can be harnessed to drive the transhydrogenase reaction is the same as that which is utilized in oxidative phosphorylation, since a competitive relationship between the two can be demonstrated (50). The energy-linked transhydrogenase has also been shown to be reversible. Under properly chosen experimental conditions, the reduction of NAD^+ by NAPH supports ATP synthesis (51) and the uptake of lipophilic anions (52).

The ATP-supported pyridine nucleotide transhydrogenation is also inhibited by the F_1 inhibitor protein. Ernster et al. (53) showed that the inhibition profile of the transhydrogenase activity of submitochondrial particles by the inhibitor protein follows the same pattern as that of the mitochondrial ATPase activity, even when the ATPase activity is not rate-limiting for the transhydrogenation. These results are in accord with the data obtained with the respiratory control and the reversed electron transfer experiments discussed in the preceding paragraphs, and provide further support for the concept that interactions of localized, individual energy transducing units in the membrane are involved in energy coupling.

ION TRANSPORT

The selectivity of the mitochondrial inner membrane to
ion transport is expressed through an awesome array of
transport systems for anions and cations; some require input of energy derived from the respiratory chain, while
others operate passively. In the last decade, much effort
has been expended with intact mitochondria on characterizing these pathways, as reflected by the unusually large
number of reviews (54-63). There have been surprisingly few
studies reported on ion transport in submitochondrial particles, however.

In this section we consider the transport activities
of submitochondrial particles with regard to anions and the
cations, K^+ and Ca^{++}. The hydrogen ion, H^+, is so central
to the mechanism concerning energy coupling and the energized state of the inner membrane (64) that consideration
of H^+ transport into and within the membrane is reserved
for the final section of this chapter.

Energy-linked K^+ transport. The ability of mitochondria
to accumulate K^+ against a concentration gradient in the
presence of the ionophore, valinomycin, is an energy-linked
reaction which functions in the presence of oligomycin and
so utilizes the energy of substrate oxidation, rather than
the ATP derived therefrom (65, also cf. ref. 54, 61). The
physiological significance of this reaction is unclear.
In the absence of valinomycin, K^+ uptake occurs, but at a
much slower rate. In addition, there is no evidence that
a K^+ gradient across the mitochondrial inner membrane exists
in the intact cell. The reaction is very useful for studying mitochondrial energy transduction, and has become very
popular in the armamentarium of mitochondriologists. Submitochondrial particles which have the inverted membrane

orientation carry out this reaction in reverse (66). Under
appropriate conditions, an energy-linked efflux of K^+ in
the presence of valinomycin is readily observed with coupled
EDTA particles (67). A nigericin-mediated K^+/H^+ exchange
similar to that seen with intact mitochondria has also been
observed in these particles. The combination of valinomycin and nigericin in the presence of K^+ induces complete
uncoupling, as revealed by the release of oligomycin-induced
respiratory control and the inhibition of energy-linked
pyridine nucleotide transhydrogenation and energy-linked
probe responses (67). The energy linked efflux of K^+ requires pre-loading of the particles with K^+ in the absence
of permeant anion. If a permeant anion is present and the
particles are not pre-loaded with K^+, an energy-linked uptake of K^+ plus the anion X^- is observed, detectable as
swelling of the particles due to an uptake of water accompanying the uptake of the salt, KX (68). This process is
markedly stimulated by nigericin. A maximal water uptake
in the presence of nigericin of 0.2 μl H_2O/mg protein was
calculated from light scattering measurements (23), while
determination of increases in dextran-inaccessible water
gave a water uptake of 0.5 μl H_2O/mg protein (68). The
two methods would be expected to yield the extreme low and
high values of water uptake. This uptake was reversed by
anaerobiosis, uncoupler, and valinomycin, and was not observed if Na^+ replaced K^+. The mechanism of K^+ transport
is still obscure. The kinetics of K^+ uptake by submitochondrial particles in the absence of valinomycin does not
suggest the existence of a carrier, although these kinetics
in intact mitochondria do suggest one (69). Valinomycin,
when present, appears to act as a carrier operating through
the phospholipid region of the membrane. The question of
whether K^+ in submitochondrial particles is sequestered in

the inner aqueous volume or acts as a gegen-ion to negatively charged groups in the membrane also remains to be
solved.

Energy-linked Ca^{++} Transport. In intact mitochondria, Ca^{++}
outcompetes ADP for the energy conserved during substrate
oxidation, thereby provoking a near maximal rate of respiration during uptake of Ca^{++} (58, 59, 62, 63). Neither stimulation of respiration by added Ca^{++} nor uptake of this
cation has been reported in a convincing manner for submitochondrial particles with inverted orientation. In
fact, submitochondrial particles which are made by cholate
treatement of rat liver mitochondria and consist of 70%
"right-side-out" particles show energy-linked Ca^{++} uptake
only if the "right-side-out" fraction is functional (30).
Maximal activity in the "right-side-out" particles is
observed with P_i or P_i plus ATP in the presence of oligomycin, and this activity is sensitive to inhibition by
both ruthenium red and uncoupler. "Inside-out" particles
made from rat liver mitochondria by sonication with EDTA
at pH 8.5 are reported (31) to take up Ca^{++} if P_i plus
ATP is present in the presence of ruthenium red. The conclusion is that Ca^{++} is being taken up by these particles
through the pathway normally used for Ca^{++} efflux in intact
mitochondria and that ruthenium red prevents energy-linked
extrusion in the opposite direction. On the other hand,
Wehrle and Pedersen (70) report that their inverted membrane vesicles, made from rat liver by sonication in HEPES
buffer, take up Ca^{++} in a reaction requiring the co-trans-
port of P_i, which is sensitive to inhibition by uncoupler,
respiratory inhibitors, and ruthenium red. In these experiments reporting Ca^{++} uptake with "inside-out" particles

(31, 70) the amounts taken up are about one order of magnitude less than those observed with "right-side-out" particles.

It is apparent that the reaction involved in energy-linked Ca^{++} uptake or release from "inside-out" submitochondrial particles has yet to be clarified. With regard to "right-side-out" particles, one question should first be answered: Are particles which retain their membrane orientation actually a form of "mini-mitoplast", containing matrix enzymes and Ca^{++} binding protein which are lost from those with inverted membrane orientation? This question is of particular importance in view of the possibility that the putative Ca^{++} carrier may be in the intermembrane space and the matrix as well as in the membrane, and, so, may be lost during disruption of the mitochondria (71). Submitochondrial particles with inverted membrane orientation would be expected to have lost this carrier, were this the case, while those which retain their membrane orientation would still have it. The question might be answered experimentally if the Ca^{++} carrier could be shown to be reconstitutively active with coupled submitochondrial particles. At present, however, there seems to be little agreement among the investigators concerning the nature of this Ca^{++} carrier (62), although the most recent candidate has already been named "calciphorin" (72). Indeed, perhaps the most convincing demonstration that the calcium carrier has been found would be by means of reconstitution experiments.

Anion Transport. There are at least 8 anion carriers which have been studied in great detail in intact mitochondria (56, 60, 61), including the adenine nucleotide (73) and phosphate carriers (74), whose isolation has been reported.

These carriers are responsible for the transport of metabolites across the inner mitochondrial membrane to fulfill the physiological demands of the cell. Most of these
carriers have yet to be demonstrated in submitochondrial
particles. The energy-linked glutamate/aspartate exchange
system has been demonstrated in rat liver and heart mitochondria sonicated in the presence of glutamate (75). The
adenine nucleotide translocase has been shown to be functional in beef heart submitochondrial particles (76, 77).
Similar reports have also been made on P_i (78) and citrate
(79) transport in liver particles. However, maximal activities reported for all these anion carriers are one to
two orders of magnitude lower than those seen with intact
mitochondria. The cause of the slow rates is not known at
present. One possibility is that carrier binding sites are
lost or damaged during the sonication of the mitochondria.
A more probable cause is the small volume of any aqueous
phase which may be trapped inside the vesicles comprising
"inside-out" submitochondrial particles. Given the water
content of submitochondrial particles (68), it appears that
this volume may be small compared to the volume of the membrane phase itself, which makes transport measurements exceedingly difficult. The apparent lack or low activity of
anion transport may, then, be due to technical difficulties
which, in turn, pose a challenge to the experimentalist.
Otherwise, it is difficult to comprehend why submitochondrial particles which retain so many mitochondrial activities should deficient in these transport activies. There
may be a fundamental reason for this deficiency, the understanding of which might lead to new insights into the functioning of the inner membrane.

THE ENERGIZED INNER MEMBRANE AND H^+

While there appears to be a general agreement among mitochondriologists that H^+ transfer in some form is a key component of the primary energy coupling reaction (64), there is as yet no agreement on how H^+ functions in this reaction.   Among the various models and mechanisms proposed, one can identify two categories which are fundamentally different. One category of such mechanisms has an underlying postulate that energy coupling is a transmembrane reaction generating an electrochemical proton gradient, $\Delta\tilde{\mu}_{H^+}$, across the inner membrane. This category includes all the variants and derivations of the original chemiosmotic hypothesis (80). The other group of mechanisms has an underlying postulate that energy coupling is an intramembrane reaction involving direct H^+ transfer between membrane proteins, or assemblies of proteins, in localized areas within the membrane.   This group includes the limited energy domain model (39,   81), the conformational hypothesis of energy coupling (82) and the concept of energization occurring through a membrane Bohr effect (83).

The transmembrane hypothesis requires that two aqueous phases be separated by the inner mitochondrial membrane of selective permeability, across which the electrochemical proton gradient, $\Delta\tilde{\mu}_{H^+}$, may be generated, maintained, and utilized. There are strict architectural requirements. The transmembrane hypothesis predicts that energy coupling can only occur in vesicular systems which have three phases: an outside aqueous phase, a membrane phase, and an inner aqueous phase. The magnitude of $\Delta\tilde{\mu}_{H^+}$ in the steady state is set by the balance between its rate of generation and rate of decay. The latter rate is dependent only on the intactness of membrane structure. The rate of generation is a function of the rate of energy input. The transmembrane hypothesis considers only the global $\Delta\tilde{\mu}_H{}^+$ across the membrane of each submitochondrial particle, and predicts that the degree of energy coupling, as measured, for instance, by respiratory control, would be a direct function of the rate of energy input from electron transport.

The intramembrane hypothesis requires only two phases for energy coupling: an aqueous phase and a membrane phase. Since the reaction of energy coupling is postulated to occur in localized regions of the membrane, the need to maintain a global $\Delta\tilde{\mu}_H{}^+$ across the membrane of the vesicle is obviated. This hypothesis predicts that the degree of energy coupling would be a function of the interaction of the localized energy-transducing units, rather than of the overall rate of electron transport.

Much experimental effort has been expended in characterizing the reactions of H^+ transfer and in differentiating between these two hypotheses. Submitochondrial particles have played a key role in these experiments (84). The intramembrane hypothesis requires experimental evidence that H^+ transfer can and does occur within the membrane in reactions directly related to energy coupling, and the transmembrane hypothesis requires demonstration of the establishment of $\Delta\tilde{\mu}_H{}^+$, including both ΔpH and $\Delta\Psi$, across the membrane. One set of useful tools in the search for this experimental evidence has been optical probes, whose absorbance and/or fluorescence spectra change in response to H^+ transfer or membrane potential. We shall concentrate only on the subject of H^+ transfer and H^+ gradients in submitochondrial particles. This is done for the sake of brevity. The subject of membrane potentials measured in

mitochondria and submitochondrial particles by means of
microelectrodes, optical probes and ion distribution is
a broad and complex one attended by much controversy, as
is evident from perusal of three recent reviews on the
topic (85-87).

Transmembrane H^+ Gradient. The evidence for energy-linked
uptake of H^+ from the suspending medium by submitochondrial
particles, particularly oligomycin-coupled EDTA particles,
is both convincing and extensive (84). The direction of
H^+ movement is opposite to that observed upon energization
of intact mitochondria. The question of how large is the
transmembrane H^+ gradient, or ΔpH, is less clear.   In
energized mitochondria, there is considerable variation
around ΔpH = 0.5, depending on both the mode of measurement and experimental conditions (88-90). With energized,
coupled EDTA particles, a ΔpH value of 3.6 was calculated
(91) from quenching of the fluorescence of the probe, 9-
aminoacridine (9AA), under the assumption that the 9AA
taken up into the particles is completely non-fluorescent.
A similar value was obtained in this study from the distribution of NH_4^+, based on electrode measurement of NH_4^+
in the external medium. A more recent report gives a
value of 2.0, using both 9AA and fluorescein-containing
dextran (FITC-dextran) "trapped" in the interior of the
particle (89). There is one reservation with these
measurements of ΔpH: the location of the probes. Are
they inside the submitochondrial particle in a vacuole
representing a true aqueous phase, or are they in the
membrane phase? If they are in a vacuole, what is the
mechanism for the fluorescence quenching of 9AA? The
protonated form of 9AA is highly fluorescent. Without
some independent evidence of location, it is not possible

to state that NH_4^+, 9AA and FITC-dextran are free solutes
in an aqueous vacuole, as opposed to being associated with
hydrated membrane proteins through electrostatic interactions in the case of NH_4^+ and 9AA, and through a combination of electrostatic and surface adsorption effects in
the case of FITC-dextran. In fact, analyses of the spectral properties of 9AA and its substituted analog, 9-amino-
6-chloro-2-methoxyacridine (9ACMA) indicate that the
energy-linked fluorescence lowering of these probes is due
to binding of the dye molecules to negatively charged
membrane components with formation of non-fluorescent
complexes (92). The question of whether a transmembrane
ΔpH exists in vesicular submitochondrial particles, and
the corollary question of its magnitude, if it does exist,
in our opinion remains to be elucidated.

Thayer and Hinkle (93) showed that ATP was synthesized when phosphorylating submitochondrial particles,
ETP_H , were pre-incubated in a low K^+ , acidic medium in
the presence of valinomycin followed by a rapid increase
in K^+ (100-fold) and in pH (2.5 units). The kinetics of
this reaction were also measured (94) and the combined
results were considered as strong evidence for the electrochemical gradient as the obligatory intermediate of energy
transduction. On the other hand, these findings can
equally be interpreted on the basis of a transient potential across the phase boundary of the membrane (23). The
latter interpretation readily accommodates the concept of
activation of localized assemblies of membrane proteins,
since only two phases, the membrane and aqueous medium
are involved. It seems to be far more compatible with
results concerning respiratory control, energy-linked
reversed electron flow, and energy-linked pyridine

nucleotide transhydrogenation in response to the rate of
energy input by either electron transfer or ATP hydrolysis
in submitochondrial particles, than is the interpretation
based on a global transmembrane $\Delta\tilde{\mu}_H{}^+$.

Intramembrane $\underline{H^+}$ $\underline{Transfer}$. In order to monitor intramembrane reactions it is necessary to have a probe which
binds tightly to the membrane and, so, reports events in
the membrane rather than in the aqueous phase. The pH indicator, bromthymol blue (BTB), was one of the first pH
probes which was shown to bind to mitochondrial membranes
(95) and to register a change of absorbance corresponding
to acidification upon energization of submitochondrial
particles (96). However, the location and distribution
of BTB molecules was subsequently called into question (97),
and little further work was done with this probe.

Quinacrine (9-{4'-diethylamino-1-methylbutyl}amino-
6-chloro-2-methoxy-acridine) (QA) has proved to be a very
useful intramembrane pH indicator for submitochondrial
particles (98). Use of this compound as a fluorescent
probe for the energized state of biological membranes
was first introduced by Kraayenhof (99) in 1970 with a
chloroplast membrane preparation. Detailed analyses of
the fluorescence properties of QA (100, 101) provided
evidence that the energy-linked fluorescence decrease of
QA is a quantitative measure of the conversion of the
monoprotonated form of the QA molecules into the diprotonated form:

$$QA \cdot H^+ \quad + \quad H^+ \longrightarrow \quad QA \cdot H_2^{++}$$

This would, in turn, furnish the number of H^+ associated
with the energized membrane in the steady state. Fluorescence polarization data indicate that the diprotonated QA

molecules associated with the energized particles are in
a highly immobilized environment. Furthermore, the temperature profile of the fluorescence polarization of QA
shows a distinctive break at 15° characteristic of membrane-
linked functions (92). These data strongly support the
idea that the diprotonated dye molecules are tightly bound
in the membrane phase of the particles.

Further evidence for an intramembrane location of QA
in energized submitochondrial particles comes from results
obtained with preparations which are functionally open
membrane fragments, the Y-fragments (22) and the rabbit
skeletal muscle submitochondrial particles (23). QA gives
responses to energization in these preparations which are
qualitatively identical to those seen in the oligomycin-
coupled EDTA particles, as described above (102). This
indicates that the binding site for the probe in the two
types of particles is similar. An additional point
supporting the intramembrane nature of the energy-linked
H^+ transfer is the energy-linked movement of H^+ in the
suspending medium monitored by H^+ electrode: an uptake
of H^+ in the case of coupled-EDTA particles, and a release
of H^+ from the skeletal muscle particles (103). These
findings further substantiate our original idea that QA
probes the intramembrane H^+ of the energized particles.

As pointed out above, the mechanism responsible for
the fluorescence decrease of QA associated with energization of submitochondrial particles is distinctively different from that of 9AA and 9AMCA, as revealed by the
detailed analyses of their spectroscopic properties (92).
The energy-linked fluorescence decrease of 9AA and 9ACMA
results mainly from the formation of non-fluorescent complexes with negatively charged ions possessing π-bonding
systems, while the energy-linked fluorescence decrease of

QA results quantitatively from the protonation of membrane-bound $QA \cdot H^+$ to $QA \cdot H_2^{++}$. The interaction of $QA \cdot H_2^{++}$ with the submitochondrial membrane does not significantly affect the fluorescence intensity of the $QA \cdot H_2^{++}$ molecules. For other kinds of biological membranes, such as those of chloroplasts and chromatophores, the interaction between $QA \cdot H_2^{++}$ and the membrane may play an important role in the fluorescence intensity of this dye as a result of the different compositions of the membrane preparations (104). It is apparent that QA on the one hand, and 9AA and 9ACMA on the other, are probing different regions of the energized mitochondrial membrane. Application of photoaffinity analogs of these dyes to the identification of the membrane components associated with energization of submitochondrial particles was made recently by Mueller et al. (105, 106) and offers great potential for continued investigation. There is little doubt that studies of this kind will provide us with information on the molecular mechanism of the energy-linked H^+ transfer and its role in energy transduction in general.

EPILOGUE

No discussion of submitochondrial particles could proceed any distance without invoking the name of Professor Lars Ernster. His contributions to understanding the workings of the inner mitochondrial membrane with the use of this powerful experimental tool are so numerous and many-faceted that we have been able to cite but a small portion in this review. We wish to compliment him on this great achievement and, of great importance to us personally, to thank him for the wisdom, guidance and friendship which has had such a profound influence on our own work.

REFERENCES

1. Depierre, J. W. and Ernster, L. (1977) Annu. Rev. Biochem. 46, 201-262.
2. Ernster, L. and Kuylensterna, B. (1970) in Membranes of Mitochondria and Chloroplasts (Racker, E., ed.), pp. 172-212, Van Nostrand Reinhold, New York.
3. Lee, C. P. and Ernster, L. (1966) BBA Libr. 7, 218-236.
4. Keilin, D. and Hartree, E. F. (1940) Proc. Roy. Soc. B 129, 277-306.
5. Cooper, C. and Lehninger, A. L. (1956) J. Biol. Chem. 219, 489-506.
6. Greenawalt, J. W. (1979) Methods Enzymol. 31, 310-323.
7. Kielly, W. W. and Bronk, J. R. (1958) J. Biol. Chem. 230, 521-533.
8. Beechey, R. B. and Cattell, K. J. (1973) Curr. Top. Bioenerg. 5, 305-307.
9. Crane, F. L., Glenn, J. L. and Green, D. E. (1956) Biochim. Biophys. Acta 22, 475-487.
10. Lee, C. P. (1979) Methods Enzymol. 55, 105-112.
11. Löw, H. and Vallin, I. (1963) Biochim. Biophys. Acta 69, 361-374.
12. Vallin, I. and Löw, H. (1968) Eur. J. Biochem. 5, 402-408.
13. Beyer, R. E. (1967) Methods Enzymol. 10, 186-194.
14. Lee, C. P. and Ernster, L. (1967) Methods Enzymol. 10, 543-548.
15. Fessenden, J. M. and Racker, E. (1967) J. Biol. Chem. 241, 2483-2489.
16. Racker, E. (1967) Fed. Proc. 26, 1335-1340.
17. Lee, C. P. and Ernster, L. (1965) Biochem. Biophys. Res. Comm. 18, 523-529.
18. Lee, C. P. and Ernster, L. (1968) Eur. J. Biochem. 3, 391-400.
19. Robertson, A. M., Holloway, C. T., Knight, I. G. and Beechey, R. B. (1968) Biochem. J. 108, 445-456.
20. Lee, C. P., Ernster, L. and Chance, B. (1969) Eur. J. Biochem. 8, 153-163.
21. Racker, E., Tyler, D. D., Estabrook, R. W., Conover, T. E., Parsons, D. F. and Chance, B. (1965) in Oxidases and Related Redox Systems (King, T. E., Mason, H. S. and Morrison, M., eds.), pp. 1077-1094, John Wiley and Sons, New York.
22. Huang, C. H., Keyhani, E. and Lee, C. P. (1973) Biochim. Biophys. Acta 305, 455-473.

23. Storey, B. T., Lee, C. P., Papa, S., Rosen, S. G. and Simon, G. (1976) Biochemistry 15, 928-933.
24. Lee, C. P., Huang, C. H. and Cierkosz, B. I. T. (1974) in Membrane Proteins in Transport and Phosphorylation (Azzone, G. F., Klingenberg, M. E., Quagliariello, E. and Siliprandi, N., eds.), pp. 161-170, Elsevier/North Holland, Amsterdam.
25. Ernster, L., Nordenbrand, K., Chude, O. and Jantii, K. (1974) in Membrane Proteins in Transport and Phosphorylation (Azzone, G. F., Klingenberg, M. E., Quagliariello, E. and Siliprandi, N., eds.), pp. 29-41, Elsevier/North Holland, Amsterdam.
26. Gomez-Puyou, M. T., Gomez-Puyou, A., Nordenbrand, K. and Ernster, L. (1979) in Function and Molecular Aspects of Biomembrane Transport (Quagliariello, E., Papa, S., Palmieri, F., Slater, E. C. and Siliprandi, N., eds.), pp. 119-133, Elsevier/North Holland, Amsterdam.
27. Hackenbrock, C. R. and Hammon, J. M. (1974) J. Biol. Chem. 250, 9185-9197.
28. Hackenbrock, C. R. (1966) J. Cell Biol. 30, 269-297.
29. Pedersen, P. L. and Hullihen, J. (1978) J. Biol. Chem. 253, 2176-2183.
30. Niggli, V., Mattenberger, N. and Gazzotti, P. (1978) Eur. J. Biochem. 89, 361-366.
31. Lötscher, H. R., Schwerzmann, K. and Carafoli, E. (1979) FEBS Lett. 99, 194-198.
32. Godinot, C. and Gautheron, D. C. (1979) Methods Enzym. 55, 112-114.
33. Scott, D. M., Storey, B. T. and Lee, C. P. (1978) Biochem. Biophys. Res. Comm. 83, 641-648.
34. Scott, D. M., Storey, B. T. and Lee, C. P. (1978) in Frontiers of Biological Energetics (Dutton, P. L., Leigh, J. S. and Scarpa, A., eds.), Vol. I, pp. 422-429, Academic Press, New York.
35. Von Jagow, G. and Klingenberg, M. (1970) Eur. J. Biochem. 12, 583-592.
36. Klingenberg, M. and Bucholz, M. (1970) Eur. J. Biochem. 13, 247-252.
37. Gomez-Puyou, A., Gomez-Puyou, M. T. and Ernster, L. (1979) Biochim. Biophys. Acta 547, 252-257.
38. Hinkle, P. C., Tu, Y. L. and Kim, J. J. (1975) in Molecular Aspects of Membrane Phenomena (Kaback, N. R., Neurath, H., Radda, G. K., Schwizer, R. and Wiley, R. R., eds.), pp. 222-232, Springer-Verlag, New York.
39. Ernster, L. (1977) Annu. Rev. Biochem. 46, 981-995.
40. Chance, B. and Williams, G. R. (1956) Advan. Enzymol. 17, 65-134.

41. Vallin, I. (1968) Biochim. Biophys. Acta 162, 477-486.
42. Klingenberg, M. (1967) in Mitochondrial Structure and Compartmentation (Quagliariello, E., Papa, S., Slater, E. C. and Tager, J. M., eds.), pp. 320-324, Adriatica Editrice, Bari.
43. Lee, C. P. (1967) in Mitochondrial Structure and Compartmentation (Quagliariello, E., Papa, S., Slater, E. C. and Tager, J. M., eds.), pp. 325-326, Adriatiaca Editricè, Bari.
44. Löw, H., Krueger, H. and Ziegler, D. M. (1961) Biochem. Biophys. Res. Comm. 5, 231-237.
45. Chance, B., Lee, C. P. and Schoener, B. (1966) J. Biol. Chem. 241, 4574-4576.
46. Baum, H., Hall, G. S. and Nalder, J. (1971) in Energy Transduction in Respiration and Photosynthesis. (Quagliariello, E., Papa, S. and Rossi, C. S., eds.), pp. 747-755, Adriatica Editrice, Bari.
47. Klingenberg, M. and Slencka, W. (1959) Biochem. Z. 331, 486-517.
48. Davidson, L. and Ernster, L. (1963) Biochem. Biophys. Res. Comm. 10, 91-96.
49. Davidson, L. and Ernster, L. (1963) Biochem. Z. 338, 188-205.
50. Lee, C. P. and Ernster, L. (1968) Eur. J. Biochem. 3, 385-390.
51. Van de Stadt, R. J., Nieuwenhuis, F. J. R. M. and Van Dam, K. (1971) Biochim. Biophys. Acta 234, 173-176.
52. Skulachev, V. P. (1971) Curr. Top. Bioenerg. 4, 127-190.
53. Ernster, L., Juntii, K. and Asami, K. (1973) J. Bioenerg. 4, 149-159.
54. Moore, C. L. (1971) Curr. Top. Bioenerg. 4, 191-236.
55. Azzone, G. F. and Massari, S. (1973) Biochim. Biophys. Acta 301, 195-226.
56. Meijer, A. J. and Van Dam, K. (1974) Biochim. Biophys. Acta 346, 213-244.
57. Fonyo, A. (1976) Horizons Biochem. Biophys. 2, 60-105.
58. Bygrave, F. L. (1977) Curr. Top. Bioenerg. 6, 260-318.
59. Mela, L. (1977) Curr. Top. Membrane Transp. 9, 322-366.
60. La Noue, K. F. and Schoolwerth, A. C. (1979) Annu. Rev. Biochem. 48, 871-922.
61. Scarpa, A. (1979) in Membrane Transport in Biology II. Transport Across Single Biological Membranes (Biesbisch, G., Tosteson, D. C. and Ussing, H. H., eds.), pp. 263-355, Springer-Verlag, Berlin.
62. Saris, N.-E. and Åkerman, K. E. O. (1980) Curr. Top. Bioenerg. 10, 104-180.

63. Nicholls, D. and Crompton, M. (1980) FEBS Lett. 111, 261-267.
64. Boyer, P. D., Chance, B., Ernster, L., Racker, E. and Slater, E. C. (1977) Annu. Rev. Biochem. 46, 955-1025.
65. Cockrell, R. S., Harris, E. J. and Pressman, B. C. (1966) Biochemistry 5, 2326-2335.
66. Chance, B. and Montal, M. (1971) Curr. Top. Membranes Transp. 2, 49-156.
67. Montal, M., Chance, B. and Lee, C. P. (1970) J. Membrane Biol. 2, 201-234.
68. Papa, S., Storey, B. T., Lorusso, M., Lee, C. P. and Chance, B. (1973) Biochem. Biophys. Res. Comm. 52, 1395-1402.
69. Diwan, J. J. and Leher, P. H. (1978) Membrane Biochem. 1, 43-60.
70. Wehrle, J. P. and Pedersen, P. L. (1979) J. Biol. Chem. 254, 7269-7275.
71. Sandri, G., Panfili, E. and Sottocasa, G. L. (1976) Biochem. Biophys. Res. Comm. 68, 1272-1279.
72. Jeng, A. Y. and Shamoo, A. E. (1980) J. Biol. Chem. 255, 6897-6903.
73. Hackenberg, H. and Klingenberg, M. (1980) Biochemistry 19, 548-555.
74. Wohlrab, H. (1979) Biochemistry 18, 2098-2102.
75. LaNoue, K. F., Duszynski, J., Watts, J. A. and McKee, E. (1979) Arch. Biochem. Biophys. 195, 578-590.
76. Lauquin, G. J. M., Villiers, C., Michejda, J. W., Hrymewiecka, L. V. and Vignais, P. V. (1977) Biochim. Biophys. Acta 460, 331-345.
77. Klingenberg, M. (1977) Eur. J. Biochem. 76, 553-565.
78. Wehrle, J. P., Cintron, N. M. and Pedersen, P. L. (1978) J. Biol. Chem. 253, 8598-8603.
79. Preizioso, G., Palmieri, F. and Quagliariello, E. (1977) FEBS Lett. 81, 249-252.
80. Mitchell, P. (1966) Biol. Rev. 41, 445-502.
81. Williams, R. J. P. (1978) FEBS Lett. 85, 9-19.
82. Boyer, P. D. (1975) FEBS Lett. 58, 1-6.
83. Chance, B. (1972) FEBS Lett. 23, 3-20.
84. Papa, S. (1976) Biochim. Biophys. Acta 456, 39-84.
85. Cohen, L. B. and Salzberg, B. M. (1978) Rev. Physiol. Biochem. Pharmacol. 83, 35-88.
86. Waggoner, A. S. (1979) Annu. Rev. Biophys. Bioenerg. 8, 47-68.
87. Tedeschi, H. (1980) Biol. Rev. 55, 171-206.
88. Tischler, M. E., Hecht, P. and Williamson, J. R. (1977) Arch. Biochem. Biophys. 181, 278-292.
89. Rottenberg, H. (1979) Methods Enzymol. 55, 547-569.
90. Dodgson, S. J., Forster, R. E., Storey, B. T. and Mela, L. (1980) Proc. Nat. Acad. Sci. USA 77, 5562-5566.

91. Rottenberg, H. and Lee, C. P. (1975) Biochemistry 14, 2675-2680.
92. Huang, C. S. and Lee, C. P. (1978) in Frontiers of Biological Energetics (Dutton, P. L., Leigh, J. S. and Scarpa, A., eds.), Vol. 2, pp. 1285-1292, Academic Press, New York.
93. Thayer, W. S. and Hinkle, P. C. (1975) J. Biol. Chem. 250, 5330-5335.
94. Thayer, W. S. and Hinkle, P. C. (1975) J. Biol. Chem. 250, 5336-5342.
95. Chance, B. and Mela, L. (1966) J. Biol. Chem. 241, 4588-4599.
96. Chance, B. and Mela, L. (1967) J. Biol. Chem. 252, 830-844.
97. Mitchell, P., Moyle, J. and Smith, L. (1968) Eur. J. Biochem. 4, 9-19.
98. Lee, C. P. (1971) Biochemistry 10, 4375-4381.
99. Kraayenhof, R. (1970) FEBS Lett. 6, 161-165.
100. Lee, C. P. (1974) BBA Libr. 13, 337-353.
101. Huang, C. S., Kopacz, S. J. and Lee, C. P. (1977) Biochim. Biophys. Acta 459, 241-249.
102. Storey, B. T., Scott, D. M. and Lee, C. P. (1980) J. Biol. Chem. 255, 5224-5229.
103. Scott, D. M., Storey, B. T. and Lee, C. P. (1979) Biochem. Biophys. Res. Comm. 87, 1058-1065.
104. Kraayenhof, R., Brocklehurst, J. R. and Lee, C. P. (1976) in Concepts in Biochemical Fluorescence (R. F. Chen and H. Edelhoch, eds.), pp. 767-809, Marcel Dekker, New York.
105. Mueller, D. M., Lee, C. P., Hudson, R. A. and Kopacz, S. J. (1979) in Membrane Bioenergetics (Lee, C. P., Schatz, G. and Ernster, L., eds.), pp. 507-520, Addison-Wesley, Boston.
106. Mueller, D. M., Hudson, R. A. and Lee, C. P. (1981) J. Amer. Chem. Soc. (in press).

THE FLAVOPROTEINS OF THE RESPIRATORY CHAIN

Thomas P. Singer, Rona R. Ramsay,
and Christian Paech

Molecular Biology Division, Veterans Administration
Medical Center, San Francisco, California 94121 and
Department of Biochemistry and Biophysics, University
of California, San Francisco, California 94143

Iron flavoproteins of the respiratory chain have been
one of the most widely investigated aspects of mitochondrial
structure and function for over twenty-five years and there
is no sign that interest in this field is declining. Given
the volume of information, it seemed logical to concentrate
on recent developments, with prior information only summarized as a framework for the newer findings. Lack of space
has also necessitated restricting this chapter to mammalian
enzymes, although much interesting and important information
on iron flavoproteins in higher plants and microorganisms has
been published in recent years (1,2). Even for mammalian enzymes only three are dealt with, since little new information
has been reported on other respiratory chain-linked iron flavoproteins (choline dehydrogenase, L-glycerophosphate dehydrogenase) since publication of the last overview of the subject
(3).

C. P. Lee, G. Schatz, G. Dallner (eds.), Mitochondria and Microsomes
in honor of Lars Ernster   ISBN 0-201-04576-1

SUCCINATE DEHYDROGENASE

1. ASSAYS. Assays of succinate dehydrogenase depend on
whether or not the enzyme is membrane-bound. "Natural" electron acceptors such as O_2 in the succinoxidase assay, are
only useful in reconstitution experiments, since they do not
measure the full activity of the dehydrogenase because the
$Q \rightarrow$ cyt. $\underline{b}$-$\underline{c}_1$ step is rate-limiting. With proper care, artificial electron acceptors can readily be used to distinguish
between native and modified preparations and even between
different conformational states (see below).

In membrane preparations, given that the solubility barrier is abolished, the spectrophotometric assays using phenazine methosulfate (PMS), Q, or TMPD·(the radical of tetra-
methylphenyldiamine) all give the same activity (4,5). The
Q analog conventionally used is 2,3-dimethoxy-5-methyl-6-
pentyl-1,4-benzoquinone (DPB). One report that the TMPD·as-
say measured 2-2.5 times more activity than PMS (6) was not
corroborated in other laboratories where the same activity
was measured with both dyes and with low concentrations of
ferricyanide (7,8).

At relatively high concentrations, ferricyanide accepts
electrons from the enzyme in both soluble and membranous
forms, but the activity shows complex temperature dependence
and may measure an uncertain fraction of the activity (4).
Low concentrations of ferricyanide measure the full activity
in soluble samples, reacting with the HiPIP Fe-S cluster (9,
10), but this "low K_m" ferricyanide activity is not seen in
membranes, presumably because the HiPIP site is inaccessible
to the non-penetrant ferricyanide (10).

The turnover number of the enzyme in any of the reductase
assays drops from $\sim$ 22,000 to $\sim$ 14,000 (V_{max} (PMS) at $38^{\circ}C$)

on extraction from mitochondria but is fully restored on re-
inserting the enzyme into the membrane. These changes are
thought to be due to different conformations in the membranous and soluble forms, which may also account for the large
difference in stability between the two (3,11). The reduction of ferricyanide ("low K_m" site) appears on solubilization and disappears on reconstitution of the system, while
the converse happens with Q reductase activity and with sensitivity to the selective inhibitors, TTF and the carboxanilides (11).

In Complex II, the turnover number of the enzyme is lower
($\sim$ 10,000-12,000) than in the best soluble preparations, so
that extraction from this source increases the activity in
the PMS and TMPD· assays, concurrently with the disappearance
of TTF-sensitive Q reductase activity (11). Reinsertion of
this enzyme into the inner membrane further increases the
turnover number to its original value ($\sim$ 22,000). The reversible changes demonstrate that the catalytic activity and
other properties of the enzyme are functions of its conformation which, in turn, hinges on its environment. An important
implication of this concept concerns the elusive "Center 2",
as discussed below.

2. STRUCTURE. Succinate dehydrogenase contains two subunits
of 70,000 and 27,000 daltons, respectively (12). In Complex
II the two subunits are present in 1:1 ratio (13) but in all
soluble heart preparations examined, regardless of starting
material or method of extraction, the 27,000 dalton subunit
seems to be present in excess, using a variety of criteria
to determine the subunit ratio (11,13). The usual molar
ratio found is 1.2 to 1.5 in favor of the 27,000 dalton unit.

It has been proposed (13) that during the extraction step
some dissociation of the quaternary structure may occur, resulting in precipitation of the relatively insoluble 70,000
dalton protein and recombination of the 27,000 dalton part
with the holoenzyme. If this is indeed the mechanism, one
would expect that dissociation of the two subunits is accompanied by loss of their Fe-S clusters (see below).

The FAD moiety, $8\alpha[N(3)-histidyl]-FAD$ is covalently bound
to the 70,000 dalton subunit. The substrate binding site,
which contains a highly reactive -SH group, is on the same
subunit but at some distance from the flavin, probably folded
back so as to give a close juxtaposition of the substrate to
the isoalloxazine ring (14). Although an extensive part of
the amino acid sequence around the flavin has been determined
(15), that around the substrate site cysteine is not known,
because of difficulties in isolating a short enough peptide
for sequencing.

The soluble enzyme tends to dimerize in all but extremely
dilute solutions (13). Electron micrographs analyzed by the
optical autocorrelation technique confirmed some dimerization
(Fiskin, A.M., Peterson, A., Singer, T.P., and Coles, C.J.,
unpublished).

3. IRON-SULFUR CLUSTERS. Both Complex II (16,17) and the
soluble enzyme (11,12,18) contain 8 g atoms each of Fe and
labile S by chemical analysis per mol of flavin. Two variants of the cluster extrusion technique have shown that
these are made up of one [4Fe-4S] and two [2Fe-2S] clusters
(21). However, considerable confusion exists as regards the
number of EPR detectable clusters in the enzyme and recently
even the amount of Fe and labile S in the enzyme has been

questioned solely on this basis (20).

There is general agreement that EPR at very low temperature reveals a tetranuclear cluster in the oxidized enzyme ("HiPIP"or Center 3) with a g = 2.01 signal which disappears on reduction with succinate, and a binuclear cluster (Center 1), g = 1.92, which appears on reduction with succinate and is not seen in the oxidized enzyme (19,20,22). Quantitation of the EPR signal shows a 1:1 ratio of both of these to the flavin in membrane preparations, including Complex II, but the HiPIP signal of this cluster seems to be very labile, so that it decays rapidly after isolation, at the same rate as the "low K_m" ferricyanide activity (23).

The focus of disagreement is the second [2Fe-2S] cluster, first described by Ohnishi's group (24). The signal (g = 1.93) is not seen either in the oxidized or in the succinate-reduced enzyme but appears on treating either of these with dithionite. The redox potential of this "Center 2" has been reported to be exceedingly low (-400 ± 15 mV (24)). A higher spin concentration of the [2Fe-2S] EPR signal, accompanied by a change in line shape, on reduction with dithionite, as compared with succinate, has also been reported by Beinert et al. (19,23) for Complex II and a variety of soluble preparations. It should be noted, however, that the signal intensity after dithionite reduction was only 1.2 - 1.5 times greater than after succinate. This behavior was interpreted as being due to spin coupling (19). In the pure enzyme (isolated by the method of Ackrell et al. (11)), which is ∿ 100% reconstitutively active and has the highest turnover number among soluble preparations, the signal intensity did not change on adding dithionite to the succinate-reduced enzyme (Albracht, S. P.J., Beinert, H., and Singer, T.P., unpublished experiments).

Albracht has been studying submitochondrial particles from heart and fragments thereof, such as succinate-cyt. c reductase, but not Complex II, and in these he did not find any change in the EPR signal of Center 1 on adding dithionite to the enzyme reduced with succinate (20). He concluded that "Center 2" is, in fact, a preparative artifact not seen in native preparations. In order to account for the binding of 8 g atoms of Fe and S per mol of flavin by analysis, he has made two assumptions: that the [4Fe-4S] cluster is in the 70,000 dalton subunit and the [2Fe-2S] in the smaller one and that partial dissociation of the enzyme occurs during solubilization, without detachment of Fe-S clusters during this event or on subsequent recombination of the small subunit with the holoenzyme to give a molecule containing a 30-50% excess of the 27,000 dalton subunit over the 70,000 dalton one. Given these assumptions, one may be led to the conclusion that the dehydrogenase contains 6 atoms of Fe and S each and that the finding of 8 g atoms per mol, as well as the results of cluster extrusion showing 2 binuclear centers per mol are the consequence of the presence of an excess of the 27,000 dalton subunit in the soluble enzyme (13).

Although we do not agree with these interpretations, the experimental part of Albracht's work (20) is among the most careful studies in the field. Some of the reasons why the interpretations of Albracht contradict existing data are as follows. First, the ratio of the two subunits is 1:1 in Complex II (13), although this particle contains 8, not 6, g atoms of Fe and S per mol of flavin (12,16,17) and invariably shows "Center 2" (i.e., more intense g = 1.93 signal with a different line shape, on reduction by dithionite than by succinate (19)). Second, in order to arrive at an Fe:S:flavin ratio of 8:8:1, on the assumptions discussed above, the ratio of the 27 K:70 K subunits would have to be 2. The excess of 27 K subunit is never this great. Third, there is no real evidence that the [4Fe-4S] cluster is in the large subunit; in fact, it seems likely that it is located in the small one (23). Finally, preparations of the enzyme extracted from acetone powders without succinate (19,22) contain both subunits but only 4 g atoms of Fe and S and lack the HiPIP EPR signal. This would suggest that this enzyme contains the two binuclear clusters but not the tetranuclear one.

We propose that a simpler explanation, compatible with all data known to us, is that the enzyme does contain 8 g atoms of Fe and S, which exist as one tetranuclear and two binuclear Fe-S clusters. One of the latter, "Center 2", is EPR silent in relatively intact membranes, becomes EPR detectable on reduction with dithionite in Complex II and in most soluble preparations (but not in that of Ackrell et al.(11)) and again becomes EPR silent on reinsertion into the inner membrane. The EPR detectability of this cluster may be an expression of the different conformations the enzyme can assume in these preparations, just as with the turnover number. Finally, the fact that both chemical analysis and cluster extrusion indicate the presence of 8 g atoms of Fe and S, despite the excess of the 27,000 dalton subunit in soluble preparations can be rationalized if, during the extraction step, the dissociated 27,000 dalton subunit loses its non-heme Fe and labile S, so that only the aposubunit recombines with the holoenzyme.

4. INHIBITORS. The many inhibitors known to affect this enzyme may be classified into three groups: inhibitors affecting

the essential -SH group at the substrate site, inhibitors
which compete with the substrate, and inhibitors which block
reoxidation of the enzyme by Q.

Among alkylating agents of -SH groups, $[^{14}C]$-N-ethyl-
maleimide has been used to label specifically the substrate
site thiol, demonstrating that it is located on the 70 K subunit (14). Of the many -SH groups of the enzyme, that in the
substrate site was most rapidly alkylated and its modification was prevented by substrates and competitive inhibitors.

To the long list of competitive inhibitors of the dehydrogenase (malonate, D-chlorosuccinate, D-methylsuccinate,
fumarate, bicarbonate, etc.) recently two new nitro analogs
have been added (25): nitraminoacetate and nitroacetate, with
K_i values of 26 μM and 68 μM for the dianion forms at 25°,
pH 8. In contrast to these the dianion of 3-nitropropionate
(26,27) is slowly oxidized by the dehydrogenase (at 0.1% of
the rate for succinate) to 3-nitroacrylate, which immediately
alkylates the -SH group at the substrate site (27) (Fig.1).
Thus, 3-nitropropionate is a suicide inhibitor of the enzyme,
1 mol inactivating each mol of succinate dehydrogenase.

Thenoyltrifluoroacetone (TTF) is a selective inhibitor
of succinate dehydrogenase in membrane preparations, where
it inhibits reoxidation of the enzyme by Q, but not in soluble ones (28). The inhibition site has been localized between the HiPIP cluster and Q (29). In recent years a series
of oxathiin derivatives ("carboxanilides") have been shown to
inhibit the enzyme in fungi, higher plants and bacteria, as
well as in animal tissues (30-32). Carboxanilides inhibit at
the same site and in the same manner as TTF but are several
orders of magnitude more effective (32). The kinetics of the
inhibition tend to be complex and depend on the type of mem-

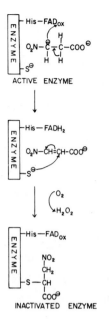

Fig. 1.   Postulated mechanism
          for the inactivation of
          succinate dehydrogenase
          by 3-nitropropionate.

brane preparation used (32). Binding studies with $[G-^3H]2,4,$
5-trimethyl-3-carboxanilinofuran have revealed the existence
of both a specific binding site, responsible for the inhibition, and of unspecific binding. TTF displaced the furan
from the specific but not from the unspecific sites, but the
specific binding at ∿100% inhibition amounts to only ∿0.5 mol
carboxanilide per mol of succinate dehydrogenase (33). An
extension of these studies using a photoaffinity labeled carboxanilide (3'-azido-5,6-dehydro-2-methyl-1,4-oxathiin-3-car-
boxanilide) and Complex II indicate that the two small "binding peptides" (C_{II-3} and C_{II-4}) and phospholipids are all
close enough to the binding site to be specifically labeled,
while the two subunits of the dehydrogenase do not bind carboxanilides specifically (34).

5.   THE "BINDING PEPTIDES" OF COMPLEX II AND THE RECONSTITU-
TION OF SUCCINATE-Q REDUCTASE.   The characteristic differences

between succinate dehydrogenase in purified, soluble form and
in membranes is that the former is labile, has "low K_m" ferricyanide activity but does not reduce Q and is not inhibited
by TTF and carboxanilides, whereas the latter is more stable
and lacks "low K_m" ferricyanide activity but reduces Q in a
TTF-sensitive reaction. The differences would suggest that
the conformation of the enzyme is different in and out of
the membrane.

The first step in defining the factors responsible for
the differences, was the demonstration by Keilin and King
(35) that recombination of an anaerobically isolated sample
of the dehydrogenase with an inner membrane sample (Keilin-
Hartree preparation) previously exposed to pH $\sim$10, which
lacks succinoxidase activity, restores the latter. The
Keilin-Hartree preparation is, however, too complex to permit definition of the binding factors. Moreover, although
inactivated by the alkali treatment, the enzyme was still
present in the membrane (36,37), so that there was uncertainty as to whether the added enzyme occupied the same site in
this "double headed" preparation as it does in intact ones.

Several years later Bruni and Racker (38) and McPhail
and Cunningham (39) combined the soluble enzyme with their
cyt. b preparations but restored only $\sim$10% of the Q reductase
activity seen in Complex II. Besides this low activity, the
peptide composition of the cyt. b preparations was unknown,
so that it was not clear whether, in fact, cyt. b was involved in the reconstitution. Yu et al. (8) isolated a peptide
of 15,000 daltons molecular weight, accompanied by $\sim$20% phospholipid,from their cyt. b-c$_1$ sample, which restored TTF-sensitive Q reductase activity to soluble succinate dehydrogenase, with disappearance of the "low K_m" ferricyanide activity.

The succinate-Q activity per mol of the dehydrogenase in the recombined preparation was rather low, however, and it was later noted (40,41) that two peptides are required for full reconstitution.

The studies of Ackrell et al. (40,42) used Complex II, in which the properties of membrane bound succinate dehydrogenase are preserved (except for a lower turnover number in the PMS assay) and which has the virtue of a simple peptide composition. PAGE shows that it contains equimolar amounts of the two subunits of succinate dehydrogenase and of two smaller peptides of 13,500 and 7,000 daltons, respectively, besides lesser amounts of cyt. b and of trace contaminants from Complex III (43). If Complex II is extracted with perchlorate to remove the enzyme (12) and the residue is further extracted with Triton X-100, the two low molecular weight peptides are obtained free from the subunits of succinate dehydrogenase and may then be further purified. When the peptide preparation is recombined anaerobically with the pure, reconstitutionally fully active dehydrogenase (11), the full succinate-Q reductase activity is restored and "low K_m" ferricyanide activity disappears (Table I). The Q reductase activity, assayed with DPB, is fully TTF- and carboxanilide-sensitive. The two small peptides, C_{II-3} and C_{II-4}, are both needed for reconstitution (42), as confirmed by others (41, 44). That C_{II-4} by itself is not sufficient for reconstitution of the activity has been shown by selective digestion of C_{II-3} with chymotrypsin, which leaves C_{II-4} unaffected. After this treatment the peptide no longer confers Q reductase activity, TTF sensitivity, nor can mask ferricyanide activity, although it still combines with the dehydrogenase, as shown by immunoprecipitation (40). Because the two small

Table I. Reconstitution of Succinate-Q Reductase Activity
with the Peptides C_{II-3} and C_{II-4}.

Sample	Reductase activities[a]			
	$Fe(CN)_6^{3-}$	PMS	DPB	Cyt. c
SDH	24	73	0	0
SDH + peptides	0	90	40	0
SDH + peptides + Complex III				9.4

[a] Succinate-acceptor reductase activities (μmol succinate
oxidized min^{-1} mg^{-1}) at fixed acceptor concentrations:
$Fe(CN)_6^{3-}$, 200 μM; PMS, 1.08 mM; DPB, 35 μM; cyt. c, 80 μM.
(Note that the results do not represent full V_{max} activity
in any of the assays).

peptides physically combine with the dehydrogenase during reconstitution, the term "binding peptides" appears appropriate.
The additional role for lipids in the reconstitution (39) has
not been ruled out since all peptide preparations contain
lipid.

6. REGULATION. The activity of succinate dehydrogenase, in
purified form as well as in intact mitochondria, is modulated
by a wide array of positive effectors and a negative effector,
oxaloacetate. The subject has been extensively reviewed (5,
45, 46) so that the present summary will be brief.

The dehydrogenase exists in active and deactivated forms,
requiring a high energy of activation for interconversion
(47). The inactive form is stabilized by and binds tightly
oxaloacetate, probably in a thiohemiacetal linkage. Oxaloacetate is also bound to the active form but much less tightly, so that dialysis or gel exclusion can remove it and succinate displaces it in the cold, with full restoration of activity. Activators which bind (and probably stabilize) the

active form are substrates, competitive inhibitors, a number
of monovalent anions, certain nucleotides, and, in mitochondria, ATP. One of the simplest ways to activate the enzyme,
i.e., change it to a form from which oxaloacetate can be
readily dissociated, is to reduce its covalently bound flavin
moiety (48,49).

Oxaloacetate is thus bound to the enzyme in two ways.
The initial reaction with the active form yields reversible
binding at the substrate site, and, hence, oxaloacetate is
competitive with succinate. The presence of oxaloacetate,
however, displaces the equilibrium in favor of the deactivated form, in which the binding is irreversible. Although
Gutman (46) believes that oxaloacetate as well as positive
effectors bind at a second (regulatory) site, the evidence
for this is unconvincing and the experimental findings can
be readily explained without assuming the existence of a
regulatory site (50).

An intriguing question is why this enzyme, which is neither rate-limiting in the Krebs cycle nor a recognized control point, is subject to such complex regulation. The hypothesis advanced by us nearly a decade ago (51), which remains
to be proven, proposes that the reason its activity is modulated to a low level in state 3, where the ADP/ATP ratio is
high, is to permit rapid ATP synthesis, which is more efficiently accomplished by permitting NADH oxidation to take over
from succinate oxidation, since the two compete for a common
respiratory chain. Conversely, in state 4, the ATP level is
high, which permits energy consuming reactions to proceed and
succinate accumulated in state 3 is metabolized by the now
fully activated succinate dehydrogenase.

7. ARRANGEMENT OF THE ENZYME IN THE INNER MEMBRANE. Since

ferricyanide, a non-penetrant, accepts electrons from the dehydrogenase in antimycin-treated inverted particles but not
mitochondria (52), it is clear that some part of the enzyme
must be exposed to the matrix side in intact membranes. The
same conclusion is reached from the fact that antibodies to
the enzyme inhibit its activity in inverted membranes (53).

The first insight into the arrangement of the peptides
of Complex II in the membrane was provided by the elegant
surface-labeling studies of Merli et al. (53). Mitochondria
incubated with $[^{35}S]$diazobenzenesulfonate were solubilized
with Triton X-100, and the succinate dehydrogenase containing fragments immunoprecipitated and analyzed by SDS-PAGE.
The subunits of succinate dehydrogenase itself were not labeled under these conditions but one of the other components
of Complex II, peptide C_{II-3}, was significantly labeled with
^{35}S. In contrast, when inverted submitochondrial particles
were labeled and analyzed, the ^{35}S label was found in the
large subunit, and, to a lesser extent, in the small subunit
but not in the binding peptides. The results are consistent
with the view that the 70,000 dalton subunit is located on
the matrix side of the membrane and the 30,000 dalton subunit
in close proximity but only partially exposed. Of the peptides, C_{II-3} is exposed to the cytoplasmic side of the membrane, but must also be contiguous with the larger subunits,
since it masks the "low K_m" ferricyanide site of the enzyme
in Complex II (54), while C_{II-4} is buried in the membrane.

NADH DEHYDROGENASE

1. FORMS OF THE ISOLATED ENZYME. NADH dehydrogenase is the
most complex flavoprotein of the respiratory chain. It is

difficult to isolate because of its high affinity for lipids originating from the membrane and for detergents used in some isolation procedures, which can entrap the enzyme and cause aggregation. These properties also interfere with molecular weight determination by physical methods (55,56). Since the native enzyme does not enter polyacrylamide gels, the molecular weight is usually approximated from the FMN content, a method with obvious limitations.

The enzyme has been isolated in three forms, namely Complex I, the high molecular weight, soluble form (type I dehydrogenase), and the low molecular weight form (type II dehydrogenase). Complex I (57), being a particulate preparation, is usually contaminated with proteins from other segments of the respiratory chain but is useful as a reference material, since, as far as it is known, it retains all the properties of the enzyme in mitochondria.

The high molecular weight form was first isolated by extraction with phospholipase A_2 (58,59). Its catalytic properties, Fe and labile S content, and EPR signals agree closely with those of Complex I, except for the absence of rotenone-sensitive NADH-Q reductase activity (60-62) and phospholipids. Reactivity with Q homologs, and interaction with rotenone (and piericidin A), depend, however, on phospholipids (62-64) which are absent from the type I enzyme regardless of how isolated, so that NADH-Q reductase activity or inhibition by rotenone would not be expected to be present. A brief report by Baugh and King (65) of the isolation of the type I enzyme free from lipids and still endowed with high Q reductase activity and sensitivity to rotenone could not be confirmed and would not be expected (64). Recent procedures for the isolation of type I dehydrogenase using either Triton X-100 (65,66) or Lubrol (67) offer no obvious advantage, since the

products obtained appear less pure than that isolated with
phospholipase, as judged from their respective compositions.
Moreover, Triton X-100, under the conditions used (65), is
known to modify the catalytic properties of the enzyme (68).

The type II dehydrogenase, first isolated in 1952 (69),
has been prepared by many procedures, all yielding enzyme
with substantially the same composition and properties (60,
70-72) but very different from the type I (55,60,71,73). The
former contains 16 to 18 g atoms of Fe and labile S per mol
of FMN and 4 EPR detectable Fe-S clusters, the latter 2 to 4
g atoms of Fe and labile S but no EPR detectable Fe-S cluster.
The turnover number of the type II enzyme is lower, its substrate specificity is different, and, most characteristically,
the type II enzyme lacks the ability to pass electrons via
its Fe-S cluster 1 to ferricyanide, but activities masked in
Complex I and in the type I enzyme have emerged, such as reactivity with cyt. c, rotenone-insensitive Q reduction, and
direct reoxidation of the flavin by ferricyanide (60). When
the type I enzyme is dissociated to the type II by heat, heat-
acid-ethanol, thiourea, chaotropic agents, or proteolysis, the
conversion is accompanied by these alterations in the catalytic properties (55,70,74,75).

The fact that two entirely different proteins can be isolated from the inner membrane of mitochondria, both of which
catalyze the oxidation of NADH but with different oxidants,
was the basis of long debates on the relation of these two
enzymes to each other.

The two main points of view concerning the interrelations
of the high and low molecular weight forms of NADH dehydrogenase may be, in the last analysis, only semantically different. The view of Hatefi (76) is that Complex I consists of
a flavoprotein (FP), an iron-protein (IP) and some structural

protein, and that electron flow from NADH to Q takes the path:

$$NADH \rightarrow FP \rightarrow IP \rightarrow Q$$

He recognizes that the properties of FP (the type II enzyme) differ materially from those of Complex I and the type I enzyme, and suggests that dissociation of the FP from the complex may cause a conformation change, responsible for the altered catalytic properties, i.e., that other components of Complex I modulate the properties of FP but that the latter is isolated as a native, unmodified protein. However, Complex I and the type I enzyme consist of a large number of subunits and even the FP (the type II enzyme) consists of 3 polypeptides, so that there is no reason to call any group of these subunits, particularly one with modified properties (FP) or of no known catalytic properties (IP) an entity in its own right. The proposed proteins should each be catalytically competent and kinetically adequate to fulfill their presumed role in Complex I. If the modification of the type II suffered during extraction is reversible, it should be possible to reconstitute the individual units to yield a multienzyme complex with the properties of the type I enzyme, as has been done with the protein components of α-keto acid oxidases.

The alternative point of view is that the type II enzyme is a breakdown product of the native, high molecular weight enzyme and is not a pristine subunit but is modified in that its catalytic and molecular properties are altered (60,70,71, 77). To quote from a paper published nearly 20 years ago(78):

"These results suggest that NADH-cytochrome reductases [the type II enzyme]...are derived from the fragmentation of NADH dehydrogenase. The transformation involves the change of a high molecular weight protein to a low molecular weight fragment... One cannot be

completely certain that the NADH dehydrogenase [type
I enzyme]... is in every regard the unmodified enzyme
in highly purified form... What seems reasonably certain is that [it is] closer to the native form of the
enzyme than preparations isolated by other methods
[i.e., the type II enzyme]."

Apart from semantic problems as to whether the products
of the action of chaotropic agents on Complex I can be called
individual proteins, the main difference between the two views
is whether the type II dehydrogenase has been reversibly or
irreversibly modified. Definitive answers to this can only
come from reconstitution of Complex I or of the type I enzyme
which has so far not been accomplished (79).

2. PROSTHETIC GROUPS. In addition to its FMN component,
Complex I (57) and the type I enzyme (58) each contains 16 to
18 g atoms of Fe per mol of FMN. In Complex I labile S and
Fe are present in equal amounts (61,80), while in the soluble
enzyme an excess of labile S over Fe is usually found (80),
an artifact for which no satisfactory explanation has been
found. On reduction with NADH, 4 clusters of the low potential ferredoxin type have been detected by EPR at $12^{o}K$ in
Complex I (81) and in the type I enzyme (82), each at about
the same concentration as the flavin, which were named Centers 1-4. With $\sim$ 16 g atoms of Fe and labile S, it would be
tempting to conclude that the enzyme contains 4 [4Fe-4S]
clusters, but this is unlikely for several reasons.

First, Center I is readily detected by EPR at temperatures where tetranuclear clusters are not expected to show
an EPR signal (83,84). Other lines of evidence point to the
probability that at least one binuclear cluster must be in

the enzyme (86). Second, potentiometric titrations of pigeon heart mitochondria with dithionite in the presence of mediators, correlated with EPR signals, led Ohnishi (87) to the conclusion that NADH dehydrogenase has 6, not 4, Fe-S centers and that the lowest potential center among these, Center 1, consists of two components, 1a and 1b, with $E_{m\ 7.2}$ values of -380 and -240 mV, respectively. The stated potential of Center 1a is far below that of the $FMN/FMNH_2$ couple which it is supposed to oxidize. A more recent report by this group (88) states the number of Fe-S centers to be 5, but the notion that Center 1 consists of two components is retained. EPR studies by Albracht et al.(89,90) on beef heart mitochondria did not confirm the existence of Center 1a, 5, or 6, but these workers also believe that Center 1 yields two overlapping EPR signals (90), although Centers 1a and 1b of Albracht are not the same as 1a and 1b of Ohnishi. The two groups do agree, however, that each of the two components of Center 1 is present at concentrations far below that of the FMN. The lack of stoichiometry of 1a and 1b with the flavin is puzzling, unless one accepts the postulate   (90) that NADH dehydrogenase is dimeric with 2 FMN but only one Center 1a and one Center 1b per molecule, for which no evidence has been adduced.

Since quantitation of the EPR signals in various laboratories failed to provide congruent results on Fe-S cluster composition of the enzyme, we turned to the cluster exclusion technique (91), using ^{19}F NMR spectroscopy for analysis of the extruded clusters (92). Table II is a summary of the findings. [2Fe-2S] and [4Fe-4S] clusters were detected and the integrated intensities of the signals correspond to the ratio [2Fe-2S]:[4Fe-4S] $\simeq$ 2:1. Thus, the enzyme appears to contain 4 binuclear and 2 tetranuclear clusters per FMN.

Table II. Cluster Extrusion of NADH Dehydrogenase

Exp. No.	Concentration before extrusion (μM)		Concentration after extrusion (μM)		Extrusion %	Ratio d^{2-}/t^{2-}
	Flavin	Fe	d^{2-}	t^{2-}		
1	43	770	37	18	95	2.1
2	55	990	49	25	100	2.0
3	43	750	35	17	92	2.0

Experimental conditions: ligand exchange was carried out in 80% (v/v) hexamethylphosphoramide/aqueous buffer (30 mM potassium phosphate, pH 7.8, in experiments 1 and 2 or 33 mM potassium phosphate/30 mM Tris-phosphate, pH 8.5, in experiment 3). The molar ratio of o-xylyl-α,α'-dithiol:Fe was 125:1 and that of p-trifluoromethylbenzenethiol:o-xylyl-α,α' -dithiol ≃ 4:1. The designations d^{2-} and t^{2-} refer to the [2Fe-2S] and [4Fe-4S] clusters, respectively. Note that the samples were diluted 5-fold in the unfolding step.

The oxidized form of NADH dehydrogenase is EPR silent and in the reduced state all signals are of the low potential ferredoxin type, with g values below 2.0, indicating that the clusters giving these signals must be of the [2Fe-2S] and [4Fe-4S] type. Although [3Fe-3S] clusters can break down to yield [2Fe-2S] clusters during extrusion, their presence is ruled out, since in all known instances [3Fe-3S] clusters have a g = 2.01 EPR signal in the oxidized state.

Since the chemical method unambiguously shows the presence of 6 Fe-S clusters in the type I enzyme, while EPR reveals only 4, it would seem that two clusters are EPR silent, which is not uncommon. It remains to be determined which of the EPR detectable components besides Center 1 is binuclear, which tetranuclear.

3. SUBUNITS AND MOLECULAR WEIGHT. Determination of the polypeptide composition of a complex enzyme, consisting of a large
number of subunits, is always difficult, particularly if the
preparations are not homogeneous and if the properties of the
enzyme do not permit reliable determination of its purity.
In the case of NADH dehydrogenase advantage can be taken of
its availability in three different forms, which are isolated
by radically different procedures, and that one of these, the
type II enzyme, has been available in essentially pure form
since 1963 (70). The constituent peptides of this low molecular weight form can serve as reference in the analysis of
the peptides in Complex I or the high molecular weight, purified enzyme, and peptides which occur at lower concentration
than the reference peptides may be considered impurities.
The procedures used in comminuting the inner membrane to
yield Complex I are unspecific and thus this particle may be
expected to contain contaminants and even the best preparations of the type I enzyme contains detectable impurities
(59). Nevertheless, considering the different methods used
for isolating the two preparations, those peptides which are
found in both in amounts at least stoichiometric to the reference peptides may be considered probable constituents of
the enzyme and a peptide present in the soluble enzyme but
not in Complex I is very likely to be an impurity. This is
the approach we have used in analyzing the data in Table III.

     Although many reports have appeared on the peptide compo-
sition of Complex I or of the type II enzyme (72,79,94-99),
the results are by no means harmonious (Table IV). Only one
previous paper has compared the subunits in all three forms
of the enzyme (94) but the data for the type I enzyme isolated by the phospholipase method were not documented. The

report concluded that Complex I and the type I enzyme (59) have identical polypeptide composition, except for the absence of the 87 K dalton from the latter. We consider an

Table III. Peptide Composition of Different Forms of NADH Dehydrogenase.

Complex I	Type I enzyme	Type II enzyme
87 K (1)		
75 K (1)	75 K (1)	
56 K (2)	56 K (2)	
53 K (2)	53 K (2)	53 K (1)
40 K (2)	40 K (2)	
33 K (2)	33 K (2)	
	30 K (2-3)	
27 K (1)	27 K (1)	27 K (1)
23 K		
21 K		
19 K		
16 K	16 K	
14 K	14 K	
12 K		
11 K	11 K	11 K
9 K	9 K	
5 K		

Data from densitometry after PAGE according to Laemmli (93). Numbers in parentheses represent estimates of the mols of each peptide per mol of flavin. No concentrations are given for the peptides < 25 K, since the analytical procedure is not reliable in this range with many components of similar m.w. Type I enzyme was prepared according to Cremona and Kearney (59) except that it was applied to the sucrose gradient immediately after the first ammonium sulfate precipitation. Unpublished data from this Laboratory.

identical peptide composition unlikely, since, by analogy
with the succinate dehydrogenase system, small "binding" peptides might be required for the interaction of NADH dehydrogenase and Q.

Our findings (Table III) agree with those of Ragan with
respect to the absence of the 87 K peptide from the type I
enzyme but also show that several small (<25 K) peptides
commonly seen in Complex I (Table IV) are absent from the
purified type I enzyme. Some of these may be, of course, impurities. The 30 K component in the type I enzyme was not
detected in our preparations of Complex I and thus may also
be an impurity, although other investigators have noted its
presence in the Complex (Table IV).

Although the type II enzyme has been available in homogeneous form, there are contradictory reports concerning its
subunit composition. Galante and Hatefi (72) and we (Table
III) find 3 subunits of 51-53 K, 24-27 K, and 9-11 K daltons.
Others (79,94) found only the two larger subunits in the preparation. Possibly, the loss of the $\sim$ 11 K dalton subunit,
like the tendency of the FMN and Fe-S clusters to be lost
(55), is a mark of the notorious lability of this form of
the enzyme.

Considering the fact that one of the three subunits of
the type II enzyme appears to be absent from some preparations,
the variations in molecular weight reported in the literature
(70 K to 80 K (85)), based on physical measurements and FMN
content, are not surprising. In principle, the most reliable
value would be the summation of the weights of the constituent
polypeptides; this would give a figure of $\sim$ 91 K from our data
(Table III) and $\sim$ 84 K from those of Galante and Hatefi (72).
The latter authors also reported a minimum molecular weight

Table IV. Comparison of Data for Peptide Composition of Complex I in Literature

Reference (94)	(95)	(79)	(96)	(97)	(98)	(99)	From Table III
			90K (1)				87K (1)
75K (1)	75K 75K	77K (1)	77K (1)	75K (1)		74K	75K (1)
			73K (1)				
					65K		
53K (2)	53K 53K	56K (2)	56K (2)	53K (1)	55K	53K	56K (2)
		52K (1)	52K (1)				53K (2)
	49K						
42K](2)	42K 42K	40K (1)	40K (2)	42K](2)	42K	43K	40K (2)
39K (2)	39K 39K			38K	36K	37K	
33K (1)	33K				33K		33K (2)
		32K (1)	32K (6)				
	30K 30K		30K (1.6)	30K (2)			
29K (1)							
26K (1)	27K 27K	28K (1)	28K (1)	27K (1)		27K	27K (1)
	26K						
25K	25K 25K		24K (1)	25K			
23.5K	23.5K 23.5K			23.5K (1)		23K	23K
22K	22K 22K	22K(~4)	22K (2.5)	22K			
20.5K	20.5K 20.5K			20K (1.5)			21K
18K	18K 18K	18K(~7)	18K (7.6)	18K		19K	19K
	16.5K			16.5K			
15.5K	15.5K 15.5K	11-16K(~16)	11-16K(~17)	15.5K		16K	16K
							14K
							12K
							11K
8K	8K 8K			8K			9K
5K	5K 5K			5K			5K

of 74 K ± 3 K, based on FMN content but a rather low value
(69 ± 1 K) on the basis of gel filtration. Since the latter
procedure is more subject to error than estimates from PAGE
of the subunits, we would consider that the most likely value
for preparations containing all three subunits to be 80 K to
90 K daltons.

Assignment of a molecular weight to the type I enzyme is
even more difficult. Based on FMN content, with correction
for ultracentrifugally detectable impurities, many years ago
we suggested a minimum molecular weight of 550 K daltons
(59). An independent estimate may be obtained from the subunit composition reported in Table III. For reasons already
discussed, the concentration of the low molecular weight peptides is uncertain, but from the best estimates, based on
their staining intensities on SDS-PAGE gels, a molecular
weight of 520 to 550 K daltons is obtained, in agreement with
earlier data.

4. SPECIFICITY, KINETICS, AND INHIBITORS. The fact that
NADPH was a slow substrate of NADH dehydrogenase, reacting
at the same site as NADH, was unambiguously demonstrated
with the highly purified type I enzyme in 1965 (100). The
evidence was based on (a) direct measurement of the oxidation of NADPH in the ferricyanide assay, (b) competitive
inhibition of the oxidation of NADH by NADPH, (c) stopped-
flow measurements of the bleaching of the chromophores of
the enzyme by NADPH, (d) demonstration of NADPH-NAD and
NADPH-acetylpyridine NAD transhydrogenation by the purified
enzyme, and (e) EPR studies showing that NADPH and NADH
elicited the Center 1 signal to the same extent, though at
different rates.

Working with Complex I, Hatefi rediscovered these facts

several years later (101-104), except that in Complex I at
neutral pH NADPH induced only the EPR signals of Fe-S Centers
2,3, and 4, not of Center 1. This is, of course, to be expected from the fact that the Complex is slightly autooxidiz-
able, so that while the rapidly oxidized substrate, NADH,
could keep all four centers reduced in the presence of O_2,
the more slowly oxidized NADPH could only reduce the high
potential components. This fact was later realized by
Hatefi's group (105) so that it is not necessary to postulate
the existence of a separate NADPH dehydrogenase (104) in
Complex I.

Continued interest in the transformation of the type I
to the type II dehydrogenase has prompted extensive studies
of the kinetics of the two forms of the enzyme in Slater's
laboratory (73,106). The type II enzyme was reported to act
by both an ordered and a ping-pong mechanism, the type I only
by a ping-pong mechanism, in accord with earlier data (107).
The very large kinetic differences between the two forms are
ascribed to a greater accessibility of certain acceptors to
the type II enzyme (71,73), to the loss of Fe-S Center 1
where ferricyanide reacts in the intact enzyme (84), and to
a very much lower rate of intramolecular electron transfer
in the type II enzyme (73). Since dissociation constants of
NADH and NAD and association of NADH are comparable in the
two forms, it was suggested that no major alteration in the
tertiary structure of the flavoprotein subunit occurs during
transformation of the type I enzyme (73). This conclusion
is not supported, however, by the observations that the relative rates of oxidation of a series of NADH analogs and the
K_m values of the enzyme for these are very different in the
two forms of the enzyme and that the FMN is much more prone

to dissociate from the type II than the type I enzyme (70).
These and other properties of the type II enzyme clearly suggest that its conformation is different than when attached
to the rest of the NADH dehydrogenase complex.

Perhaps the potentially most interesting inhibitor of
NADH dehydrogenase is rhein (4,5-dihydroxyanthraquinone-2-
carboxylic acid). Although no structural resemblance to NADH
or NAD is apparent, it is an extremely effective inhibitor
of the enzyme in either soluble (type I) or membrane-bound
form, with a K_i value of ~ 2 μM at 30° (107). At similarly
low concentrations it also inhibits mitochondrial transhydrogenase (108) and Ernster's DT-diaphorase (109), as if rhein
recognized some basic similarity in the structures of the
substrate sites of these three enzymes. Other NAD-linked
enzymes are inhibited only at higher concentrations or are
unaffected. An important aspect of Kean's studies (107) is
that rhein does not inhibit significantly the type II soluble
enzyme. The conclusion of Dooijewaard and Slater (73) that
during isolation of the type II enzyme no gross changes of
its tertiary structure occurs is in direct conflict with
these observations, as also with earlier comparisons of the
substrate specificities of the two forms (70).

5. COUPLING SITE 1 AND ARRANGEMENT IN THE MEMBRANE. The first
localization of energy coupling in the NADH dehydrogenase
region of the respiratory chain came from the observation
that in submitochondrial particles inhibited with piericidin
A, reduction by NADH causes major absorbance changes which
may be monitored at 470 minus 500 nm (110) and that only a
part of this bleaching is reversed by the slow electron flux
to O_2 which incomplete inhibition by piericidin permits. The
chromophore which remains bleached is an Fe-S center (111).

When this experiment is performed with phosphorylating particles, the addition of ATP causes rapid reoxidation of this
chromophore by NAD present in the sample. Freeze-quench EPR
studies have established that the Fe-S cluster reoxidized on
adding ATP is Center 2 and that the reaction is oligomycin
and dinitrophenol sensitive (112). It was, therefore, concluded that coupling site 1 is at or near Center 2 of the
dehydrogenase. On the basis of these studies, Mitchell (113)
and Garland et al. (114) proposed a chemiosmotic scheme for
the arrangement of the dehydrogenase in the membrane. An
alternate scheme has been proposed by Skulachev (115) and
De Vault (116).

Although there appears to be a concensus as regards the
involvement of Center 2 in site 1 energy conservation (88,
117), this has not always been the case. Thus, Albracht and
Slater (118) reported that in anaerobiosis the addition of
ATP to phosphorylating particles induces reoxidation of Center 1 previously reduced by NADH. It was later shown, however, that the reoxidation was due to O_2 introduced with
the ATP, not to ATP itself (119). Earlier, Ohnishi et al.
(120) reported, on the basis of potentiometric titrations
with mediator dyes present, that coupling occurs at Center 1,
more specifically at the low potential component of that
center. Later she (117) proposed that both Centers 1a and 2
are involved in site 1. The reasons why Center 1 is not
likely to participate in site 1 energy coupling have been
presented elsewhere (119). A very recent report by Ingledew
and Ohnishi (88) describes the uncoupler-sensitive oxidation
of Center 4 on adding ATP and poses the possibility that both
Centers 2 and 4 play a role in energy coupling.

ETF DEHYDROGENASE: THE LINK BETWEEN β-OXIDATION AND
THE RESPIRATORY CHAIN

Fatty acid oxidation in mitochondria is initiated by
acyl-CoA dehydrogenases, a family of soluble flavoproteins
(121-124), which donate electrons to another flavoprotein,
the electron transferring flavoprotein (ETF). ETF also
accepts electrons from sarcosine dehydrogenase but reacts
only slowly with Q or cytochrome (125,126). ETF, a dimeric
protein, accepts 2 electrons from its dehydrogenase substrates, being reduced to the anionic semiquinone (127,128).
Thus, the catalytic cycle does not seem to involve full reduction to the hydroquinone (128).

The protein linking ETF to the respiratory chain, ETF
dehydrogenase (also called ETF-Q oxidoreductase) was recently isolated from heart mitochondria (129). It contains
1 mol of FAD and a tetranuclear Fe-S cluster per min. molecular weight of 70,000. The cluster shows an EPR spectrum
typical of reduced Fe-S proteins with a characteristic signal at g = 2.08 (129). Reduction of this cluster by ETF
and its dehydrogenase ($t_{\frac{1}{2}}$ < 10 ms) and reoxidation by Q-1
($t_{\frac{1}{2}}$ ~ 20 ms) has been reported (129) but the catalytic properties of the enzyme were not investigated, owing to lack
of a suitable assay.

$$\left.\begin{array}{l} \text{Acyl-CoA} \\ \text{Enoyl-CoA} \end{array}\right\} \begin{array}{l}\text{Acyl-CoA}\\\text{dehydrogenase}\end{array} \to \text{ETF} \to \begin{array}{l}\text{ETF}\\\text{dehydrogenase}\end{array} \to Q \to cyt.\underline{b}\text{-}\underline{c}_1 \to cyt.\underline{c}$$

We devised an assay in which the reduction of cyt. $\underline{c}$ is
measured in the presence of excess octanoyl-CoA, its dehydrogenase, ETF, a soluble Q analog, and Complex III as a

source of cyt. $\underline{b}$ and $\underline{c}_1$, and cyt. $\underline{c}$, with limiting ETF dehydrogenase (Ramsay, R. R., McIntire, W., and Singer, T. P., unpublished). As expected from the many components used, an appreciable blank rate was obtained without ETF dehydrogenase, although its addition to the system caused significant acceleration of cyt. $\underline{c}$ reduction in crude or partially purified preparations. The highly purified enzyme, isolated by the procedure of Ruzicka and Beinert (129), with the expected FAD and Fe-S content and EPR spectrum, was virtually inactive, however, in the spectrophotometric assay. On incorporation of the dehydrogenase into phospholipid micelles some catalytic activity was recovered, suggesting that phospholipids may be important for rapid interaction of the components in this system.

Despite the difficulty in reproducing the proper catalytic environment required for catalysis with the soluble components, the system merits intensive study, since it is unique in that a chain of three flavoproteins appear to be required before electrons from the substrate reach the iron-sulfur cluster and, thence, the respiratory chain.

ACKNOWLEDGMENTS. Original data from this Laboratory presented here were obtained with the support of Program Project HL-12651, The Veterans Administration, and a grant from the National Science Foundation (PCM 78-23716).

REFERENCES

1. Storey, B.T. (1980) in The Biochemistry of Plants, Vol.2, (D.D. Davies, ed.), pp. 125-195, Academic Press, New York.
2. Ainsworth, P.J., Ball, A.J.S., and Tustanoff, E.R. (1980) Arch. Biochem. Biophys. $\underline{202}$, 187-200.
3. Singer, T.P., and Edmondson, D.E. (1978) Methods Enzymol. $\underline{53}$, 397-418.
4. Singer, T.P. (1974) Methods of Biochem. Anal. $\underline{22}$, 123-175.
5. Ackrell, B.A.C., Kearney, E.B., and Singer, T.P. (1978) Methods Enzymol. $\underline{53}$, 466-483.
6. Vinogradov, A.D., Goloveshkina, V.G., and Gavrikova, E.V. (1977) FEBS Lett. $\underline{73}$, 235-238.
7. Ackrell, B.A.C., Coles, C.J., and Singer, T.P. (1977) FEBS Lett. 75, 249-253.
8. Yu, C.A., Yu, L., and King, T.E. (1977) Biochem. Biophys. Res. Commun. $\underline{79}$, 939-946.
9. Ackrell, B.A.C., Kearney, E.B., Mowery P., Singer, T.P., Beinert, H., Vinogradov, A.D., and White, G.A. (1976) in Iron and Copper Proteins, (K.T. Yasunobu, H.F. Mower, and O. Hayaishi, eds.), pp. 161-181, Plenum Publishing Corporation, New York.
10. Vinogradov, A.D., Gavrikova, E.V., and Goloveshkina, V.G. (1975) Biochem. Biophys. Res. Commun. $\underline{65}$, 1264-1269.
11. Ackrell, B.A.C., Kearney, E.B., and Coles, C.J. (1977) J. Biol. Chem. $\underline{252}$, 6963-6965.
12. Davis, K.H. and Hatefi, Y. (1971) Biochemistry $\underline{10}$, 2507-2516.
13. Coles, C.J., Tisdale, H.D., Kenney, W.C., and Singer, T.P. (1972) Physiol. Chem. Phys. $\underline{4}$, 301-316.
14. Kenney, W.C., Mowery, P.C., Seng, R.L., and Singer, T.P. (1976) J. Biol. Chem. $\underline{251}$, 2369-2373.
15. Kenney, W.C., Walker, W.H., and Singer, T.P. (1972) J. Biol. Chem. $\underline{247}$, 4510-4513.
16. Lusty, C.J., Machinist, J.M., and Singer, T.P. (1965) J. Biol. Chem. $\underline{240}$, 1804-1810.
17. Baginsky, M.L. and Hatefi, Y. (1969) J. Biol. Chem. $\underline{244}$, 5313-5319.
18. King, T.E. (1964) Biochem. Biophys. Res. Commun. $\underline{16}$, 511-516.
19. Beinert, H., Ackrell, B.A.C., Kearney, E.B., and Singer, T.P. (1975) Eur. J. Biochem. $\underline{54}$, 185-194.
20. Albracht, S.P.J. (1980) Biochim. Biophys. Acta $\underline{612}$, 11-28.
21. Coles, C.J., Holm, R.H., Kurtz, D.M., Orme-Johnson, W.H., Rawlings, J., Singer, T.P., and Wong, G.B. (1979) Proc. Natl. Acad. Sci. U.S.A. $\underline{76}$, 3805-3808.

22. Ohnishi, T., Lim, J., Winter, D.B., and King, T.E. (1976) J. Biol. Chem. 251, 2105-2109.
23. Beinert, H., Ackrell, B.A.C., Vinogradov, A.D., Kearney, E.B., and Singer, T.P. (1977) Arch. Biochem. Biophys. 182, 95-106.
24. Ohnishi, T., Winter, D.B., Lim, J., and King, T.E. (1973) Biochem. Biophys. Res. Commun. 53, 231-237.
25. Alston, T.A., Seitz, S.P., Porter, D.J.T., and Bright, H.J. (1980) Biochem. Biophys. Res. Commun. 97, 294-300.
26. Alston, T.A., Mela, L., and Bright, H.J. (1977) Proc. Natl. Acad. Sci. U.S.A. 74, 3767-3771.
27. Coles, C.J., Edmondson, D.E., and Singer, T.P. (1979) J. Biol. Chem. 254, 5161-5167.
28. Ziegler, D.M. (1961) in Biological Structure and Function (T.W. Goodwin and O. Lindberg, eds.) Vol. 2, pp. 253-260, Academic Press, New York.
29. Ackrell, B.A.C., Kearney, E.B., Coles, C.J., Singer, T.P., Beinert, H., Wan, Y.P., and Folkers, K. (1977) Arch. Biochem. Biophys. 182, 107-117.
30. White, G.A. (1971) Biochem. Biophys. Res. Commun. 44, 1212-1219.
31. Ulrich, J.T. and Mathre, D.E. (1972) J. Bacteriol. 110, 628-632.
32. Mowery, P.C., Steenkamp, D.J., Ackrell, B.A.C., Singer, T.P., and White, G.A. (1977) Arch. Biochem. Biophys. 178, 495-506.
33. Coles, C.J., Singer, T.P., White, G.A., and Thorn, G.D. (1978) J. Biol. Chem. 253, 5573-5578.
34. Ramsay, R.R., Ackrell, B.A.C., Coles, C.J., Singer, T.P., White, G.A., and Thorn, G.D. (1980) Proc. Natl. Acad. Sci. U.S.A., in press.
35. Keilin, D. and King. T.E. (1958) Nature 181, 1520-
36. Kimura, T., Hauber, J., and Singer, T.P. (1963) Nature 198, 362-366.
37. Hanstein, W.G., Davis, K.A., Ghalambor, M.A., and Hatefi, Y. (1971) Biochemistry 10, 2517-2524.
38. Bruni, A. and Racker, E. (1968) J. Biol. Chem. 243, 962-971.
39. McPhail, L. and Cunningham, C. (1975) Biochemistry 14, 1122-1131.
40. Ackrell, B.A.C., Ball, M.B., and Kearney, E.B. (1980) J. Biol. Chem. 255, 2761-2769.
41. Yu, L. and Yu, C.A. (1980) Biochim. Biophys. Acta. 593, 24-38.
42. Ackrell, B.A.C., Ball, M.B., and Kearney, E.B. (1979) Abstr. 11th Int. Cong. Biochem., Toronto, Canada, p.273.

43. Capaldi, R.A.. Sweetland, J., and Merli, A. (1977) Biochemistry 16, 5707-5710.
44. Hatefi, Y. and Galante, Y.M. (1980) J. Biol. Chem. 255, 5530-5537.
45. Singer, T.P., Kearney, E.B., and Kenney, W.C. (1973) Advan. Enzymol. 37, 189-272.
46. Gutman, M. (1978) Mol. Cell. Biochem. 20, 41-60.
47. Ackrell, B.A.C., Kearney, E.B., and Mayr, M. (1974) J. Biol. Chem. 249, 2021-2027.
48. Salach, J.I. and Singer, T.P. (1974) J. Biol. Chem. 249, 3765-3767.
49. Ackrell, B.A.C., Kearney, E.B., and Edmondson, D.E. (1975) J. Biol. Chem. 250, 7114-7119.
50. Coles, C.J. and Singer, T.P. (1977) FEBS Lett. 82, 267-268.
51. Singer, T.P., Gutman, M., and Kearney, E.B. (1971) FEBS Lett. 17, 11-13.
52. Klingenberg, M. and Buchholz, M.(1970) Eur. J. Biochem. 13, 247-252.
53. Merli, A., Capaldi. R.A., Ackrell, B.A.C., and Kearney, E.B. (1979) Biochemistry 18, 1393-1400.
54. Ackrell, B.A.C., Ramsay, R.R., Kearney, E.B., Singer, T.P., White, G.A., and Thorn, G.D. (1981) in Function of Quinones and Energy Conserving Systems (B.L. Trumpower, ed.). Academic Press, New York, in press.
55. Singer, T.P. and Gutman, M. (1971) Advan. Enzymol. 34, 79-153.
56. Dooijewaard, G.,DeBruin, G.J.M., Van Dijk, P.J., and Slater, E.C. (1978) Biochim. Biophys. Acta 501, 458-469.
57. Hatefi, Y., Haavik, A.G., and Griffiths, D.E. (1962) J. Biol. Chem. 237, 1676-1680.
58. Ringler, R.L., Minakami, S., and Singer, T.P. (1963) J. Biol. Chem. 238, 801-810.
59. Cremona, T. and Kearney, E.B. (1964) J. Biol. Chem. 239, 2328-2334.
60. Singer, T.P. (1966) in Comprehensive Biochemistry (M. Florkin and E.H. Stotz, eds.) Vol. 14, pp. 127-198, Elsevier, Amsterdam.
61. Machinist, J.M. and Singer, T.P. (1965) Proc. Natl. Acad. Sci. U.S.A. 53, 467-474.
62. Machinist, J.M. and Singer, T.P. (1965) J. Biol. Chem. 240, 3182-3190.
63. Singer, T.P., Horgan, D.J., and Casida, J.E. (1968) in Flavins and Flavoproteins (K. Yagi, ed.), pp. 192-213, University of Tokyo Press, Tokyo.

64. Ragan, C.I. and Racker, E. (1973) J. Biol. Chem. 248, 6876-6884.
65. Baugh, R.F. and King, T.E. (1972) Biochem. Biophys. Res. Commun. 49, 1165-1173.
66. Kaniuga, Z. (1967) in Flavins and Flavoproteins (H. Kamin, ed.) pp. 649-658, University Park Press,Baltimore.
67. Huang, P.C. and Pharo, R.L. (1971) Biochim. Biophys. Acta 245, 240-244.
68. Gutman, M. (1970) Physiol. Chem. Physics 2, 9-14.
69. Mahler, H.R., Sarkar, N.K., Vernon, L.P., and Alberty, R.A. (1952) J. Biol. Chem. 199, 585-597.
70. Watari, H., Kearney, E.B., and Singer, T.P. (1963) J. Biol. Chem. 258, 4063-4073.
71. Cremona, T., Kearney, E.B., Villavicencio, M., and Singer, T.P. (1963) Biochem. Z. 338, 407-442.
72. Galante, Y.M. and Hatefi, Y. (1979) Arch. Biochem. Biophys. 192, 559-568.
73. Dooijewaard, G. and Slater, E.C. (1976) Biochim. Biophys. Acta 440, 1-15.
74. Kaniuga, Z. (1963) Biochim. Biophys. Acta 73, 550-564.
75. Salach, J., Singer, T.P., and Bader, P. (1967) J. Biol. Chem. 242, 4555-4562.
76. Hatefi, Y. (1968) Proc. Natl. Acad. Sci. U.S.A. 60, 733-740.
77. Singer, T.P. (1963) in The Enzymes (P.D. Boyer, H. Lardy, and K. Myrbäck, eds.) Vol. 7, pp.345-381, Academic Press New York.
78. Singer, T.P. and Kearney, E.B. (1962) in Redoxfunktionen cytoplasmatischer Strukturen (Th. Bücher, ed.) pp.251-265, Austrian Biochemical Society, Vienna.
79. Dooijewaard, G., Slater, E.C., Van Dijk, P.J., and De Bruin, G.J.M. (1978) Biochim. Biophys. Acta 503, 405-424.
80. Lusty, C.J., Machinist, J.M., and Singer, T.P. (1965) J. Biol. Chem. 240, 1804-1810.
81. Orme-Johnson, N.R., Hansen, R.E., and Beinert, H. (1974) J. Biol. Chem. 249, 1922-1927.
82. Gutman, M., Singer, T.P., and Beinert, H. (1971) Biochem. Biophys. Res. Commun. 44, 1572-1578.
83. Beinert, H. and Sands, R.H. (1960) Biochem. Biophys. Res. Commun. 3, 41-47.
84. Beinert, H., Palmer, G., Cremona, T., and Singer, T.P. (1965) J. Biol. Chem. 240, 475-480.
85. Albracht, S.P.J. and Subramanian, J. (1977) Biochim. Biophys. Acta 462, 36-48.
86. Salerno, J.C., Ohnishi, T., Blum, H., and Leigh, J.S. (1977) Biochim. Biophys. Acta 494, 191-197.

87. Ohnishi, T. (1975) Biochim. Biophys. Acta 387, 475-490.
88. Ingledew, W.J. and Ohnishi, T. (1980) Biochem. J. 186,
     111-117.
89. Albracht, S.P.J., Dooijewaard, G., Leeuwerik, F.J., and
     Van Swol, B. (1977) Biochim. Biophys. Acta 459, 300-317.
90. Albracht, S.P.J., Leeuwerik, F.J., and Van Swol, B.(1979)
     FEBS Lett. 104, 197-200.
91. Wong, G.B., Kurtz, D.M., Jr., Holm, R.H., Mortenson, L.
     E., and Upchurch, R.G. (1979) J. Am. Chem. Soc. 101,
     3078-3090.
92. Paech, C., Reynolds, J.G., Singer, T.P., and Holm, R.H.
     (1981) J. Biol. Chem.,submitted for publication.
93. Laemmli, U.K. (1970) Nature 227, 680-685.
94. Ragan, C.I. (1976) Biochem. J. 154, 295-305.
95. Crowder, S.E. and Ragan, C.I. (1977) Biochem. J. 165,
     295-301.
96. Dooijewaard, G., De Bruin, G.J.M., Van Dijk, P.J., and
     Slater, E.C. (1978) Biochim. Biophys. Acta 501, 458-469.
97. Hatefi, Y., Galante, Y.M., Stiggall, D.L., and Ragan, C.
     I. (1979) Methods Enzymol. 56, 577-602.
98. Capaldi, R.A. (1974) Arch. Biochem. Biophys. 163, 99-105.
99. Hare, J.F. and Crane, F.L. (1974) Sub-Cell. Biochem. 3,
     1-25.
100. Rossi, C., Cremona, T., Machinist, M., and Singer, T.P.
     (1965) J. Biol. Chem. 240, 2634-2643.
101. Hatefi, Y. (1974) in Dynamics of Energy-Transducing Mem-
     branes (L. Ernster, R.W. Estabrook, and E.C. Slater,
     eds.) pp. 125-141, Elsevier, Amsterdam.
102. Hatefi, Y. (1973) Biochem. Biophys. Res. Commun. 50,
     978-984.
103. Hatefi, Y. and Hanstein, W.G. (1973) Biochemistry 12,
     3515-3522.
104. Djavadi-Ohaniance, L. and Hatefi, Y. (1975) J. Biol.
     Chem. 250, 9397-9403.
105. Hatefi, Y. and Bearden, A.J. (1976) Biochem. Biophys.
     Res. Commun. 69, 1032-1038.
106. Dooijewaard, G. and Slater, E.C. (1976) Biochim. Biophys.
     Acta 440, 16-35.
107. Kean, E.A., Gutman, M., and Singer, T.P. (1971) J. Biol.
     Chem. 246, 2346-2353.
108. Ernster, L. and Lee, C.P. (1967) Methods Enzymol. 10,
     738-744.
109. Ernster, L. (1967) Methods Enzymol. 10, 309-317.
110. Bois, R. and Estabrook, R.W. (1969) Arch. Biochem.
     Biophys. 129, 362-369.

111. Gutman, M. and Singer, T.P. (1970) Biochemistry 9, 4750-4758.
112. Gutman, M., Singer, T.P., and Beinert, H. (1972) Biochemistry 11, 556-562.
113. Mitchell, P. (1972) in Mitochondria/Biomembranes (S.G. van den Bergh, P. Borst, L.L.M. Van Deenen, J.C. Riemersma, E.C. Slater, and J.M. Tager, eds.) pp. 358-370, North Holland, Amsterdam.
114. Garland, P.B., Clegg, R.A., Downie, J.A., Gray, T.A., Lawford, H.G., and Skyrme, J. (1972) in Mitochondria/ Biomembranes (S.G. van den Bergh, P. Borst, L.L.M. Van Deenen, J.C. Riemersma, E.C. Slater, and J.M. Tager, eds.) pp. 105-117, North Holland, Amsterdam.
115. Skulachev, V.P.(1971) in Current Topics in Bioenergetics, Vol. 4, (D.R. Sanadi, ed.) pp. 127-185, Academic Press, New York.
116. De Vault, D. (1976) J. Theor. Biol. 62, 115-139.
117. Ohnishi, T. (1976) Eur. J. Biochem. 64, 91-103.
118. Albracht, S.P.J. and Slater, E.C. (1971) Biochim. Biophys. Acta 245, 508-511.
119. Gutman, M., Beinert, H., and Singer, T.P. (1975) in Electron Transfer Chains and Oxidative Phosphorylation (E. Quagliariello, S. Papa, F. Palmieri, E.C. Slater, and N. Siliprandi, eds.) pp. 55-62, North Holland, Amsterdam.
120. Ohnishi, T., Wilson, D.F., and Chance, B. (1972) Biochem. Biophys. Res. Commun. 49, 1087-1092.
121. Beinert, H. (1962) Methods Enzymol. 5, 546-557.
122. Hall, C.L. (1978) Methods Enzymol. 53, 502-518.
123. Frerman, F.E., Kim, J-J. P., Huhta, K., and McKean, M.C. (1980) J. Biol. Chem. 255, 2195-2198.
124. Engel, P.C. (1980) in Flavins and Flavoproteins (K. Yagi and T. Yamano, eds.) pp. 423-430, University Park Press, Baltimore.
125. Crane, F.L. and Beinert, H. (1956) J. Biol. Chem. 218, 717-731.
126. Hall, C.L. and Kamin, H. (1975) J. Biol. Chem. 250, 3476-3486.
127. Hall, C.L. and Lambeth, J.D. (1980) J. Biol. Chem. 255, 3591-3595.
128. Reinsch, J.W., Feinberg, B.A., and McFarland, J.T.(1980) Biochem. Biophys. Res. Commun. 94, 1409-1416.
129. Ruzicka, F.J. and Beinert, H. (1977) J. Biol. Chem. 252, 8440-8445.

MITOCHONDRIAL FLAVO-IRON-SULFUR CLUSTERS[1]

Tomoko Ohnishi

Department of Biochemistry and Biophysics
School of Medicine
University of Pennsylvania
Philadelphia, PA 19104

INTRODUCTION

Detailed review articles (1-3) on the mitochondrial iron-sulfur clusters have recently appeared. Therefore in this mini-review article on the dehydrogenase iron-sulfur clusters, I have focused on further progress with attention to some controversial topics. I have also placed some emphasis on the spatial organization of iron-sulfur clusters relative to the neighboring redox components as well as to the mitochondrial membrane, since I believe that this quantitative information will greatly facilitate our efforts to understand the mechanism of electron transfer and energy coupling.

SUCCINATE-UQ REDUCTASE SYSTEM

Although the first solubilized and purification of succinate dehydrogenase (SDH) was reported in 1955 by Kearney and Singer (4) and Wang et al. (5), its molecular composition, subunit structure, identification of multiple iron-sulfur

[1]Supported by NIH grant GM-12202 and NSF grant PCM 78-16779.

C. P. Lee, G. Schatz, G. Dallner (eds.), Mitochondria and Microsomes
in honor of Lars Ernster   ISBN 0-201-04576-1

components and their distribution in the two subunits have become clear only within the last several years. The first soluble SDH contained one covalently bound FAD and 2-4 atoms each of non-heme iron and acid labile sulfur per flavin; it only catalyzed electron transfer from succinate to non-physiological electron acceptors. By improving the method of Wang et al. (5), King (6) succeeded in isolating a more intact SDH preparation which can reconstitute the respiratory chain (i.e., can bind to SDH-depleted particles and restore electron transfer activity from succinate to oxygen) in addition to the non-physiological electron transfer. This enzyme was found to contain 8 non-heme iron and 8 acid labile sulfides per flavin, but it was only 30-40% pure. The next advance was made in 1971 by Hatefi and his colleagues (7,8), who isolated an essentially 100% pure SDH preparation by extracting the enzyme from the purified particulate succinate-ubiquinone (UQ) reductase (Complex II)[1], using chaotropic agents. These investigators demonstrated that SDH has a molecular weight of approximately 97,000 daltons and contains 7-8 gram atoms each of non-heme iron and acid labile sulfur and one molecule of flavin. They also showed that the SDH molecule is composed of 2 non-identical subunits; namely, the flavo-iron-sulfur protein (FP) and the iron-sulfur protein (IP), with molecular weights of 70,000 and 27,000, respectively. The FP contains one covalently bound FAD, 4 non-heme iron and 4 acid labile sulfides; IP contains 3-4 non-heme iron and 3-4 acid labile sulfides. This was the first pure SDH preparation. It was, however, only about 20% active in reconstituting respiratory chain electron transport.

[1]Complex II consists of SDH and two additional polypeptides of smaller molecular weights (9-11).

In 1977, Ackrell et al. (12) isolated a SDH preparation which is essentially pure and fully active in reconstituting the physiological electron transport. The strategy for this success was to use purified particulate enzyme, namely, Complex II, as a starting material for the extraction of SDH as in Hatefi's preparation and apply King's extraction method which seems to be somewhat milder than chaotropic treatment, namely, alkaline extraction (pH 9) together with butanol treatment. Yu and Yu (13) subsequently isolated a similar high quality SDH preparation from succinate-cytochrome c reductase (SCR) (14) utilizing the same King procedure. Hatefi and Galante (11) also reported a fully active and pure SDH preparation by their original method (7) under more strictly anaerobic conditions. These pure and reconstitutively fully active bovine heart SDH contain, on the average, 1 covalently bound FAD, 8 non-heme iron, and 8 acid labile sulfides per molecule.

From the EPR analysis of paramagnetic redox centers in older type SDH preparations and in Complex II, conducted mostly in Beinert's and Ohnishi's laboratories, the following composition of SDH redox components was proposed: FP subunit contains two spin-coupled binuclear $[2Fe-2S]^{+1 (+1, +2)}$ iron-sulfur clusters, namely, S-1 and S-2, in addition to a covalently bound FAD; IP contains one tetranuclear iron-sulfur cluster, S-3, as will be described in more detail in the following section. Coles et al. (15) demonstrated that two binuclear and one tetranuclear iron-sulfur clusters are present in the SDH molecule as a whole, based on a completely independent experimental approach, iron-sulfur core extrusion and core displacement method, supporting the above proposed three iron-sulfur cluster composition of SDH.

Albracht (16, 17) reported that the EPR absorption responsible for the second binuclear iron-sulfur cluster, S-2, is not detectable in the SDH molecule in the intact state; this finding was confirmed by other investigators. Based on this evidence together with some additional data from his own laboratory (16), Albracht concluded that the apparent existence of two binuclear clusters in earlier SDH preparations is due to an isolation artifact and the intact SDH molecule contains only one each of the binuclear and tetranuclear clusters; he assigned one $[4Fe-4S]^{+3(+2,+3)}$ to the FP and one $[2Fe-2S]^{+1(+1,+2)}$ to the IP subunit, opposite to the proposed assignment by other investigators (1-3).

SDH Iron-Sulfur Clusters

Since all investigators in the field agree on the identification of Clusters S-1 and S-3, I will first summarize the current views on these two clusters and then discuss Cluster S-2 which has remained a controversial component of SDH. EPR signals of Cluster S-1 in the purified SDH preparation were detected as early as 1960 by Beinert and Sands (18). Cluster S-1 is a binuclear iron-sulfur component (19, 20) and gives an EPR spectrum of rhombic symmetry with g-values of $g_z=2.025$, $g_y=1.93$, and $g_x=1.905$ in the reduced state (21). Its spin relaxation is relatively slow so that EPR signals are readily saturated at temperatures below 30 K (22, 23). The midpoint redox potential of this cluster is around 0 mV in the soluble SDH (23); thus it is quantitatively reduced with a high concentration of succinate. Its spin concentration is approximately equivalent to that of flavin.

Cluster S-3 is paramagnetic in the oxidized state (HiPIP-type cluster) and shows relatively isotropic highly temperature sensitive EPR signals with g-values of $g_z=2.015$, $g_y=2.014$, and g_x=1.990 (24). The midpoint redox potential of Cluster S-3 is +65 mV in Complex II and is above 120 mV in the intact mitochondrial system (3). This cluster is generally accepted as a tetranuclear iron-sulfur species (15). Cluster S-3 is stable in the membrane-bound state and the spin concentration was found to be equal to that of the flavin in Complex II preparations (22, 25). Cluster S-3 becomes extremely labile towards oxidants once the enzyme is solubilized and its EPR signals are detectable only in the reconstitutively active form of SDH (26). It has been accepted as an integral component of the SDH molecule based on various indirect evidence (3).

The presence of the second Fd-Type iron-sulfur cluster (S-2) was pointed out when EPR spectra of King-type reconstitutively active SDH preparations reduced with succinate and with dithionite, respectively, were compared at temperatures below 20°K (3, 21). As exemplified in Fig. 1 succinate reduced and dithionite reduced dehydrogenase show a similar spin intensity above 40 K, but in the lower temperature range, the dithionite reduced SDH gives 1.5-1.75 spins per flavin compared to a smaller value (0.8-1.0) for the succinate reduced enzyme (3). Beinert et al. (25) also obtained a similar spin concentration for the succinate reduced enzyme, but 1.2-1.75 equivalents for the dithionite reduced enzyme relative to Cluster S-1. These two groups agreed on the existence of an EPR active Cluster S-2 in older type SDH preparations and in Complex II; however, the S-2 spin concentrations appeared to vary   considerably depending on the preparations used. The measured midpoint potential of S-2 is extremely low, approximately -400 mV in all soluble dehydrogenase preparations and -260 mV in particulate preparations (3). In the reconstitutively inactive SDH, Cluster S-2, has a slower relaxation, similar to Cluster S-1, in the higher temperature range.

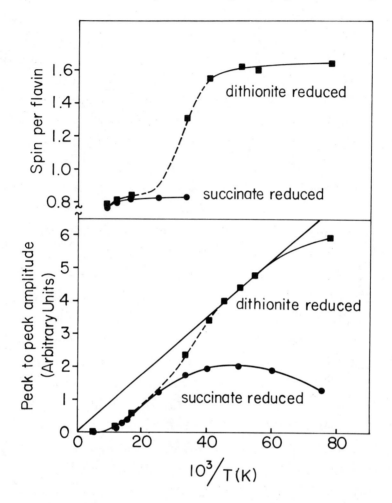

Fig. 1   Spin concentration and the g=1.93 signal amplitude of King-type reconstitutively active succinate dehydrogenase as a function of reciprocal temperature. The enzyme was reduced with succinate and dithionite, respectively.   EPR measurements for the lower figure were conducted with a microwave power level of 2 mW.

The Cluster S-2 lineshape was estimated from the difference

between the spectra of the dithionite reduced and succinate reduced reconstitutively inactive enzyme at temperatures above 25 K, with a minimal effect from the spin coupling between S-1 and S-2. The lineshape and the principal g-values of S-1 and S-2 were found to be almost identical (21). It was also found that at extremely low temperatures (<6.5 K), reconstitutively inactive SDH exhibits a splitting of the central resonance of 23 gauss and broadening of both the g_z and g_x peaks. In contrast, in King's reconstitutively active enzymes, only broadening of the central signal (splitting was not resolved) was observed in the same temperature range. The reversible conversion of the enzyme from the reconstitutively active to the inactive form and vice versa accompanied the change in the low temperature spectral pattern described above (23, 27). Thus in both extremes of the temperature range, a quite remarkable difference between reconstitutively active and inactive forms of the enzyme was demonstrated using the older type SDH preparations. The low tempertature spectrum was well simulated assuming a weak dipolar interaction between Clusters S-1 and S-2 and their spatial arrangement was reported by Salerno et al. (21). On the other hand, in all SDH preparations, a dynamic spin-spin interaction between Clusters S-1 and S-2 is observable, i.e., an enhancement of S-1 spin relaxation upon S-2 reduction. This observation was supported by other investigators (25). As described in the preceding section, Albracht et al. reported that the Cluster S-2 spin is not EPR detectable in SDH in the intact state, namely, in beef heart SMP, SCR (16, 17) and in a fully active and pure preparation of SDH (15). This was confirmed recently in the author's laboratory (Ohnishi, Blum, Yu, and Yu, unpublished data). It is important, however, to emphasize the fact that a dramatic enhancement of S-1 spin relaxation is also seen in the most

intact form of SDH preparation [two orders of magnitude enhancement in the half-saturation parameter[2] ($P_{\frac{1}{2}}$): from 0.07
mW to 6.0 mW at 12 K (Ohnishi et al., unpublished results)].
We propose that the apparent non-detectability of the S-2 signal in the intact state of SDH is most likely caused by the
spin coupling between S-1 and S-2 which renders the total EPR
detectable spin concentration of S-1 plus S-2 not higher than
the flavin concentration in this system. Coles et al. (15)
also proposed an EPR silent Cluster S-2. Albracht on the other
hand interprets the fact that Cluster S-2 is not EPR detectable to imply its non-existence in the intact SDH molecule.
He proposed that the spin relaxation enhancement is caused by
a protein conformational change induced by the reduction of
the flavin to the fully reduced state. The enhancement of the
S-1 spin relaxation, however, is caused by an n=1 redox component with midpoint potential of approximately -400 mV in the
soluble SDH preparations. Direct redox titration of the SDH
flavin free radical indicates that the fully reduced form of
flavin titrates with a midpoint potential of -81 mV at pH 7
and an n-value close to two (29). Thus neither midpoint potential nor n-value support the hypothesis proposed by Albracht
and the dramatic enhancement of S-1 spin relaxation seems to
be most consistent with the cross relaxation of S-1 via S-2
spins. Albracht (17) discredited the results of the core extrusion and core transfer data, because these experiments were
conducted on partially impure preparations of SDH. Coles et
al. (15) used two different kinds of preparations, one was

---

[2]$P_{\frac{1}{2}}$ is a half saturation parameter; a quantitative definition was given in ref. (28). The faster the spin relaxation
the greater the microwave power $P_{\frac{1}{2}}$ needed for saturation.

about 85% and the other about 65% pure based on either flavin
or on non-heme iron content. This implies that the impurity
of both enzyme preparations was not due to modified iron-sulfur subunit (IP) of SDH, contrary to the interpretation of
Albracht (16, 17). In both enzyme preparations an average
of 1.8 binuclear and 1.0 tetranuclear cluster per flavin was
detected, supporting the presence of both binuclear clusters,
S-1 and S-2, as well as a tetranuclear Cluster S-3 in the SDH
molecule.

Spatial organization of redox components. Topographical relationships between adjacent redox components can be studied
by analyzing their spin-spin interactions which are manifested as either spectral splitting, broadening, or spin relaxation enhancement. This is exemplified by a dipole-dipole interaction between Clusters S-1 and S-2 which show various interesting features as described in the preceding section. The
distance between S-1 and S-2 was estimated to be approximately 10 Å (21, 23). Proximity of Cluster S-1 to flavin has also been demonstrated based on the faster spin relaxation of
SDH flavin free radical ($P_{\frac{1}{2}}$ of 2 mW at 233 K) than that of
flavodoxin ($P_{\frac{1}{2}}$ of 0.25 mW); the latter contains no paramagnetic metal ions in the molecule. The faster relaxation is
seen irrespective of the presence or absence of EPR detectable S-3 cluster in the SDH molecule: destruction of Cluster S-1 and S-2 by acid treatment lowers the spin relaxation
rate of the flavin to the level of flavodoxin. The distance
to S-1 was estimated to be in the range of 12-18 Å (29).

Another interesting spin-spin interaction (24) is seen
as low temperature split signals near g=2.04, 1.98, and 1.96
which arise from a specific UQ species which most likely
functions as the electron acceptor of SDH. Computer line

shape simulation of the spectra of Complex II trapped kinetically during the reoxidation process by addition of an oxidized UQ analogue (24) as well as computer analysis of potentiometric titrations of these signals (30) have shown that
the above signals arise from the spin-spin interaction between
a pair of electrochemically similar ubisemiquinones (designated as SQ_S) located at a distance of about 8 Å from each
other. Their close proximity to Cluster S-3 was also suggested from a very rapid spin relaxation of low temperature SQ-SQ
spin coupled signals (31) and of the g=2 signal of SQ_S species
detected subsequently at higher temperatures as well as from
their response to various inhibitors (30-33). In addition to
the spatial relationship between neighboring redox components,
one can examine the topographical distribution of redox components relative to the mitochondrial inner membrane based on
the analysis of spin-spin interactions between intrinsic redox
components and extrinsic paramagnetic probes bound to the surface of the system, for example, isolated proteins, isolated
complexes, SMP, or mitochondria (the last two systems have
opposite membrane polarity). A quantitative procedure to
estimate distance to the intrinsic redox center has been
developed using dysprosium complexes which are very potent
paramagnetic probes (34, 35). For example, Cluster S-3 is
located approximately 12 Å away from the Complex II surface
and about 16 Å from the matrix surface of the mitochondrial
membrane, but greater than 33 Å from the cytosolic surface.
These data are consistent with a somewhat buried location of
the SDH IP-subunit, containing S-3, on the matrix side of the
mitochondrial inner membrane (36).

Another useful strategy to study the spatial organization
of redox components is the analysis of EPR signals of redox

components in oriented multilayered preparations of mitochondrial membranes (37). For example, the iron-iron axes
of binuclear clusters are found to be within the membrane
plane, and the g_z axis of tetranuclear iron-sulfur clusters
is either directed normal to the membrane or parallel to the
plane depending on individual clusters (38). Analysis of the
EPR spectra of the spin-coupled UQ pair in oriented multilayered preparations of mitochondria demonstrated that the vector
connecting the ubiquinone pair is perpendicular to the mitochondrial membrane (38).

Information based on the different approaches described
above combined with surface labelling studies reported by
Capaldi's group (39) is summarized schematically in Fig. 2.

NADH-UQ REDUCTASE SYSTEM

Based on two-dimensional electrophoresis and specific
immuno-precipitation studies, Ragan and his co-workers have
described the NADH-UQ oxidoreductase complex (Complex I) (40)
as "a single enzyme of enormous intricacy, comprising some 26
polypeptides" (41). NADH-UQ reductase is much more complicated than succinate-UQ reductase because in addition to the
NADH dehydrogenase segment which is analogous to succinate
dehydrogenase, it contains components which are required for
energy coupling at site I. Not all of the redox components
or polypeptides may be directly involved in these functions.
The situation may be even more complicated than it appears
if EPR signals of some iron-sulfur clusters are hidden as a
result of spin-spin interaction with nearby components.

Resolution of this complex enzyme using chaotropic reagents in combination with ammonium sulfate fractionation has
greatly facilitated the structural and functional analysis of
the complex. Such treatment separates the hydrophilic flavo-

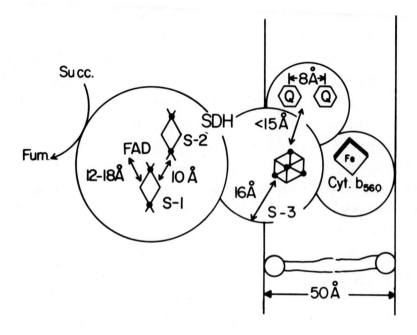

Fig. 2   Tentative scheme for the spatial relationship
          of redox components associated with succinate
          dehydrogenase.

iron-sulfur protein (FP) and the iron-sulfur protein (IP)
in soluble form and a hydrophobic fraction in insoluble form
(42).   The FP fraction   corresponds to soluble NADH dehydrogenase and contains three polypeptides (subunits I-III) (43-
46).   Their molecular weights have recently been estimated to
be approximately 51,000, 24,000, and 9,000, respectively (44).
The FP fraction was further resolved into two subfractions,
namely, subunit I and II + III by a stronger chaotropic treatment combined with freeze-thawing.   Subunit I seems to contain 4 non-heme iron and equivalent acid labile sulfides and
subunit II + III contains about two each of non-heme iron and
acid labile sulfide (47).   The hydrophilic iron-sulfur subunit

(IP) was also further resolved into two subfractions: IP-I
is mostly composed of a 75,000 dalton polypeptide and IP-II
contains 49,000, 29,000 and 51,000 dalton polypeptides. Both
IP subfractions contain non-heme iron and acid labile sulfur
at equivalent concentrations (47). About 80% of the total
protein and 50% of the non-heme iron and acid labile sulfide
of Complex I are located in the hydrophobic polypeptide subunits. Further resolution of the hydrophobic fraction has
not been conducted to date. Recent EPR analysis of soluble
NADH dehydrogenase and the hydrophilic subunits will be described in the following section.

The distribution of these constituent polypeptides in
the mitochondrial inner membrane has been studied by the
surface labeling of Complex I, SMP, or whole mitochondria with
the impermeable probes, diazobenzene [^{35}S] sulfonate or lactoperoxidase-catalyzed [^{125}I] iodination, followed by SDS/poly-
acrylamide-gel electrophoresis (48). None of the three FP
subunits are exposed to the surface even in isolated Complex
I, although the substrate (NADH) binding site resides on the
51,000 dalton subunit (49). The large molecular weight hydrophilic subunits (75,000, 49,000 and 29,000 daltons) of the
IP fraction are transmembraneous. Several hydrophobic protein subunits are not exposed to either side of the mitochondrial inner membrane surface, but are in contact with the
hydrophobic interior of the membrane (48).

Iron-Sulfur Clusters in the NADH-UQ Reductase Segment

In the NADH-UQ reductase (Complex I) segment of the respiratory chain, only the N-1 type clsuter has remained controversial among the 5 or 6 EPR detectable iron-sulfur clusters, even though the EPR spectrum of Cluster N-1 was detected as early as 1960 (18). Orme-Johnson et al. (50) identified

a "g=1.94" type EPR spectrum from Cluster N-1 in Complex I
(40) with rhombic symmetry and g-values of 2.022, 1.938,
1.923. It has a resonance absorption approximately equal to
the flavin concentration. This cluster is completely reducible with NADH in Complex I. Ohnishi and her colleagues po-
tentiometrically resolved the "g=1.94" species in intact mitochondrial membrane systems, into two n=1 components with
midpoint potentials of -240 $\pm$ 20 mV and -380 $\pm$ 20 mV at pH
7.2 and designated these two clusters as N-1b and N-1a, respectively (3). These two clusters differ in their response
to phosphate potential (51) and in the pH dependence of their
midpoint potentials (52). Both clusters exhibit EPR spectra
with rhombic symmetry in beef heart mitochondria while N-1a
in the pigeon heart system exhibits a spectrum of axial
symmetry. In early studies (53) using Complex I prepared in
Racker's laboratory, both N-1a and N-1b type iron-sulfur clusters were detected with similar properties as those in mitochondria. Subsequent analysis of these clusters using Complex
I isolated in King's laboratory showed the presence of the
two N-1 type clusters with $E_{m8.0}$ values of about -150 and
-390 mV. A more quantitative analysis of these clusters was
hampered by the variability of the higher midpoint potential
component in its relative quantity and line shape, depending
on the preparation (3, 46). Albracht and Slater (54) suggested earlier that the EPR spectrum of the "g=1.94" species in
Complex I can be resolved into two rhombic species, by utilizing their different lability towards NADH. Subsequently,
Albracht et al. proposed the presence of two clusters of axial
symmetry, based on the computer simulation of their EPR line
shapes using Complex I (55) and beef heart SMP (56). Each
cluster was reported to be present at only 0.25 spin equiva-

lents per FMN, fully reducible with NADH, and were designated
Clusters 1a and 1b. These investigators also reported the
absence of any further lower midpoint potential components.
More recently Ohnishi et al. (57) have studied the N-1 type
iron-sulfur species quantitatively by the combined application
of potentiometric analysis and computer line shape simulations
using Complex I preparations isolated in Hatefi's laboratory
(40). The results obtained have clarified some discrepancies
among the data obtained with Complex I prepared in different
laboratories.

Potentiometric titration of the g=1.94 signal of Cluster
N-1 at pH 8.0 showed only one major component with a midpoint
redox potential of -335 mV and with an n-value equal to one.
In order to display the actual line shape of the $E_{m8.0}$=-335 mV
component, the difference (Fig. 3C) between the spectra of
Complex I poised at -407 mV (Fig. 3A) and -286 mV (Fig. 3B)
was obtained using a Nicolet model 1024 instrument computer.
This component shows a typical binuclear "g=1.94" type EPR
spectrum which can be simulated as a single rhombic component
with g-values of g_z=2.019, g_y=1.937, and g_x=1.922 and with line
widths of Lz=7.5 x 10^{-4}, Ly=7.6 x 10^{-4}, Lx=1.05 x 10^{-3} tesla
(Fig. 3D). This component was tentatively assigned to Cluster
N-1b. Our g-values from the spectral simulation and relative
spin concentration are almost identical to those reported
earlier by Orme-Johnson et al. (50) in Complex I reduced with
NADH.

Complex I isolated in Hatefi's laboratory contains only
very low concentrations of an impurity which is reducible only
with dithionite and EPR detectable at temperatures higher than
30 K. NADH alone or dithionite followed by NADH reduces N-1b
type cluster giving rise to a spectrum almost identical to

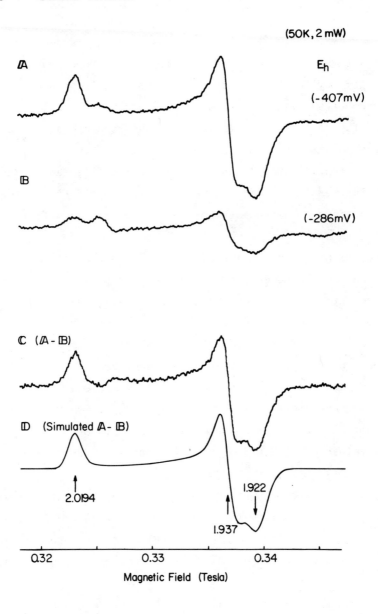

Fig. 3   EPR spectrum of iron-sulfur cluster N-1b obtained
from potentiometric resolution in Complex I and
its computer simulation.   Simulation parameters
used were g-values shown in the figure and line
widths of Lz=7.5 x 10^{-4} tesla, Ly=7.6 x 10^{-4}
tesla, and Lx=1.05 x 10^{-3} tesla (57).

that obtained from the potentiometric analysis (Fig. 3, spectrum C), inconsistent with the observation reported by Albracht
et al. (55).

Upon further lowering the redox poise of Complex I in
the presence of redox mediators with very low midpoint potentials, namely, benzylviologen ($E_{m7.0}$=-311 mV), methyl viologen
(-430 mV), triquat (-500 mV), and ICI (-730 mV), the line
shape of the spectrum is gradually altered; the g=2.02 peak
shifts towards lower magnetic field, the positive peak of the
central resonance absorbance is diminished with concurrent
emergence of a g=1.95 peak, and at the same time the g=1.91
negative peak appears.

The spectral line shape of the lower midpoint potential
component was obtained as a difference spectrum between the
lowest E_h (-518 mV) sample and that of -448 mV: a spectrum
with an apparent rhombic symmetry with g-values of 2.03, 1.95,
and 1.91 was obtained. Spin quantitation of spectra poised
at these two potentials has revealed that 1.5 times more spins
are seen in the lower E_h sample. We tentatively identified
this very low E_m component as Cluster N-1a. The detailed
spectral line shape of this component, however, is difficult
to simulate as a single component; it appears that either
Cluster N-1a was partially modified during the isolation of
Complex I or that there is a spin-spin interaction between
Cluster N-1a and N-1b (or with another component) similar to
the case of Clusters S-1 and S-2 in succinate dehydrogenase.
The existence of Cluster N-1a has been questioned by Albracht
et al. (55, 56) because no additional resonance absorbance
was detected upon addition of dithionite together with methyl
viologen to the NADH reduced Complex I. However, in order to
obtain this very low potential (E_h) range, an extremely re-

ducing environment is required with mediator dyes such as
triquat and ICI.   In the earlier studies using Complex I prepared in King's laboratory, a high potential component ($E_{m8.0}$=
-150 mV) was assigned to a modified Cluster N-1b and the -390
mV component to N-1a, because the latter had the same E_m value
as N-1a in the intact system (3) as well as in Complex I prepared in Racker's laboratory (53).   The high potential component is now considered to be an extraneous impurity (not
N-1b) because this component has been found to be almost completely removable from Complex I.   The -390 mV component in
Complex I isolated in King's laboratory seems to be N-1b with
an E_m value about -60 mV lower than that in the Complex I
isolated in Hatefi's laboratory and the N-1a component appears
to be not seen due to its extremely low E_m value.   A recent
Complex I preparation from King's laboratory gave almost the
same N-1b potentiometric titration and the difference spectrum
(Ohnishi, Salerno, Blum, King, and Widger, manuscript in
preparation), as that from Hatefi's laboratory seen in Fig. 3.

These results identify Cluster N-1b as an intrinsic component of Complex I present in an equivalent concentration to
that of FMN, and displaying a rhombic spectrum.   Cluster 1a
and 1b reported by Albracht (55) appear to correspond to
Cluster N-1b designated by Ohnishi et al. (57).

Cluster N-2 is the most established component in NADH-UQ
reductase (50) having the highest midpoint redox potential
among iron sulfur clusters in this segment of the respiratory
chain (3).   The cluster is present in approximately equal
concentration to that of FMN and displays an EPR spectrum of
axial symmetry with g-values of g=2.054 and g=1.922 (50).

The final identification of the EPR spectra of Clusters
N-3, N-4, and N-5 was achieved by computer simulations con-

ducted by Albracht et al. (55). Orme-Johnson et al. discovered
EPR signals from two distinct iron-sulfur species and assigned
g-values of (2.100, 1.886 and 1.862) for N-3 and (2.103,
? , and 1.864) for N-4. Subsequent potentiometric titrations
of these clusters in pigeon heart SMP by Ohnishi (58) indicated
that the g=1.87 resonance (equivalent to the g=1.86 resonance
of beef heart Complex I) corresponds to the g_x peak of N-3
and the 1.89 signal to the g_x peak of N-4, based on their
different midpoint potentials. She, however, attributed the
g_z peak position shift from 2.10 to 2.11 (2.103 in Complex I),
in parallel with the g_x titrations, to the overlapping of two
components with slightly different g-values, giving the g_z=2.10
to N-3 and the g_z=2.11 to N-4, in the same fashion as Orme-
Johnson et al. (50); the slight shift of the g_z peak is presently considered to be associated with some spin-spin interaction. More recently, Albracht et al. (55) conducted computer
simulations of the spectral line shape of individual clusters,
assuming that each cluster has a spin concentration equal to
that of the flavin. He came to the conclusion that the g_z
peak of N-3 is 2.036 rather than 2.10. The g=2.036 signal
was initially missed by other workers (50, 58)[3]; its identity as the g_z peak of N-3 was confirmed by the demonstration
of concurrent power saturation of the 2.036 and 1.86 signals
of Complex I poised at an appropriate redox potential (3, 46).

The revised assignment of the N-3 and N-4 spectra necessitated a re-interpretation of the spectra of Clusters N-5 and

[3]Footnote: Albracht et al. (55) designated Cluster N-4 for
a component with g-values of (2.036, 1.92-1.93, 1.863) component and N-3 for (2.103, 1.93-1.94, 1.884). We retained
for simplicity, however, our original nomenclature of
N-3 and N-4, with revised g_z value for N-3.

N-6. Signals observed at very low temperatures could be attributed to partially saturated N-4 signals and an additional cluster N-5 with an extremely fast spin relaxation, discarding Cluster N-6 (3, 55). The spin relaxation rates of Clusters N-1 to N-5 increase according to the numerical order of the clusters. All clusters except N-5 seem to be present at a concentration equivalent to FMN in Complex I. All the above described clusters except N-1a and N-1b appear to be of tetranuclear structure (20), although the recently identified trinuclear structure (59, 60) has not yet been excluded.

Spatial Organization of Redox Components

Spin-spin interaction between intrinsic redox components in the NADH-UQ reductase segment was first reported in 1965 by Beinert (61). He demonstrated an interaction between flavin and a transition metal component (iron-sulfur cluster) in close proximity, which was revealed as an enhanced spin relaxation of the flavin free radical. A similar spin coupling between SDH flavin radicals and S-1 spins has been observed (29).

Subsequently, spin-spin interaction between Cluster N-3 and flavin was detected in the author's laboratory (3, 62). Potentiometric titration of the signal amplitude of Cluster N-3 at g=1.86 gives an n=1 Nernst curve with a midpoint redox potential of about -240 mV. Further lowering of the ambient redox potential (E_h) gives rise to an interesting anomaly; the signal decreases in amplitude, reaches a minimum at about -330 mV at pH 8 (recent data obtained using HEPES buffer) and then increases again to its maximum value. Concomitant titration of the flavin g=2.00 signal recorded at -50°C shows a bell-shaped titration curve with the peak at -330 mV, mirroring the trough of the titration curve of Cluster N-3.

This strongly indicates the spin-spin interaction of flavin free radical with N-3, which is presently under investigation in more detail. Spin-coupling was also suggested between Cluster N-1a and N-1b (or another iron-sulfur cluster), based on the spectral line shape alteration at extremely low redox potentials (57). A slight shift of the field position of the g_z peak of Cluster N-4 during redox titrations (50, 58) is also indicative of some interaction between Cluster N-4 and another redox component, as mentioned in the preceding section.

It has been postulated that the iron-sulfur clusters in the site I segment of the respiratory chain are distributed transmembranously, with Cluster N-1 close to the cytosolic side and N-2 close to the matrix side of the inner mitochondrial membrane, forming an electron transferring arm of the chemiosmotic site I loop (63, 64).

The topographical distribution of iron-sulfur clusters relative to the mitochondrial inner membrane in this segment of the respiratory chain is presently being analyzed, utilizing spin-spin interactions between intrinsic redox components and membrane impermeable extrinsic paramagnetic dysprosium complex probes. Preliminary results indicate that Clusters N-2 and N-3, for example, are approximately 40 Å and 30 Å, respectively, from the cytosolic surface of the mitochondrial membrane (Ohnishi et al., unpublished results).

Another promising approach for unravelling the structural intricacy of Complex I is EPR analysis of iron-sulfur clusters in subfractions of the complex, for example, soluble low molecular weight NADH dehydrogenase, an isolated hydrophilic iron-sulfur fraction, and their further resolved subunits, all of which contain equivalent amounts of non-heme iron and

acid labile sulfides. These preparations were obtained by
chaotropic treatment combined with ammonium sulfate fractionations (44, 47). EPR spectra of reduced soluble NADH dehydro-
hydrogenase recorded at different temperatures exhibit signals
from at least two distinct species of iron-sulfur cluster;
one is a binuclear N-1 type (g values of 2.03, 1.94, and 1.92)
with slow spin relaxation (57), the other is a rapidly relaxing species with g values of 2.05, 1.95 and 1.87 (50, 57).
The latter species cannot be assigned to a specific iron-
sulfur cluster identified in Complex I, due to the line shape
modification during the enzyme isolation; identification of
the cluster structure requires further investigation.

CONCLUDING REMARKS

Steady progress has been achieved in the identification
and spatial organization of iron-sulfur clusters present in
the mitochondrial dehydrogenases, particularly succinate and
NADH dehydrogenase, as discussed in this review.

In addition to flavins, which have an intermediate
stability for the free radical state and function as converters
from n=2 to n=1 electron transfer steps, there is also evidence that bound ubiquinone species (Q_S and Q_N)[4] (30, 33, 46,
65), play a converter role in the reversed direction (from
n=1 to n=2), namely from iron-sulfur clusters to the bulk ubiquinone pool in the mitochondrial inner membrane.

In this chapter, I have not reviewed studies of the site
I energy conservation mechanism, because no substantial new
information is available beyond that which I summarized in my
previous review (3). Although it appears certain that site I

[4]Footnote: Q_S and Q_N; specifically bound ubiquinone species
in the succinate-UQ and NADH-UQ reductase segments, respectively, of the respiratory chain.

energy coupling operates, in principle, by a chemiosmotic mechanism (redox driven proton ejection), studies on the structural organization of the redox components and subunit polypeptides in site I tend to rule out FMN as a mobile hydrogen carrier as formulated in the original loop mechanism (63, 64). Experiments using paramagnetic probes indicate that none of the iron-sulfur clusters are located close to either surface of the mitochondrial inner membrane. This information favors the transductase type proton pump mechanism formulated by DeVault (66), exemplified by a simple model of Skulachev (67) or the multi-component pump models of Salerno (68). A detailed comparison of these proton pump models and the local loop-type pump model proposed by Mitchell (69) will be presented elsewhere (70).

ACKNOWLEDGMENTS

The author wishes to express her thanks to Drs. John Bowyer and Haywood Blum for critically reading the manuscript and their stimulating discussions. She is also grateful to Drs. H. Beinert and S.P.J. Albracht for providing their manuscripts before publication and their helpful comments. Thanks are also due to Dr. John Salerno for helpful discussions.

REFERENCES

1. Beinert, H. (1977) in Iron-Sulfur Proteins (W. Lovenberg ed.) Vol. III, pp. 61-100, Academic Press, New York.
2. Beinert, H. (1978) Methods in Enzymology 53, 133-150.
3. Ohnishi, T. (1979) in Membrane Proteins in Energy Transduction (R.A. Capaldi, ed.) pp. 1-87, Marcel Dekker, Inc., New York.
4. Kearney, E.B. and Singer, T.P. (1956) J. Biol. Chem. 219, 963-975.
5. Wang, T.Y., Tsou, C.L. and Wang, Y.L. (1956) Sic. Sinica Peking 5, 73-90.
6. King, T.E. (1963) J. Biol. Chem. 238, 4036-4051.
7. Davis, K.A. and Hatefi, Y. (1971) Biochemistry 10, 2509-2516.
8. Hanstein, W.G., Davis, K.A., Ghalamber, M.A. and Hatefi, Y. (1971) Biochemistry 10, 2517-2524.
9. Bell, R.J., Sweetland, J., Ludwig, B. and Capaldi, R.A. (1979) Proc. Natl. Acad. Sci. USA 76, 741-745.

10. Ackrell, B.A., Ball, M.B., and Kearney, E.B. (1980) J. Biol. Chem. 255, 2761-2769.
11. Hatefi, Y. and Galante, Y.M. (1980) J. Biol. Chem. 255, 5530-5537.
12. Ackrell, B.A.C., Kearney, E.B. and Coles, C.J. (1977) J. Biol. Chem. 252, 6963-6965.
13. Yu, C.A. and Yu, L. (1980) Biochim. Biophys. Acta 591, 409-420.
14. Yu, C.A., Yu, L., and King, T.E. (1974) J. Biol. Chem. 249, 4905-4910.
15. Coles, C.J., Holm, R.H., Kurtz, D.M., Orme-Johnson, W.H., Rawlings, J., Singer, T.P., and Wong, G.B. (1979) Proc. Natl. Acad. Sci. USA 76, 3805-3808.
16. Albracht, S.P.J. (1980) Biochim. Biophys. Acta 612, 11-28.
17. Albracht, S.P.J. (1980) Abstract of First European Bioenergetic Conference, pp. 39-40, Urbino, Italy.
18. Beinert, H. and Sands, R.H. (1960) Biochem. Biophys. Res. Commun. 3, 41-46.
19. Salerno, J.C., Ohnishi, T., Blum, H., and Leigh, J.S. (1977) Biochim. Biophys. Acta 494, 191-197.
20. Albracht, S.P.J. and Subramanian, J. (1977) Biochim. Biophys. Acta 462, 36-48.
21. Salerno, J.C., Lim, J., King, T.E., Blum, H. and Ohnishi, T. (1979) J. Biol. Chem. 254, 4828-4835.
22. Beinert, H., Ackrell, B.A.C., Kearney, E.B. and Singer, T.P. (1975) Eur. J. Biochem. 54, 185-194.
23. Ohnishi, T., Salerno, J.C., Winter, D.B., Lim, J., Yu, C.A., Yu, L. and King, T.E. (1976) J. Biol. Chem. 251, 2094-2104.
24. Ruzicka, F.J., Beinert, H., Schepler, K.L., Dunham, W.K. and Sands, R.H. (1975) Proc. Natl. Acad. Sci. USA 72, 2886-2890.
25. Beinert, H., Ackrell, B.A.C., Vinogradov, A.D., Kearney, E. and Singer, T.P. (1977) Arch. Biochem. Biophys. 182, 95-106.
26. Ohnishi, T., Lim, J., Winter, D.B., and King, T.E. (1976) J. Biol. Chem. 251, 2105-2109.
27. Ohnishi, T., Leigh, J.S., Winter, D.B., Lim, J., and King, T.E. (1974) Biochem. Biophys. Res. Commun. 61, 1026-1035.
28. Blum, H. and Ohnishi, T. (1980) Biochim. Biophys. Acta 621, 9-18.
29. Ohnishi, T., King, T.E., Salerno, J.C., Blum, H., Bowyer, J.R. and Maida, T. (1981) J. Biol. Chem., in press.
30. Salerno, J.C. and Ohnishi, T. (1980) Biochem. J. 192, 769-781.
31. Ingledew, W.J., Salerno, J.C. and Ohnishi, T. (1976) Arch. Biochem. Biophys. 177, 176-184.
32. Konstanchinov, A.A. and Ruuge, K.E. (1977) FEBS Letts. 81,

33. Ohnishi, T. and Trumpower, B.L. (1980) J. Biol. Chem. 255, 3278-3284.
34. Blum, H., Leigh, J.S. and Ohnishi, T. (1980) Biochim. Biophys. Acta 626, 31-40.
35. Blum, H., Cusanovich, M.A., Sweeney, W.V. and Ohnishi, T. (1981) J. Biol. Chem. 256, 2199-2206.
36. Ohnishi, T., Blum, H., Harmon, H.J. and Hompo, T. (1981) in Interaction Between Iron and Proteins in Oxygen and Electron Transport (C. Ho & W.A. Eaton, eds.), Elsevier/North-Holland, New York, in press.
37. Blasie, J.K., Erecinska, M., Samuels, S. and Leigh, J.S. (1978) Biochim. Biophys. Acta 501, 33-52.
38. Salerno, J.C., Blum, H., and Ohnishi, T. (1979) Biochim. Biophys. Acta 547, 270-281.
39. Girdlestone, J., Bisson, R. and Capaldi, R.A. (1981) Biochemistry 20, 152-156.
40. Hatefi, Y., Haavik, A.G. and Griffiths, D.E. (1962) J. Biol. Chem. 237, 1676-1680.
41. Heron, C., Smith, S. and Ragan, C.I. (1979) Biochem. J. 181, 435-443.
42. Hatefi, Y. and Stempel, K.E. (1969) J. Biol. Chem. 244, 2350-2357.
43. Dooijewaard, G., Slater, E.C., van Dijk, P.J. and deBruin, G.J.M. (1978) Biochim. Biophys. Acta 503, 405-424.
44. Galante, Y.M. and Hatefi, Y. (1979) Arch. Biochem. Biophys. 192, 559-568.
45. Heron, C., Smith, S., and Ragan, C.I. (1979) Biochem. J. 181, 435-443.
46. Widger, W. (1979) Ph.D. Thesis, Department of Chemistry, State University of New York at Albany.
47. Ragan, C.I., Galante, Y.M., Hatefi, Y. and Ohnishi, T., submitted for publication.
48. Smith, S. and Ragan, C.I. (1980) Biochem. J. 185, 315-326.
49. Chen, S. and Guillory, R.J. (1980) Fed. Proc. 39, 2057.
50. Orme-Johnson, N.R., Hansen, R.E. and Beinert, H. (1974) J. Biol. Chem. 249, 1922-1927.
51. Ohnishi, T. (1976) Eur. J. Biochem. 64, 91-103.
52. Ingledew, W.J. and Ohnishi, T. (1980) Biochem. J. 186, 111-117.
53. Ohnishi, T., Leigh, J.S., Ragan, C.I. and Racker, E. (1974) Biochem. Biophys. Res. Commun. 56, 775-782.
54. Albracht, S.P.J. and Slater, E.C. (1970) Biochim. Biophys. Acta 223, 454-459.
55. Albracht, S.P.J., Dooijewaard, G., Leeuwerik, F.J. and van Swol, B. (1977) Biochim. Biophys. Acta 459, 300-317.
56. Albracht, S.P.J., Leeuwerik, F.J. and van Swol, B. (1979) FEBS Letts. 104, 197-200.

57. Ohnishi, T., Blum, H., Galante, Y. and Hatefi, Y., submitted for publication.
58. Ohnishi, T. (1975) Biochim. Biophys. Acta 387, 475-490.
59. Emptage, M.H., Kent, J.A., Huyuh, B.H., Rawlings, J., Orme-Johnson, W.H. and Munck, E. (1980) J. Biol. Chem. 255, 1793-1796.
60. Stout, D., Ghosh, W., Pattahhi, V. and Robbins, A.H. (1980) J. Biol. Chem. 255, 1797-1800.
61. Beinert, H. (1965) in Oxidases and Related Redox Systems (T.E. King, H.S. Mason, & M. Morrison, eds.), Vol. I, pp. 198, Plenum Press, Elmwood, NY.
62. Salerno, J.C., Ohnishi, T., Lim, J., Widger, W.R. and King, T.E. (1977) Biochem. Biophys. Res. Commun. 75, 618-624.
63. Mitchell, P. (1972) in Mitochondria/Biomembranes (S.G. van der Bergh, P. Borst, L.L.M. van Deenen, J.C. Riemersma, E.C. Slater, J.M. Tager, eds.) Vol. 28, North-Holland, Amsterdam, pp. 358-370.
64. Garland, P., Clegg, R.A., Downie, J.A., Gray, T.A., Lawford, H.G. and Skyrme, J. (1972) in Mitochondria/Biomembranes, (S.G. van der Bergh, P. Borst, L.L.M. van Deenen, J.C. Riemersma, E.C. Slater, J.M. Tager, eds.), Vol. 28, North-Holland, Amsterdam, pp. 105-117.
65. King, T.E., Yu, L., Nagaoka, S., Widger, W.R., Yu, C.A. (1978) in Frontiers of Biological Energetics (P.L. Dutton, J.S. Leigh, & A. Scarpa, eds.), Academic Press, NY, pp. 174-182.
66. DeVault, D. (1976) J. Theor. Biol. 62, 115-139.
67. Skulachev, V.P. (1975) in Proceedings of the 10th FEBS Meeting, Paris (Y. Raoul, ed.), Associated Scientific, Amsterdam, pp. 225-238.
68. Salerno, J.C. (1981) submitted for publication.
69. Mitchell, P. (1981) in Oxidases and Related Redox Systems, (T.E. King, H.S. Mason & M. Morrison, eds.), Plenum Press, Elmwood, NY, in press.
70. Ohnishi, T. and Salerno, J.C. (1981) in Metal Ions in Biology (T.G. Spiro, ed.), John Wiley & Sons, Inc. New York, Vol. 4, in press.

THE COMPOSITION, STRUCTURE, TOPOLOGY, AND

FUNCTION OF COMPLEX III
(UBIQUINOL–CYTOCHROME c REDUCTASE)

B. Dean Nelson

Department of Biochemistry, Arrhenius Laboratory, University
of Stockholm, 106 91 Stockholm, Sweden

This paper is dedicated to Lars Ernster with deep gratitude for the scientific stimulation, guidance, and support
he has provided over the years.

INTRODUCTION

It is now agreed that Complex III (ubiquinol-cytochrome
c reductase) catalyzes a redox-driven translocation of
protons across the mitochondrial inner membrane. In view of
this property, isolated Complex III should provide an excellent model with which to study the molecular basis for
these events.

Complex III is, however, a complicated enzyme. In addition to 4 separate redox centers (Table 1), it contains a
minimum of 8 polypeptides. Five of these do not contain
prosthetic groups, and their functions are unknown. In

C. P. Lee, G. Schatz, G. Dallner (eds.), Mitochondria and Microsomes
   in honor of Lars Ernster   ISBN 0-201-04576-1

addition, the catalytic mechanism is complicated by the
fact that the reducing substrate (ubiquinol) is a two
electron donor, while the electron acceptor (cytochrome c)
receives only one electron. During the past decade a number
of models have been proposed which account for the pathway
of electron fluxes through Complex III and for the mechanism of proton translocation. Although these provide important working models, the final solution to these problems requires a great deal more information on the molecular organization of Complex III.

The purpose of the present review is to summarize what
is known about the composition and structure of isolated
Complex III, with emphasis being placed upon the peptides.
A cursory attempt will be made to relate structure to function. However, due to space limitation this discussion will
be limited to those instances where structural arguments
can, or are thought to be able to, shed light on the mechanism of electron transport and proton translocation through
the Complex. The reader is referred to (1-5) for earlier
reviews of this subject.

Table 1. Redox Components of Complex III

Component	Molar ratio	E_m (Mv)
Cytochrome c_1	1	+ 230
Iron-sulfur protein	1	+ 230
Cytochrome b-562	1	+ 90
Cytochrome b-565	1	− 30
QH_2/Q	−	+ 30
$QH_2/QH\cdot$	−	+ 300[a]
$QH\cdot/Q$	−	− 300[a]

[a] estimated midpoint potentials (8). For additional
information see ref. (40,110).

MODELS OF ELECTRON TRANSPORT AND PROTON TRANSLOCATION

The mechanisms of proton translocation and electron transport in Complex III remain topics of controversy. For sake of orientation, a brief description of the two most general models are presented.

Electron transport from QH_2 to cytochrome c can take place via two general mechanisms involving linear (6,7) or cyclic (8-10) transfer sequences. The first of these is illustrated by the Wikström–Berden model (7) in Fig. 1A, or by a modification (Fig. 1B) of the Wikström–Berden model which takes into account the currently accepted electron transfer sequences from QH_2 (8,10) and the role of the iron sulfur protein (11). According to this model, Complex III operates as a QH_2 oxidase with both electrons moving linearly through the complex to cytochrome c.

The cyclic model of electron flow through Complex III is illustrated by the Q-cycle of Mitchell (8) (Fig. 2). In contrast to the linear model, the Q-cycle predicts that only one of the two electrons extracted from QH_2 flows directly to cytochrome c, the second is recycled through Complex III.

The consequences of these two types of electron fluxes are important with respect to the mechanism of proton translocation. Both the linear and cyclic models can account for the experimentally observed $H^+/2e^-$ value of 4 (9,12,13). Since the linear models predict a $H^+/2e^-$ of 2 from the oxidation of QH_2, the 2 extra protons must be translocated via a proton pump mechanism (5,13,14). In contrast, the Q-cycle postulates that protons are translocated as hydrogen on QH_2 and that the $H^+/2e^-$ of 4 is accounted for by re-cycling of one electron (8).

A

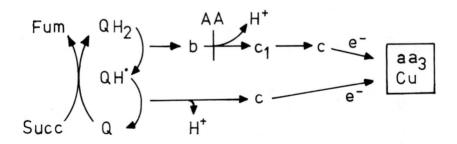

B

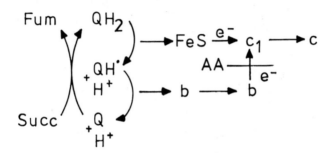

Fig. 1. Linear pathways of electron transport in Complex III
A) Wikström–Berden model from ref. (7)
B) Modified Wikström–Berden model.

Both models appear to carry specific predictions, as
well as certain constraints, regarding the molecular organization of Complex III. In the past, the limited structural and topological information which was available on Complex III has been interpreted rather freely in support of
various kinetic models. The remainder of this paper will
summarize what is currently known about the composition,
topology and, to some extent, function of the subunits of
Complex III.

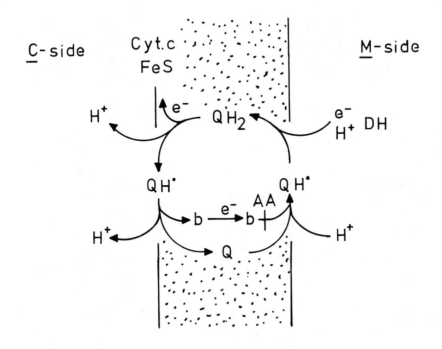

Fig. 2. Cyclic pathway of electron transport in Complex III
        (ref. 8).

COMPARATIVE PEPTIDE COMPOSITION

    Complex III has been isolated from Neurospora crassa
(15,16), Saccharomyces cerevisiae (17,18), beef heart (19-
21) and rat liver (22). Although different isolation procedures have been used, the polypeptide composition of the
various preparations are strikingly similar (Table 2). The
characteristics of the 8 peptides comprising Complex III are
summarized in Table 3.

Table 2. Apparent molecular weights of the subunits of Complex III from different sources.

Peptide	Beef heart ref. (36)	Rat ref. (22)	Yeast ref. (17)	Neurospora ref. (15)
		molecular weight x 10^{-3}		
I	48	50	44	50
II	46	47	40	45
III	30	33	32	31
IV	28	33	32	27
V	25	25	-	25
VI	14	12	17	14
VII	11	10	14	10
VIII	5	5	11	8

CHARACTERISTICS OF THE INDIVIDUAL PEPTIDES OF COMPLEX III

Core proteins. The two core proteins are the largest peptitides in Complex III (Table 3). Neurospora core proteins have been isolated as a water soluble complex (23). In contrast, the beef heart core proteins appear to be physically integrated into the complex (1,24,25). Furthermore, the available evidence suggest that detergents are required for removal of beef heart core proteins, although this has not been studied methodically.

Functions for the core proteins have not been established. Alkylation of core protein I leads to the partial inactivation of duroquinol-cytochrome c reductase activity (64), suggesting a role for this peptide in either the catalytic mechanism or in the structural assembly of the Complex. Specific antibodies against core proteins I and II are,

however, without effect on electron transport in either submitochondrial particles or isolated Complex III (26). Furthermore, these antibodies did not exert an oligomycin- or
uncoupler-like effect on NADH oxidase in submitochondrial
particles.

It has been suggested that yeast core proteins are
electrophoretic artifacts (27). Although this would explain
the difficulties in establishing a function for these peptides, several observations make this proposal untenable:
1) core proteins are consistently found in the same relative amounts in Complex III independent of the source of the
complex or the method of isolation, 2) additional peptides
are not generated upon re-electrophoresis of isolated core
proteins (25,26,28,29), 3) immunoreplica experiments with
antibodies against the core proteins reveal no low molecular

Table 3. Subunit Composition of Beef Heart Complex III

Peptide	Molecular weight x 10^{-3}	designation	molar ratio	function
I	48	core protein I	1	unknown
II	46	core protein II	2	unknown
III	30	cytochrome b	2	electron transport
IV	28	cytochrome c_1	1	electron transport
V	25	iron sulfur protein	1	electron transport
VI	14	cytochrome c_1-associated	1-2	unknown
VII	11	cytochrome b-associated	1-2	unknown
VIII	5-9	antimycin-binding protein	1-2	unknown

weight, unaggregated forms of the peptide (26), 4) immunological studies also indicate that core proteins in intact
mitochondria are associated only with Complex III (26), and
that they are present in amounts stoichiometric with the
cytochromes of the b-c_1 region (23,26,30). Thus, core proteins are specifically associated with Complex III and, by
implication, probably have a specific function in this
complex.

Cytochrome b. Cytochrome b has been isolated from beef
heart (31), Neurospora (32,33) and yeast (34) mitochondria.
The characteristics of these preparations have been recently reviewed (3). Isolated cytochrome b behaves as a dimer of
approximately 60,000 molecular weight in Triton micelles
(31,33). Molecular weights of 27,000-33,000 have been reported for the monomer (31-34). Earlier assignments of
molecular weights of 37,000 (35) and 14-17,000 (1,28,29,35,
36) for cytochrome b probably resulted from either anomalous migration of this peptide during SDS-polyacrylamide gel
electrophoresis (29) or to the isolation of a Complex II-
associated b-cytochrome (36,37,38). Although some evidence
suggests that the dimer is composed of two non-identical
monomers (33), genetic evidence points to a single peptide
(for review see ref. 39). Apo-cytochrome b is encoded on
mitochondrial DNA and, as expected, contains a high per-
centage of nonpolar amino acid residues (31).

The redox heterogeneity of the heme centers is retained during purification of beef heart cytochrome b (31),
even though the midpoint potentials are 100 mV lower than
in Complex III (40,41). The midpoint potentials of these
centers (-5 mV and -100 mV (31)) exhibit a pH dependency
similar to those reported for isolated Complex III (21) and

mitochondria (42). Unlike mitochondrial cytochrome b, the
isolated protein exhibits no absorption maximum in the α
region at 566 nm (31).

Cytochrome c_1. Cytochrome c_1 from beef heart (43–45) and
yeast (46) exhibit molecular weights of approximately
31,000 on SDS-polyacrylamide gels. They are sensitive to
proteolytic attack, and peptides of 2–3,000 daltons are
released (44,46). The complete amino acid sequence of beef
heart cytochrome c_1 was recently reported (47). This peptide contains 241 amino acid residues. The calculated molecular weight is 27,874. Heme is attached to cysteine residues 37 and 40 located near the N-terminus (47). The N-terminus is not blocked in this preparation (47), in contrast
to earlier reports (44). The midpoint potential of the heme
(approximately + 230 mV) (48,49) is similar in the isolated
peptide, in isolated Complex III (40) and in mitochondria
(42), as are the spectral properties (40,49,42).

It is now abundantly clear that cytochrome c binding
to Complex III takes place through amino acid residues on
cytochrome c_1. Evidence supporting this conclusion include:
1) the formation of kinetically competent 1:1 molar complexes of cytochrome c_1 and cytochrome c (48,49), 2) cytochrome c binding sites on Complex III and isolated cytochrome c_1 which are indistinguishable on the basis of
different ionic conditions (50), 3) blockage of acetylation
of the same lysine residues on cytochrome c bound to either
Complex III or isolated cytochrome c_1 (50), 4) kinetic
studies using chemically modified cytochrome c (51–53), and
5) the binding of photoaffinity labeled cytochrome c to
Complex III (54).

It is not known which amino acid residues on cyto-

chrome c_1 are involved in the binding of cytochrome c. In
contrast, the amino acid residues on cytochrome c which are
required for binding to the reductase have been thoroughly
studied (55-57). These residues are concentrated on the
upper left hand side of cytochrome c, near the heme cleft.
The same residues are required for cytochrome c binding
to cytochrome oxidase, implying rotational movement of
cytochrome c during oxidation/reduction.

Rieske iron sulfur protein. The isolated iron sulfur protein
from beef heart mitochondria has a molecular weight of approximately 25,000 on SDS-polyacrylamide gels (58). The peptide contains a single 2S:2Fe center per mole of enzyme
(18,59), the midpoint potential of which is close to that
of cytochrome c_1 (18,60).

Reconstitution studies by Trumpower and colleagues
(10,11,58) show that the Rieske protein is identical to the
oxidation factor (61). The BAL-sensitive factor of Slater
(62) has also been identified as the Rieske protein (63).
The Rieske protein is required for transfer of electrons
from succinate or quinols to cytochrome c_1 (18,58-62). In
the presence of antimycin, the Rieske protein is also required for transfer of electrons from succinate to cytochrome b (10,61). It is located on the oxygen side of the
antimycin block in the electron transport chain.

Low molecular weight peptides. In addition to the peptides
mentioned above, Complex III contains at least 3 peptides
of lower molecular weight. Considerable variation has been
reported in the apparent molecular weights of these peptides (see ref. 28), and even in their numbers (29,36,65).
This is due to the use of different electrophoretic systems,

which influence both the resolution and the relative mobilities (29,66) of the small peptides. For example, peptides VI
and VII (Table 3), which in the early literature were often
associated with cytochrome c_1 and cytochrome b, respectively
(1,28,29,36,43,67), exchange places on polyacrylamide gels
in the presence of different electrophoretic buffer systems
(66, cf. ref. 29).

Peptide VIII was identified as the antimycin binding
protein (68). However, recent experiments (69) show that
antimycin does not bind to yeast mitochondria containing a
mutation in the structural gene for cytochrome b. It was
suggested that antimycin binds to cytochrome b, in accordance with earlier reports (for review see ref. 70). However, these experiments (69) do not eliminate the possibility that cytochrome b is needed only for proper insertion
of subunit VIII into Complex III, and that the antimycin
binding site is located on subunit VIII.

Two quinone binding proteins have been identified in
different preparations of Complex III (71-75). One of these,
with a reported molecular weight of 15,000, is a component
of Complex II (71,73,76). It is found only in preparations
of Complex III which are reconstitutively active with purified succinate dehydrogenase (71,73). The second quinone-
binding protein is indigenous to Complex III (quinol–cytochrome c reductase). It has been identified by kinetic experiments (76,75). Photoaffinity labeling of Complex III
with quinone analogues leads, however, to the labeling of
two peptides which exhibit apparent molecular weights of
37,000 and 17,000 (74) upon electrophoresis in the Weber-
Osborn buffer system. These peptides have earlier been
associated with the b-cytochromes (35). Their relationships
to the peptides listed in Table 3 are not clear. The dis-

covery of two quinone-binding proteins appears to fulfill
one of the predictions of the proton motive Q-cycle (8).

STOICHIOMETRY OF COMPLEX III PEPTIDES

The stoichiometry of Complex III subunits has been determined by Coomassie blue binding (28,29), total leucine
content (30), total amino acid analysis (77), radioimmuno-
logical assay (26) and quantification of prosthetic groups.
It is generally accepted that the electron transport proteins are present in molar ratios of 2:1:1 (cytochrome b,
cytochrome c_1 and the iron sulfur protein, respectively).
In contrast, large variations in the stoichiometries of the
non-redox proteins have been reported, particularly for the
3 smallest subunits. The latter is due, in part, to difficulties arising from the use of different electrophoretic
methods. Furthermore, since no functions are established
for these smaller peptides it is difficult to determine
which are bona fide members of the complex.

Estimates of the molar ratios of the core proteins also
vary. Recent immunological experiments suggest a 2:1 molar
ratio for core protein II: core protein I (26). This result
agrees with those obtained for Neurospora (30), and those
earlier obtained for beef heart (24) core proteins. Exclud-
ing the 3 smallest subunits, the most probably stoichiometry of the remaining subunits of Complex III is 1:2:2:1:1
for core protein I: core protein II: cytochrome b: cytochrome c_1: and the iron sulfur protein, respectively.

THE MOLECULAR ORGANIZATION OF COMPLEX III

Complex III is a dimer in the presence of Triton X-100 (78) and a monomer in the presence of taurocholate (79). The molecular weight of the monomer is approximately 230,000 to 290,000 (79). Electron microscopy of membrane crystals or ordered membrane sheets of Complex III indicate that both the beef heart (1) and Neurospora enzymes (23,80, 81) have the dimension of 70 x 90 Å. Image reconstruction of the Neurospora enzyme indicates that this is the dimeric form (23,82). Beef heart Complex III was considered to be a monomer with a 15 Å wide and 25 Å deep well in its center (1). It seems reasonable, however, that this represents the gap between the two monomers.

Holo-Complex III. The exact arrangements of the peptides within Complex III or within the membrane are not known (for Review see ref. 96). Based upon image reconstruction data, Weiss and colleagues (23) have drawn the following general conclusions for the Neurospora enzyme. The enzyme is approximately 150 Å long and it extends through the bilayer. The complex is 50 Å wide in the bilayer. It projects approximately 70 Å into the matrix, and this projection is suggested to contain the core proteins. The membrane sector contains the b-cytochromes and certain domains of the iron sulfur protein and cytochrome c_1. Cytochrome c_1 projects 30 Å into the cytoplasmic compartment. Locations of the three smallest subunits were not specified. In more recent experiments, Weiss and coworkers studied the membrane crystals of a subcomplex preparation which contains only cytochromes b and c_1 and the 3 smallest peptides (81). The dimensions of this preparation in the direction perpendicul-

ar to the membrane is about 70 Å, approximately half of
that of the intact complex. It is assumed that the greatest
portion of this sub-complex is submerged in the bilayer.

In the following paragraphs we summarize more specific
information available on the topology of the individual
subunits of Complex III.

Topology of core proteins. Neurospora core proteins can be
isolated as water soluble complexes (23). This result, plus
image reconstruction data (23,81), led to the conclusion
that core proteins are loosely attached to the matrix surface of the inner membrane (23).

Topology of the beef heart core proteins has been studied with non-penetrating protein reagents. Core protein II
is labeled on the matrix surface of the membrane with both
^{35}S-DABS (83,84) and lactoperoxidase + ^{125}I (85). It has
been claimed, however, that beef heart core protein II is
also labeled on the cytoplasmic surface, and thus penetrates the bilayer (84). Core protein II was not labeled, however, in the bilayer with hydrophobic membrane probes (86).
Although the bulk of evidence suggests that core protein II
is highly exposed on the matrix surface of the inner membrane, more work is clearly needed.

The location of core protein I in beef heart is even
less clear, since some groups have observed its labeling
only on the matrix surface (83) while others observed
labeling only on the cytoplasmic surface (84). In an other
type of experiment, insertion of Complex III into phospholipid vesicles resulted in the disappearance of ^{35}S-DABS
and ^{125}I labeling sites on both core proteins (87). These
sites were re-exposed upon treatment of the vesicles with
detergents. Such experiments suggest that the core proteins

are either buried in the bilayer or located on the inside of
the vesicle (mitochondrial orientation).

Topology of cytochrome b. Cytochrome b is hydrophobic and is
therefore assumed to be buried in the lipid bilayer. However, it is not known if certain domains of the protein
project from the bilayer. The cytochrome b hemes (cytochrome
b-565 and b-562) have been located to within 10 Å of the
inner membranes cytoplasmic surface by electron spin resonance techniques (88,89). If this is the case, models
proposing a cytochrome b transmembrane electron conducting
pathway (Fig. 2) should account such heme asymmetry. No information is available on the heme binding domains on the
apoprotein. Thus, it is not known if heme binds to a single
peptide or if binding is shared between two peptide chains
in the dimeric structure. This information will help to determine the molecular significance of the two spectrally
distinguishable b-cytochromes in Complex III (Table 1).

Topology of cytochrome c_1. The location of cytochrome c_1 on
the cytoplasmic surface of the inner membrane is firmly established by the demonstration that cytochrome c_1 contains
the binding sites for cytochrome c (see above). This conclusion is also confirmed by studies in which specific antibodies (90) and surface labeling reagents (83,84) were used
as probes. Cytochrome c_1 heme has also been located to within 10 Å of the cytoplasmic surface using paramagnetic cations (Ni^+ and Gd^{3+}) bound to the surface of the membrane as
pertubing agents (88,89).

Little is known about the domains of the apoprotein
which are anchored in the bilayer. Weiss reported that
Neurospora cytochrome c_1 projects as far as 30 Å from the

outer surface of the membrane (23) suggesting that the peptide is anchored by a relatively small domain.

An approach to this problem was initiated by Trumpower
and Kati (44) who demonstrated that a peptide fragment of
2-3,000 molecular weight was released upon treating isolated cytochrome c_1 with trypsin. Cytochrome c_1 integrated
into Complex III was insensitive to trypsin, suggesting that
the fragment was also integrated into the holoenzyme. In
these experiments the tryptic fragment was released from
the N-terminal end of the peptide chain. However, amino
acid sequence analysis of cytochrome c_1 also show a hydrophobic core between residues 199 and 241 at the carboxy
terminus which is presumed to represent a membrane binding
domain of the peptide (47). Finally, evidence indicating
that cytochrome c_1 might span the bilayer comes from studies showing that yeast cytochrome c_1 is translated in the
cytoplasm as a larger molecular weight precursor molecule
(for review see ref. 91). Processing is thought to occur
in the matrix (91).

Topology of the iron sulfur protein. The iron sulfur protein
is partially exposed on the surface of isolated Complex III
since trypsin treatment of the complex results in the specific digestion of this peptide (25,92,93) and a concomitant loss of enzyme activity. The surface location of the
iron sulfur protein is also confirmed by labeling isolated
Complex III with ^{35}S-DABS (87).

The exact location of the Rieske protein within the intact membrane is not known. In mitochondria, this peptide
is weakly labeled by ^{35}S-DABS (83). However, a large number
of DABS binding sites on the iron sulfur protein are shielded when Complex III is inserted into the lipid bilayer (87).

Under these conditions trypsin sensitivity is also lost
(93). These findings suggest that a large part of the peptide is submerged in the bilayer or is located on the inner
surface of the liposomes. In yeast, the iron sulfur protein
is translated as a larger precursor peptide which undergoes
processing (91). This strongly suggests that the peptide
might extend through the bilayer. The iron sulfur centers
of the protein have been placed 19-24 Å from the outer surface, i.e., in the middle of the bilayer (94).

Topology of the small molecular weight peptides. Little is
known about the location of these peptides in the complex
or in the membrane. Subunit VIII (5,000 Mr) is labeled with
^{35}S-DABS only after dissociating Complex III with detergents
(87), suggesting that it is buried within the complex. This
agrees with reports that subunit VIII is extractable in methanol (25), and that it is labeled with reagents believed
to probe the hydrophobic core of the bilayer (86).

The two largest of these 3 subunits (subunits VI and
VII, Table 3) are labeled with ^{35}S-DABS on the surface of
isolated Complex III (84,87), as well as on the cytoplasmic
surface of intact mitochondria (84) and the surface of reconstituted Complex III-vesicles (87).

Cross linking studies. Cross linking agents have been used
in order to assess the distances between the individual
subunits of isolated Complex III (23,95). It should be
stressed, however, that an inherent difficulty in such experiments is the state of aggregation of the complex. If
the enzyme exists as a dimer or in higher aggregation
states, it can be difficult to determine if cross linking
occurs between peptides within the complex or peptides of

different complexes.

Capaldi and associates (95) conducted a detailed study using conditions in which the complex was judged to be in the monomeric state. Bifunctional reagents spanning 6, 11 and 18 Å were used. Apocytochromes b and c_1 were both linked to the iron sulfur protein and to subunits VI and VII (cytochrome b- and cytochrome c_1-associated peptides, respectively) by reagents spanning 6 and 18 Å. They did not cross link to the core proteins. The two core proteins were linked to each other by all three reagents, as were core proteins I + iron sulfur protein and core protein II + subunit VI (cytochrome c_1 associated). No cross linking of cytochrome b and cytochrome c_1 was observed, and no dimers of cytochrome b or of cytochrome c_1 were found. It would be interesting to know if these reagents penetrate the hydrophobic domains of the enzyme or if cross linking occurs primarily at the surface of the complex.

Weiss et al. (23) conducted cross linking experiments using a sub-complex free of core proteins and the iron sulfur peptide. Dimers of cytochrome c_1 were formed, as were dimers of cytochrome b. In agreement with (95), Weiss et al. (23) also showed extensive cross linking of cytochrome b and c_1 to subunits VI-VIII.

STRUCTURAL REQUIREMENTS FOR CYCLIC AND LINEAR ELECTRON TRANSPORT PATHWAYS

At this point it is important to ask if cyclic and linear electron transfer pathways can be characterized by certain predictable structural features, and, if so, does our current knowledge of the structure of Complex III provide any clues as to the type of electron transport system present.

Proton translocation step. A major structural difference
predicted by the two models of electron transport (Figures
1 and 2) is related to the mechanism of proton translocation. The Q-cycle predicts two diffusible forms of ubiquinone and two quinone binding sites on either side of the
membrane. The function of ubiquinone in the membrane has
received considerable attention, and the available information is vast (see ref. 96-100). However, no clear picture
has yet emerged regarding the location or movements of ubiquinone in the membrane.

In linear electron transport pathways (Fig. 1), redox
linked proton pump mechanisms are predicted. Prerequisite
to this are redox linked conformational changes in Complex
III. Although conformational changes can be demonstrated
(1,25,101-103), there is no proof that they are linked to
the translocation of protons. Furthermore, there is little
information suggesting which peptide(s) could be involved
in proton pumping, if such a mechanism exists. To date, the
most likely candidate for this role is cytochrome b (3,104).
The pH dependency of the midpoint potentials in mitochondria
(42) and isolated cytochrome b (31), plus the dimeric structure of cytochrome b and the presence of two heme centers
(31), make this peptide suitable as a component in a proton
pump. Cytochrome b is also hydrophobic, and could, thus,
provide the necessary transmembrane channel. A detailed
model for the role of cytochrome b in proton translocation
has been proposed (3,104).

Ubiquinone-binding proteins. The Q-cycle requires two Q-
binding proteins for stabilizing the semiquinones at the
inner and outer surfaces of the coupling membrane (8). Two
Q-binding proteins have been identified, one a component of

Complex II (73), and a second which is intrinsic to Complex III (72,74). Although the discovery of these peptides appears to fulfill a prediction of the Q-cycle, they can also be accommodated by linear mechanisms. If, for example, Complex II catalyzes a two step reduction of Q to QH_2 (see Fig. 1), the semiquinone generated in this reaction might also require stabilization prior to transfer of the second electron. The ubiquinone binding protein in Complex III is required at the same site (near cytochrome b) in both the Q-cycle and the linear mechanism (cf. Fig. 1B and Fig. 2). Thus, the mere presence of two Q-binding proteins is not sufficient to allow a choice between the cyclic and linear electron transport models shown in Figures 1 and 2.

Finally, it should be emphasized that even though the Q-cycle appears to require Q-binding proteins on either side of the membrane, the distances between the Q-binding domains on these peptides could be short. Thus, the location of Q-binding sites in the membranes, and the distances through which Q and QH_2 are required to diffuse (8), must await detailed structural analysis of the enzyme.

Interaction of the primary dehydrogenases with Complex III.
There are two general mechanisms by which primary dehydrogenases can interact with ubiquinone and Complex III. In linear electron transport pathways the dehydrogenases function as quinone reductases, supplying both electrons directly to Q ($Q + 2H^+ + 2e^- \rightarrow QH_2$). In this case, ubiquinone should behave as a mobile pool of reducing equivalents for Complex III. In contrast, in the Q-cycle the primary dehydrogenase functions as a semiquinone reductase ($QH\cdot + H^+ + e^- \rightarrow QH_2$) (8,10,100). Thus, a careful analysis of the catalytic mechanism of Complex II should provide support

for either cyclic or linear electron transport in Complex
III. A number of important papers have recently appeared on
the interaction of ubiquinone with succinate dehydrogenase
(99,100,105–110). A thorough discussion of these is beyond
the scope of the present review.

Clearly, the same physical integration of Complexes II
and III is not needed for the participation of ubiquinone
in linear or cyclic electron transport pathways. In Bacillus
subtilis, binding of succinate dehydrogenase to the membrane
requires the presence of a Complex II-associated b-cytochrome (11). Binding occurs to holocytochrome b, but not to
the apoprotein, indicating a need for heme (112). A similar
b-cytochrome is also present in beef heart Complex II (38),
although no role of this cytochrome in succinate dehydrogenase binding has been reported. These findings (111,112)
indicate, however, that a careful study of the Complex II-
associated cytochrome b, and its structural and functional
relationships to cytochrome b of Complex III, is warranted.

Rieske iron sulfur protein. The recent studies of Trumpower
and coworkers (10,58) have helped greatly in elucidating
the function of the iron sulfur protein. This protein is
required for reduction of cytochrome c_1 by duroquinol and
succinate (10,61), and was therefore assigned the role of a
QH_2/cytochrome c_1:semiquinone/cytochrome b oxido-reductase
(100). Based upon this assignment, Trumpower was able to
integrate a number of independent kinetic observations, including: 1) the role of the Rieske protein in the antimy-
cin-induced reduction of cytochrome b (7), 2) inhibition
of this phenomena upon prereduction of Complex III with
ascorbate (113,114) and 3) antimycin inhibition of cytochrome c_1 reduction during the second, but not the first,

turnover of the complex (115). However, it seems that most of these kinetic observations can be explained in the same manner if the iron sulfur protein functions as shown in the modified Wikström-Berden model (Fig. 1B). Neither the Q-cycle nor the model in Fig. 1B explain, however, the reduction of cytochrome b in particles depleted of the iron sulfur protein (10).

The Q-cycle in its most general form appears to predict the location of the iron sulfur protein near the cytoplasmic surface (10,100). However, even though some domains of the peptide might be exposed to the cytoplasmic surface, much evidence indicates that a large part of the peptide is buried in the membrane (see above). The significant point, however, is the location of the FeS cluster, which appears to be 25 Å from the cytoplasmic surface, i.e., near the center of the bilayer (94).

POSSIBLE FUNCTIONS OF NON REDOX PEPTIDES

With the exception of possible roles in antimycin- or ubiquinone-binding, little is known about the functions of the non-redox peptides (Table 3). Three possible functions are available: 1) catalytic, 2) structural or 3) regulatory. The first of these is normally rejected out of hand since these peptides lack known prosthetic groups. However, the role of sulfhydryl groups or of electron tunneling through amino acid residues of Complex III have not been evaluated, much less eliminated. Structural involvement of these peptides can occur at two levels. One could be to aid in the alignment of the redox proteins during electron transport and/or proton translocation. A second could be in the recognition, binding and assembly of newly synthesized sub-

units of Complex III approaching the membrane from either
the matrix or cytoplasmic sides.

Regulatory functions of these peptides, at least in
the regulation of electron transport, is more difficult to
imagine. Electron transport in the cytochrome b-c$_1$ region
is in near equilibrium with the phosphorylation reactions
(42), and there is no need for a regulatory mechanism.
However, it is well documented that one of the early effects
of glucagon on liver cells is activation of the respiratory
chain (116-118), apparently in the cytochrome b-c$_1$ region
(116,117). Since glucagon acts via a cAMP mechanism, one
must consider the possibility that some components of
Complex III undergo phosphorylation.

OUTLOOK

It should be clear from the present review that studies
on the structure and topology of Complex III are still in
their infancies. Interest in this subject has been greatly
stimulated by the recent introduction of several working
hypotheses describing possible pathways of electron flow
through Complex III and the mechanisms of proton translocation. Although our present knowledge of the organization of
Complex III does little to help elucidate these mechanisms,
progress is rapid and the future prospects appear promising.
Three areas of research seem particularly promising: 3 dimensional structural studies of the enzyme, isolation and
reconstitution of the individual subunits, and the use of
mutants for analyzing the functional roles of the subunits.

Methods for producing ordered membrane sheets and
crystals of various respiratory chain proteins have only
recently developed, and will certainly expand rapidly.

Presently, both cytochrome oxidase and Complex III have
been prepared in various forms which permit 3 dimensional
structural analysis. Increased resolution, coupled to the
analysis of subcomplex preparations, should prove very useful in elucidating the topological organization of these
complicated proteins.

Rapid advances have also been made in the area of isolation and reconstitution. Examples are: studies on the
role of the iron sulfur protein in electron transport, the
identification of quinone-binding proteins and their roles
in reduction of ubiquinone, and the identification of cytochrome c_1 as the binding site for substrate (cytochrome c).
The isolation of the above mentioned redox proteins, plus
the isolation of relatively intact cytochrome b were necessary prerequisites for reconstitution of electron transport starting from the individual peptides of Complex III.
It now seems only a matter of time before this reconstitution is achieved. Establishing roles for the non-redox proteins of Complex III is more problematic. Although reconstitution studies represent a means to this end, the difficulties are clearly greater than when starting with proteins
for which functions are, within limits, already known.

To date, little advantage has been taken of mutants as
tools for investigating subunit functions. A relatively
large collection of genetically well-defined yeast and
Neurospora mutants are now available. These include both
mitochondrial and nuclear mutants of Complex III and cytochrome oxidase. The gene products of many of these mutants
have been thoroughly characterized, particularly those encoded in mitochondrial DNA. Methodical studies are now
needed on the effects of these mutations on electron transport and proton translocation in intact membranes and in

isolated enzymes. Since the available mutants range between those in which individual peptides are absent to those in which the same peptide is translated as a larger molecular weight form, the relationship between molecular structure and enzyme function can be assessed. In addition, mutants lacking specific peptides should be useful as starting material for reconstitution studies. Thus, the use of mutants represents the most potentially important tool available for elucidating the molecular functions of the subunits of Complex III.

Acknowledgements

The author wishes to thank B. Trumpower and L. Hederstedt for making available manuscripts which were in press. The work quoted from our own laboratory was supported by funds from the Swedish Natural Sciences Research Council and the Swedish Cancer Society.

REFERENCES

1. Rieske, J.S. (1976) Biochim. Biophys. Acta 456, 195-247.
2. Crane, F.L. (1977) Ann. Rev. Biochem. 46, 439-469.
3. von Jagow, G. and Sebald, W. (1980) Ann. Rev. Biochem. 49, 281-314.
4. Wikström, M.F.K. (1973) Biochim. Biophys. Acta 301, 155-193.
5. Papa, S. (1976) Biochim. Biophys. Acta 456, 39-84.
6. Baum, H., Rieske, J.S., Silman, H.I. and Lipton, S.H. (1967) Proc. Natl. Acad. Sci. (USA) 57, 798-805.
7. Wikström, M.K.F. and Berden, J.A. (1972) Biochim. Biophys. Acta 283, 403-420.
8. Mitchell, P. (1976) J. Theoret. Biol. 62, 327-367.
9. Mitchell, P. (1980) Annal. New York Acad. Sci. 341, 564-584.
10. Trumpower, B.L. (1976) Biochem. Biophys. Res. Commun. 70, 73-80.
11. Trumpower, B.L., Edwards, C.A. and Ohnishi, T. (1980) J. Biol. Chem. 255, 7487-7493.
12. Papa, S., Guerrieri, F., Lorusso, M., Izzo, G., Capuano, F. and Boffoli, D. (1979) in Functions and Molecular Aspects of Biomembrane Transport (Quagliariello et al., eds.) Elsevier/North-Holland BioMedical Press, pp. 197-207.
13. Alexandre, A., Galiazzo, F. and Lehninger, A.L. (1980) J. Biol. Chem. 255, 10721-10730.
14. Ernster, L. (1975) FEBS Symp. 35, 253-276.
15. Weiss, H. and Kolb, H.J. (1979) Eur. J. Biochem. 99, 139-149.
16. Weiss, H., Juchs, B. and Ziganke, B. (1978) Methods in Enzymology 53, 98-112.
17. Kaṭan, M.B., Pool, L., and Groot, G.S.P. (1976) Eur. J. Biochem. 65, 95-105.
18. Siedow, J.N., Power, S., De La Rosa, F.F. and Palmer, G. (1978) J. Biol. Chem. 253, 2392-2399.
19. Hatefi, Y., Haavik, A.G., Fowler, L.R., Griffiths, D.E. (1962) J. Biol. Chem. 273, 1681-1685.
20. Rieske, J.S., Zaugg, W.S. and Hansen, R.E. (1964) J. Biol. Chem. 239, 3023-3030.
21. Engel, W.D., Schägger, H. and von Jagow, G. (1980) Biochim. Biophys. Acta 592, 211-222.
22. Gellerfors, P. and Nelson, B.D. (1981) Eur. J. Biochem. in press.
23. Weiss, H., Wingfield, P. and Leonard, K. (1979) in Membrane Bioenergetics (Lee, C.P., Schatz, G. and Ernster, L., eds.), Addison-Wesley Publishing Co., pp. 119-132.

24. Silman, H.I., Rieske, J.S., Lipton, S.H. and Baum, H. (1967) J. Biol. Chem. 242, 4867-4875.
25. Baum, H., Silman, H.I., Rieske, J.S. and Lipton, S.H. (1967) J. Biol. Chem. 242, 4876-4887.
26. Mendel-Hartvig, I. and Nelson, B.D., (1981) Biochim. Biophys. Acta, in press.
27. Augen, J., Power, S. and Palmer, G. (1979) Biochem. Biophys. Res. Commun. 86, 271-277.
28. Gellerfors, P. and Nelson, B.D. (1975) Eur. J. Biochem. 52, 433-443.
29. Morres, C.A.M. and Slater, E.C. (1977) Biochim. Biophys. Acta 462, 531-548.
30. Weiss, H. and Juchs, B. (1978) Eur. J. Biochem. 88, 17-28.
31. von Jagow, G., Schägger, H., Engel, W.D., Machleidt, W. and Machleidt, I. (1978) FEBS Lett. 91, 121-125.
32. Weiss, H. and Ziganke, B. (1974) Eur. J. Biochem. 41, 63-71.
33. Weiss, H. (1976) Biochim. Biophys. Acta 456, 291-313.
34. Lin, L.-F.H. and Beattie, D.S. (1978) J. Biol. Chem. 253, 2412-2418.
35. Yu, C., Yu, L. and King, T.E. (1975) Biochem. Biophys. Res. Commun. 66, 1194-1200.
36. Bell, R.L. and Capaldi, R.A. (1976) Biochemistry 15, 996-1001.
37. Davis, K.A. and Hatefi, Y. (1971) Biochem. Biophys. Res. Commun. 44, 1338-1344.
38. Hatefi, Y. and Galante, Y.M. (1980) J. Biol. Chem. 255, 5530-5537.
39. Borst, P. and Grivell, L.A. (1981) Nature 289, 439-440.
40. Nelson, B.D. and Gellerfors, P. (1974) Biochim. Biophys. Acta 357, 358-364.
41. Riccio, P., Schägger, H., Engel, W.D. and von Jagow, G. (1977) Biochim. Biophys. Acta 459, 250-262.
42. Wilson, D.F., Erecinska, M. and Dutton, P.L. (1974) Ann. Rev. Biophys. Bioengineering 3, 203-230.
43. Yu, C.A., Yu, L. and King, T.E. (1972) J. Biol. Chem. 247, 1012-1019.
44. Trumpower, B.L. and Katki, A. (1975) Biochemistry 14, 3635-3642.
45. König, B.W., Schilder, L.T.M., Tervoort, M.J. and Van Gelder, B.F. (1980) Biochim. Biophys. Acta 621, 283-295.
46. Ross, E. and Schatz, G. (1976) J. Biol. Chem. 251, 1991-1996.
47. Wakabayashi, S., Matsubara, H., Kim, C.H., Kawai, K. and King, T.E. (1980) Biochem. Biophys. Res. Commun. 97, 1548-1554.

48. Yu, C.A., Yu, L. and King, T.E. (1973) J. Biol. Chem. 248, 528-533.
49. Chiang, Y.-L. and King, T.E. (1979) J. Biol. Chem. 254, 1845-1853.
50. Bosshard, H.R., Zürrer, M., Schägger, H. and von Jagow, G. (1979) Biochem. Biophys. Res. Commun. 89, 250-258.
51. Rieder, R. and Bosshard, H.R. (1978) FEBS Lett. 92, 223-226.
52. König, B.W., Osheroff, N., Wilms, J., Muijsers, A.O., Dekker, H.L. and Margoliash, E. (1980) FEBS Lett. 111, 395-398.
53. Rieder, R. and Bosshard, H.R. (1980) J. Biol. Chem. 255, 4732-4739.
54. Broger, C., Nalecz, M.J. and Azzi, A. (1980) Biochim. Biophys. Acta 592, 519-527.
55. Speck, S.H., Koppenol, W.H., Osheroff, N., Dethmers, J.K., Kang, C.H., Margoliash, E. and Ferguson-Miller, S. (1979) in Membrane Bioenergetics (Lee, C.P., Schatz, G. and Ernster, L., eds.) Addison-Wesley Publishing Co., Inc., pp. 31-43.
56. Speck, S.H., Ferguson-Miller, S., Osheroff, N. and Margoliash, E. (1979) Proc. Natl. Acad. Sci. USA 76, 155-159.
57. Osheroff, N., Jemmerson, R., Speck, S.H., Ferguson-Miller, S. and Margoliash, E. (1979) J. Biol. Chem. 254, 12717-12724.
58. Trumpower, B.L. and Edwards, C.A. (1979) J. Biol. Chem. 254, 8697-8706.
59. Rieske, J.S., Zaugg, W.S. and Hansen, R.E. (1964) J. Biol. Chem. 239, 3023-3030.
60. Rieske, J.S., Hansen, R.E. and Zaugg, W.S. (1964) J. Biol. Chem. 239, 3017-3022.
61. Nishibayashi-Yamashita, H., Cunningham, C. and Racker, E. (1972) J. Biol. Chem. 247, 698-704.
62. Slater, E.C. (1950) Biochem. J. 46, 484-
63. Slater, E.C. and de Vries, S. (1980) Nature 288, 717-718.
64. Gellerfors, P., Lundén, M. and Nelson, B.D. (1976) Eur. J. Biochem. 67, 463-468.
65. Nelson, B.D. and Gellerfors, P. (1978) Methods in Enzymology 53, 80-91.
66. Capaldi, R.A., Bell, R.L. and Branchek, T. (1977) Biochem. Biophys. Res. Commun. 74, 425-433.
67. Yu, C.A., Yu, L. and King, T.E. (1974) J. Biol. Chem. 249, 4905-4910.
68. Das Gupta, U. and Rieske, J.S. (1973) Biochem. Biophys. Res. Commun. 54, 1247-1253.

69. Roberts, H., Smith, S.C., Marzuki, S. and Linnane, A. (1980) Arch. Biochem. Biophys. 200, 387-295.
70. Slater, E.C. (1973) Biochim. Biophys. Acta 304, 129-154.
71. Yu, C.A., Yu, L. and King, T.E. (1977) Biochem. Biophys. Res. Commun. 78, 259-265.
72. Yu, C.A., Nagaoka, S., Yu, L. and King. T.E. (1978) Biochem. Biophys. Res. Commun. 82, 1070-1078.
73. Yu, C.A. and Yu, L. (1980) Biochemistry 19, 3579-3585.
74. Yu, C.A. and Yu, L. (1980) Biochem. Biophys. Res. Commun. 96, 286-292.
75. Yu, C.A., Nagoaka, S., Yu, L. and King. T.E. (1980) Arch. Biochem. Biophys. 204, 59-70.
76. Ackrell, B.A., Ball, M.B. and Kearny, E.B. (1980) J. Biol. Chem. 255, 2761-2769.
77. Yu, L., Yu, C.A. and King, T.E. (1977) Biochim. Biophys. Acta 495, 232-247.
78. von Jagow, G., Schägger, H., Riccio, P., Klingenberg, M. and Kolb, H.J. (1977) Biochim. Biophys. Acta 462, 549-558.
79. Tzagaloff, A., Yang, P.C., Wharton, D.C. and Rieske, J.S. (1965) Biochim. Biophys. Acta 96, 1-8.
80. Wingfield, P., Arad, T., Leonard, K. and Weiss, H. (1979) Nature 280, 696-697.
81. Hovmöller, S., Leonard, K. and Weiss, H. (1981) FEBS Lett. 123, 118-122.
82. Leonard, K.R., Arad, T., Wingfield, P. and Weiss, H. (1980) First Eur. Bioenergetics Conf. 83-84.
83. Mendel-Hartvig, I. and Nelson, B.D. (1978) FEBS Lett. 92, 36-40.
84. Bell, R.L., Sweetland, J., Ludwig, B. and Capaldi, R.A. (1979) Proc. Natl. Acad. Sci. USA 76, 741-745.
85. Low, D.C., Szary, C. and Boxer, D.H. (1980) Biochem. Soc. Transact. 8, 1980.
86. Montecucco, C., Gutweniger, H., Bisson, R., Dabbeni-Sala, F., Pitotti, A., Zaccolin, P., Mascia, A., Santato, M. and Arslan, P. (1980) First Eur. Bioenergetics Conf. 85-86.
87. Gellerfors, P. and Nelson, B.D. (1977) Eur. J. Biochem. 80, 275-282.
88. Case, G.D. and Leigh, J.S. (1976) Biochem. J. 160, 769-783.
89. Case, G.D., Ohnishi, T. and Leigh, J.S. (1976) Biochem. J. 160, 785-795.
90. Schneider, D.L. and Racker, E. (1971) Proc. Exptl. Biol. 30, 1190.
91. Schatz, G. (1979) FEBS Lett. 103, 203-211.

92. Ball, M.B., Bell, R.L. and Capaldi, R.A. (1977) FEBS Lett. 83, 99-102.
93. Mendel-Hartvig, I. and Nelson, B.D., unpublished results.
94. Ohnishi, T. and Blum, H. (1980) First. Eur. Bioenergetics Conf. Patron Editore, Bologna, pp. 71-72.
95. Smith, R.J., Capaldi, R.A., Muchmore, D. and Dahlquist, F. (1978) Biochemistry 17, 3719-3723.
96. DePierre, J.W. and Ernster, L. (1977) Ann. Rev. Biochem. 46, 201-262.
97. Trumpower, B.L. and Landeen, C.E. (1977) The Ealing Review 1, 4-7.
98. Ernster, L., Glaser, E. and Norling, B. (1978) Methods in Enzymology 53, 573-579.
99. Gutman, M. (1980) Biochim. Biophys. Acta 594, 53-84.
100. Trumpower, B.L. (1981) J. Bioenerget. Biomembrane, in press.
101. Reed, J., Reed, T.A. and Hess, B. (1978) Eur. J. Biochem. 91, 255-261.
102. Das Gupta, U., Wharton, D.C. and Rieske, J.S. (1979) J. Bioenerget. Biomembranes 11, 79-95.
103. Rieske, J.S., Baum, H., Stoner, C.D. and Lipton, S.H. (1967) J. Biol. Chem. 242, 4854-4866.
104. von Jagow, G. and Engel, W.D. (1980) FEBS Lett. 111, 1-5.
105. Salerno, J.C. and Ohnishi, T. (1980) Biochem. J. 192, 769-781.
106. Konstantinov, A.A. and Ruuge, E.K. (1977) FEBS Lett. 81, 137-141.
107. Salerno, J.C., Harmon, H.J., Blum, H., Leigh, J.S. and Ohnishi, T. (1977) FEBS Lett. 82, 179-182.
108. Ingledew, W.J. and Ohnishi, T. (1977) Biochem. J. 164, 617-620.
109. Trumpower, B.L. and Simmons, Z. (1979) J. Biol. Chem. 254, 4608-4616.
110. De Vries, S., Berden, J.A. and Slater, E.C. (1980) FEBS Lett. 122, 143-148.
111. Hederstedt, L. and Rutberg, L. (1980) J. Bacteriology 144, 941-951.
112. Holmgren, E., Hederstedt, L. and Rutberg, L. (1979) J. Bacteriol. 138, 377-382.
113. Eisenbach, M. and Gutman, M. (1975) Eur. J. Biochem. 59, 223-230.
114. Eisenbach, M. and Gutman, M. (1976) FEBS Lett. 61, 247-250.
115. Bowyer, J.R. and Trumpower, B.L. (1981) J. Biol. Chem. in press.

116. Titheradge, M.A., Binder, S.B., Yamazaki, R.K. and Haynes, R.C. (1978) J. Biol. Chem. 253, 3357-3360.
117. Titheradge, M.A. and Haynes, R.C. (1979) FEBS Lett. 106, 330-334.
118. Siess, E.A. and Wieland, O.H. (1980) Eur. J. Biochem. 110, 203-210.

CYTOCHROME OXIDASE

Mårten Wikström

Department of Medical Chemistry, University
of Helsinki, SF-00170 Helsinki 17, Finland

INTRODUCTION

Cytochrome oxidase (ferrocytochrome $\underline{c}$: O_2 oxidoreductase; EC 1.9.3.1; also, cytochrome $\underline{c}$ oxidase) is the principal respiratory enzyme in biology, which reduces dioxygen to water by reducing equivalents derived from the oxidation of foodstuffs. Cytochrome oxidase is probably responsible for more than 90% of the O_2 consumption by living organisms on Earth. It is present not only in the inner membrane of mitochondria, but also in the cell membrane of some bacteria capable of aerobic growth (1).

Presently there is strong interest in cytochrome oxidase and current research on this enzyme is truly multidisciplinary. Cytochrome oxidase is a multisubunit membrane protein, three polypeptides of which are coded by the mitochondrial DNA and synthesised on mitochondrial ribosomes (see 2). It is both a haem and a copper protein, which has stimulated many studies with sophisticated physical techniques. Moreover, the cytochrome oxidase reaction is coupled to energy conservation, which has been subject of much recent research.

C. P. Lee, G. Schatz, G. Dallner (eds.), Mitochondria and Microsomes
in honor of Lars Ernster   ISBN 0-201-04576-1

The latter will be the main topic of this paper, but the intimate relationships between structure, electron transfer, O_2
reduction and energy transduction demand some attention to
all these topics. The picture obtained from this paper may be
completed by consultation of recent reviews (2-5). Some provocative hypotheses will also be presented to provide frames
for the data. With this I wish to honour Lars Ernster, whose
remarkable contributions to membrane bioenergetics have never
been beset with encyclopedic dullness.

THE OVERALL REACTION AND ITS TOPOGRAPHY

Cytochrome oxidase catalyses electron transfer from
cytochrome c, located on the outside (C-side) of the mitochondrial membrane (6,7), to O_2. The enzyme complex spans the
membrane completely (see 3-5). O_2 may primarily diffuse into
the active site from the phospholipid bilayer (8), in which
it is 5-10 times more soluble than in the aqueous phases.
One "substrate" proton is taken up per e^- transferred in the
generation of water from reduced oxygen. The previous evidence for uptake of this proton from the matrix (M-) phase (9,
10) was rendered ambiguous by the discovery that the enzyme
functions as a proton pump (11-13). Fig. 1 shows two extreme
interpretations of these data, differing with respect to the
"sidedness" of the substrate proton. Note, however, that this
proton may be taken from the M-phase (as in A) also without
postulating translocation of the electrons (as in B). In such
a case (12) a "channel" is required to conduct the proton to
the site of O_2 reduction. The earlier data (9,10) only showed
that protons are taken up from the M-phase, which clearly
does not distinguish between A and B (Fig. 1). Only very recent findings (see below) might provide such a distinction.

In contrast to what is generally believed, there is no

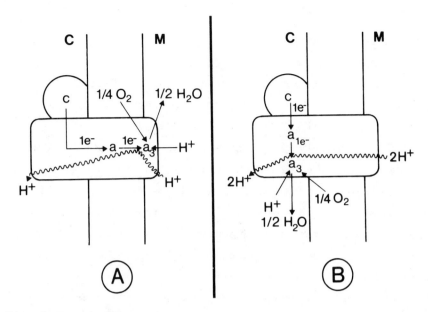

Fig. 1.<u>Two extreme arrangements of the proton-translocating
cytochrome oxidase</u>

Cytochrome oxidase (rectangula) is "plugged through"
the membrane (vertical lines), in between the aqueous
C- and M-phases. "c" denotes cytochrome c and wavy lines indicate H^+ translocation.

unambiguous evidence for an "electron-translocating" function
of the enzyme (6,9,10). Since the position of the O_2-reducing
site is not known with respect to the dielectric of the protein, it is difficult to determine to what extent electron
and proton translocation may contribute to the overall elec-
togenic function (5,13). It is important to note that the
function depicted in Fig. 1 is <u>thermodynamically</u> equivalent
to translocation of $2H^+/e^-$, quite independently of the two
extreme mechanisms (5).

The function in Fig. 1 is not yet unanimously accepted. Two groups claim that the enzyme has no H^+-translocating
activity (14-18). In contrast, two other groups claim that
the pump releases a net of $2H^+/e^-$ on the C-side (19,20; but

see also 5,13,21,22).

REDOX CENTRES

The cytochrome aa_3 unit contains two haems and two coppers, all with different properties (3,4,23). Haem a_3 and Cu_B (usually "invisible" by EPR spectroscopy, but see 24) form a binuclear centre of O_2 reduction (25-27). Haem a and Cu_A are often thought merely to act as electron carriers. Haem a may be located fairly close to the C-phase, whereas a_3 is inaccessible to "probes" from either side of the membrane (28,29). The planes of both haems are oriented perpendicularly to the plane of the membrane (3). Subunits (see 30) I and II might "carry" haems a_3 and a, respectively (31-33), and II might "carry" copper (31,34), but this matter is not yet settled.

The aa_3 unit accepts four electrons. All four redox centres have been regarded as one-electron carriers (35-37, but see below). The haems, but not the coppers, have pH-dependent redox potentials (36,37). There are strong anti-cooperative redox interactions between the haems (38-40).

Spectroscopic assignment of individual centres has long been ambiguous. The evidence is now strong for the original proposal (41) that haem a is the main contribuent to the reduced minus oxidised band at 605 nm (39,40). The band at 830 nm is due to Cu_A^{II} to at least 85% (42,43).

THE REDUCTION OF DIOXYGEN TO WATER

Several intermediate states of the O_2 reaction have been described using the ingenious low-temperature "triple-trapping" technique (44). Yet, only the "oxy" Compound A is clearly identified with an $Fe^{II}-O_2$ structure of haem a_3.

Structural, functional and thermodynamic arguments suggest that O_2 is reduced to bound peroxide in a concerted two-electron step (3,5,27,37). Compound C, which is formed when the half-reduced enzyme (a_3 and Cu_B reduced) reacts with O_2 (44-46) might thus be a peroxidic species. Its unique spectrum is characterised by a strong band at 607 nm, and a very weak band near 445 nm (relative to the oxidised enzyme). The absence of the 655 nm band of the oxidised enzyme has led to the conclusion that a_3 may be ferrous in Compound C (45, 46). However, ferric a_3 can occur in the absence of this band (47) in certain states, as shown by EPR spectroscopy.

It was recently shown that the O_2 reaction may be partially reversed in mitochondria at a high phosphate potential (48). The oxidised a_3/Cu_B centre apparently reacts with water with stepwise transfer of two electrons to cytochrome c. The product after such "reversed electron transfer" is spectrally identical with Compound C, but the experimental conditions make a ferrous state of a_3 very unlikely (48). Two alternative structures for Compound C are shown in Fig. 2 (no. 1), viz. a μ-peroxo species (3,27,48) and a species with ferryl haem iron. The latter might be more consistent with the spectrum of Compound C. In contrast to Clore et al. (46), I find this spectrum similar to ferryl iron haemoproteins, such as the ES compound of cytochrome c peroxidase (49) and Compound II of horseradish peroxidase (50), when it is taken into account that there is haem A in the former but iron protoporphyrin IX in the two latter cases. Owing to its electron-withdrawing properties, the formyl group of haem A is expected to shift the spectra to longer wavelengths.

The evidence for Cu_B^{II} in Compound C is not very strong. The near infrared transitions observed (51) could also arise from haem.

Compound A might not be a significant intermediate in the catalytic cycle (3,48; Fig. 2), but may accumulate in

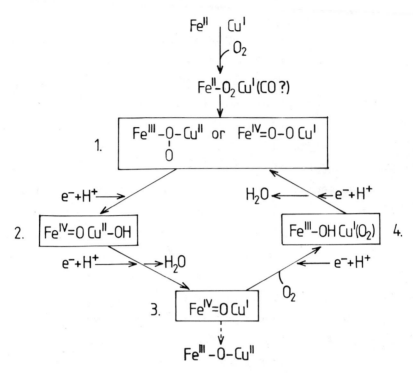

Fig. 2. <u>Tentative mechanism of O_2 reduction at the $\underline{a}_3/Cu_B$</u>
<u>centre of cytochrome oxidase</u>

the low-temperature experiments (44) due to binding of CO to
Cu_B^I (Fig. 2; see 48). A species corresponding to no. 2 (Fig.
2) was observed after "reversed" transfer of one electron
(48). Two isoelectronic alternatives are suggested for the
oxidised $\underline{a}_3/Cu_B$ centre (no. 3). An important feature of Fig.
2 is that O_2 reduction occurs by concerted two-electron steps
whereas the $\underline{a}_3/Cu_B$ centre receives electrons in discrete one-
electron steps. This may be of importance for the function of
the proton pump, which may be linked to oxidoreduction of cytochrome $\underline{a}$ (see below).

THE PROTON PUMP AND ELECTRON TRANSFER

In 1977 it was discovered (11,12) that cytochrome oxi-

dase is a redox-linked proton pump (Fig. 1; see 13 for a review). This has been quantitatively confirmed in reconstituted oxidase liposomes (52-55). Objections to this contention
(14-20) have been tested and found unwarranted (13,21,22,52,
56-58). More recent contradictory data of Papa et al. (18)
have also been criticised (22). These authors also did not
consider the known interactions between ascorbate and haemo-
globin (see 59) in their spectrophotometric assay of oxygen
consumption. In view of this, and of results from previous
experiments (13,56,60,61), I would tend to regard these
authors' negative conclusion with respect to the existence of
the proton pump somewhat premature.

A GENERAL MODEL. Three basic molecular elements are of importance in a redox-linked proton pump (5,13), viz. (i) a
redox centre to which (ii) at least one acid/base group is
linked functionally, and (iii), proton-conducting structures
("channels"). It is essential that both (i) and (ii) can
exist in two different states of (electron and proton) input
and output, respectively. If only one redox-linked acid/base
group is postulated, and if the electronic input (and output)
state merges with the protonic one, this model may be described as shown in Fig. 3. Any further simplifications considerably diminish the usefulness of the model (5,13). As it is,
it provides some specific predictions that may be tested experimentally (see below).

THE REDOX CENTRE. The redox centre of the proton pump is expected to exhibit a pH-dependent midpoint redox potential (5,
13,62). Cytochromes a and a_3, but not the coppers, have this
property (36,37,40). The energy-dependent spectral shift in
haem a_3 was previously thought to indicate a direct involvement in H^+ translocation (11-13). However, this (cf. Fig. 3)

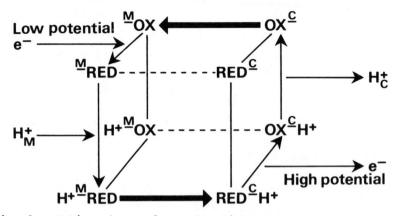

Fig. 3. Cubic scheme of a redox-linked proton pump

RED and OX denote reduced and oxidised states of the
redox centre. Bars with M or C denote a redox-linked
acid/base group, in protonic contact with the respective aqueous phases (via "H$^+$-channels" ?). Thick
arrows show reorientations between electronic and
protonic input and output states (left vs. right).
Reaction along broken lines lead to futile cycles.
Thin plus thick arrows show one of many possible
reaction paths (see refs. 5 and 13).

is difficult to reconcile with the involved function of a_3 in
O_2 reduction. A solution to this dilemma was recently provided by the finding that the spectral shift is due to a partial reversal of the O_2 reaction (48; cf. above). There is therefore no longer any basis for the idea that cytochrome a_3
would be directly involved in energy transduction.

The obligatory shuttling of the redox centre between
input and output states during turnover (Fig. 3) may render
it unique kinetically, and therefore identifiable. Since this
shuttling may be slower than electron transfer due to its
"conformational" character, overall kinetics may become heterogeneous, particularly at low temperatures.

O_2-pulse experiments. When the reduced enzyme is pulsed with
O_2 at low temperatures, a fairly stable state is reached

(Compound B; 44,45), in which 30-50% of haem $\underline{a}$ and Cu_A are oxidised, based on both optical and EPR spectroscopy (63,64). Oxidation of the rest of haem $\underline{a}$ takes place only at much higher temperatures, and then in conjunction with oxidation of cytochrome $\underline{c}$ (65). This is precisely the kind of kinetic heterogeneity expected from the redox element of the proton pump (fig. 3 and above).

These biphasic kinetics are not an artefact of the low-temperature technique, but provide, in fact, a long-sought-for explanation of the kinetics of haem $\underline{aa}_3$ oxidation at higher temperatures (66). Here the O_2-induced absorption decrease at 605 nm is biphasic, but the extent of the rapid phase is much too large to be explained by haem $\underline{a}_3$ oxidation alone (see also 1,67). It was thought subsequently (68) that the spectral contribution of $\underline{a}_3$ to the 605 nm band is much larger than anticipated from ligand-binding studies (1,41), as proposed on the basis of redox titrations (36). However, the original spectral assignments of $\underline{a}$ and $\underline{a}_3$ are correct, as mentioned above, and the diverging results from redox titrations are due to the strong redox interactions between the haems (23,38-40,64). The low-temperature data (see above) now suggest that the extent of the rapid phase is large simply because it includes 30-50% oxidation of haem $\underline{a}$ in addition to the oxidation of $\underline{a}_3$. Also the heterogeneity of Cu_A oxidation was observed at room temperature (66), and is no longer subject to ambiguity with respect to spectroscopic interpretation (42,43).

Cytochrome $\underline{c}$-pulse experiments. When the oxidised enzyme is pulsed with ferrocytochrome $\underline{c}$ there is a rapid phase of electron transfer, but electrons reach the $\underline{a}_3/Cu_B$ centre only very slowly (69-74). It is agreed that haem $\underline{a}$ is the primary

electron acceptor, but there has been some controversy as to
whether Cu_A is reduced as well in the fast phase. Andréasson
(72) proposed that only haem $\underline{a}$ is reduced, because the oxidase accepts only a maximum of $le^-/\underline{aa}_3$ unit present. In contrast, Wilson $\underline{et}$ $\underline{al}$. (74) concluded that also Cu_A is reduced
on the basis of absorption changes at 830 nm and EPR data
(75). Due to the established spectral assignments of haem $\underline{a}$
and Cu_A (see above), it may be safely concluded that the data of both groups show reduction of haem $\underline{a}$ as well as of Cu_A.
Yet, Andréasson's finding remains highly significant (and is
supported by data in 74). We must thus simultaneously explain
why the oxidase accepts only $le^-/\underline{aa}_3$ unit present no matter
how high the ratio between ferrous $\underline{c}$ and enzyme, and that
this results in about half-maximal reduction of both haem $\underline{a}$
and Cu_A. Since there are no indications of redox interactions
between these centres, the simplest explanation is that only
half of the $\underline{aa}_3$ units can accept electrons (about equally to
haem $\underline{a}$ and Cu_A), whereas the other half cannot. Although this
is highly anomalous $\underline{a}$ $\underline{priori}$, it is again expected if haem $\underline{a}$
is directly involved in energy transduction, as discussed
above. If this interpretation is correct, these data suggest
that the input/output transition (Fig. 3) may be anomalously
slow in the fully oxidised "resting" enzyme. Moreover, as also suggested by the O_2-pulse data above, Cu_A appears to equilibrate electronically with haem $\underline{a}$ whether it is in the input
or the output state.

Steady state data. Yonetani (76) titrated cytochrome oxidase
aerobically with ascorbate in the presence of cytochrome $\underline{c}$.
He observed that the 605 and 445 nm bands "saturated" at 60
and 30% of full reduction, respectively, whereas the steady
state level of reduction of cytochrome $\underline{c}$ increased linearly

up to 100%. With a predicted rate-limitation at the input/
output transitions (Fig. 3 and cf. above) only the input state would become reduced in the aerobic steady state, while
the output state remains highly oxidised. These data hence
provide further support to the contention that cytochrome a
may shuttle between two states as described above.

I conclude that three independent kinetic tests suggest cytochrome a as a likely candidate for the redox centre
of the proton pump.

THE ACID/BASE GROUP. The redox-linked acid/base group should
also exist in two distinct states, one in which protonic
equilibration occurs with the M-phase (input), and one in
which it occurs with the C-phase (output; cf. Fig. 3). In
the simplified model of Fig. 3, transfer of electrons from
cytochrome c to the input form of cytochrome a should specifically be associated with H^+ uptake from the M-phase.
Artzatbanov et al. (77; cf. 40) showed that the redox equilibrium between cytochromes c and a is uniquely dependent on
pH in the M-phase in mitochondria where electron transfer to
a_3/Cu_B is blocked by cyanide. This is in remarkable agreement with the model.

Reversed electron transfer, observed by the energy-
linked spectral shift in ferric haem a_3 (48), is associated
with uptake of H^+ from the C-phase (78). This provides indirect support for the converse prediction (Fig. 3) that the
output state of the redox centre should communicate protonically with the aqueous C-phase.

A GATED PROTON CHANNEL. After discovery of the proton pump
it was suggested, largely by analogy to the H^+-ATPase (12,13)
that cytochrome oxidase may also be equipped with an H^+-
channel, which might be formed by one of the mitochondrially

synthesised subunits (I, II or III). This proposal has recently received experimental support.

A subunit III-free oxidase preparation (79,80) is
otherwise similar to the original enzyme, but exhibits no net
H^+ translocation after reconstition into liposomes (contrast 81). Yet, full respiratory control is retained, showing
that the membranes remain impermeable to H^+. Since the rate
of electron transfer is unaffected, it follows that the proton pump must be "locally uncoupled". Such local or intrinsic uncoupling of the pump may be the result of "slipping" of
the coupling between electron and proton transfer. This could
occur, for instance, if the reaction path is along the dotted
steps in Fig. 3. This suggests that subunit III might provide
a "gating" function, equivalent to the input/output transitions in Fig. 3 (thick arrows), where the acid/base group is
rendered sequentially in protonic contact with the C- and M-
phases. Preliminary data by Timo Penttilä in our group have
shown, moreover, that the pH-dependence of the E_m of cytochrome a is much reduced in the subunit III-free enzyme. It is
therefore possible that one of the essential acid/base groups
that are linked to the redox reaction might be located on
subunit III.

Casey et al. (82,83) and Steffens et al. (84) showed
evidence suggesting that dicyclohexyl carbodiimide (DCCD) may
block the proton pump by specific binding to a single site in
subunit III. This again indicates a specific involvement of
this subunit in the function of the proton pump, and particularly favours the idea that it may constitute a proton channel.

Ludwig and Schatz (32) showed that cytochrome oxidase
isolated from Paracoccus contains only two different polypep-

tides, most closely corresponding to nos. I and II of the mitochondrial enzyme (33). Reconstitution into liposomes yields
excellent respiratory control (32), but little or no net H^+
translocation (B. Ludwig, personal communication; and cf. the
subunit III-free enzyme above). In view of the finding (85)
that the Paracoccus enzyme catalyses H^+ translocation in bacterial membranes, it is conceivable that a putative third
subunit (corresponding to no. III of the mitochondrial enzyme) may easily be detached on purification of the bacterial
enzyme. This might be analogous to the relative ease by which
subunit III is removed from beef heart cytochrome oxidase
(79-81).

       If our proposal for the function of subunit III above
is correct, then the observed respiratory control in its absence and during putative "slipping" of the proton pump, can
only be explained if "substrate" protons are indeed taken up
from the N-phase (see Fig. 1A). An intrinsic uncoupling of
the proton pump in Fig. 1B must lead to complete loss of respiratory control, since in this case the reaction would no
longer be associated with translocation of electrical charge
or generation of a pH gradient.

       A RECIPROCATING SITE MECHANISM
       All three sets of experiments showing kinetic hetero-
geneity of cytochrome a (see above) also suggested that the
two proposed states of input and output may exist with about
equal "occupancy" at all times. Although this might be due
to a coincidental distribution of enzyme molecules in the two
states, there is the intriguing possibility that the input
and output states occur simultaneously in an $(aa_3)_2$ dimer,
which would then be the functional entity of the enzyme.

       Recent structural data provide support for the contention of a dimeric enzyme (80,86,87). The finding by Bisson and Capaldi (88) that covalent attachment of 1 cytochrome c per 2 aa$_3$ units blocks enzymatic activity almost completely, provides strong evidence for functioning of the enzyme as a dimer. However, this possibility is alien for most oxidase enzymologists, who have almost traditionally regarded the aa$_3$ unit as the catalytic entity. If the enzyme indeed functions as a dimer, it must surely mean that interactions between the monomers occur, and may be of catalytic significance. However, apart from the finding in (88), such interactions have not been reported, but might simply have gone unnoticed.

In Fig. 4 a schematic model is presented of how the dimeric enzyme might function in the catalytic processes of electron transfer and proton translocation. This model is designed to provide one possible structural rationalisation of the more abstract model in Fig. 3. However, Fig. 4 should not be interpreted in terms of extensive structural changes. These changes have been drawn large merely for illustrative purposes. The actual transitions between input and output modes may involve only very subtle structural adjustments (5,13). The essence of the proposed model may be described as follows:

The dimer (aa$_3$)$_2$ is the minimal functional unit of the enzyme. During turnover there is a reciprocal (or anticooperative) interaction between the monomers so that one is always in the input state and the other in the output state. These states correspond precisely to the analogous states in Fig. 3. Thus electrons are transferred from cytochrome c to haem a, and H$^+$ is taken up from the M-phase, exclusively in the input state. No electron transfer is possible to the

$\underline{a}_3/Cu_B$ centre in this state. The latter takes place specifically in the output state, coupled to release of H^+ into the
C-phase and stepwise reduction of O_2 to water. Here no electron transfer is possible between cytochromes $\underline{c}$ and $\underline{a}$.

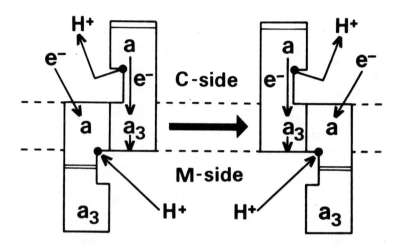

Fig. 4. Tentative reciprocating site mechanism of electron
transfer and proton translocation by dimeric cytochrome oxidase

The dimeric enzyme is depicted highly schematically in
the mitochondrial membrane (dotted lines). Monomeric
aa_3 units attain reciprocal states with respect to
input/output of electrons and protons. After completion of the specific function of each monomer (left),
there is a switch (thick arrow) by which the monomer
previously in the input state (downward position) is
converted to the output state (upward position), and
vice versa for the companion monomer. The large intermonomeric upward/downward movement is depicted only
for illustrative purposes. Double lines in monomers
indicate blockage of electron transfer between $\underline{a}$ and
$\underline{a}_3$, and between $\underline{c}$ and $\underline{a}$, in input and output states,
respectively.

The overall cytochrome oxidase reaction is thus divided into two parts, each of which take place simultaneously
on the two monomers. After completion of these reactions,

there is a transition (analogous to the steps denoted by thick arrows in Fig. 3) by which the monomers switch catalytic functions reciprocally.

Reciprocal or anticooperative interactions between the monomeric aa_3 units are of fundamental importance in this model. This could, for the first time, provide a clue for the functional relevance of the strong negative interactions previously observed in the enzyme (see above), but which to date have been assumed to take place between haems a and a_3. It may thus be motivated to re-evaluate the experimental data in view of the possibility of intermonomeric effects.

Although the present model is consistent with several experimental findings indicated above, I must stress that it is highly hypothetical. However, it may stimulate further research and certainly poses several new questions (see 5,64). One of these is the interesting analogy to the "alternating site" model proposed for the mitochondrial and the chloroplast H^+-ATPase (89). Anticooperativity between similar catalytic units (90) could indeed find particularly important applications more generally in biological energy-transducing systems (5).

CONCLUSION

Research on various aspects of function and structure of cytochrome oxidase is currently very active. A detailed molecular mechanism of cellular respiration is emerging, although much more work is still needed for its complete elucidation. Progress has been made recently beyond the much debated stoicheiometric aspects of proton translocation by the enzyme. A general model has helped to identify cytochrome a as directly involved in this process, based largely on electron transfer kinetics previously available in the literature. The recent demonstration of a partial reversal of the O_2 reaction of cytochrome oxidase has helped this identification, but also provided some insight into the mechanism of reduction of dioxygen. There are also indications for a specific function of subunit III in proton translocation, possibly as a gated proton channel. Finally, there are both structural and functional indications for the $(\underline{aa}_3)_2$ dimer being the catalytic entity of cytochrome oxidase. A speculative reciprocating site mechanism was presented in an attempt to rationalise a catalytic function as a dimer with electron transfer, proton translocation and anticooperative properties of the enzyme.

ACKNOWLEDGEMENTS

I wish to thank Ms. Hilkka Vuorenmaa for expert technical assistance, as well as my collaborators, Drs. Klaas Krab, Timo Penttilä and Matti Saraste for stimulating discussions. I further wish to acknowledge several useful suggestions by Dr. Clyde H. Barlow and by Prof. R.J.P. Williams. This work is supported by the Sigrid Jusélius Foundation and by the Finnish Academy.

REFERENCES

1. Lemberg, M.R. (1969) Physiol. Rev. 49, 48-121.
2. Borst, P. and Grivell, L.A. (1978) Cell 15, 705-723.
3. Erecińska, M. and Wilson, D.F. (1978) Arch. Biochem. Biophys. 188, 1-14.
4. Malmström, B.G. (1979) Biochim. Biophys. Acta 549, 281-303.
5. Wikström, M., Krab, K. and Saraste, M. (1981) Annu. Rev. Biochem. Vol. 50, in the press.
6. Hinkle, P. and Mitchell, P. (1970) J. Bioenerg. 1, 45-60.
7. DePierre, J.W. and Ernster, L. (1977) Annu. Rev. Biochem. 46, 201-262.
8. Sharrock, M. and Yonetani, T. (1977) Biochim. Biophys. Acta 462, 718-730.
9. Mitchell, P. and Moyle, J. (1967) in Biochemistry of Mitochondria (E.C. Slater et al. eds.) pp. 53-74, Academic Press/PWN, London/Warsaw.

10.   Papa, S. (1976) Biochim. Biophys. Acta 456, 39-84.
11.   Wikström, M. (1977) Nature (Lond.) 266, 271-273.
12.   Wikström, M. and Saari, H.T. (1977) Biochim. Biophys.
      Acta 462, 347-361.
13.   Wikström, M. and Krab, K. (1979) Biochim. Biophys.
      Acta 549, 177-222.
14.   Moyle, J. and Mitchell, P. (1978) FEBS Lett. 88,
      268-272.
15.   Moyle, J. and Mitchell, P. (1978) FEBS Lett. 90,
      361-365.
16.   Mitchell, P. and Moyle, J. (1979) Biochem. Soc. Trans.
      7, 887-894.
17.   Lorusso, M., Capuano, F., Boffoli, D., Stefanelli, R.
      and Papa, S. (1979) Biochem. J. 182, 133-147.
18.   Papa, S., Guerrieri, F., Lorusso, M., Izzo, G., Boffo-
      li, D., Capuano, F., Capitanio, N. and Altamura, N.
      (1980) Biochem. J. 192, 203-218.
19.   Alexandre, A. and Lehninger, A.L. (1979) J. Biol.
      Chem. 254, 11555-11560.
20.   Azzone, G.F., Pozzan, T. and Di Virgilio, F. (1979)
      J. Biol. Chem. 254, 10206-10212.
21.   Wikström, M. and Krab, K. (1979) Biochem. Soc. Trans.
      7, 880-887.
22.   Wikström, M. and Krab, K. (1980) Curr. Topics Bioenerg.
      10, 51-101.
23.   Malmström, B.G. (1974) Q. Rev. Biophys. 6, 389-431.
24.   Malmström, B.G., Karlsson, B., Aasa, R., Andréasson,
      L.-E., Clore, G.M. and Vänngård, T. (1980) Proc. Symp.
      Interaction Between Iron and Proteins in Oxygen and
      Electron Transport (C. Ho and W.C. Eaton eds.) Else-
      vier, New York, in the press.
25.   Falk, K.-E., Vänngård, T. and Ångström, J. (1977)
      FEBS Lett. 75, 23-27.
26.   Tweedle, M.F., Wilson, L.J., Garcia-Iniguez, L., Bab-
      cock, G.T. and Palmer, G. (1978) J. Biol. Chem. 253,
      8065-8071.
27.   Reed, C.A. and Landrum, J.T. (1979) FEBS Lett. 106,
      265-267.
28.   Ohnishi, T., Blum, H., Leigh, J.S., Jr. and Salerno,
      J.C. (1979) in Membrane Bioenergetics (C.P. Lee et al.
      eds.) pp. 21-30, Addison-Wesley, London.
29.   Saari, H., Penttilä, T. and Wikström, M. (1980) J.
      Bioenerg. Biomembr. 12, 325-338.
30.   Downer, N.W., Robinson, N.C. and Capaldi, R.A. (1976)
      Biochemistry 15, 2930-2936.
31.   Winter, D.B., Bruyninckx, W.J., Foulke, F.G., Grinich,
      N.P. and Mason, H.S. (1980) J. Biol. Chem. in press.

32. Ludwig, B. and Schatz, G. (1979) Proc. Natl. Acad. Sci. U.S.A. 77, 196-200.
33. Ludwig, B. (1980) see ref. 24.
34. Steffens, G.J. and Buse, G. (1979) Hoppe-Seyler's Z. Physiol. Chem. 360, 613-619.
35. Van Gelder, B.F. and Muijsers, A.O. (1966) Biochim. Biophys. Acta 118, 47-57.
36. Wilson, D.F., Lindsay, J.G. and Brocklehurst, E.S. (1972) Biochim. Biophys. Acta 256, 277-286.
37. Lindsay, J.G., Owen, C.S. and Wilson, D.F. (1975) Arch. Biochem. Biophys. 169, 492-505.
38. Nicholls, P. and Petersen, L.C. (1974) Biochim. Biophys. Acta 357, 462-467.
39. Wikström, M., Harmon, H.J., Ingledew, W.J. and Chance, B. (1976) FEBS Lett. 65, 259-277.
40. Wikström, M. (1980) see ref. 24.
41. Keilin, D. and Hartree, E.F. (1939) Proc. Roy. Soc. (London) Ser. B. 127, 167-191.
42. Beinert, H., Shaw, R.W., Hansen, R.E. and Hartzell, C.R. (1980) Biochim. Biophys. Acta 591, 458-470.
43. Boelens, R. and Wever, R. (1980) FEBS Lett. 116, 223-226.
44. Chance, B., Saronio, C. and Leigh, J.S., Jr. (1975) J. Biol. Chem. 250, 9226-9237.
45. Chance, B., Saronio, C. and Leigh, J.S., Jr. (1979) Biochem. J. 177, 931-941.
46. Clore, G.M., Andréasson, L.-E., Karlsson, B., Aasa, R. and Malmström, B.G. (1980) Biochem. J. 185, 155-167.
47. Shaw, R.W., Hansen, R.E. and Beinert, H. (1978) Biochim. Biophys. Acta 504, 187-199.
48. Wikström, M. (1981) Proc. Natl. Acad. Sci. U.S.A. in the press.
49. Yonetani, T. (1970) Advan. Enzymol. 33, 309-335.
50. George, P. (1953) J. Biol. Chem. 201, 413-426.
51. Chance, B. and Leigh, J.S., Jr. (1977) Proc. Natl. Acad. Sci. U.S.A. 74, 4777-4780.
52. Krab, K. and Wikström, M. (1978) Biochim. Biophys. Acta 504, 200-214.
53. Casey, R.P., Chappell, J.B. and Azzi, A. (1979) Biochem. J. 182, 181-188.
54. Sigel, E. and Carafoli, E. (1979) J. Biol. Chem. 254, 10572-10574.
55. Coin, J.T. and Hinkle, P.C. (1979) ref. 28, pp. 405-412.
56. Krab, K. and Wikström, M. (1979) Biochim. Biophys. Acta 548, 1-15.

57. Wikström, M. and Krab, K. (1978) in Frontiers of
     Biological Energetics (P.L. Dutton et al. eds.) pp.
     351-358, Academic Press, New York.
58. Krab, K. and Wikström, M. (1980) Biochem. J. 186,
     637-639.
59. Lemberg, R and Legge, J.W. (1949) Hematin Compounds
     and Bile Pigments, Interscience Publ. Inc., New York.
60. Sigel, E. and Carafoli, E. (1978) Eur. J. Biochem.
     89, 119-123.
61. Sigel, E. (1980) Functional Studies on Mitochondrial
     Cytochrome c Oxidase in Natural and Reconstituted
     Membrane Systems, Dissertation, Swiss Federal Institu-
     te of Technology, Zürich.
62. Wikström, M. (1980) Curr. Topics in Membr. and Trans-
     port (C. Slayman, ed.) Academic Press, New York, in
     the press.
63. Clore, G.M., Andréasson, L.-E., Karlsson, B., Aasa,
     R. and Malmström, B.G. (1980) Biochem. J. 185, 139-
     154.
64. Wikström, M., Krab, K. and Saraste, M. (1981) A Syn-
     thetic Treatise on Cytochrome Oxidase, Academic Press,
     London. In the press.
65. Chance, B., Saronio, C., Waring, A. and Leigh, J.S.,
     Jr. (1978) Biochim. Biophys. Acta 503, 37-55.
66. Gibson, Q.H. and Greenwood, C. (1965) J. Biol. Chem.
     240, 2694-2698.
67. Nicholls, P. and Chance, B. (1974) in Molecular Mecha-
     nisms of Oxygen Activation (D. Hayashi, ed.) pp. 479-
     534, Academic Press, New York.
68. Erecińska, M. and Chance, B. (1972) Arch. Biochem.
     Biophys. 151, 304-315.
69. Gibson, Q.H., Greenwood, C., Wharton, D.C. and Palmer,
     G. (1965) J. Biol. Chem. 240, 888-894.
70. Antonini, E., Brunori, M., Greenwood, C. and Wilson,
     M.T. (1973) Biochem. Soc. Trans. 1, 34-35.
71. Van Buuren, K.J.H., Van Gelder, B.F., Wilting, J. and
     Braams, R. (1974) Biochim. Biophys. Acta 333, 421-
     429.
72. Andréasson, L.-E. (1975) Eur. J. Biochem. 53, 591-597.
73. Andréasson, L.-E., Malmström, B.G., Strömberg, C. and
     Vänngård, T. (1972) FEBS Lett. 28, 297-301.
74. Wilson, M.T., Greenwood, C., Brunori, M. and Antonini,
     E. (1975) Biochem. J. 147, 145-153.
75. Beinert, H., Hansen, R.E. and Hartzell, C.R. (1976)
     Biochim. Biophys. Acta 423, 339-355.
76. Yonetani, T. (1960) J. Biol. Chem. 235, 3138-3143.

77.   Artzatbanov, V. Yu., Konstantinov, A.A. and Skula-
       chev, V.P. (1978) FEBS Lett. 87, 180-185.
78.   Wikström, M. (1975) in Electron Transfer Chains and
       Oxidative Phosphorylation (E. Quagliariello et al.
       eds.), pp. 97-103, North-Holland, Amsterdam.
79.   Penttilä, T., Saraste, M. and Wikström, M. (1979)
       FEBS Lett. 101, 295-300.
80.   Saraste, M., Penttilä, T. and Wikström, M. (1981)
       Eur. J. Biochem. in the press.
81.   Saraste, M., Penttilä, T., Coggins, J.R. and Wik-
       ström, M. (1980) FEBS Lett. 114, 35-38.
82.   Casey, R.P., Thelen, M. and Azzi, A. (1980) J. Biol.
       Chem. 255, 3994-4000.
83.   Azzi, A., Casey, R.P. and Thelen, M. (1980) 1st Eur.
       Bioenerg. Conf. Short Reports, pp. 21-22, Pàtron Edi-
       tore, Bologna, Italy.
84.   Steffens, G.C.M., Prochaska, L. and Capaldi, R.A.
       (1980) see ref. 83, pp. 95-96.
85.   Van Verseveld, H.W., Krab, K. and Stouthamer, A.H.
       (1981) Biochim. Biophys. Acta, in the press.
86.   Robinson, N.C. and Capaldi, R.A. (1977) Biochemistry
       16, 375-381.
87.   Henderson, R., Capaldi, R.A. and Leigh, J.S., Jr.
       (1977) J. Mol. Biol. 112, 631-648.
88.   Bisson, R. and Capaldi, R.A. (1980) see ref. 83,
       pp. 103-104.
89.   Kayalar, C., Rosing, J. and Boyer, P.D. (1977) J.
       Biol. Chem. 252, 2486-2491.
90.   Levitzki, A. and Koshland, D.E., Jr. (1976) Curr.
       Topics Cell Regul. 10, 1-40.

STRUCTURE, FUNCTION, AND CHARGE SEPARATION IN CYTOCHROME OXIDASE, AN EXAFS STUDY

B. Chance

Johnson Research Foundation
University of Pennsylvania
Philadelphia, PA 19104

and

L. Powers and Y. Ching

Bell Laboratories
600 Mountain Avenue
Murray Hill, NJ 07974

INTRODUCTION

The function of cytochrome oxidase in electron transfer, oxygen reduction, charge separation, and energy conservation has been cited as one of the unsolved problems
of energy transduction (1). The possibility that the
redox cycle is coupled to structural changes at the active
site of the protein that are propagated to the periphery
with attendant asymmetric charge separation comes under
the general category of "membrane Bohr effects" (2, 3).
Since cytochrome oxidase does not crystallize appropriately for X-ray diffraction, direct evidence for the
location and magnitude of a conformation change has been
lacking; synchrotron X-ray studies (4), particularly extended X-ray absorption fine structure (EXAFS), have been
required for this study. This new technique has been

Abbreviations: EXAFS, extended X-ray absorption fine
structure; SSRL, Stanford Synchrotron Radiation Laboratory; rrdf, relative radial distribution function.

C. P. Lee, G. Schatz, G. Dallner (eds.), Mitochondria and Microsomes
    in honor of Lars Ernster   ISBN 0-201-04576-1

applied to cytochrome oxidase for the past several years
at Stanford Synchrotron Radiation Laboratory (SSRL) under
Project 341/423B (5). After coping with a number of
technical difficulties involving the intensity and positional stability of the light beam, the integrity of the
sample under X-irradiation, and the preparation of cytochrome oxidase in fully occupied derivatives at maximal
concentration (1 mM), it has been possible recently to
obtain sets of data on the four states of cytochrome
oxidase (oxidized, reduced, and the mixed valence states),
and, through detailed data analysis involving comparison
with a number of appropriate models, to derive the most
probable structure for the four centers themselves (6-11).
This technique, which has also provided a precise measurement of Fe-N bond length in hemoglobin (12) and the
nature of the binuclear O_2-bearing complex in hemocyanin
(13), now affords key data on the nature of the active
site of the enzyme which explain kinetic results on the
formation and inter-conversion of oxy- and peroxy-com-
pounds observed by low temperature kinetic techniques.
Of great importance is the result that structure-based
reaction mechanisms involve bond rupture and reformation
in the redox cycle, generating conformation changes
appropriate to the Bohr charge separation hypothesis.
This contribution reviews the experimental bases for the
structural changes and points to their possible application to the energy coupling problem.

The intermediates of greatest interest in the study
of the function of cytochrome oxidase are an oxy-compound
formed, as most recently observed, at -130° (14, 15)
and found to have properties very nearly identical to

those of the ferrous cuprous carboxy compound, as might
indeed have been expected from Warburg's pioneering
photochemical action spectrum (16). However, the oxygen
compound differs distinctly from the CO compound in that:
a) it is not demonstrably light sensitive; and b) it is
labile, undergoing an oxido-reduction reaction beginning
at -120° that leads to the donation of electrons from
both reduced Fe_{a_3} and Cu_{a_3} (one electron each) (14), and
to the formation of what is proposed to be a ferric-
cupric peroxide compound (15).   The Cu could be close
enough to the Fe in these complexes to donate electrons
directly to oxygen and possibly to bridge between Fe and
Cu as in hemocyanin, where the cupric-cupric peroxide
has a Cu-Cu distance of 3.65 $\overset{\circ}{A}$ (13) and presumably a
corresponding distance in the Fe-Cu binuclear complex
of cytochrome oxidase. Thus, distance measurements from
Fe to Cu and, indeed, Cu to Fe are of greatest importance
to the mechanism of cytochrome oxidase oxygen reduction.

Anomalies relevant to the structure of the redox
center are found in the chemistry of cytochrome oxidase
(11): a) a slow reaction with cyanide in the oxidized
form and a rapid reaction in the reduced state; b) spin
coupling between Fe_{a_3} and Cu_{a_3} that causes difficulties
in the observation of their epr signals; and c) the
addition of mercury (and urea) causes enhancement of
the epr signals which appears to be due to ablation of
spin coupling between $Fe^{3+}_{a_3}$ and $Cu^{2+}_{a_3}$.

Approaches to the atomic structure of the active site
of cytochrome oxidase are restricted by the fact that cytochrome oxidase is a membrane protein insoluble in water,

very large in molecular weight (the dimer is 240 kilo-
daltons), has neither been crystallized in a form suitable for X-ray examinations nor has been obtained in
the necessary heavy metal derivatives. The X-ray synchrotron technique is directly applicable to amorphous
solids, such as cytochrome oxidase, and is precise enough
to record interatomic distances with a high accuracy
(0.05 $\overset{o}{A}$) and small changes therein. However, the synchrotron technique has basic drawbacks, for example,
the method is insensitive, usually only 10^5 counts per
second can be obtained from the highest concentration
of cytochrome oxidase available (1.5 mM). The method
is therefore slow, and data recording of any intermediates
must take place under trapped conditions where stability
of intermediates, for several hours at least, is ensured.
Furthermore, X-radiation can damage the protein, either
directly or indirectly, through the production of hydrated
electrons which have been shown to alter the redox state
of the cytochrome oxidase (16). And thirdly, beam time
at optimal intensities and beam stability (dedicated
synchrotron operation) is a precious commodity and one
for which only a feasible and important project should
be undertaken.

 With these problems in mind, we have attempted continuously to improve techniques for X-ray synchrotron
observation of the structures of cytochrome oxidase
compounds and intermediates, and to increase the sensitivity and authenticity of measurements of cytochrome
oxidase a) by increasing the efficiency of photon capture
by using detectors of improved sensitivity, count rate,

and solid angle; b) by recording data when precise position and intensity stabilization of the X-ray synchrotron
beam is available (for which dedicated electron synchrotron operation is the sine qua non); c) by online optical
spectroscopy and offline epr spectroscopy of the samples
in order to verify that they are unaltered by the photon
beam; and d) by the use of low temperatures to reduce
the diffusivity and reactivity of hydrated electrons
generated by the intense photon beams (5, 15, 16).

These improvements in the precision of data and the
authenticity of the sample has permitted acquisition of
sets of EXAFS data that are suitable for detailed analysis and structure determinations (11).

THE EXAFS METHOD

The principle of structure determination by EXAFS
as it applies to cytochrome oxidase is illustrated in
Fig. 1, which indicates that irradiation of the Fe atoms
of cytochrome oxidase above K - edge, i.e., at ≥ 7 KeV,
will cause, as a function of increased energy of excitation, a series of oscillations of the emitted fluorescence intensity which are the Fourier transforms of the
radial distribution of surrounding atoms around the
central one; for Fe, the first shell pyrrole nitrogens
and, in the second or third shell, the Cu_{a_3} atom of the
redox center of cytochrome oxidase. Similarly, excitation
and emission from the Cu center at ≥ 9.0 KeV gives, with
increasing energy, oscillations of the fluorescence signal
in accordance with the first, second, and third shell
atoms surrounding both Cu atoms and presumably would

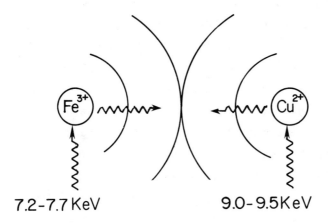

7.2-7.7 KeV   9.0-9.5 KeV

Fig. 1.   The principle of the EXAFS method. Irradiation
          of the metal atoms above the ionization potential
          causes ejection of a photoelectron which is
          back-scattered from nearest and next-to-nearest
          neighbors giving characteristic interference
          effects as a function of increasing energy of
          the incident X-rays.

thereby indicate the Fe_{a_3} atom of the redox center. The
nature of the experiment immediately signifies the need

for precision exceeding that required in hemoglobin EXAFS,

where the problem was simply to determine the distances

to the pyrrole nitrogens (12). In hemocyanin, the third

shell was found to be occupied by nearest neighbor Cu,

but hemocyanin concentrations of 5-10 mM were available

(13). In cytochrome oxidase, the problem is about ten

times more difficult since the maximal concentration of

enzyme that can be obtained and manipulated for chemical

modifications is approximately 1 mM.

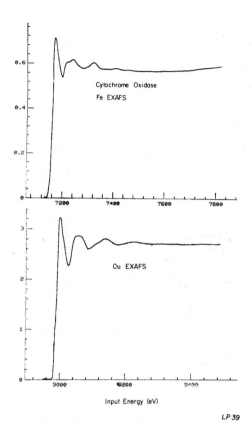

Fig. 2. Typical EXAFS recordings for the sum of the two
         cooper and the two iron atoms of cytochrome oxy-
         dase. Note that the energy scales are different
         for the two ecordings. The traces represent
         the average of ~25, 7 min. runs. Data taken on
         beam lines II-3 and I-5 at SSRL.

Figure 2 shows EXAFS data for cytochrome oxidase, in
the top trace for Fe and in the bottom trace for Cu. The
traces illustrate the characteristic oscillations of
intensity of the fluorescence signal as a function of
energy for 1 mM cytochrome oxidase in the oxidized state.
These data comprise the average of ~25, seven min. runs

with energy intervals of 1 eV as obtained at ~1 x 10^{11}
photons/sec and with ~3 eV resolution on beam line II-3
at SSRL (5).

The arrangement for obtaining these data is illustrated in Fig. 3, where the X-ray beam impinges upon the
sample in a double mylar walled cryostat maintaining the
sample in an atomosphere of chilled nitrogen at -130°,
yet allowing viewing the sample over a large solid angle.

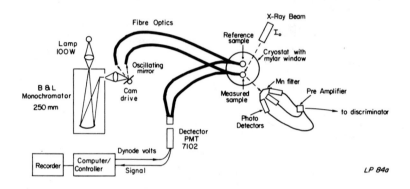

Fig. 3. Diagram of low temperature sample cryostat,
photodetectors and optical monitor as used
for the registration of cytochrome oxidase
EXAFS.

The photodetectors (Fig. 3) are constructed of pilot
B plastic scintillation material and are shielded from
light with aluminized mylar film (11).   The elastic
scattering signal was diminished by a magnanese filter
similar to that suggested by Stern and Heald (11, 17).
The solid angle viewed by 7 of these detectors is about
50% of the solid angle available from a single side of
the sample.   The count rate of total photons for 1 mM

cytochrome oxidase on beam line II-3 is ~5 x 10^5/sec of
which less than 1/4 is unwanted elastically scattered
radiation and over 3/4 are the desired fluorescent
photons.

In order to more clearly differentiate the EXAFS
from the initial edge jump (Fig. 4), a linear background
is subtracted which sets the absorption below the edge
to 0.   The EXAFS signal (a function of the photo-electron

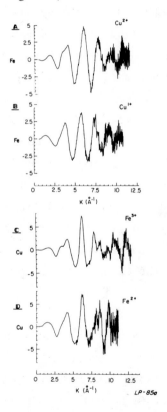

Fig. 4.   The corrected background and the normalized EXAFS
data, in this case multiplied by k^3 for iron and
copper.

wave vector (k)) was isolated from the data by cubic B
spline fit to remove the "isolated atom" contribution and
was then compensated for the inverse k^3 dependence of the
EXAFS data of Fig. 1. This equalizes the amplitude of
the oscillations over the observed range of k. The
function was then normalized to one absorbing atom by
the magnitude of the edge (this corresponds to 2 Fe or
2 Cu atoms). The signal-to-noise ratio of the data was
sufficiently great so that no smoothing was used. In
fact, the data showed a signal-to-noise ratio three
times that of the hemoglobin data used to determine the
Fe-N bond length in that case (12). Thus, Fig. 4 shows
the corrected EXAFS data for Cu and Fe in the oxidized
and reduced states over a range of k over 11 $\overset{o}{A}^{-1}$. The
contributions to the individual shells were analyzed in
detail with Fourier filters about each shell, the results
of which were compared with the corresponding shells of
appropriate Fe, Cu models (11). The Fourier transform
of the functions of Fig. 4 gives the desired radial distribution function plus an absorber-scatterer phase
shift $\alpha(k)$ (Fig. 5). The Cu and Fe data for oxidized
and reduced states show clearly that the noise level is
very small.

Considering first the relative radial distribution
function (rrdf) for oxidized Cu, it is seen to consist
of one large peak and two minor ones, corresponding
respectively to the first, second, and third shells.
The contrast of these data to the fully reduced state
is remarkable; no peak above the first shell is seen to
exceed the noise level. Similarly, in the case of Fe,

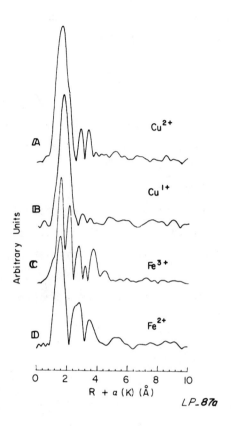

Fig. 5.   The Fourier transform of the data of Fig. 4
          gives the relative radial distribution function
          (rrdf) for copper and iron in the oxidized and
          fully reduced states.

there is one large, bifurcated peak and two smaller ones
above the noise level. The third peak occurs at the
same radial distance as the third peak in the Cu rrdf.
Unlike the Cu rrdf of the reduced enzyme, the Fe rrdf
of the reduced state still contains three peaks. However,
the third peak is shifted in the reduced state (see below).

Thus, without further analysis, one can identify a large
conformation change in the oxidized-reduced transitions
for Cu and, to a smaller extent, Fe as well.

DATA ANALYSIS

The principle of the analysis chosen here is to model
each site with a single compound, utilizing the mixed
valence states to alter one site while the other remains
unaltered, thus affording specific perturbations of each
structure and its relationship to the model compound (18).
This affords as unique a solution as permitted by the
available accuracy for identifying structural data on
each of the four redox centers of cytochrome oxidase.
The data presented here are a brief synopsis of the data
on Fe_{a3} and Cu_{a3}. Full details are available elsewhere
(11).

STRUCTURE OF THE REDOX CENTERS

First Shell Atoms. The results of the detailed analysis
procedure (11) are illustrated in Fig. 6, and the bifurcated peak of the first shell of the oxidized oxidase is
interpreted as a sulfur atom at 2.6 Å distance from Fe.

The back-transformed data from the first shell
(Fourier filter) is consistent only with a sulfur atom
and not with an oxygen atom as postulated for the μ-oxo
structure of the oxidized enzyme (19). The sulfur atom
cannot be seen in the reduced state of the Fe, but can
be seen in the reduced state of Cu. It is concluded
therefore that it is one of the sulfur ligands of Cu_{a3}, for
which the appropriate model is stellacyanin, a type I "blue"
Cu.

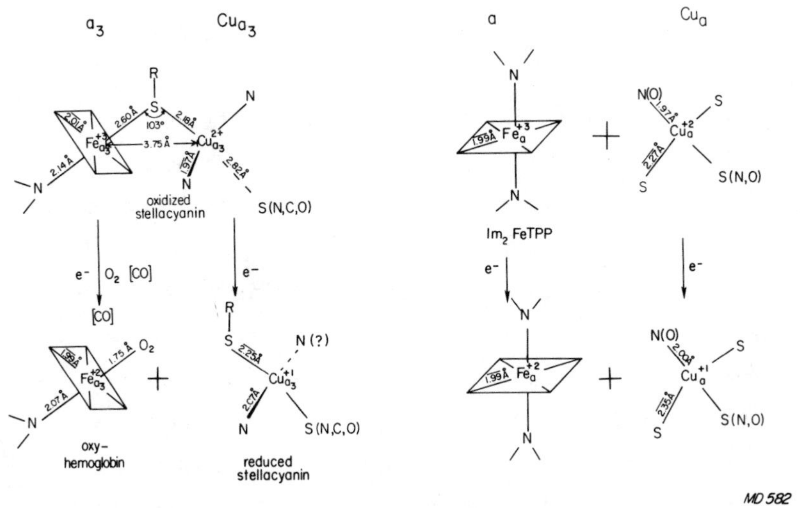

Fig. 6.   Results of the data analysis procedure giving
          ligands and bond length for oxidized and re-
          duced cytochrome oxidase using appropriate
          models as indicates in the figure.

The Interatomic Distance.   The higher shell contributions
of both Fe and Cu, as already pointed out above, contain
information on their interatomic distances at ≤ 4 Å.   The
observation of spin pairing in Fe_{a3} and Cu_{a3} is consistent with this distance.   This interatomic distance is
observed from both iron and copper.   The phase shift
could not be satisfied by N, C, O, S, or any combination
of these except a Cu atom.   Model compounds having
appropriate iron-copper distances are lacking, although
copper metal seems appropriate, as are several other
models containing additional copper atoms.   Appropriate
corrections to these give signals similar to those of
the higher shells in both the oxidized and mixed valence

formate states. On the basis of these comparisons, only
the third shells of the Cu EXAFS contain iron at the distances indicated in Fig. 6. Similarly, the Fe higher
shells show the contribution in the fourth shell which
includes nitrogen and carbon contribution of the heme
groups at 4.12 $\overset{o}{A}$. Subtraction of the latter contribution
for both hemes gives filtered data, which, by phase
comparison, shows that this remainder contains only Cu
at the distance shown in Fig. 6. Thus, the two contributions meet our criteria and can be identified as an
Fe-Cu distance of 3.75 $\pm$ 0.5 $\overset{o}{A}$ between the components
of the $\underline{a}_3$ redox center. The possibility that this might
be due to the heme $\underline{a}$ and its associated Cu_a is extremely
unlikely because there is no antiferromagnetic coupling
exhibited by Fe_a/Cu_a.

Examination of the iron fourth shell of the reduced
and CO mixed valence states in the same manner does not
produce results uniquely identified as a single Cu, and
it is apparent that either the distance between these
two atoms in these states is larger than that observable,
or an unusual Debye-Waller factor makes the distance
determination inadequately accurate (11).

STRUCTURAL CHANGES IN THE REDOX CYCLE

Figure 6 gives the summary of the metal atom distances and the bond lengths for the four centers of
cytochrome oxidase; the models are listed as applied
and the dashed lines indicate ligands that are possibly
too long to be real bonds. In several cases, alternatives between nitrogen and oxygen are given where

these are possibilities. The error bars are not included
for simplicity, and a line over the distance indicates an
average for those with the same ligands in the metal
center. Obviously, charge delocalization can occur in
such complex systems and the formal valence listed can
be so interpreted.

The redox site, Fe_{a_3}-Cu_{a_3}, employs the cysteine
sulfur of the stellacyanin model to form the bridging
ligand between the two metal atoms. This sulfur remains
three-liganded and has the same bond length between Cu
and S as in stellacyanin. However, the Fe-S distances
are longer than those usually observed for octahedral
coordination, as in cytochrome c or cytochrome P_{450}.
The addition of the sulfur ligand makes iron 6-coordinate
and high spin in accordance with the observations of the
Raman spectroscopy of Babcock et al. (20). The dimensions
of the iron-sulfur-copper redox center permit a calculation of the included bond angle at the S atom to be ~103°,
appropriate to sp^3 bonding required of a 3-liganded
sulfur.

The role of the S atom in the spin coupling between
Fe and Cu is not resolved, but is consistent with the
early observation (21) that mercuration (and urea denatura-
tion) destroys the spin coupling and renders the invisible
Cu "visible", presumably by displacing it outside a distance appropriate for antiferromagnetic coupling with
Fe and Cu.

Relation to the Reaction Mechanism . A simplified reaction scheme for the oxido-reduction cycle of cytochrome oxidase is illustrated in Fig. 7.   This diagram emphasizes the structural changes that occur in the reduction of the redox center by an electron pair from Fe$_a$ and Cu$_a$ and the reoxidation of the center by molecular oxygen.

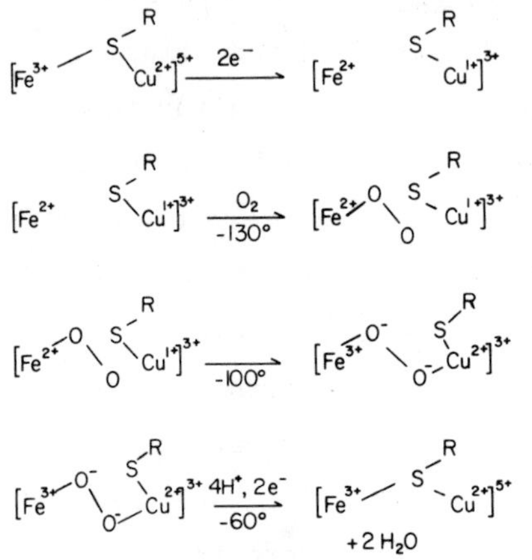

Fig. 7.   Steps of cytochrome oxidase reaction mechanism.

The oxidized state is characterized by a low reactivity towards the usual ligands: F$^-$, SH$^-$, CN$^-$, and especially H_2O_2. This is explained as a consequence of the bridging sulfur atom as the sixth ligand of the high-spin iron. The first step in conformation change in the cycle is caused by the acceptance of electrons by Fe$_{a_3}$, Cu$_{a_3}$ to give, either in two one-step electron transfer reactions or in a one-step, two-electron transfer reaction, the reduction of the iron and the copper. A sulfur radical may appear in this reaction and, in fact, the location of the charge on the copper-sulfur moiety is uncertain.

The reduction of the iron greatly enhances its activity
towards ligands, particularly O_2, CO, etc., and in fact
the reaction with oxygen is as fast as that of hemoglobin
with oxygen, suggesting an "unfettered" approach of dioxygen to the reactive site. Thus, the formation of an
oxy-cytochrome oxidase compound at -130° (14, 15) is now
also explained by the structural data, which show no
evidence of the sulfur bridge and the copper atom at
3.75 $\overset{\circ}{A}$ from the reduced iron. Oxygen reduction, by both
iron and copper, nevertheless requires a reestablishment
of an appropriate iron-copper distance, for example, as
obtained in a hemocyanin model at 3.6-3.7 $\overset{\circ}{A}$ (13) or in
the oxidized state of cytochrome oxidase (11). Thus,
the presence of a bridge peroxide as the first electron
transfer intermediate in oxygen reduction is consistent
with our previous proposals for the structure of Compound B--peroxy cytochrome oxidase (14). Further experiments are required to verify this structure, but the evidence that the iron and copper can approach one another
sufficiently close to form the sulfur bridge makes the
probability of a peroxide bridge structure for Compound B
very high, and represents a second conformation change.

The reaction thereafter would presumably follow the
pathways clearly identified for peroxidases, namely, the
formation of quadrivalent, or indeed, pentav lent iron
and appropriate ferryl ion forms, as well, following
rupture of the peroxide bond. Thus, the iron may become
deligated and appropriate to a reestablishment of the iron
sulfur bond in the ferric cupric state, a third conformation change. Thus, a cycle of conformation changes occurs

during the redox activity of cytochrome oxidase and is
appropriate as the energy concerving mechanism involving
energy-linked structural changes.

Relation to Charge Separation.   The separation of charge
across the membrane is of greatest interest in consider-
tion of the three biological functions of cytochrome
oxidase (namely, electron transport, oxygen reduction
and charge separation across the membrane).   One mechanism for causing this charge separation to occur is that
previously proposed for chromatophores, namely that a
structural change in the redox center activates peripheral
carboxy and amino groups to change their pK values, much
in the way as there is coupling between the binding of O_2
in hemoglobin and the pK's grouped within the crevice of
the tetramer (22,23).This results in the exchange of one or
more protons so that at usual pH apparently one proton
is taken up per oxygen bound (23).   Similarly, in the
purple membrane protein, the stoichiometry of light-
induced hydrogen uptake may vary from zero to two,
depending upon the ionic strength the pH, and other
parameters which would be expected to influence the pK's
of peripheral groups on the protein (24).

    In the case of cytochrome oxidase, the linkage be-
tween the structural changes in the redox cycle and the
pumping of protons from one side of the membrane ot hte
other, via a membrane Bohr effect (22), is illustrated in
Fig. 8.   In this case the effects at the redox center
vastly exceed those in liganding of hemoglobin and may
be directly involved in the translocation.   State A

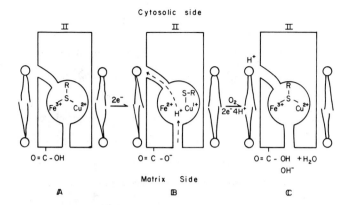

Cytochrome oxidase as a membrane protein with Bohr effect.

Fig. 8.   Diagram of the conformation coupled charge sepa-
          ration mechanism in the cytochrome oxidase.

is the resting or oxidized state, in which the Fe-S linkage
is present and the peripheral group(s) on the matrix side
is (are) protonated. The channel is blocked, as suggested
here, by the Fe-S bonding and no proton translocation via
the redox site is possible. In state B, two electrons
have been accepted by the redox center, and the Fe-S-Cu
bond ruptures. This results in a propagated structural
change causing the peripheral protonated group(s) to
deprotonate, releasing one (or more) hydrogen ions on
the matrix side (a number in any case always in excess
of those released on the cytosolic side). The proton
is indicated to be translocated through the open gate
to the cytosolic side through a channel that leads
through the redox site. Obviously, other channels and
gates are possible and the diagram affords only a simplified example (25). Energy must be provided in this

translocation when the gradient has already been established by previous transport activities, and is presumably imparted to the gate by the bond breaking and making, or in the intermediate reaction steps such as peroxide bond breaking in Compound B. In step C the reaction with oxygen occurs; the remaining two electrons and four hydrogen ions are transferred, resulting in the reestablishment of the Fe-S-Cu bond, therby closing the gate. A molecule of water reprotonates the carboxyl group, leaving a hydroxyl ion on the matrix side and a hydrogen ion on the cytosolic side. The cycle may be repeated. Obviously, many variations of this scheme are possible; the diagram only attempts to establish the principle of the membrane Bohr effect as applied to cytochrome oxidase.

This mechanism is in distinct contrast to that of Mitchell, whose hypothesis is based on cytochrome oxidase as a vectorial electron conductor (26), operating perpendicularly to the plane of the membrane; the charge separation across the membrane being due to the protonation of oxygen intermediates in water reduction (27). It is obvious that the observation of protons in excess of those required for water reduction violates the "Mitchell Loop" as the sole source of charge separation across the membrane (25). A more extreme view is that none of the water protons is active in the transmembrane charge separation, but instead that all occurs through a linked-function (Bohr) mechanism as described above (22). In this case, the asymmetry of proton binding and translocation is greater than that cited above, but nevertheless well within the capability of charge separation in large

molecules which contain numerous peripheral groups, as
may indeed be the case in 120 kilodalton cytochrome
oxidase.

REFERENCES

1. Chance, B., Boyer, P., Ernster, L., Mitchell, P.,
    Racker, E. C. (1977) Ann. Rev. Biochem. 46, 955-1026.
2. Chance, B. (1972) FEBS Lett. 23, 3-20.
3. Chance, B., Crofts, A. R., Nishimura, M. and Price, B.
    (1970) Eur. J. Biochem. 13, 364-374.
4. Winick, H. and Doniach, S., eds. (1980) Synchrotron
    Radiation Research, Plenum Press, New York.
5. Bienenstock, A. and Winick, H. (1979, 1980, 1981)
    Stanford Synchrotron Radiation Laboratory Activity
    Reports, SSRL Reports 79/03, 80/01 and 81/in press.
6. Powers, L., Chance, B., Ching, Y. and Blumberg, W.
    (1981) Fed. Proc., in press.
7. Chance, B. and Powers, L. (1981) Biophys. J. 33, 94a.
8. Powers, L., Chance, B., Ching, Y. and Angiolillo, P.
    (1981) Biophys. J. 33, 94b.
9. Peisach, J., Blumberg, W. and Chance, B. (1981)
    Fed. Proc., in press.
10. Powers, L., Chance, B. and Ching, Y. (1981) VIIth
    IUPAB Congress, Mexico City, in press.
11. Powers, L., Chance, B., Ching, Y. and Angiolillo, P.
    (1981) Biophys. J. 33 , 123-144.
12. Eisenberger, P., Shulman, R. G., Brown, G. and
    Ogawa, S. (1976) Proc. Natl. Acad. Sci. USA 73,
    491-495.
13. Brown, J., Powers, L., Kincaid, B., Larrabee, J.
    and Spiro, T. (1980) J. Am. Chem. Soc. 102 4210-4216.
14. Chance, B., Saronio, C. and Leigh, J. S., Jr. (1975)
    Proc. Natl. Acad. Sci. USA 72, 1635-1640.
15. Yang, E. K. and Chance, B. (1981) Biophys. Biochim.
    Acta, submitted.
16. Warburg, O. (1949) Heavy Metal Prosthetic Groups and
    Enzyme Action (Translated by A. Lawson), Oxford
    University Press, Oxford.
17. Stern, E. and Heald, S. M. (1979) Rev. Sci. Instr.
    50, 1579-1582.

18. Powers, L., Blumberg, W., Chance, B., Barlow, C., Leigh, J.S., Smith, J., Yonetani, T., Vik, S. and Peisach, J. (1979) Biochim. Biophy. Acta 546, 520-538.

19. Chance, B., Waring, A.and Powers, L. (1979) in Cytochrome Oxidase (King, T., Orii, Y., Chance, B. and Okunuki, K., eds.) Elsevier/North Holland, Amsterdam, pp.353-360.

20. Babcock, G., Callahan, P., McMahon, J., Ondrias, M. and Salmeen, I. (1980) in Symposium on Interaction Between Iron and Proteins in Oxygen and Electron Transport. Arlie House, Virginia, Abstract V-11.

21. Beinert, H. and Palmer, G. (1965) in Oxidase and and Related Redox Systems (King, T., Mason, H. and Morrison, M., eds.), John Wiley and Sons, New York, pp. 257-264.

22. Chance, B., Crofts, A.R.,Nishimura, M., and Price, B., Eur J Biochem (1970) 13 364-374

23. Perutz, M. (1969) The Croonian Lecture, Proc. Royal Soc. B173, 130-140.

24. Hess, B. and Kurschmitz, D. (1978) in Frontiers of Biological Energetics (Dutton, P., Leigh, J.S. and Scarpa, A., eds.), Academic Press, New York, pp. 257-264.

25. Krab, K. and Wikstrom, M. (1978) Biochim. Biophys. Acta 504, 200-214.

26. Mitchell, P. (1966) Chemiosmotic Coupling and Photosynthetic Phosphorylation Glynn Res. Bodmin Cornwall England.

27. Keynes, R. D. and Davies, R. E. (1961) in Membrane Transport and Metabolism (Kleinzeller, A. and Kotyk, A., eds.), Academic Press, New York, pp 336-340.

THE ADP,ATP TRANSLOCATION SYSTEM OF MITOCHONDRIA

M. Klingenberg

Institute for Physical Biochemistry, University of Munich,
Goethestrasse 33, 8000 Munich 2, F.R.G.

INTRODUCTION

In 1964 at the time of greatest activity and fertility
in research on oxidative and photosynthetic energy transduction, Ernster and Lee (1) published their review article
which was a landmark of comprehensive analysis of our
knowledge in bioenergetics. It was around this time that
the role of the inner mitochondrial membrane as an osmotic
barrier became recognized and thus also as the barrier for
the various solutes or substrates for electron transport and
oxidative phosphorylation. In these exciting times at the
International Congress in New York and at the satellite
"compostium" we first reported on the findings of an ADP,ATP-
specific and AMP-excluding exchange system between the inner-
and extramitochondrial adenine nucleotides (2,3,4).

This research started from two approaches already
systematically exploited in our laboratory (5), the investigation of mitochondrial bound adenine- and other nucleotides
and the development of methods for directly measuring transport in mitochondria (2). The recognition of a specific

C. P. Lee, G. Schatz, G. Dallner (eds.), Mitochondria and Microsomes
    in honor of Lars Ernster   ISBN 0-201-04576-1

ADP,ATP transport emerged from a general investigation of
nucleotide transport through the mitochondrial membrane, in
particular under the viewpoint of the metabolic interaction
of the intra- and extramitochondrial hydrogen transfer and
phosphate transfer systems (6).

In this relatively brief review attention will be
focussed on a few highlights. For further information the
reader may be referred to a number of longer or shorter
reviews dealing with this subject (7-12).

GENERAL CHARACTERISTICS OF THE ADP,ATP EXCHANGE IN MITOCHONDRIA.

Mitochondria from all sources tested so far have an
ADP,ATP translocation system. Most studies have been performed with the "classical" mitochondria from rat liver and
beef heart. And therefore data given will be limited to
mitochondria from these sources.

For understanding the studies on the ADP,ATP transport it is important to realize that the isolated "intact"
mitochondria contain an endogenous pool of adenine nucleotides of ATP,ADP and AMP (14,15). A function of endogenous
adenine nucleotides in oxidative and substrate level phosphorylation was determined (5,13). However, the state of the
nucleotides remained unclear, for example, they were
supposed to be adsorbed in the mitochondrial matrix. Only
after the transport in and out of this pool had been found did
it become clear that the nucleotides are trapped in the
inner mitochondrial space by virtue of the impermeability
of the inner mitochondrial membrane (4,13). Whether the
internal nucleotide pool is on the pathway of the oxidative
phosphorylation of external nucleotides has been a

controversy until recent years (16,17). This question has been resolved by new techniques that in fact these nucleotides are part of the main path of ATP synthesis (18). Only in this case the measured rates of ADP and ATP translocation would reflect the interaction of external adenine nucleotides with the ATP synthesis system. Arguments that the endogenous nucleotides are bypassed have not been fully aware of this reasoning.

The translocation of ADP and ATP to the mitochondria operates in general as a 1:1 mol exchange between the endogenous and exogenous ADP or ATP (4) (Fig.1). A specific net release of ATP from the mitochondria by the same path can be induced, however, at about a 100-fold slower rate (19). A slow net uptake of adenine nucleotides is observed in foetal (20) and plant mitochondria (21). The first, however, may use another pathway since it is not sensitive to specific inhibitors.

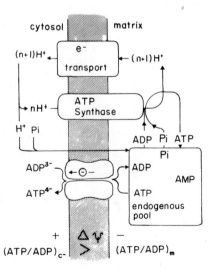

Fig. 1. General scheme for the role of the ADP-ATP exchange system in oxidative phosphorylation by mitochondria.

A most striking feature is the high specificity of
the exchange only for ADP and ATP, in particular concerning
the base portion (22). Apart from adenine, all other bases are
excluded. Of great significance for metabolic compartmentation is the exclusion of AMP from transport. It requires
that AMP is transphosphorylated in the two compartments
separately and that there is no connection between the
adenylate equilibrium in both compartments. Obviously, the
link between the phosphate transfer systems of the mitochondria to the cytosol has been concentrated to a single
system, the transport of ADP and ATP only. All other
phosphate transfer reactions have to be hooked up to the
ADP,ATP system in order to "communicate" between both compartments. Only the "free" ATP or ADP are translocated,
the Mg^{++} complexes being inactive.

Basic kinetic properties of the ADP,ATP exchange such
as concentration and temperature dependence could only be
determined under great experimental efforts to resolve the
relatively short time interval until the added radioactive
nucleotides have been equilibrated with the endogenous
nucleotides. So far, relatively accurate kinetic data have
been collected only for mitochondria from rat liver and beef
heart, using advanced rapid sampling techniques (8). The
exchange rates at 37° have to be extrapolated from values at
lower temperature or from measurements in mitochondria where
the transport has been partially inhibited. There is a clear
temperature break at about 14°, the high activation energy
changes from 124 kJ to only 36 kJ. In beef heart mitochondria
this break is absent. The strong temperature dependence of
the transport is responsible for the very low rate of synthesis of external ATP in rat liver mitochondria. As a result

at low temperatures, below $5^{o}C$, phosphorylation of endogenous
ATP is five times more active than of exogenous ATP (5,23).

The dependence of the translocation rate on the concentration of ADP or ATP does not follow a simple hyperbolic
(Michaelis) type relation, but can be broken up into about
two hyperbolic relations, each characterized by one K_m value
(8). The K_m's are considerably higher for ATP than for ADP
in normal "energized" mitochondria. The value for ADP
reflects a K_m of mitochondria in oxidative phosphorylation.

At this point it should be stressed that numerous
parameters concerning the ATP synthesis in whole mitochondria
are reflecting the parameters of the translocation rather
than the ATP synthesis per se. This is true for the temperature dependence, the high specificity with respect to
the base portion of the nucleotide, insensitivity to external
Mg^{++}, the K_m for ADP, etc. In all these cases the ATP synthase, being masked, has different properties than when it
is directly accessible as in submitochondrial particles.

Of great metabolic importance is the differentiation
between ADP and ATP by the energization of the mitochondrial
membrane. Basically the ADP,ATP exchange is not energy
dependent and in uncoupled mitochondria ADP and ATP are
translocated in both directions with about equally high
activity. In energized mitochondria, however, the influx of
ATP is strongly depressed as compared to that of ADP and
conversely, in the release from the inside ATP is strongly
preferred to ADP as the counterexchanging species (8,22).
For example, when equal amounts of ATP and ADP are offered
to liver mitochondria, ADP is preferred about eight times
for the uptake, and vice versa, ATP is preferred about twelve
times to ADP for the efflux (8,24). This strong asymmetry,

induced by energization of the membrane, has far-reaching
consequences. Firstly, it makes the ADP,ATP carrier select
a combination between exchanging partners ADP and ATP which
is required for producing external ATP in oxidative phosphorylation. In other words, out of the four possible combinations, the one which takes up ADP and releases ATP is preferred to such an extent that it may account for 90% of the
exchange activity. Secondly, the asymmetry of the exchange
causes accumulation of ATP outside and of ADP inside. As a
consequence the ATP/ADP ratio is considerably higher outside
than inside the mitochondria, by a factor which may reach 30
(25,26). Despite the low intramitochondrial ATP concentration, by buffering with Mg^{++}, the ATP content remains
sufficiently high for effective functioning in intramitochondrial transphosphorylation reactions.

The mechanism for the energy control of the ADP,ATP
exchange has been shown to be a membrane potential-driven
electrophoretic effect which utilizes the charge difference
between ATP^{4-} and ADP^{3-} anions (27,28,29) (see Fig. 1).
Evidence specifies that ATP rather than ADP is controlled.
It can be visualized that the binding center in the carrier-
ATP complex carries one negative charge excess whereas the
carrier-ADP complex is electroneutral. More detailed data
are obtained from the reconstituted system (see below).
Evidence for the electrophoretic mechanism comes from correlation of the ADP/ATP ratio to the membrane potential, a one
to one cotransport of cations (K^+ or H^+) with uptake of ATP
but not of ADP (27,28,29) and from studies with ATP analogues.

Electrophoretic ADP,ATP exchange requires energy with-
drawn from the electrochemical H^+ potential. As a consequence
the phosphorylation potential of external ATP is raised by

the transport by about 10 kJ above the inner mitochondrial
potential (30). The great metabolic significance for separation of the intramitochondrial and cytosolic ATP systems
in the cell is evident (27,30,31). Extensive investigations
using new methods to separate the metabolites in the cytosolic and intramitochondrial compartments for liver and
heart clearly concurred with the results obtained in the
mitochondria. In well fed liver, the ratio of cytosolic to
intramitochondrial ATP/ADP is found in the range of 25,
whereas in starved livers it decreases to about 3 (31,32).
As a result the cytosolic phosphorylation potential is calculated to be higher by about 12 kJ compared to the intramitochondrial one.

The fact that in oxidative phosphorylation part of
the phosphorylation energy of ATP is generated by the export
from the mitochondria, came as a surprise and therefore was
at first not easily accepted (33,34). In particular it
raised considerable problems for the "stoichiometries"
between H^+ production by electron transport and H^+ consumption by ATP synthesis, according to the chemiosmotic hypothesis (34). It required for the production of one mol ATP
an additional mol H^+ for the transport (Fig. 1). Up to that
stage the P/O ratio could be reconciled with the postulate
that 2 H^+ are generated per coupling site and are required
for the synthesis of one mol ATP. This balance now became
unsettled (30). In order to maintain the observed P/O
ratios, more than 2 H^+ are to be generated per coupling site,
for example 3 H^+, if 2 H^+ are used for the synthesis of one
mol ATP. In fact, there is now widespread agreement that 3
or even 4 H^+ are generated per coupling site, which resolves
the previous dilemma (34,35,36). It may be mentioned,

however, that these higher ratios require a fundamentally different mechanism for H^+ generation than proposed in the original chemiosmotic theory (33).

CHARACTERIZATION OF CARRIER BINDING SITES IN THE MEMBRANE.

The specific ligands of the ADP,ATP carrier such as substrates ADP and ATP and the inhibitors have been used in a series of binding studies to elucidate not only the number of binding sites but also its function in the translocation process. Binding data were collected from measuring the concentration dependence of the ligands and the interaction of the substrates with "inhibitor ligands." Using atractylate (ATR) or carboxyatractylate (CAT), the specific binding of ADP and ATP could be discriminated from total binding to the mitochondria. The analysis of the binding data in the classical manner yielded two families of binding sites with high and low affinity, at a ratio of 1:3 (37) (Table I). More penetrant and fruitful analyses of the binding data became possible by introducing the concept that the sites are flexible and can be directed to the inner or outer face of the membrane. With this novel approach, binding sites with apparently high affinity were identified as those on the inside, and the ones with low affinity as those looking to the outside. Only the latter are unmasked with a K_D (1.2 x 10^{-5}M) which is not too low to permit sufficiently high dissociation rate during the translocation process.

The "inhibitor ligands" bind with much higher affinity (Table I) and are therefore very useful for identifying the maximum number of binding sites and also for tracing the isolation of the carrier protein. The atractylates are relatively impermeant whereas bongkrekate can penetrate to

Table I. Summary of Binding Data to ADP,ATP Translocator
          in Beef Heart Mitochondria

| Ligand | Number of Binding Sites | | Diss.Constant |
	μmol/g prot.	mol/mol cyt.a	K_d (M)
ADP	0.70	1.3	0.3; 6.5×10^{-6}
ATP	0.75	1.4	0.6; 12×10^{-6}
ATR	1.2 to 1.6	2.2 to 2.9	1×10^{-7}
CAT	1.3 to 2.0	2.3 to 3.6	2×10^{-8}
BKA	1.2	2.2	$<10^{-8}$

the inside. All inhibitor ligands exclude each other, indicating that they all bind to the same site.

        Most revealing is the influence of these inhibitors
on the binding of ADP. Whereas atractylate (ATR) displaces
ADP or ATP, bongkrekate (BKA) appears to tighten ADP or ATP
(38). Dismissing the classical approach of "allosteric"
effects of the inhibitors on the substrate binding sites,
the novel approach of flexible, either inside or outside
oriented binding sites produced an unexpected treasure of
knowledge on the carrier translocation mechanism (39)
(Fig. 2). A single binding site can be assumed to exist in
the carrier which can bind the substrates ADP and ATP or the
inhibitors. This site can be opened either to the outside
or the inside of the membrane. The transition between the
two sides is identical with the central translocation step
of the overall transport and is possible only when the
carrier is loaded with ADP or ATP. Facing the outer side,
the carrier binds tightly atractylate and becomes fixed in
the outside orientation, while the permeant BKA can bind at
the inside and correspondingly fixes the binding site in
that state. Thus the population of carrier sites, being in

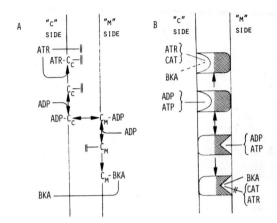

Fig. 2. A. The reorientation mechanisms of the ADP,ATP
          carrier. The impermeant ligand atractylate (ATR)
fixes the carrier at the "c"-side and the permeant ligand
bongkrekate (BKA) at the "m"-side of the inner mitochondrial
membrane. The transition between "c"- and "m"-state (C_c and
C_m) is only possible for the carrier-ADP- or ATP-complex.

          B. The difference between both states is illustrated
to reflect the opening of the binding site either to the "c"-
or "m"-surface. It changes specificity towards the inhibitors
but not to substrates, as a result of the directional translocation of the substrate through the stationary carrier (43).

the steady state usually distributed partially inside and

outside, can be concentrated either to the outer or the

inner face by binding either the ATR or BKA type inhibitors.

Without added ADP,mitochondria extensively bind atractylate

and carboxyatractylate because "empty" binding sites are

trapped outside. For binding bongkrekate, addition of ADP or

ATP is necessary in order to reorientate the sites to the

inner face. At the same time ADP molecules are internalized

at the amount as carrier sites are reoriented by picking up

ADP to the inside. This explains that ADP is apparently

tightly bound to the mitochondria under the influence of BKA.

          The discrimination of the outer and inner location of

binding sites by these inhibitors permitted for the first

time to penetrate to the molecular level of translocation by
the orientation of the binding sites (Fig. 2A). Other
mechanisms, e.g. reciprocating "flip-flop" exchange, can
thus be eliminated. Taking into account that the carrier is
actually a dimer, further restrictions on the carrier can be
made, as will be described below. In assuming an identical
binding site for substrate and both inhibitors, it seems
necessary to postulate an important conformational change in
order to accommodate the strong specificity change on the
transition between the"c"-and"m"-state. A different plausible
interpretation could postulate two distinct binding sites for
BKA and ATR which would correspond to two consecutive substrate sites on the translocation path (40). This interpretation has been revived recently (41) on the basis of
results which, however, can be equally well accommodated
with a single-site model (42). Moreover, according to all
available evidence, never do two ligands bind at the same
time to the carrier.

From a single site reorientation mechanism strong
support can be derived for the stationary versus the rotational model (8,43). In the stationary carrier, relatively
large asymmetric molecules such as ADP can be assumed to
enter and leave the binding site in the same orientation,
according to the translocation direction. As a result the
interface between the asymmetric substrate and the binding
site is reoriented in the two states and this is reflected
in the drastic difference in the specificity towards BKA
and ATR (Fig. 2B). In a rotational carrier, the mutual
orientation between substrate and carrier has not changed
and consequently the binding site remains unaltered.

Quite unexpected support and stimulus was received for the reorientation concept from "microscopic" changes of the mitochondrial configuration parallel to the changes of the carrier state (44,45). "Low amplitude" contraction and swelling were recorded by absorption change on addition of ADP or ATP which was fully abolished by ATR and enhanced by BKA. Therefore, the absorption changes can be specifically discriminated as reflecting the interaction of the ligands with the carrier. The recording facilitates the observation of kinetics and more subtle change in the interaction of these ligands and its dependence on various parameters, such as pH, etc. Surprisingly good agreement with the binding data emerges if one assumes that the contraction is caused by the "m" (matrix)-state and decontraction by the "c" (cytosolic)-state. Full decontracted and contracted states as well as intermediate distribution states are induced by varying the amount of ADP addition and are easily read off the absorption changes. The morphological changes in the contraction/decontraction states were analyzed by electron microscopy and could be interpreted to be caused by the enlargement of one membrane surface relative to the others. Enlargement of the "m" surface on addition of BKA would correspond to an accumulation of carriers on the "m"-face, and vice versa. It seemed at that time quite far-fetched to assume that a single type of protein molecule can induce such gross morphological changes in the membrane. Retrospectively this idea seems more feasible in view of the high density of carriers in the membrane. It still remains to be established whether the transition between the "c"- and "m"-state is accompanied by slight repositioning of the carrier between the "c"- and "m"-side of the membrane (see also 46).

Biochemical conformation changes in support of the reorientation mechanism have been detected using amino acid reagents and proteases. Unmistakable is the unmasking of an SH-group, which is essential for carrier activity, on addition of ADP or ATP (47,48). The group remains masked when first ATR is added or is reactive in the presence of BKA (8,49). By differential incorporation of labeled NEM, a protein band with M_r = 30 000 was identified in SDS gel electrophoresis as a major target of ADP stimulated and ATR sensitive NEM incorporation (50). As will be shown below, also by other studies this band was identified after isolation as the ADP,ATP carrier.

Lactoperoxidase catalyzed iodine incorporation into mitochondria has the 30 000 M_r band, identified as the ADP, ATP carrier, as a major target (51). While BKA largely inhibited, ATR stimulated the iodination of the 30 000 M_r band. The same direction of effects was found in submitochondrial particles. Therefore, it seems to indicate a more open exposure of thyrosins in the "c"-state than in the "m"-state, rather than the withdrawal of the carrier from the membrane surface.

The degradation of the 30 000 M_r protein by various proteases is strongly protected by carboxyatractylate, both in mitochondria and submitochondrial particles (52). The protection also against endogenous proteases by carboxyatractylate is an important provision for isolation and purification of the ADP,ATP carrier as shown below.

THE ISOLATED ADP,ATP CARRIER.

The isolation of the ADP,ATP carrier met first with considerable difficulties due to the very limited require-

ments in the choice of detergents and its unusually high
sensitivity towards endogenous proteases (53,54,55). The
aim of the purification was to obtain an undenatured protein,
as defined by the ability for specific ligand binding.Labeled
carboxyatractylate was the most useful ligand as a specific
detector and marker of the ADP,ATP carrier and a most effective protector against proteolysis. The best purification
procedure therefore starts by loading the ADP,ATP carrier in
the mitochondria with carboxyatractylate, followed by an
extraction of soluble and membrane adherent proteins and subsequent solubilization of the membrane and ADP,ATP carrier
with high amounts of Triton X-100 in the presence of salt.
Being a deeply imbedded membrane protein, the carrier
requires full disintegration of the membrane for solubilization. A subsequent adsorption chromatography on hydroxyl-
apatite adsorbs the majority of proteins while the carboxy-
atractylate-protein complex passes through. The following
gel chromatography can result in a 95% pure preparation, as
judged from carboxyatractylate binding. For removing excess
Triton X-100 and phospholipid, sucrose gradient centrifugation
is applied. The unusually simple and short purification procedure has a yield of nearly 40% of the protein in the
original mitochondria.

Requirement for detergents is quite limited. For
example, cholate denatures the protein and Brij or Lubrol do
not extract the carrier (53). Aminoxide type detergent is
useful but yields less stable preparations. The same is
true for Emulphogen and octylglucoside (56). For this and
further reasons,other published procedures (57,58) for the
purification of the ADP,ATP carrier cannot be reproduced to
render pure preparations with appreciable yields.

Table II.   Isolated ADP,ATP Carrier.

CAT binding	18 μmol/g protein
BKA binding	17 μmol/g protein
M_r in SDS-gel-electrophoresis	30 000
Isoelectric point	$\simeq$ 10
Amino acid composition:	39% polarity

Properties of protein-Triton X-100 micelle:

Triton X-100 binding:	150 mol/$(P_{30\ 000})_2$
Phospholipid binding:	16 mol/$(P_{30\ 000})_2$

Stoke's radius: 65 Å, sed. coefficient: 3.9 S

M_r (sed.equil.) 184 000, M_r protein portion 64 000

Frictional ratio: 1.56

Circular dichroism (180-220 Å):   41% helicity

The purification starting from mitochondria is about 14-fold, as based on the CAT/protein ratio. Accordingly 9% of the mitochondrial protein consists of the ADP,ATP carrier. It amounts to 13% of the total membrane protein and is the most abundant single protein in the membrane in mitochondria. In SDS-gel-electrophoresis a single protein band of M_r 30 000 is seen (Table II). The binding of CAT gives an effective M_r of about 60 000 which indicates that the carrier is a dimer of two 30 000 subunits. A high content of lysine and arginine makes the protein relative basic and raises the isoelectric point to about 10. The polarity index of the amino acid composition is 39%. More pronounced than this index would indicate, the very hydrophobic properties of this protein are reflected in the high amount of detergent binding (59). The purified CAT-protein complex forms a Triton-phospholipid protein micelle which contains about

1.5 g Triton/g protein. The large Stoke's radius with a
relatively small sedimentation coefficient is typical for
polyoxyethylene detergent micelles with a large hydration
shell. From sedimentation equilibrium runs, total M_r 184 000
is calculated which after subtraction of the Triton and
phospholipid portions yield for the protein M_r 64 000. Also
sedimentation velocity runs yield similar values.
These data substantiate that the ADP,ATP carrier is a dimer,
in agreement with the CAT binding value.

The hydrodynamic data can be best fitted by assuming
for the shape of the mixed micelle an asymmetric oblate
ellipsoid with the radii 65, 57 and 27 Å, due to a broad
ring of Triton molecules around the protein, extending along
the short axis. Other detergents, such as lauroyl-dimethyl-
aminopropyl aminoxide (LAPAO) also have similar high binding.
In this detergent, ultraviolet circular dichroism spectrum of
the CAT-protein complex was evaluated, yielding a helix
content of 41% (S. Knof and M. Klingenberg, unpublished).

Although being much more labile against endogenous
proteases, the BKA-protein complex was also isolated according to the same methods as the CAT-protein complex (60).
The protein turned out to be identical with that of the CAT-
protein complex and the BKA binding (16 μmol/g protein)
corresponds to a dimer of two 32 000 M_r subunits. Besides
the much larger instability towards endogenous and exogenous
proteases, other properties indicate a different conformation
of the BKA protein, in accordance with the conclusion drawn
from the mitochondrial studies (Fig. 2B). Thus, in the BKA
protein one SH-group per subunit, i.e., two SH-groups per
carrier molecule are easily accessible to SH-reagents whereas
they are blocked in the CAT-protein complex. In fact, it is

possible to isolate from the membrane the ethyl-maleylated
form in adequate purity only when BKA is added to the mitochondria and thus BKA,NEM-protein complex is purified (H.
Aquila and M. Klingenberg, unpublished).

Also immunological differences between the two protein
forms are detectable. Antibodies raised against the CAT-
protein complex react only poorly with the BKA-complex (61).

If the two isolated protein inhibitor complexes constitute the "c"- and "m"-state as originally postulated from
the mitochondrial studies, an interconversion of the complexes should be possible and require ADP or ATP. The transition from the BKA to the CAT-protein complex is favored by
the tighter binding of CAT than BKA. Thus, starting with the
BKA-protein complex on addition of CAT, BKA is removed and
replaced by CAT (60). Only when ADP or ATP are present in
appreciable amounts a displacement is found. However, the
mutual displacement of ^{3}H-BKA by BKA is not ADP-dependent.
The catalytic effect of the nucleotide is highly specific
and the exchange requires only μM ADP or ATP. The transition into the opposite direction from the "c"- to the "m"-
state has been possible by starting with the ATR-protein
complex which can be purified quite easily and does not bind
ATR as tightly as CAT (49). Furthermore, the appearance of the SH-group can be elegantly used for following the
appearance of the "m"-state. Another advantage is the trap
set by the SH-reagents for the "m"-state because BKA alone
would not be a sufficiently tight ligand. A clear stoichiometric relation between removal of ATR, incorporation of SH-
reagents such as NEM, DTNB, etc., and a strict dependence on
ADP or ATP of the "c"-"m" state transition is observed. The
absorption change recorded for DTNB incorporation reflected

fully the kinetics and also conveniently the specificity, as illustrated in Fig. 3.

The mutual displacement of the inhibitors and strict dependence on the specific substrates ADP and ATP are undoubtedly representations of the "c"- and "m"-state transition, as it was first observed in the mitochondrial membrane. The reproduction of this transition in a solubilized system therefore constitutes a first reproduction in solution of the translocation process. It should permit to study on the solubilized protein the actual molecular events in the transport catalysis analogous to the study of catalysis in isolated enzymes. Also the fixation of carrier in the two translocation states in solution should facilitate to investigate in more detail the structural changes which are associated with the translocation process.

RECONSTITUTION OF THE ISOLATED ADP,ATP CARRIER by incorporation into artificial phospholipid has been successful on various levels. As a prerequisite the ligand-free protein has to be partially purified in spite of its lability and then to be incorporated into the phospholipid systems. At a first stage of incorporation into liposomes the protein is stabilized against degradation (62). The binding of the inhibitor ligands is increased and the specific dependence of BKA binding on ADP shows that the incorporated carrier is mostly in the "c"-state.

The interaction of the ADP,ATP carrier protein with the liposomes produced new insight into the purely empirical reconstitution procedures (63). Firstly, on addition of the protein small liposomes merge to larger multilamellar aggregates which then on second sonication and freezing are disaggregated to single bilayer vesicles. Only these vesicles,

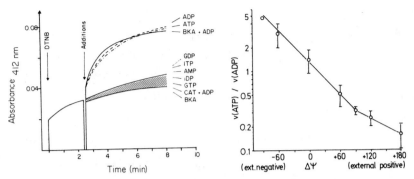

Fig. 3.   The transition from the "c"-state and "m"-state in
          a solution of the purified ADP,ATP carrier. The
          Triton X-100 solubilized and purified atractylate-
          protein-complex is exposed to the SH-reagent DTNB.
          Interaction of DTNB, which corresponds to transi-
          tion into the "m"-state, occurs only on addition
          of the substrates ADP and ATP (49).

Fig. 4.   The influence of the membrane potential on the
          uptake of ATP versus ADP in a reconstituted system
          of purified ADP,ATP carrier in egg yolk phospho-
          lipid vesicles (63).

loaded by ADP or ATP during sonication, are transport active.
A relatively high phospholipid: protein ratio dilutes Triton
sufficiently.

        The reconstituted ADP,ATP carrier system permits a
quantitative evaluation of the number of carrier molecules
incorporated and furthermore by titration with specific
ligands to discriminate active from inactive molecules (63).
This quantitative correlation proved that the isolated ADP,
ATP carrier protein is fully competent for transport and
there is no other cofactor or subunit required besides the
dimeric 32 000 M_r peptides. By titrating the reconstituted
transport and the binding sites with BKA and CAT, the orientation of the incorporated molecules could be determined.
All fully active (about 6%) molecules are sensitive to CAT

with "c"-side outside, about 25% have only very weak transport activity and expose the "m"-side outside, as shown with
BKA, and about 70% molecules are non-binding and inactive.

In the reconstituted system various parameters of the
transport could be readily measured, independent of the
complications, and side reactions in intact mitochondria.
Thus the temperature dependence and K_m for ADP and ATP were
determined and found to be in quite good agreement with
those in mitochondria. Mg^{++}-nucleotide complexes were found
to be inactive in transport and even in binding, as they do
not compete with transport (64). A particular success was
the elucidation of the membrane potential effect on the ADP,
ATP transport (65). Thus by simply creating a diffusion
potential with a K^+-gradient and valinomycin, the exchange of
ATP against ADP is selectively influenced. This eliminates
all speculations of an ATPase-ADP,ATP carrier interaction
for the energy-linked relation of the exchange. It could
be demonstrated that there is a simple relation between the
membrane potential and the control of the ATP versus ADP
transport such that ATP is preferentially driven towards the
positive side of the membrane. As shown in Fig. 4 the transport rates for ATP and ADP are equal in the absence of
membrane potential and can even be opposed to the reaction
prevailing in mitochondria, by inverting the membrane
potential. This is the more remarkable, as the active
carrier molecules are asymmetrically inserted with "c"-side
outside. Obviously the membrane potential influences primarily translocation rates. A more detailed analysis of the
four possible exchange combinations of ATP and ADP in- and
outside demonstrates the great efficiency of the membrane

potential control, as the ATP efflux plus ADP influx is
thirty times more active than the opposite case.

SUMMING UP

The short review stresses the vital role of the ADP,
ATP carrier in the interaction between mitochondria and
cytosol and in all expressions of oxidative phosphorylation
as studied on isolated mitochondria. In addition, the ADP,
ATP carrier provides unique possibilities to penetrate to
the molecular level of the translocation mechanism by
investigating either the original membrane, the purified
protein in solution or the protein after its reincorporation
into artificial vesicles.

REFERENCES

1. Ernster, L. and Lee, C.P. (1964) Ann. Rev. Biochem. 33,
   729-788.
2. Klingenberg, M., Pfaff, E. and Kröger, A. (1964) in
   Rapid Mixing and Sampling Techniques in Biochemistry
   pp. 333-337, Academic Press, New York.
3. Pfaff, E., Klingenberg, M. and Heldt, H.W. (1965)
   Biochim. Biophys. Acta 104, 312-315.
4. Klingenberg, M. and Pfaff, E. (1966) in Regulation of
   Metabolic Processes in Mitochondria (J. M. Tager, et
   al., eds.) pp. 180-201, Elsevier, Amsterdam.
5. Heldt, H.W. and Klingenberg, M. (1965) Biochem. Z. 343,
   433-451.
6. Klingenberg, M. (1964) in Abstracts of VIth IUB
   Congress, New York, Vol. 32, pp. 699-700.
7. Klingenberg, M. (1970) in Essays in Biochemistry (P.N.
   Campbell, ed.) Vol. 6, pp. 119-159, Academic Press,
   New York.
8. Klingenberg, M. (1976) in The Enzymes of Biological
   Membranes: Membrane Transport (A.N. Martonosi, ed.)
   pp. 383-438, Plenum Publishing Corp., New York.
9. Klingenberg, M. (1979) in Trends in Biochemical
   Sciences, Vol. 4, pp. 249-252.

10. Vignais, P.V. (1976) Biochim. Biophys. Acta 456, 1-38.
11. LaNoue, K.F. and Schoolwerth, A.C. (1979) Ann. Rev. Biochem. 48, 871-922.
12. Scarpa, A. (1979) in Membrane Transport in Biology (G. Giebisch et al., eds.) Vol. II, pp. 263-355, Springer Verlag, Berlin.
13. Heldt, H.W., Jacobs, H. and Klingenberg, M. (1965) Biochem. Res. Comm. 18, 174-179.
14. Siekevitz, P. and Potter, V.R. (1955) J. Biol. Chem. 215, 221-255.
15. Pressman, B.C. (1958) J. Biol. Chem. 232, 967-978.
16. Vignais, P.V., Vignais, P.M. and Doussiere, J. (1975) Biochim. Biophys. Acta 376, 219-230.
17. Out, T.A., Velaton, E. and Kemp, Jr. A. (1976) Biochim. Biophys. Acta 440, 697-710.
18. Klingenberg, M. (1977) in Structure and Function of Energy-Transducin Membranes (K.van Dam and B.F. van Gelder, eds.) BBA Library Vol. 14, pp. 275-282, Elsevier, Amsterdam.
19. Meisner, H. and Klingenberg, M. (1968) J. Biol. Chem. 243, 3631-3639.
20. Pollak, J.K. and Sutton, R. (1980) Biochem. J. 190, 1-9.
21. Aboukhalil, S. and Hanson, J.B. (1979) Plant Physiol. 64, 276-280.
22. Pfaff, E. and Klingenberg, M. (1968) Eur. J. Biochem. 6, 66-79.
23. Heldt, H.W. (1966) in Regulation of Metabolic Processes in Mitochondria (J. M. Tager et al., eds.) pp. 51-63.
24. Klingenberg, M. (1980) J. Membrane Biol. 56, 97-105.
25. Heldt, H.W., Klingenberg, M. and Milovancev, M. (1972) Eur. J. Biochem. 30, 434-440
26. Klingenberg, M. and Rottenberg, H. (1977) Eur. J. Biochem. 73, 125-130.
27. Klingenberg, M., Wulf, R., Heldt, H.W. and Pfaff, E. (1969) in Mitochondria: Structure and Function (L. Ernster and Z. Drahota, eds.) pp. 59-77, Academic Press, New York.
28. Wulf, R.,Kaltstein, A. and Klingenberg, M. (1978) Eur. J. Biochem. 82, 585-592.
29. LaNoue, K.F., Mizani, S.M. and Klingenberg, M. (1978) J. Biol. Chem. 253, 191-198.
30. Klingenberg, M., Heldt, H.W. and Pfaff, E. (1969) in The Energy Level and Metabolic Control in Mitochondria (S. Papa et al., eds.) pp. 237-253, Adriatica Editrice, Bari.

31. Soboll, S., Schol, R. and Heldt, H.W. (1978) Eur. J.
     Biochem. 87, 377.390.
32. Soboll, S., Scholz, R., Freisl, M., Elbers, R. and
     Heldt, H.W. (1976) in Use of Isolated Liver Cells and
     Kidney Tubules in Metabolic Studies (J. M. Tager, et
     al., eds.) pp. 29-40, North-Holland Publishing Co.,
     Amsterdam.
33. Mitchell, P. (1979) Eur. J. Biochem. 95, 1-20.
34. Wikström, M.K.F. and Saari, H.T. (1977) Biochim.
     Biophys. Acta 462, 347-361.
35. Brand, M.D., Chen, C.-H., Lehninger, A.L. (1976) J.
     Biol. Chem. 251, 968-972.
36. Alexandre, A., Reynafarje, B. and Lehninger, A.L.
     (1978) Proc. Natl. Sci. U.S. 75, 5296-5300.
37. Weidemann, M.J., Erdelt, H. and Klingenberg, M. (1970)
     Eur. J. Biochem. 16, 313-335.
38. Klingenberg, M. and Buchholz, M. (1973) Eur. J. Biochem.
     38, 346-358.
39. Klingenberg, M., Scherer, B., Stengel-Rutkowski, L.,
     Buchholz, M. and Grebe, K. (1973) in Mechanisms in
     Bioenergetics (G. F. Azzone et al., eds.) pp. 257-284,
     Academic Press, New York.
40. Klingenberg, M., Buchholz, H., Erdelt, H., Falkner, G.,
     Grebe, K., Kadner, H., Scherer, B., Stengel-Rutkowski,
     L. and Weidemann, M.J. (1971) in Biochemistry and
     Biophysics of Mitochondrial Membranes (G. F. Azzone
     et al., eds.) pp. 465-486, Academic Press, New York.
41. Block, M.R., Lauquin, J.M. and Vignais, P.V. (1979)
     FEBS Lett. 104, 425-430.
42. Klingenberg, M. and Appel, M. (1980) FEBS Lett. 119,
     195-200.
43. Klingenberg, M., Riccio, P., Aquila, H., Buchanan, B.B.
     and Grebe, K. (1976) in The Structural Basis of Membrane
     Function (Y. Hatefi and L. Djavadi-Ohaniance, eds.) pp.
     293-311, Academic Press, New York.
44. Scherer, B. and Klingenberg, M. (1974) Biochemistry 13,
     161-170.
45. Stoner, C.D. and Sirak, H.D. (1973) J. Cell Biol. 56,
     51-64.
46. Brandolin, G., Doussiere, J., Gulik, A., Gulik-
     Krzywicki, T., Lauquin, G.J.M. and Vignais, P.V. (1980)
     Biochim. Biophys. Acta 592, 592-614.
47. Leblanc, P. and Clauser, H. (1972) FEBS Lett. 23,
     107-113.
48. Vignais, P.V. and Vignais, P.M. (1972) FEBS Lett. 26,
     27-31.

49. Aquila, H. and Klingenberg, M. (1979) in Function and Molecular Aspects of Biomembrane Transport (E.Quagliariello et al., eds.) pp. 305-308, Elsevier, Amsterdam.
50. Klingenberg, M., Riccio, P., Aquila, H., Schmiedt, B., Grebe, K. and Topitsch, P. (1974) in Membrane Proteins in Transport and Phosphorylation (G. F. Azzone et al., eds.) pp. 229-243, North-Holland Publishing Co., Amsterdam.
51. Boxer, D.H., Feckl, J. and Klingenberg, M. (1977) FEBS Lett. 73, 43-46.
52. Hofmann, W. (1979) Diploma Thesis, University of Munich.
53. Riccio, P., Aquila, H. and Klingenberg, M. (1975) FEBS Lett. 56, 129-132.
54. Riccio, P., Aquila, H. and Klingenberg, M. (1975) FEBS Lett. 56, 133-138.
55. Klingenberg, M. Aquila, H. and Riccio, P. (1979) in Methods in Enzymology (S. Fleischer and L. Packer,eds.) Vol. LVI, pp. 407-414, Academic Press, New York.
56. Klingenberg, M., Hackenberg, H., Eisenreich, G. and Mayer, I. (1979) in Function and Molecular Aspects of Biomembrane Transport (E. Quagliariello et al., eds.) pp. 291-303, Elsevier, Amsterdam.
57. Brandolin, G., Meyer, C., Defaye, G., Vignais, P.M. and Vignais, P.V. (1974) FEBS Lett. 46, 149-153.
58. Bojanowski, D., Schlimme, E., Wang, C.S. and Alaupovic, P. (1976) Eur. J. Biochem. 71, 539-548.
59. Hackenberg, H. and Klingenberg, M. (1980) Biochemistry 19, 548-555.
60. Aquila, H., Eiermann, W., Babel, W. and Klingenberg, M. (1978) Eur. J. Biochem. 85, 549-560.
61. Buchanan, B.B., Eiermann, W., Riccio, W., Aquila, H. and Klingenberg, M. (1976) Proc. Natl. Acad. Sci. U.S. 73, 2280-2284.
62. Krämer, R. and Klingenberg, M. (1977) Biochemistry 16, 4954-4961.
63. Krämer, R. and Klingenberg, M. (1979) Biochemistry 18, 4209-4215.
64. Krämer, R. (1980) Biochim. Biophys. Acta 592, 615-620.
65. Krämer, R. and Klingenberg, M. (1980) Biochemistry 19, 556-560.

MITOCHONDRIAL NICOTINAMIDE NUCLEOTIDE TRANSHYDROGENASE

Jan Rydström

Department of Biochemistry, Arrhenius Laboratory,
University of Stockholm, S-106 91 Stockholm, Sweden

It is with a deep sense of gratitude that I dedicate
this paper to my friend, teacher and former supervisor
Lars Ernster, who introduced and addicted me to the transhydrogenase enzyme.

INTRODUCTION

The properties and function of nicotinamide nucleotide
transhydrogenase have intrigued a number of investigators
since its discovery in 1952 (1), and especially after the
discovery of the energy-linked transhydrogenase (2). In the
course of the early studies of the organisation of oxidative and photosynthetic phosphorylation and electron transport systems energy-linked transhydrogenase was extensively
used as an energy probe (3). The thermodynamic and kinetic
properties of transhydrogenase in combination with its linkage to the energy pool of the phosphorylating system provided a simple and highly sensitive device for probing the
"energy pressure" at the level between electron transport
and the oligomycin sensitive site of the ATPase. During this
period of time the detailed functional relationships between transhydrogenase and ATPase received considerable
attention (4). However, very little progress was reported
regarding the detailed molecular properties of energy-linked

C. P. Lee, G. Schatz, G. Dallner (eds.), Mitochondria and Microsomes
in honor of Lars Ernster   ISBN 0-201-04576-1

transhydrogenase although a soluble factor was isolated
from Rhodospirillum rubrum chromatophores which indeed proved to be a true component of the enzyme (5). Further progress was apparently hampered by the extreme lability of the
membrane component from these chromatophores or the enzyme
from mitochondrial inner membranes. Around 1975 major and
eventually successful attempts to purify mitochondrial transhydrogenase were made independently by two investigators,
Dr. Ronald Fisher at the University of South Carolina (USA)
and myself in collaboration with Dr. Bo Höjeberg. In spite
of the fact that completely different methods of purification were employed both procedures resulted in apparently
identical and homogeneous transhydrogenase preparations
(6,7). This progress proved to be a major breakthrough in
the characterization of the enzyme which subsequently has
allowed a number of important conclusions. The first of these was that transhydrogenase was composed of a single polypeptide with a molecular weight of 100.000-120.000, the
largest polypeptide sofar identified in the mitochondrial
inner membrane (6,7). Two recent reviews have covered the
progress up to this point (8,9). The present review will
therefore attempt to describe important findings from 1978
and onwards including some unpublished data and will be
focussed mainly on the functional and molecular properties
of mitochondrial transhydrogenase, a selection which is motivated by the lack of space as well as by the comparatively slow progress with energy-linked transhydrogenases from
other sources. Consequently, in order to limit the scope
of this review it is unavoidable that some reports which
may be important are not cited. Various aspects of the
coupling mechanism of energy-linked transhyd-

rogenase will be the subject of a separate forthcoming review (10).

METHODS OF PURIFICATION

Both of the available procedures for purification of transhydrogenase from beef heart mitochondria (6,7) have been developed independently of previous attempts to purify the enzyme (cf. 8). The starting material is submitochondrial particles which is preferable to use as compared to intact mitochondria, not only because particles represent a purification of almost two times but also because fractionation procedures applied to particles have proven to be more efficient than with mitochondria. Subsequent steps of the procedure first published (6) involve a crude fractionation of the particles using cholate-ammonium-sulphate followed by ion-exchange chromatography on a DEAE-Sepharose CL-6B column in the presence of a medium containing glycerol and Triton X-100 for increased stability and dispersion. This step is then followed by hydroxyapatite chromatography which yields the final homogenous transhydrogenase. The stability of the preparation is limited to about 6-12 hours and immediate reconstitution of the protein in liposomes provides a convenient way of storing the enzyme in the cold for several months. An important factor in this context is that SS-reducing agents must be present during storage which means that dithiothreitol or an equivalent compound must be added to reduce autooxidized SH groups. Activation of transhydrogenase activity by such agents is an indication that autooxidation of the enzyme has occurred. Prolonged storage of the enzyme under oxidizing conditions leads to irreversible inactivation.

The procedure described above is suitable for preparations of up to 0.5 mg of enzyme. If larger amounts of enzyme are required e.g., 3-10 mg, the procedure may be scaled up and modified to also include additional steps. The most important of these are hydrophobic absorption on phenyl-
-Sepharose CL-6B, washing with buffer A (cf. 11) and subsequent elution with a mixture of 2% (v/v) Triton X-100 and 1% (w/v) lysolecithin in buffer A, and, if required, gel-filtration of reconstituted vesicles on Sepharose 4B (12). Recovery of solid transhydrogenase suitable for amino acid determinations and sequencing is achieved by precipitating the enzyme with 60% acetone or 60% saturation of ammonium sulphate plus 2% sodium cholate at 0-4°C followed by centrifugation for 2 hours at 10.000 x g; in the latter case the precipitated enzyme is still active (12).

Purification of mitochondrial transhydrogenase by the method of Anderson and Fisher (7) employs extraction of submitochondrial particles from beef heart with sodium perchlorate to remove loosely bound NADH dehydrogenase. The next step is extraction of transhydrogenase from the particlex with lysolecithin as previously described by Rydström (13) in the presence of a low and critical concentration of protein. After alumina Cγ gel and calcium phosphate gel chromatography the last chromatography step utilizes an NAD affinity column (AG-NAD-type). In this step the calcium phosphate gel eluate is adsorbed and specifically eluted with an NADH-containing buffer after several washing steps with buffers containing salt and $NADP^+$. If necessary, the transhydrogenase may be concentrated on a hydroxyapatite column and eluted with 100 mM sodium phosphate (pH 7.4). Like the first procedure by Höjeberg and Rydström (6) it appears that

the method is suitable for the preparation of small amounts
(0.2-0.6 mg) of transhydrogenase protein. Recently, it was
reported (14) that Lubrol WX, which was the detergent originally used after the lysolecithin extraction step, is replaced by cholate.

A comparison of the two methods described by Höjeberg
and Rydström (6) and Anderson and Fisher (7) reveals only

TABLE I

A comparison of different methods for purification
of mitochondrial transhydrogenase from beef heart.

Method	Protein[1] (final, mg)	spec.act.	Purification factor	Yield (%)
Höjeberg Rydström (6)	0.19	4.4^2	40	8.7
Anderson Fisher (7)	0.6	7.00^3	23.3	3.1

1. The starting material is submitochondrial particles in
   both cases.
2. spec. act. is defined as μmoles NADH formed/min/mg protein.
   protein.
3. spec.act. is defined as μmoles AcPyADH formed/min/mg protein.

minor differences (Table 1). The extent of purification calculated on the basis of specific activity is 40 and 23.3,
respectively. Since different assay conditions were used the
specific activities can not be compared directly although
they appear to be similar under similar assay conditions.
For instance, the specific activity is 7 μmoles/min/mg protein with acetylpyridine-NAD$^+$ as hydrogen acceptor at 25°C

and in the presence of 80 mM sodium phosphate (pH 6.8)(7) but
only 4.4 µmoles/min/mg protein with NAD^+ as hydrogen acceptor
at 30^oC and in the presence of 100 mM trisacetate (pH 7.4)(6).
However, this difference is eliminated by the fact that activities measured by the two assay systems differ by a factor
of 2-3 as judged by the activities obtained with the same type of submitochondrial particles, 0.30 and 0.11 µmoles/min/mg
protein, respectively. Recently, Earle and Fisher (14) reported activities of the order of 16 µmoles/min/mg protein. However, both pH and temperature were changed which certainly
contributed substantially to the increase in activity. With
regard to yield the method of Höjeberg and Rydström (6) gives
more than twice that of the other method. Also, it appears
that the method of Höjeberg and Rydström (6) is comparatively
faster, easier and relatively more simple to scale up for
larger preparations. However, an important advantage with the
preparation of Anderson and Fisher (7) is its superior efficiency with respect to proton pumping capacity. The reason
for this discrepancy is presently not known (but see the last
chapter in this review).

MOLECULAR PROPERTIES

Mitochondrial transhydrogenase has been shown to be composed of a single polypeptide of a molecular weight of about
97.000 to 120.000 (6,7). A recent reinvestigation of the minimal molecular weight indicates that, as judged by SDS-polyacrylamide gel electrophoresis, the molecular weight varies
between 108.000 and 112.000 depending on whether delipidated
enzyme or lipid-saturated reconstituted enzyme is used as
sample. Provided that the enzyme may be regarded as a globular protein in both cases the difference is interpreted to

represent loosely bound phospholipids amounting to 5 moles/
/mole transhydrogenase. Tightly bound phospholipids were determined to about 9 moles/mole transhydrogenase. The lower
molecular weight, i.e., 108.000, thus seems to be the more
reliable value and is also close to that originally suggested by Höjeberg and Rydström, 97.000 (6). The previous underestimation of the molecular weight of transhydrogenase was
probably due to an erroneous approximation of the molecular
weight of the largest reference protein i.e., β-galactosidase, which only recently has been sequenced and found to have
a molecular weight of 116.349 (15). A true molecular weight
will have to await the determination of the complete amino
acid sequence of the enzyme. An important issue is the hypothetical existence of subunits other than that with a molecular weight of 108.000, which may not have been detected by
the conventional Coomassie Blue staining of polyacrylamide
gels. Lactoperoxidase-catalyzed iodination of purified transhydrogenase after electrophoresis on a polyacrylamide gel in
the presence of SDS followed by autoradiography clearly showed that the Coomassie Blue-stainable band, representing the
108.000 molecular weight component, was the only labeled
band on the gel (12). Since it is very unlikely that a hypothetical and presumably hydrophobic additional peptide would
not be labeled by the lactoperoxidase system using I^{125} with
a high specific activity, the existence of such a peptide
may be excluded.

Recently, the amino acid composition of transhydrogenase from beef heart was determined (Table 2). Since tryptophan was not determined the minimal content of hydrophobic
amino acids is about 45% (12).

TABLE II

Amino acid composition of nicotinamide
nucleotide transhydrogenase from beef heart

amino acid	Composition (mol %)
Ala	10.63
Arg	3.37
Asx (Asp + Asn)	8.11
Cys[1]	-
Glx (Glu + Gln)	8.48
Gly	.11.56
His	1.84
Ile	6.13
Leu	11.10
Lys	5.65
Met	2.26
Phe	3.68
Pro	4.97
Ser	6.70
Thr	5.86
Trp[1]	-
Tyr	3.42
Val	6.86

[1]not determined

Estimation of the molecular weight of the active transhydrogenase is more complicated. A dimeric form has been reported (16,17) which was obtained after cross linking of the
native enzyme and precipitation of the complex with antibodies. Although promising, a full account of these experiments
has not yet appeared. The isoelectric point of active trans-

hydrogenase is about 6.8 (12).

One of the most intriguing questions regarding the function of transhydrogenase concerns the mechanism of hydride transfer during the catalytic reaction. Presumably the mechanism involves a direct redox process since the existence of a reduced intermediate is made unlikely by the apparent

TABLE III

Molecular properties of isolated mitochondrial transhydrogenase (TH) from beef heart.

Property	Number or value
Molecular weight of denatured delipidated TH (monomer)	108.000
Molecular weight of denatured reconstituted TH (monomer)	112.000
Number of subunits in the active form of TH	2 (?)
Loosely bound phospholipids (moles/mole monomer TH)	5
Tightly bound phospholipids (moles/mole monomer TH)	9
Isoelectric point	6.8
Flavine	n.d.[1]
Nonheme iron	n.d.[2]

[1] not detectable spectrophotometrically

[2] not detectable spectrophotometrically or by EPR.

lack of flavine (6,7) and nonheme ion (12) in the purified
transhydrogenase. However, it is interesting to note that
evidence for the existence of a reduced intermediate is accumulating in the case of Rhodospirillum rubrum transhydrogenase (18). Another key problem involves the mechanism of redox-driven proton translocation, i.e., whether it is direct
or occurs through a conformational change connecting redox
and proton translocation events occurring in different parts
of the protein. In this context it should be recalled that
trypsin has been shown to produce two relatively high molecular weight fragments (7). Preliminary results suggest, however, that neither of these have uncoupling, i.e., proton-
-conducting, activity (12). A summary of the molecular properties of isolated mitochondrial transhydrogenase is presented in Table 3.

PROPERTIES OF RECONSTITUTED TRANSHYDROGENASE
Conditions and effect of reconstitution

Incorporation of purified transhydrogenase into liposomes by cholate-dialysis (19) results in a decrease of
the activity independently of the direction of the reaction
catalyzed (6,16). This inhibition is accompanied by generation of a membrane potential (6) and a pH gradient (16), both
of which are eliminated by the addition of uncouplers. The
addition of uncouplers also stimulates the transhydrogenase
activity, thus suggesting that transhydrogenase is a proton
pump and that, under coupled conditions, its activity is limited by a combination of a membrane potential and pH gradient (20,21). The original cholate-dialysis method (cf, 19)
appears to be the most efficient technique for incorporating
the enzyme in a lipid bilayer (20-22). Other methods, e.g.,

sonication in the absence of detergent of addition of enzyme
to preformed empty liposomes, results in inactivation and/or
a poor extent of coupling, i.e., extent of stimulation by
uncouplers (21). It has been noted though that in the presence of low amounts of Triton X-100 transhydrogenase, as prepared by the method of Höjeberg and Rydström (6), does enter
preformed vesicles (12). Also, removal of cholate by gel filtration using e.g., Sephadex G-50 (12), has proven to be a
faster method than dialysis but otherwise there are no significant differences.

The lipid composition of the liposomes is important for
the function of reconstituted transhydrogenase (20-22). Various natural lecithins are about equally efficient in reconstitution (21) whereas dioleoylphosphatidylcholine is the
most efficient among the synthetic lecithins (22). With dioleoylphosphatidylcholine or egg lecithin the resulting liposomes are small (21) with an outer diameter of approximately
300 Å (20); the optimal phospholipid/protein ratio (μmole/mg)
obtained with dioleoylphosphatidylcholine is about 200-300
(20,21). Slightly acidic phospholipids, e.g., dioleoylphos-
phatidylethanolamine at pH 7.4-8.0, also give an efficient
reconstitution (20, but see 21), whereas acidic phospholipids,
e.g., cardiolipin, phosphatidylserine or phosphatidylinositol,
are strongly inhibitory (9,20,21). All phospholipids tested
produce liposomes in which transhydrogenase is asymmetrically oriented with its active sites directed towards the surrounding medium (20,21).

Kinetic properties

The pH optimum for reduction of NAD^+ or acetylpyridine-
-NAD^+ by NADPH (reverse direction) is about 6.0-6.5 (7,20),
i.e., similar to that of transhydrogenase in submitochondrial

particles (7,9), and may vary somewhat depending on the type
of oxidized substrate used. The opposite (forward) direction
of the reaction, i.e., reduction of NADP$^+$ or thio-NADP$^+$ by
NADH, has an optimum below pH 5.0 (20,21). Regarding the stimulation by uncouplers at different pH it is interesting to
note that, at low pH, the reconstituted transhydrogenase catalyzing the forward reaction tends to be uncoupled (21). This
is not the case with the reverse reaction and may indicate
different properties of two separate proton entry sites, each
active in only one direction of the reaction.

The steady-state kinetics of purified and reconstituted
transhydrogenase has been largely neglected since the enzyme
was first purified. Rydström et al. (8,9) were the first to
study the kinetics of mitochondrial transhydrogenase using
submitochondrial particles and it was concluded that the enzyme followed a Theorell-Chance mechanism, ie., an ordered,
ternary complex mechanism with NAD(H) as the first substrate.
It was stressed that a random mechanism could also explain
the data (9) but studies with site-specific inhibitors appeared to eliminate this possibility (9). However, several recent reports have challenged the Theorell-Chance mechanism
and favoured a random type of mechanism (23-25). It appears
that the data from different laboratories are almost identical but that the interpretation of these differs (24). A
reinvestigation of the steady-state kinetics of reconstituted purified transhydrogenase from beef heart (26) reveals
an essentially identical pattern as compared to that obtained
previously with the enzyme in beef heart submitochondrial
particles (8,9). One of the most crucial points is the effect
of site-specific inhibitors, e.g., 2'-AMP and 5'-AMP, which
are specific for the NADP(H)- and NAD(H)-binding site, res-

pectively (26), which is consistent with a random mechanism
(20,23). Together with product inhibition data which show
that the reduced and oxidized form of the same nicotinamide
nucleotide are competing for the same active site and exert
a noncompetitive product inhibition with respect to the second
substrate, it is concluded that the kinetic mechanism is random with dead-end ternary complexes formed by the enzyme and
the oxidized or reduced forms of both nicotinamide nucleo-
tides, i.e., $NADP^+ \cdot E \cdot NAD^+$ or $NADPH \cdot E \cdot NADH$. This conclusion
is identical to that reached by Hanson (20) and Bragg (21)
using non-energy-linked transhydrogenase from E-coli. Concerning inhibitors other than 5-AMP and 2'-AMP it may be noted
that palmitoyl-CoA, the most potent and specific inhibitor of
transhydrogenase in submitochondrial particles with a K_i of
0.8 μM (8,9), also is a potent inhibitor of purified and soluble (not reconstituted) transhydrogenase (7) as well as reconstituted transhydrogenase (26). Inhibition of reconstituted transhydrogenase by acidic phospholipids affects only
V_{max} and not the affinity for the substrates, i.e., the effect is equivalent to a noncompetitive inhibition (26). A
summary of the kinetic properties of reconstituted transhydrogenase is shown in Table 4.

Antibodies against purified transhydrogenase (17) have
been very useful for elucidation of the various NADPH-dependent reactions catalyzed by beef heart submitochondrial particles (cf. 9). Aerobic oxidation of NADPH in the absence of
NAD^+ is insensitive to the antibody (17) and therefore most
likely attributed to the unspecific action of NADH dehydrogenase. Oxidation of NADPH by dichlorophenolindophenol
(DCPIP), which has been suggested (25,9) to be catalyzed by
transhydrogenase, was not inhibited (17) indicating that the

reaction is unrelated to transhydrogenase and probably catalyzed by NADH dehydrogenase.

TABLE IV

Kinetic properties of isolated and reconstituted[1]
mitochondrial transhydrogenase·from beef heart

Substrate varied	Inhibitor	Pattern/plots
NAD^+		convergent;
NADPH	–	intersections be-low or on the ab-scissa
NAD^+	–	K_m = 100 µM
NADPH	–	K_m = 17 µM
NAD^+	NADH	competitive
NAD^+	$NADP^+$	noncompetitive[2]
NAD^+	2'-AMP	noncompetitive[2]
NAD^+	5'-AMP	competitive
NADPH	NADH	noncompetitive[2]
NADPH	$NADP^+$	competitive
NADPH	2'-AMP	competitive
NADPH	5'-AMP	noncompetitive[2]
NAD^+ NADPH	cardiolipin	noncompetitive[3]
NAD^+	palmitoyl-CoA	noncompetitive[2]
NADPH	palmitoyl-CoA	competitive

[1]Reconstitution was carried out with dioleoylphosphatidylcholine

[2]The concentration of the second substrate was nonsaturating.

[3]Cardiolipin was included in the reconstitution mixture at a
relative lipid concentration of 5%.

Energy-linked functions

Purified transhydrogenase is reconstutively active, i. e., its properties when incorporated in liposomes with respect to energy-linked functions are similar or identical to those of the enzyme in the intact mitochondrial inner membrane. Like several other purified and reconstituted membrane proteins (cf. 27) the activity of transhydrogenase per se is considerably more influenced by the liposomal membrane than by the natural membrane, presumably because of the impermeability of the artificial membrane with respect to ions which in turn is caused partially by the lack of other proteins as well as by the necessary excess of phospholipids. Thus, it appears established that the activity of reconstituted transhydrogenase is limited by a counteracting membrane potential, positive inside the vesicles, and a pH gradient, low pH inside the vesicles, with NAD^+ (or acetylpyridine-NAD^+) and NADPH as substrates (6,14,16,20,21). The generation of a membrane potential has also been established separately and indirectly by the influence on the enzyme activity by a potassium gradient in the presence of valinomycine (20,21). Similarly, pH gradients have been shown to influence the activity of reconstituted transhydrogenase (20,21).

That reconstituted transhydrogenase indeed behaves essentially as the native membrane-bound mitochondrial enzyme was demonstrated by Rydström (20) who showed that the simultaneous incorporation of transhydrogenase and oligomycin--sensitive ATPase from beef heart (Complex V) produces vesicles capable of ATP-driven energy-linked transhydrogenase. These ATPase-transhydrogenase vesicles catalyze both the kinetic effect of ATP, i.e., the stimulation of the rate of reduction of $NADP^+$ by NADH, as well as the thermodynamic effect

of energy, i.e., the shift of the equilibrium towards formation of NAD$^+$ and NADPH; both effects were, as expected, oligomycin sensitive (20). In this context it is important to
stress that what may be erroneously taken as an ATP effect
may partly be an uncoupling of the coupled reconstituted
transhydrogenase by ATP. It should be recalled that the transhydrogenase activity of submitochondrial particles is not
stimulated by uncouplers, but of course, still stimulated by
ATP. It is therefore important to check that the thermodynamic effect of ATP also is expressed with the reconstituted
system. How these differences between the kinetic effect of
ATP in the reconstituted system and in the intact mitochondrial membrane system may be interpreted will be dealt with
elsewhere (10).

Until very recently the proton-pumping capacity of reconstituted transhydrogenase was probed indirectly with various pH sensitive dyes, e.g., 9-amino-acridine (16) and 9-ami-
no-6-chloro-2-methoxy-acridine (ACMA) (20), which are believed to be taken up by vesicles acidified on the inside, fol-
lozed by concentration and stacking of the dye and subsequent
quenching. A more direct way of measuring proton uptake was
obtained with pyranine (20) and dextran-linked fluoresceine
(21). Earle and Fisher (14) first succeeded in demonstrating
externally with a pH electrode direct proton uptake by reconstituted transhydrogenase vesicles driven by reduction of
acetylpyridine-NAD$^+$ by NADPH. The ratio between protons taken
up by the vesicles and hydrogen transferred between acetyl-
pyridine-NAD$^+$ and NADPH was found to be close to 1 and the
rate of proton translocation was markedly increased by the
presence of valinomycin (14). Needless to say (but cf. 14),
the conditions used by Earle and Fisher (14) have been tried

for measuring direct proton translocation by reconstituted
transhydrogenase prepared as described by Rydström (20) but
without success and in spite of the fact that the two preparations appear identical and that the latter preparation extensively quenches ACMA. The possibility that contaminating
Triton X-100 or glycerol is causing this problem is presently being investigated.

OUTLOOK

Mitochondrial transhydrogenase constitutes the only
energy-linked mammalian protein sofar isolated which is composed of a single polypeptide. The possibilities to clarify
the molecular basis for energy transduction, i.e., in this
case the conversion of redox energy into a proton motive force, are therefore likely to be more favourable with transhydrogenase than with other systems.

Due to the ease by which transhydrogenase may be incorporated in liposomes it is also suitable for studies of energy transfer between different energy-linked proteins in mul-
ticomponent reconstituted liposome systems. Studies along
these two different lines of approaches aiming at the elucidation of the function of transhydrogenase and its interaction with other proteins have already been initiated.

ACKNOWLEDGEMENTS

This work was supported by the Swedish Cancer Society
and the Swedish Natural Science Research Council. The author
is indebted to Dr. Hans Jörnvall, Karolinska Institute, for
valuable advice and help with amino acid determinations, and
to Miss Elisabeth Carlenor for excellent technical assistance.

REFERENCES

1. Colowick, S.P., Kaplan, N.O., Neufeld, E.F. and Ciotti, M.M. (1952) J. Biol. Chem. 195, 95-105.
2. Danielson, L. and Ernster, L. (1963) Biochem. Z. 338, 188-205.
3. Ernster, L. and Lee, C.P. (1964) Ann. Rev. Biochem. 33, 729-788.
4. Ernster, L. (1977) Ann. Rev. Biochem. 46, 981-995.
5. Fisher, R.R. and Guillory, R.J. (1971) J. Biol. Chem. 15, 4687-4693.
6. Höjeberg, B. and Rydström, J. (1977) Biochem. Biophys. Res. Commun. 78, 1183-1190.
7. Anderson, W.M. and Fisher, R.R. (1978) Arch. Biochem. Biophys. 187, 1183-1190.
8. Rydström, J., Hoek, J.B. and Ernster, L. (1976) The Enzymes 13, 51-88.
9. Rydström, J. (1977) Biochim. Biophys. Acta 463, 155-184.
10. Rydström, J. and Ernster, L. (1981) in The Proton Cycle (V.P. Skulachev and P.C. Hinkle, eds.), Addison-Wesley, Advanced Book Program, in preparation
11. Höjeberg, B. and Rydström, J. (1979) Methods Enzymol. 55F, 275-283.
12. Persson, B. and Rydström, J., in preparation.
13. Rydström, J. (1976) Biochim. Biophys. Acta 455, 24-35.
14. Earle, S.R. and Fisher, R.R. (1980) J. Biol. Chem. 255, 827-830.
15. Fowler, A.V. and Zabin, I. (1978) J. Biol. Chem. 253, 5521-5525.
16. Earle, S.R., Anderson, W.M. and Fisher, R.R. (1978) FEBS Lett. 91, 21-24.
17. Anderson, W.M., Fowler, W.T. and Fisher, R.R. (1979) 11th Intern. Congress in Biochem., abstr. nr. 05-2-R109.
18. Jacobs, E. and Fisher, R.R. (1979) Biochemistry 18, 4315-4322.
19. Racker, E. (1979) Methods Enzymol. 55F, 699-711.
20. Rydström, J. (1979) J. Biol. Chem. 254, 8611-8619.
21. Earle, S.R. and Fisher, R.R. (1980) Biochemistry 19, 561-569.
22. Rydström, J. and Fleischer, S. (1979) Methods, Enzymol. 55F, 811-816.
23. Blazyk, J.F. and Fisher, R.R. (1975) FEBS Lett. 50, 227-232.
24. Hanson, R.L. (1979) J. Biol. Chem. 254, 888-893.
25. Homyk, M. and Bragg, P.D. (1979) Biochim. Biophys. Acta 571, 201-217.

26. Enander, K. and Rydström, J., in preparation.
27. Eytan, G.D. and Kanner, B.I. (1978) in Receptors and
    Recognition (P. Cuatrecasas and M.F. Greaves, eds.),
    Chapman and Hall, London, pp. 64-105.

SUBUNIT FUNCTIONS OF ATP-DRIVEN PUMPS[1]

Efraim Racker

Section of Biochemistry, Molecular and Cell Biology
Division of Biological Sciences
Cornell University, Ithaca, New York 14853

INTRODUCTION

The decision as to what subject to choose for this special occasion was not an easy one. Lars Ernster, my close friend for many years, has always amazed me with his broad knowledge and the variety of his research activities. They have ranged from human diseases (1) to toxic agricultural products (2), from physical-chemical studies on purified enzymes (3), mitochondria (4), and microsomes (5) to intact cells (6). All of his papers have one common denominator: a sharp focus on the fundamental questions. I have therefore chosen the fundamental problem of the subunit functions of ATPases because it allows me to touch upon

[1]Supported by NIH grant CA-08964, CA-14454, NSF grant BMS 7517887, and ACS grant BC-156.

Abbreviations: NEM, N-ethylmaleimide; NBD-Cl, 7-chloro-4-nitrobenzo-2-oxa-1,3-diazole; DCCD, N,N'-dicyclohexyl-carbodiimide.

C. P. Lee, G. Schatz, G. Dallner (eds.), Mitochondria and Microsomes
in honor of Lars Ernster   ISBN 0-201-04576-1

337

many areas of interest to our birthday child, including the
bioenergetics of Ehrlich ascites tumor cells (6).

There is a most remarkable similarity between the
structure and subunit composition of H^+ ATPases from animals,
plants and microorganisms (7-12). Yet there are some striking differences which I believe are relevant to our understanding of the structure and function of the pump and may
be relevant to our interpretation of evolutionary changes.
The subunit composition of the H^+ ATPase from bovine heart
mitochondria (13) is more complex than that of the enzyme
isolated from thermophilic bacteria (11). Significant differences in SH function of these two complexes have also
been observed which will be discussed later.

I shall refer to the catalytic ATPase as F_1 (14) and
to the other part of the complex as $F_{oligomycin}$ or F_0 (15)
as originally defined. Although this nomenclature has been
generally adopted (some bacteriologists refer to it as F_{zero}),
there is some confusion in the literature. For example, where
does OSCP belong? It is a water soluble protein which can be
stripped from the membrane by treatment with alkali (16) or
salt (17). It may be the stalk, or more likely, it is only
part of the stalk (13). Actually a stalk is not visible in
F_1-depleted membranes, but is readily visible in some F_1 preparations (18) which suggests that the δ subunit of F_1 is the

major contributor to the structure of the stalk. Moreover,

the chloroplast complex appears to be lacking a water-soluble

equivalent of OSCP (19) yet there is a visible stalk. Thus,

comparative biochemistry has aided us in drawing conclusions

with respect to the stalk.

SUBUNITS OF H^+ ATPases

The hydrophilic ATPase protein (F_1). The subunit structure

of mitochondrial F_1 (8,20), of chloroplast CF_1 (9,10,21), and

of bacterial F_1 (11,12) have been extensively discussed in

excellent reviews and I shall emphasize only some of the inter-

esting differences that are emerging. Let us start with

square 1 in Japan and look at data reproduced in Table I on

the thermophilic bacteria (11). For ATP hydrolysis, the β

subunit emerges as the center of activity. This is consistent

with data from chloroplasts (22), mitochondria (23), and other

bacteria (12). But the β subunit alone is not active as an

ATPase and evidence from reconstitution and genetic experi-

ments point to a role by the α and perhaps also by the γ sub-

unit as contributors either directly to the catalytic site or

to the reconstitution of an active catalytic site. There is

a clear distinction between these two possibilities which has

not as yet been resolved.

It becomes increasingly apparent that the active site

of the ATPase is complex. Chemical modifications with NBD-Cl

Table I.   Structures and Functions of Subunits of TF_1

Subunit	α	β	γ	δ	ε
Mol wt	54,600	51,000	30,200	21,000	16,000
α-Helix content (%)	31	34	49	65	33
β-Sheet content (%)	19	23	4	15	24
Cysteine content (mol/subunit)	1.0	0.0	0.0	0.0	0.0
Net ATP synthesis and H^+-transport required in reconstitution	(+)	(+)	(+)	(+)	(+)
P_i-ATP exchange Required in reconstitution	(+)	(+)	(+)	(+)	(+)
Inhibition by each antibody	(+)	(+)	(−)	(−)	(−)
ATP hydrolysis Required in reconstitution	(±)[a]	(+)	(±)[a]	(−)	(−)
Inhibition by each antibody	(+)	(+)	(−)	(−)	(−)
Nucleotide binding to isolated pure subunits					
ATP and ADP	(+)	(+)	(−)	(−)	(−)
ITP and IDP	(−)	(+)	(−)	(−)	(−)
CTP	(+)	(−)	(−)	(−)	(−)
H^+-gate activity	(−)	(−)	(+)	(+)	(+)
Direct binding to TF_o	(−)	(−)	(−)	(+)	(+)
N_3-sensitivity	(−)	(−)	(+)	(−)	(−)

[a]Not always required.

point to tyrosine (22,23), with phenylglyoxal to arginine (24, 25), with pyridoxal phosphate to lysine (26), with N-bromo-succinimide to tryptophan (27).   The β subunit is the target for NBD-Cl, but a clear picture for the unique location of the other chemical modifiers has not emerged as yet.

The function of the γ subunit has received considerable attention. The important observations of McCarty and Fagan (28) of the light-dependenct inhibition by NEM pointed to a key role of the γ subunit of CF_1 in photophosphorylation. On the other hand, the γ subunit of TF_1 does not have an SH group. Convincing evidence for a role of the γ subunit as a gate for the proton channel of F_O was presented by Kagawa (11). I shall return later to the role of a gate in the mechanism of ATP generation and the possible participation of the γ subunit. The δ subunit interacts with the F_O part of the complex. In the case of the chloroplasts (9,18), CF_1 without the δ subunit does not bind to the membrane, whereas in the case of bacterial F_1 the ϵ subunit has an attachment site as well (12).

I would like to focus now on the role of the ATPase inhibitor first isolated by Pullman and Monroy (29) from bovine heart mitochondria, which Lars Ernster has intensively studied (30). We have chosen to call a similar ATPase inhibitor from chloroplasts the ϵ subunit (31) because there is less ambiguity about its role in CF_1 than in either MF_1 or BF_1. The ϵ subunit was dissociated from CF_1 with detergents and shown to inhibit the ATPase activity of activated CF_1 (31).

We therefore propose to also call the inhibitor of bovine mitochondria (29) the ε subunit and to reserve this name for regulatory subunits with inhibitory activity. In the case of MF_1, the inhibitor (ε subunit) can be dissociated by treatment with salt or heat (32) and is therefore lost in the course of some purification procedures. Purified MF_1 contains another low molecular weight polypeptide referred to by some investigators as the ε subunit (20). It should be pointed out, however, that it has not been established that this polypeptide fulfills a specific function in the case of MF_1. If this polypeptide turns out to be an integral component of MF_1 and not a contaminant, we must conclude that this enzyme has 6 rather than 5 subunits. This was the conclusion drawn by Nelson (33) who proposed that this polypeptide should be referred to as ε' subunit, and that it might participate in the binding of F_1 to the membrane. In E. coli the "ε subunit" was actually shown to participate in the binding of F_1 to the membrane but has some other confusing properties. It inhibits the ATPase activity of E. coli F_1 in solution, but fails to inhibit when it is bound to the membrane (cf 12). If under physiological conditions this subunit indeed does not control ATP hydrolysis, I propose we should call it ε' and perhaps look further for a dissociable subunit ε which acts as an effective inhibitor of membrane-bound ATPase.

The major theme of my contribution to this birthday volume is the emphasis on the role of a regulatory subunit in the function of the ATPase complex. Such a role has been repeatedly emphasized by Ernster, his collaborators (30) and others (34,35). Of particular interest is their concept that the interaction of the inhibitor with F_1 is under the control of the electrochemical proton gradient.

It is rather confusing to students, who are well-indoctrinated with the laws of thermodynamics, that an inhibitor should inhibit an enzyme in one direction (hydrolysis) and not the other (synthesis). They have to learn about the pitfalls encountered in experiments in which the kinetic properties of the systems appear to contradict these laws. Many years ago we resolved such a mystery. An inhibitor (threose-2,4-diphosphate) inhibited the forward reaction catalyzed by glyceraldehyde-3-phosphate dehydrogenase but not the reverse reaction (36), because the substrate of the back reaction interfered with the formation of the enzyme-inhibitor complex.

What is the role of the mitochondrial and chloroplast inhibitor? Why is the inhibitor readily dissociated from the ATPase of mitochondria but not of chloroplasts? Perhaps there is a valid, functional reason for this difference. In mitochondria the supply of oxidizable substrates is continuous;

in light-dependent chloroplasts it is discontinuous. In the dark, little ATPase activity was detected in chloroplast thylakoids, while ATPase activity appeared on illumination in the presence of SH compounds and uncouplers (37). Dithiothreitol activates ATPase activity of soluble CF_1 (38). In intact chloroplasts, illumination mediated by a thioredoxin system activates the ATPase activity of CF_o-F_1 which is expressed in the presence of NH_4Cl (39). Although it is now clear that the activation process and expression of ATPase activity are two separable processes, everywhere we look we see a complex and reversible mechanism controlling ATPase activity.

Clearly, pumps and turbines should not be leaky if they are expected to function efficiently. Yet, when we look at MF_1 in solution this certainly looks like a very leaky enzyme. In the presence of ATP, water has free access to the active site and the precious pyrophosphate bond is cleaved delivering heat to its medium. On the other hand, this protein is part of a very efficient machine which either pumps protons at the expense of ATP hydrolysis (in some bacteria) or generates ATP by using an electrochemical proton gradient. Indeed it is the function of the ε subunit to stop the leak which could render the process of ATP synthesis inefficient. Why did nature design a separate polypeptide to shield the

active site from the illicit entry of water? Wouldn't it
have been more efficient to design a covalent structure that
is not dissociable? I shall return to this question in the
discussion of other pumps.

The "hydrophobic" sector (F_o). The least ambiguous component
of this sector is the proteolipid. Its chemical properties
have been extensively reviewed (8,11,12) and there is general
agreement that it functions as a proton channel. The most
persuasive demonstration of this function was achieved by
Nelson et al. (40) who isolated a pure polypeptide from
chloroplasts which was inserted into liposomes where it
serves as a DCCD-sensitive proton channel. But the same
investigators, as well as colleagues in my laboratory, have
failed thus far to isolate such a functional unit from mitochondria and similar failures were reported for TF_o from
thermophilic bacteria (11). In fact, we have performed experiments that show that another component of 28,000 daltons
(13) is required for a functional proton channel, and similar
conclusions were drawn from experiments with TF_o (11). What
is the role of these other components and what is the role
of OSCP and F_6 which are two well-established "coupling
factors" in mitochondria? Without going into detail, we can
assign to these latter two components a function for the

attachment of the extramembranous F_1 to the surface of the inner mitochondrial membrane.

We have recently described some of the properties of the 28,000 dalton polypeptide. It is required for the reconstitution of an active complex which catalyzes ATP synthesis driven by the electrochemical proton gradient generated by bacteriorhodopsin in the presence of light (13). We have concluded that this polypeptide is located at the surface of the membrane because of its sensitivity to trypsin. We have proposed that it functions as a gate that opens and closes the proton channel in a cyclic manner. This proposal is not based on hard data, since such a cyclic process is difficult to demonstrate experimentally. But we have identified the 28,000 dalton component (28 K) as the inhibitor of ATPase activity which we have known to exist ever since we observed that a preparation of F_oF_1 from mitochondria cannot hydrolyze ATP unless phospholipids are added (16). Thus, a reversible interaction between 28 K in the ATPase complex and phospholipid could be visualized as a gating mechanism. A reversible opening and closing of the channel is part of a mechanism of ATP synthesis according to a hypothesis which has been described elsewhere (16a). Finally, I want to return to the question raised earlier about the gate function of the γ subunit of TF_1. It is quite possible that in the more primitive thermophilic bacteria, the γ subunit fulfills the function

of a gate. It does this without SH groups. But how effi-

cient is this process and how is it controlled? The more

complex H^+ ATPase in mitochondria and chloroplasts may

utilize the γ subunit as a gate in addition to the 28 K com-

ponent of the hydrophobic sector. Perhaps there is a double

door in these complex structures. Why should the gating

function be so much simpler in the thermophilic ATPase com-

plex? I believe that increase in efficiency and regulation

are the two major reasons for the increasing complexity we

see in evolution. This thesis is subject to an experimental

test. Is the thermophilic H^+ ATPase less efficient in ATP

synthesis or less rigidly controlled than its mammalian or

chloroplast counterpart?

SUBUNITS OF THE Ca^{2+} ATPase

Purified Ca^{2+} ATPase of sarcoplasmic reticulum con-

tains 2 proteins: a 100,000 dalton protein which contains

the active site of the ATPase and a proteolipid (41). The

latter has been reported to increase the efficiency of pump-

ing (42) but it has been difficult to demonstrate this con-

vincingly because, as in the case of the mitochondrial proteo-

lipid, we have not succeeded in preparing a pure proteolipid

in a biologically active form. Recently, we have isolated

the 100,000 dalton protein free of proteolipid (43). After

reconstitution into liposomes, the protein catalyzed ATP-

driven Ca^{2+} uptake. However, the efficiency of pumping as

measured by the ratio of Ca^{2+} translocated/ATP hydrolyzed

was low (ca 0.6). We proposed (43) that "the proteolipid

may increase the efficiency of pumping by contributing to

the formation of the transmembranous channel either by aiding

in its assembly or by participating as a functional subunit."

Proteolipids isolated from sarcoplasmic reticulum by solvent

extraction served as ionophores in reconstituted vesicles,

but were not functional as subunits of Ca^{2+} ATPase complex

involved in ATP-dependent Ca^{2+} translocation (43). This

again is reminiscent of our experiences with the proteolipid

from bovine heart mitochondria. Although it is inactivated,

even by short exposure to solvents, it retains protonophore

activity which is insensitive to oligomycin (44).

SUBUNITS OF THE Na^+K^+ ATPase

Numerous reviews and books have been published on the

Na^+K^+ ATPase of the plasma membrane (see 45). It is generally

agreed that there are two subunits present in purified prepar-

ations: A 100,000 dalton protein (α subunit) which contains

the active site and a smaller subunit (37,000 - 56,000) which

is a glycoprotein (β subunit). There is no information

available about the function of this subunit. Virtually all
preparations of the ATPase also contain a proteolipid which
was reported to interact with a radioactive ouabain derivative
(49) and to undergo phosphorylation (50).

In the course of studies on the mechanism of the high
aerobic glycolysis in malignant tumor cells, we observed that
the Na^+K^+ pump of the Ehrlich ascites tumor cells operates
inefficiently (46,47) thus supplying ADP and P_i required for
glycolysis. Quercetin and other bioflavonoids appeared to
repair the defect. Recently, the ATPase from Ehrlich ascites
tumor cells was isolated and incorporated into liposomes. The
pump operated inefficiently and quercetin greatly increased
the Na^+ transported/ATP hydrolyzed ratio (48). The reason
for the low efficiency of pumping was traced to the phosphorylation of the β subunit of the Na^+K^+ ATPase by a protein
kinase (PK_M) which in the presence of ATP and Mg^{2+} phosphorylated a tyrosine residue of the protein (51). PK_M had to be
phosphorylated to be active and the kinase responsible for
its activation was called PK_S. This enzyme was dependent on
phosphorylation by PK_L which in turn was phosphorylated by
PK_F. Thus, a cascade of 4 enzymes that all became phosphorylated on a tyrosine residue was needed for the phosphorylation
of the β subunit of the Na^+K^+ ATPase (52). PK_F was shown to
be immunologically related to the sarc gene product (53) of

Rous sarcoma virus transformed cells which had been shown to
be a protein kinase (54).

How does phosphorylation of the β subunit affect its
efficiency? Why does quercetin repair the defect? What is
the physiological meaning of such regulatory mechanisms?

Although we do not have direct answers to these questions, there are numerous clues particularly from studies of
the proton pump of mitochondria and chloroplasts. The ε subunit of these ATPases prevents the hydrolysis of ATP. When
the enzymes were treated with trypsin, hydrolysis took place
at a rapid rate. In the case of CF_1 the addition of the ε
subunit to the trypsin-treated enzyme failed to inhibit ATP
hydrolysis because the γ subunit, which is required for the
interaction with the ε subunit, was digested also. However,
quercetin inhibited even the digested enzyme (22). Thus,
the bioflavonoids seem to have a direct access to the site
where ATP is hydrolyzed. We propose that the β subunit of
the Na^+K^+ ATPase shields the active site of the α subunit
against the illicit entry of water and is physically displaced when it becomes phosphorylated. Thus, a leak is introduced which permits hydrolysis of ATP that is not associated
with ion transport. As in the case of the H^+ ATPase, quercetin
still has access to the active site of the Na^+K^+ ATPase and
inhibits the hydrolysis of the phosphorylated intermediate

$E_2 \sim P$. This was illustrated directly in studies of the Na^+-K^+ ATPase from electric eel (55). Quercetin at low concentrations had no effect on the formation of $E_1 \sim P$ from ATP, but served as a potent inhibitor of the hydrolysis of $E_2 \sim P$.

Why has nature designed a regulatory subunit that can be damaged by phosphorylation? Why do normal as well as tumor cells contain PK_M, an enzyme which renders the pump inefficient? Again, we might take a clue from the mitochondria. It is well known that in some animals uncoupled brown fat mitochondria serve as heating ovens during cold exposure. Could an inefficient Na^+ pump serve a similar function, indeed be a much more effective heating device since all mammalian cells contain this pump? A role for the Na^+ pump in thermogenesis was proposed earlier (cf 56) but without a thermostat. The β subunit, finely regulated by phosphorylation and dephosphorylation, could serve in such a function.

Now we have to analyze the fine tuning mechanism which allows for a high efficiency of pumping in normal cells and which appears to be damaged in tumor cells. Although the details of this tuning mechanism need further study, it appears that cyclic nucleotides play a critical role (57).

CONCLUSIONS

There are several functions associated with the

operation of ATP-driven ion pumps. 1) There is an active
site which serves as energy transformer. 2) There is a channel that allows transmembranous movements of ions. 3) There
may be a cyclic opening and closing of the channel. 4) There
is regulation of water entry to the active site of the enzyme.

The delineation of these functions with reference to
specific subunits of the complex protein is sharp in the case
of the ATPase-driven H^+ pumps. The proteolipid is the channel;
the α and β subunits form the active site; the γ subunit and
the 28,000 dalton protein appear to function as gate; and the
ε subunit is involved in regulation. The other coupling factors are connecting components between the transformer and
the channel.

In the case of the Ca^{2+} and Na^+K^+ pump, these functions are not as clearly separated; the unit is more compact.
The role of the proteolipid, which is usually found in association with the pump, is not clearly defined.

The major emphasis of this review is the role of
regulatory subunits. I propose that a major function of the
ε subunit of the H^+ ATPase and the β subunit of the Na^+K^+
ATPase is the shielding of the active site from illicit entry
of water. In the case of the Na^+K^+ pump, phosphorylation of
the β subunit at a tyrosine residue renders the pump inefficient. The inefficient pump operation is responsible for the

high aerobic glycolysis of some tumor cells and therefore

also for the excess heat production. In normal cells the

phosphorylation of the β subunit is inhibited but it is con-

ceivable that this control may be released during cold ex-

posure or in diseases.

REFERENCES

1. Ernster, L., Ikkos, D. and Luft, R. (1959) Nature 184, 1851-1854.
2. Heijkenskjöld, L. and Ernster, L. (1975) Acta Med. Scand. 585, Supplem. 75-83.
3. Rase, B., Bartfai, T. and Ernster, L. (1976) Arch. Biochem. Biophys. 172, 380-386.
4. Ernster, L. (1974) 9th FEBS Mtg., Plenary Lecture, Budapest, Aug. 24, 1974.
5. Ernster, L., Siekevitz, P. and Palade, G.E. (1962) J. Cell Biol. 15, 541-562.
6. Gordon, E.E., Ernster, L. and Dallner, G. (1967) Cancer Res. 27, 1372-1377.
7. Racker, E. (1979) in Membranes of Mitochondria and Chloroplasts (E. Racker, ed.) pp. 127-171, Van Nostrand Reinhold Co., New York.
8. Senior, A.E. (1973) Biochim. Biophys. Acta 301, 249-277.
9. Baird, B.A. and Hammes, G.G. (1979) Biochim. Biophys. Acta 549, 31-53.
10. Nelson, N. (1976) Biochim. Biophys. Acta 456, 314-338.
11. Kagawa, Y. (1980) J. Memb. Biol. 55, 1-8.
    "   (1978) Biochim. Biophys. Acta 505, 45-93.
12. Futai, M. and Kanazawa, H. (1980) in Current Topics Bioenerg. (Sanadi, D.R. ed.) in press.
13. Alfonzo, M. and Racker, E. (1979) Canadian J. Biochem. 57 1351-1358.
14. Pullman, M., Penefsky, H., Datta, A. and Racker, E. (1960) J. Biol. Chem. 235, 3322-3329.
15. Racker, E. (1963) Biochem. Biophys. Res. Commun. 10, 435-439.
16. Kagawa, Y. and Racker, E. (1966) J. Biol. Chem. 241, 2461-2482.
16a. Racker, E. (1977) in Calcium Binding Proteins and Calcium Function (R.H. Wasserman et al. eds.) pp. 155-163, Elsevier North-Holland, Amsterdam.

17. MacLennan, D.H. and Tzagoloff, A. (1968) Biochem. 7, 1603-1610.
18. Younis, H.M., Telford, J.N. and Koch, R.B. (1978) Pestic. Biochem. Physiol. 8, 271-277.
19. Pick, U. and Racker, E. (1979) J. Biol. Chem. 254, 2793-2799.
20. Penefsky, H.S. (1974) in The Enzymes (P.D. Boyer, ed.) pp. 375-394, Acad. Press, New York.
21. McCarty, R.E. (1978) in Current Topics Bioenerg. (D.R. Sanadi, ed.) 7, 245-278.
22. Deters, D.W., Racker, E., Nelson, N. and Nelson, H. (1975) J. Biol. Chem. 250, 1041-1047.
23. Ferguson, S.J., Lloyd, W.J. and Radda, G.K. (1974) FEBS Lett. 38, 234-236.
24. Markus, F., Schuster, S.M. and Lardy, H.A. (1976) J. Biol. Chem. 251, 1775-1780.
25. Schmid, R., Jagendorf, A. and Hulkaver, S. (1977) Biochim. Biophys. Acta 462, 177-186.
26. Sugiyama, Y. and Mucohata, Y. (1979) FEBS Lett. 98, 276-280.
27. Risi, S., Höckel, H., Hulla, F.W. and Dose, K. (1977) Eur. J. Biochem. 81, 103-109.
28. McCarty, R.E. and Fagan, J. (1973) Biochem. 12, 1503-1507.
29. Pullman, M.E. and Monroy, G.C. (1963) J. Biol. Chem. 238 3762-3769.
30. Gómez-Puyou, A., Gómez-Puyou, M.T. and Ernster, L. (1979) Biochim. Biophys. Acta 547, 252-257.
31. Nelson, N., Nelson, H. and Racker, E. (1972) J. Biol. Chem. 247, 7657-7662.
32. Horstman, L.L. and Racker, E. (1970) J. Biol. Chem. 245, 1336-1344.
33. Nelson, N. (1980) in Current Topics of Bioenerg. (D.R. Sanadi, ed.) in press.
34. Van de Stadt, R.J., de Boer, B.L. and van Dam, K. (1973) Biochim. Biophys. Acta 392, 338-349.
35. Harris, D.A. and Crofts, A.R. (1978) Biochim. Biophys. Acta 502, 87-102.
36. Racker, E., Klybas, V. and Schramm, M. (1959) J. Biol. Chem. 234, 2510-2516.
37. Petrack, B., Craston, A., Sheppy, F. and Farron, F. (1965) J. Biol. Chem. 240, 906-914.
38. McCarty, R.E. and Racker, E. (1966) Brookhaven Symposia in Biology: No. 19, New York.
39. Mills, J.D., Mitchell, P. and Schürmann, P. (1980) FEBS Lett. 112, 173-177.
40. Nelson, N., Eytan, E., Notsani, B., Sigrist, H., Sigrist-Nelson, K. and Gitler, C. (1977) Proc. Natl. Acad. Sci. USA 74, 2375-2378.

41. MacLennan, D.H., Yip, C.C., Iles, G.H. and Seaman, P. (1972) Cold Spring Harbor Sym., Quant. Biol. 37, 469-478.

42. Racker, E. and Eytan, E. (1975) J. Biol. Chem. 250, 7533-7534.

43. Knowles, A., Zimniak, P., Alfonzo, M., Zimniak, A. and Racker, E. (1980) J. Membrane Biol. 55, 233-239.

44. Racker, E. (1975) in Proceedings of International Symposium on Electron Chains and Oxidative Phosphorylation (E. Quagliariello, et al. eds.) pp. 401-406, Fasano, Italy, North-Holland Publishing Co., Amsterdam.

45. Skou, J.C. and Nørby, J.G., eds., (1979) "Na,K-ATPase Structure and Kinetics," pp. 1-549, Academic Press, New York.

46. Scholnick, P., Lang, D. and Racker, E.(1973) J. Biol. Chem. 248, 5175-5182.

47. Suolinna, E-M., Buchsbaum, R.N. and Racker, E. (1975) Cancer Research 35, 1865-1872.

48. Spector, M., O'Neal, S. and Racker, E. (1980) J. Biol. Chem. 255, 5504-5507.

49. Forbush, B., Kaplan, J.H. and Hoffman, J.F. (1978) Biochem. 17, 3667-3676.

50. Reeves, A.S., Collins, J.H. and Schwartz, A. (1980) Biochem. Biophys. Res. Commun. 95, 1591-1598.

51. Spector, M., O'Neal, S. and Racker, E. (1980) J. Biol. Chem. 255, 8370-8373.

52. Spector, M., O'Neal, S. and Racker, E. (1980) J. Biol. Chem., in press.

53. Hanafusa, H. (1977) in Comprehensive Virology (H. Fraenkel-Conrat and R.R. Wagner, eds.) Vol. 10, pp. 401-483, Plenum Press, New York.

54. Collett, M.S. and Erikson, R.L. (1978) Proc. Natl. Acad. Sci. USA 75, 2021-2024.

55. Kuriki, Y. and Racker, E. (1976) Biochemistry 15, 4951-4956.

56. Guernsey, D.L. and Stevens, E.D. (1977) Science 196, 908-910.

57. Spector, M. and Racker, E., in preparation.

THE UPTAKE AND THE RELEASE OF CALCIUM BY MITOCHONDRIA

Ernesto Carafoli

Laboratory of Biochemistry, Swiss Federal Institute of
Technology (ETH), 8092 Zurich, Switzerland

The process of mitochondrial Ca^{2+} transport was discovered (1,2) at a time when much less was known on the role of Ca^{2+} in intracellular biochemical regulation than is known today. As a result, the interest in the process remained somewhat limited for a number of years. With the explosive growth of interest in the regulatory and messenger function of Ca^{2+}, and on ways and means to transport it across membrane boundaries, a revival of interest and activity in the mitochondrial transport process soon developed. Observations that had remained obscure for a long time were rapidly understood, important variables of the reaction were defined with precision, novel concepts were proposed that made the general picture of the process more rational.

In this overview, the travel of Ca^{2+} across the inner membrane will be decomposed into its uptake and release directions. This is somewhat artificial, since the uptake and release of Ca^{2+} are necessarily linked together in what has been termed the "Ca^{2+} cycle" (3). But the separation will help the understanding of the properties of the two oppo-

C. P. Lee, G. Schatz, G. Dallner (eds.), Mitochondria and Microsomes
  in honor of Lars Ernster   ISBN 0-201-04576-1

sing portions of the process, and it will stress the concept that the entrance and the exit of Ca^{2+} occur via different routes. The emergence of the latter concept (4,5) is one of the most important recent development in the field.

THE UPTAKE OF Ca^{2+} BY MITOCHONDRIA

The penetration of Ca^{2+} into mitochondria can be conveniently described as a passive process, in which Ca^{2+} travels across the membrane electrophoretically, pulled by the negative electrical potential maintained inside mitochondria by coupled respiration. That the process is purely electrophoretic, i.e., that Ca^{2+} enters mitochondria with 2 positive charges, without electrical compensation, has been first proposed by Rottenberg and Scarpa (6) on the basis of experiments in which the equilibrium distribution of Ca^{2+} across the inner membrane of energized mitochondria was compared with that of K+, which, in the presence of valinomycin, penetrates electrophoretically. The conclusion by Rottenberg and Scarpa has been supported by evidence from different experimental lines (7,8), and is now generally accepted. Partial charge compensation by H^+ antiport (9) or by anion symport (10,11) is negated by experimental evidence, although the possibility of a symport of Ca^{2+} with β-hydroxybutyric acid (11) has not been ruled out.

The penetration of Ca^{2+} into mitochondria is inhibited specifically by ruthenium red (12,13), a polycation that interacts with an essential site in the Ca^{2+} uptake system (14). It is also inhibited by lanthanides (15). The lantha-

nide inhibition is ionic-radius related, and is maximal
with Tm $^{3+}$ (16). Of interest from a physiological standpoint is the inhibition of the uptake process by concentrations of Mg^{2+} which are normally present in most cytosols
(17 - 20). When present in the reaction medium, Mg^{2+} transforms the kinetics of the uptake process from hyperbolic to
sigmoidal (20), i.e., it markedly inhibits the uptake at
low Ca^{2+} concentrations. In liver mitochondria, however,
the kinetics of the uptake process is still sigmoidal in
the absence of Mg^{2+} indicating that the operation of the
Ca^{2+} uptake carrier may in this case be limited by factors
different from Mg^{2+}. When the "carrier" is studied under
non-limiting conditions, its kinetics becomes hyperbolic
also in liver mitochondria (21).

   Related to the kinetics of the uptake process is the
matter of the affinity of mitochondria for Ca^{2+}. There is
now general agreement that the uptake reaction operates,
under optimal conditions, with a K_m of between 2 and 13µM,
and a V_{max} of between 3 and 10 nmol per mg of protein per
sec (4). These values indicate that mitochondria may not be
adequate to control processes that require the rapid mobili-
zation of Ca^{2+} from ambients where its activity is lower
than 1 µM, e.g., the contraction-relaxation of muscle. In
addition, it must be kept in mind that, under the conditions
presumably prevailing in situ, the rate of Ca^{2+} uptake is
likely to be slowed down by natural inhibitors like Mg^{2+}.
It appears more probable that the function of mitochondria
is to buffer cytosolic Ca^{2+} so that its level will never
deviate from the limits compatible with the proper functioning of the cell (other concepts related to the overall
Ca^{2+}-buffering role of mitochondria will be discussed below).

Efforts have been made in several Laboratories to resolve the Ca^{2+} uptake system into its molecular components. At the present date, two fractions appear to have been characterized in detail, and to be involved in the uptake reaction. One is a water soluble glyco-lipoprotein (22), which is loosely bound, at least in part, to the inner membrane. This protein binds Ca^{2+} with high affinity, and becomes associated with phospholipid bilayers only in the presence of Ca^{2+} (23). It has been shown that an antibody specific for this protein abolishes the uptake of Ca^{2+} by mitochondria (24), and it has been proposed that the function of this glycoprotein is to provide a superficial "recognition site" for Ca^{2+} (25).

The other fraction is a small molecular weight polypeptide (calciphorin) (26), which binds the Ca^{2+}-analogue Mn^{2+} with high affinity, and specifically increases the permeability to Ca^{2+} of artificial phospholipid bilayers. It has also been shown that the effects of the polypeptide on bilayers is counteracted by ruthenium red.

Whether the two protein factors mentioned are related to one another is not known. Unknown is also their exact role in the permeation of Ca^{2+}, particularly with respect to the transport of the cation through the apolar phase of the membrane. Recent evidence (27) indicates that the Ca^{2+} binding glycoprotein may be involved <u>also</u> in the release of Ca^{2+} from mitochondria (see below).

THE RELEASE OF Ca^{2+} FROM MITOCHONDRIA

The electrophoretic Ca^{2+} uptake pathway operates against a potential of 180 m v, and must thus be regarded as essentially irreversible. Indeed, its reversal would require substantial oscillations of the membrane potential, a highly improbable event, since the maintenance of the membrane potential is central to the functioning of the organelle. Since Ca^{2+} _must_ neverthless be released from mitochondria to prevent their calcification, the postulation of (an) indepent Ca^{2+} release pathway(s) has in effect no alternative. As mentioned, the concept of separate uptake and release pathways came relatively late. Independent pathways have been first demonstrated by Carafoli and Crompton (4) and Puskin et al. (5) in 1976, but were already implicit in an earlier experiment by Rossi et al.(28), in which ruthenium red was shown to permit, and indeed to promote, the efflux of Ca^{2+} from mitochondria. Clearly, under these conditions Ca^{2+} _had_ to leave mitochondria on a route different from the electrophoretic uptake uniporter, which was blocked by ruthenium red. Thus, the observation by Rossi et al. (28) has provided the experimental tool which is routinely used today whenever it becomes necessary to decide whether Ca^{2+} leaves mitochondria on the (reversed) uptake uniporter, or independently. Insensitivity to ruthenium red has now become accepted as the decisive argument in favor of an independent route.

THE Na^+-PROMOTED RELEASE ROUTE

In 1973, Carafoli et al (29) reported that Na^+ could release the accumulated and the endogenous Ca^{2+} from heart mitochondria. In these early observations rather high concentrations of Na^+ were employed, and thus no conclusions were possible on the physiological significance of the phe-

nomenon. The experiment, however, clearly indicated a specific role of Na^+ in releasing Ca^{2+}. A variety of other monovalent cations were indeed tested, but they had no releasing
effects.

The original observation by Carafoli et al. (29) was studied in considerable detail in a series of investigations
carried out by Carafoli, Crompton, and their associates (8,
30, 31-33), and in a number of other Laboratories (34-37).
The process appears now established as an important and well
characterized means to release Ca^{2+} from the majority of the
mitochondrial species, and its physiological significance
appears very probable. Here, a summary will be given of its
most significant features.

It is seen with particular evidence in mitochondria from
heart, brain, adrenal cortex, and is almost absent from some
mitochondrial types, notably kidney and liver. Half maximal
rate of Ca^{2+} release is observed at 6-8 mM external Na^+, and
the maximal rate of release is about 0.25 n moles of Ca^{2+}
per mg of protein per sec. This rate is about 10 times slower than that of the uptake process: to achieve a balance
between uptake and release, either the Na^+-induced release
route must thus be stimulated, or the electrophoretic uptake
uniporter must be inhibited. The latter possibility seems
more realistic, since it has been seen already that Mg^{2+} does
indeed slow down substantially the uptake route. It is perhaps useful to stress again at this point that the uptake
and release routes must in the long run be balanced,to prevent excess accumulation or deprivation of Ca^{2+}. This is the
concept of the "mitochondrial Ca^{2+} cycle" (3) which has been
mentioned avove, and which is, the most logical way of describing the process of mitochondrial Ca^{2+} transport.

The Na-activated pathway is insensitive to ruthenium red, inhibited by lanthanides (maximal inhibition by La^{3+}) (16) and by Mg^{2+} (35), and stimulated by K^+ (33). As required of a route that operates against a negative membrane potential, it is electroneutral (38). As for its mechanism, it could either be explained with a Ca^{2+}/Na^+ antiporter, or with a H^+/Ca^{2+} antiporter, in both cases coupled to the operation of the H^+/Na^+ antiporter, which is known to exist in mitochondria (39). The experimental evidence available, however, (34), strongly suggests the first possibility, i.e., the existence of a Ca^{2+}/Na^+ exchange-diffusion carrier. Indeed, in the case of a H^+/Ca^{2+} exchange the role of added Na^+ would be to dissipate the excess matrix H^+. However, substitution of the natural Na^+/H^+ antiporter with the artificial K^+/H^+ antiporter nigericin, in a K^+ ambient, fails to elicit the rapid release of Ca^{2+} that would have been expected in the case of a direct H^+Ca^{2+} exchange.

THE RELEASE OF Ca^{2+} FROM Na^+-INSENSITIVE MITOCHONDRIA

As mentioned, some mitochondrial types, notably liver, kidney cortex, lung, smooth muscle, are insensitive to Na^+. Actually, a recent report (36) has shown Na^+-induced Ca^{2+} release also in some of these mitochondria, but the activities observed are marginal with respect to heart or brain. Yet, Na^+-insensitive mitochondria must release Ca^{2+} by way of a membrane-potential-independent process, like their heart and brain counterparts. Several mechanisms have been described. It is fair to say, however, that none of them has been characterized in nearly as much detail, or has reached the same degree of probability as a physiological route, as the Na^+-promoted pathway. A synthetic description will now be given of those, among the mechanisms proposed, that have been better documented experimentally. For the reasons mentioned above, processes linked to the obvious collapse of the transmembrane potential, for example the Ca^{2+} release induced by phosphoenolpyruvate (10), will not be considered.

a- FATTY ACIDS AND THE RELEASE OF Ca^{2+}. That fatty acids, especially polyunsaturated fatty acids, can release Ca^{2+} from mitochondria, is to be expected from their uncoupling action. In three studies, however, sub-uncoupling concentrations of fatty acids have been used, and Ca^{2+} release has still been observed. Two of these studies have employed prostaglandins (41,42), and have lead the Authors to conclude that the Ca^{2+} releasing action is unrelated to the hormonal properties of the compounds, but related to their fatty acid character. The suggestion has been made that prostaglandins would act as H^+ and Ca^{2+} ionophores, crossing the membrane in one direction with the former species, and in the other direction with the other. The third study (43) which was performed on both kidney and liver mitochondria, has employed a large number of saturated and unsaturated fatty acids, and has shown a very evident Ca^{2+}-releasing effect. The potency of the Ca^{2+}-releasing action was higher in the more unsaturated fatty acids, e.g., arachidonic acid. Most interestingly, it was found that Na^+ counteracted their Ca^{2+}-releasing action. This observation differentiates very clearly the process from that promoted by Na^+ in heart and brain mitochondria. The possible physiological role of fatty acids in releasing Ca^{2+},however,remains an open problem   Free fatty acids are likely to be present in the inner mitochondrial membrane, which contains a very active, Ca^{2+}-stimulated, phospholipase activity (44). They could, therefore, theoretically act as Ca^{2+}-releasers in vivo. One evident difficulty is the Ca^{2+}-specificity of the releasing effect, which is not easily rationalized by mechanisms based on a "mobile ionophore" effect of fatty acids. Other possibilities, based for example on changes of the fluidity of the membrane induced by fatty acids, can be envisaged.Clearly, more experiments are necessary to place the role of unsaturated fatty acids in the correct physiological perspective.

b- Ca^{2+}-RELEASE INDUCED BY ALTERATIONS IN THE REDOX BALANCE

OF PYRIDINE NUCLEOTIDES. Recent experiments by Lehninger et al. (45) have indicated that the redox state of the mitochondrial pyridine nucleotides might regulate the Ca^{2+} balance between liver mitochondria and medium. Conditions leading to the oxidation of NAD(P)H, e.g., the addition of acetoacetate, induce Ca^{2+} release, re-reduction of $NAD(P)^+$, e.g., by addition of β-hydroxybutyrate, induce re-uptake of the lost Ca^{2+}, and its retention. These observations have recently been extended by Lötscher et al. (46, 47) who have found that a possibly important physiological factor in the modulation of the redox state of pyridine nucleotides,and thus of the Ca^{2+} balance between liver mitochondria and medium, is the concerted action of the 2 matrix enzymes,glutathione peroxidase and glutathione reductase. They have also found that the loss of Ca^{2+} from mitochondria consequent upon oxidation of the pyridine nucleotides is linked to the disappearance of the latter from the organelle, and to the specific appearance of nicotinamide in the external medium. The Ca^{2+} release induced by the oxidation of NADP(H) is reversible, and so is, under specified experimental conditions, the related loss of nicotinamide. This suggests that the phenomenon may be physiologically relevant, and not linked to irreversible damage to the organelles. This view, however, has recently been challenged by Nicholls and Brand (48), on the basis of experiments in which a series of agents known to protect the mitochondrial structure against non-specific damage (BSA,ADP), abolish the release of Ca^{2+} induced by the oxidation of pyridine nucleotides.

c- THE ROLE OF INORGANIC PHOSPHATE IN RELEASING Ca^{2+} FROM

LIVER MITOCHONDRIA. That phosphate can promote the loss of accumulated Ca^{2+} is known since 1964, when Rossi and Lehninger (49) observed that the respiration of liver mitochondria became irreversibly activated in the presence of Ca^{2+} and phosphate, whereas in the absence of phosphate the stimulation of respiration was reversible. They correctly interpreted their observation as a phosphate-promoted, continuous loss of the accumulated Ca^{2+}, and attributed it to a "deleterious" effect of inorganic phosphate. The nature of the "deleterious" effect of phosphate remained unspecified, but in retrospect it is interesting to note that the conditions employed by Rossi and Lehninger had not damaged mitochondria too profoundly, because Ca^{2+} could evidently still be continuously taken up. This is an important point, because it has frequently been assumed in the literature that inorganic phosphate, when presented to liver mitochondria together

with Ca^{2+}, damages the organelles irreversibly and unspeci-
fically. While this may be true under extreme conditions,
in which mitochondria are gorged with massive amounts of
Ca^{2+} and phosphate, the assumption that phosphate invariably releases Ca^{2+} via structural damage, and not because it
activates a specific release pathway, is most likely an
oversimplification. And it may be interesting to note at
this point that even after the accumulation of massive
amounts of Ca^{2+} and phosphate liver mitochondria may remain
relatively intact structurally and functionally, if they
are permitted to accumulate in the matrix net amounts of
adenine nucleotides (50).

The nature of the release pathway activated by inorganic
phosphate under the conditions of Rossi and Lehninger (49)
remained unspecified. Inhibitors of the uptake of Ca^{2+},
which would not affect the energy state of the mitochondrion,
were unavailable in 1964. As a result, it could not be de-
cided whether Ca^{2+} left mitochondria by a specific release
route, or on the electrophoretic uptake uniporter, reversed
by a limited decrease of the transmembrane potential.

The problem of the release of Ca^{2+} by liver mitochondria
has recently been investigated in detail by Roos et al.(51)
who have used phosphate-depleted organelles to avoid possible complications arising from the presence of the rather
large amounts of inorganic phosphate which are normally
present. They have observed that phosphate-depleted mitochondria accumulate pulses of Ca^{2+}, and retain them, whereas they lose the accumulated Ca^{2+} rapidly when presented
with both Ca^{2+} and phosphate. Evidently, then, phosphate
is instrumental in promoting the release of Ca^{2+}, and it
was also shown by Roos et al. (51) that the transport of

phosphate across the inner membrane is necessary for the Ca^{2+}-releasing effect, since the inhibitor of phosphate transfer NEM eliminates it. Other observations of interst were made in the study by Roos et al. (51). Ruthenium red did not inhibit the release of Ca^{2+}, implying the operation of a route different from the uptake uniporter. Phosphate and Ca^{2+} did not move out of mitochondria simultaneously, ruling out earlier tentative suggestions (52) of a Ca-phosphate symport as a possible mechanism for the release of Ca^{2+} from liver mitochondria. The most likely possibility thus remain that of an independent Ca^{2+} releasing route (a Ca^{2+}-H^+ exchange?), somehow activated by inorganic phosphate. Another interesting aspect of the study by Roos et al. (51), possibly the most interesting in view of the above mentioned historical controversy on the damaging effects of phosphate on liver mitochondria, is that during the cycle of Ca^{2+} uptake and release in the presence of phosphate the membrane potential remains approximately the same as in the absence of phosphate. It can be thus safely concluded that the releasing effect of inorganic phosphate, albeit obscure in mechanism, is not mediated by unspecific structural damage to the mitochondria.

OTHER POSSIBLE MECHANISMS FOR Ca^{2+}-RELEASE. Heart mitochondria release Ca^{2+} via the specific Ca^{2+}-Na^+ antiporter described above. They can, however, also release it by way of a Ca^{2+}-induced membrane transition (53,54), which is a Ca^{2+}-dependent sudden increase in the permeability to Ca^{2+} promoted by incubation in the presence of a variety of agents, among them unsaturated fatty acids and phosphate.

The membrane transition, and the associated loss of Ca^{2+}, are associated with induction of ATPase, uncoupling of oxidative phosphorylation, and loss of respiratory control. It is not inhibited by ruthenium red, suggesting once again a pathway independent of the uptake mechanism, and evidence has been presented that it is not related to the Na^+-promoted release process. Whether the loss of Ca^{2+} induced by the membrane transition is related to the other, above described, mechanisms for releasing Ca^{2+} from Na^+-insensitive mitochondria, e.g., the release mediated by unsaturated fatty acids, or by the oxidation of mitochondrial pyridine nucleotides, is an open problem. It is certainly interesting that the Ca^{2+}- dependent membrane transition can be induced by oleic acid, and can be accelerated by conditions leading to the oxidation of NADH (53,54).

The point has been made above that the release of Ca^{2+} induced by the collapse of the transmembrane potential, i.e., by reversal of the electrophoretic uptake uniporter, is uninteresting physiologically, since it is unlikely that the transmembrane potential will ever be allowed to collapse in vivo. However, as Nicholls has pointed out (55) the activities of the uptake and release pathways, whatever the latter may be in liver mitochondria, are finely balanced, producing a final steady state for the transmembrane Ca^{2+}-distribution across the inner membrane, which results in the maintenance of an activity of Ca^{2+} outside mitochondria of about 0.8 μM (55). This is a true steady state, since deviations of the external Ca^{2+} activity result in net uptake or release of Ca^{2+}, to restore the original steady state balance. As long as the transmembrane electrical potential is above a minimal value, calculated at about 130 m v by

Nicholls (55), the transmembrane Ca^{2+} distribution does not vary. It the transmembrane potential decreases below this level, the external activity of Ca^{2+} becomes variable, and is now determined by the thermodynamic equilibrium of Ca^{2+} across the uptake uniporter. Whether the transmembrane potential will ever be allowed to fluctuate in vivo below 130 mv is, however, an open question.

CONCLUDING REMARKS. A considerable amount of phenomenology is now available on the process(es) by which Ca^{2+} traverses the inner membrane, and it can certainly be said that much more is understood on the matter, now, than it was only 3 or 4 years ago. As already pointed out in the Introduction section, the field has experienced a surge of progress in recent times, after a period of relative dormancy. Yet, some very fundamental questions still remain unanswered. Some pertain to purely molecular aspects of the process, like the nature of the Ca^{2+} carrier, its possible multiplicity, the mechanism of its functioning. Other unanswered questions are more general in nature, for example whether the various release pathways described have physiological relevance, and whether they are indeed completely different or converge into a final common mechanism. But the Key unanswered question is undoubtedly that of the physiological significance of the mitochondrial Ca^{2+} transport process, and, as a corollary, of its regulation in vivo. This problem has occupied investigators from the very beginning, and has continued to do so ever since. The opinion has been expressed in this chapter that mitochondria act as long-term, Ca^{2+}-buffering structures, and do not play a <u>direct</u> role in the regulation of very fast Ca^{2+} requiring processes, which are turned on and off by variations in Ca^{2+} activity below the μM range. This seems to be nowadays the prevailing view, but it is fair to say that it rests mostly on experiments and conclusions relating to the contraction and relaxation of (heart) muscle. A variety of other cytoplasmic processes require modulations of Ca^{2+} within a time scale which is less rapid, and possibly in an activity range which is higher, than muscle contraction and relaxation. Here, mitochondria may well play a <u>direct</u> role.

Finally, a novel concept, which is now beginning to emerge, deserves to be emphasized. Mitochondria themselves possess a series of enzymatic reactions that are Ca^{2+} sensitive in the μM range (56-62), and which eventually provide the reducing equivalents to the respiratory chain. It may well be that an important function, perhaps the most important function, of the mitochondrial Ca^{2+} transporting process, is that of regulating the Ca^{2+} activity <u>within</u> the mitochondrial space (matrix and/or inner membrane) rather than in the cytosol (59,63). The cytosol possesses other systems,in addition to mitochondria,which regulate the Ca^{2+} activity in the environment of Ca^{2+} sensitive enzymes. The mitochondrion, however, relies only on its transport process(es) to accomplish the same goal within its matrix and membrane spaces.

AKNOWLEDGMENT. Parts of the original work described has been made possible by the financial contribution of the Swiss Nationalfonds (Grants nos. 3.597.0-75 and 3.282-0.78).

REFERENCES

1. Saris, N.E. (1963) Soc.Sci.Fennica, Comment.Physico-
    Mathem. 28. 1-59.
2. Vasington, F.D., and Murphy, J.V., (1962) J. Biol.
    Chem. 237, 2670-2677.
3. Carafoli, E., (1979) FEBS Letters, 104, 1-5.
4. Carafoli, E., and Crompton, M. (1978) Current Top.
    Membr. Transp. 10, 152-216.
5. Puskin, J.S., Gunter, T.E., Gunter, K.K., and Russell,
    P.R. (1976) Biochemistry, 15, 3834-3842.
6. Rottenberg, H., and Scarpa, A. (1974) Biochemistry,
    13, 4811-4819.
7. Heaton, G.M., and Nicholls, D.G. (1976) Biochem. J.,
    156, 635-646.
8. Crompton, M., and Heid, I. (1978) Eur. J. Biochem.
    91, 599-608.
9. Reed, K.C., and Bygrave, F.L. (1975) Eur. J. Biochem.,
    55, 497-504.
10. Moyle, J., and Mitchell, P. (1977) FEBS Letters, 73,
    131-136.
11. Moyle, J., and Mitchell, P. (1977) FEBS Letters 77,
    136-145.
12. Moore, C.L. (1971) Biochem. Biophys. Res. Commun.
    42, 298-305.
13. Vasington, F.D., Gazzotti, P., Tiozzo, R., and
    Carafoli, E. (1972) Biochim. Biophys. Acta, 256, 43-
    54.
14. Niggli, V., Gazzotti, P., and Carafoli, E. (1978)
    Experientia, 34, 1135-1137.
15. Mela, L. (1968) Arch. Biochem. Biophys. 123, 286-291
16. Crompton, M., Heid, I., Baschera, C., and Carafoli, E.
    (1978) FEBS Letters, 104, 352-354.
17. Jacobus, W.E., Tiozzo, R., Lugli, G., Lehninger, A.L.,
    and Carafoli, E. (1975) J. Biol. Chem. 250, 7863-7870.
18. Sordahl, L.A. (1974) Arch. Biochem. Biophys. 167,
    104-115.
19. Åkerman, K.E.O., Wikström, M.K.F., and Saris, N.E.
    (1977) Biochim. Biophys. Acta, 464, 287-294.
20. Crompton, M., Sigel, E., Salzmann, M., and Carafoli,
    E. (1976) Eur. J. Biochem. 69, 429-434.
21. Affolter, H., and Carafoli, E., manuscript in prepa-
    ration.

22. Sottocasa, G.L., Sandri, G., Panfili, E., de Bernard, B., Gazzotti, P., Vasington, F.D., and Carafoli, E. (1972) Biochem. Biophys. Res. Commun. 47, 808-813.

23. Prestipino, G., Ceccarelli, D., Conti, F., and Carafoli, E. (1974) FEBS Letters, 45, 99-103.

24. Panfili, E., Sandri, G., Sottocasa, G.L., Lunazzi, G., Liut, G., and Graziosi, G. (1976) Nature, 264, 185-186.

25. Carafoli, E. (1975) Molec. Cell.Biochem., 8,133-140.

26. Jeng, A.Y., and Shamoo, A.E. (1980) J. Biol. Chem. 255, 6897-6903.

27. Sottocasa, G.L. (1980) First European Bioenergetics Conference, Short Reports, 267-268.

28. Rossi, C.S., Vasington, F.D., and Carafoli, E.(1973) Biochem. Biophys. Res. Commun., 50, 846-852.

29. Carafoli, E., Tiozzo, R., Lugli, G., Crovetti, F., and Kratzing, C. (1974) J. Molec. Cell. Cardiol., 6, 361-371.

30. Crompton, M., Capano, M., and Carafoli, E. (1976) Eur. J. Biochem., 69, 453-462.

31. Crompton, M., Moser, R., Lüdi, H., and Carafoli, E. (1978) Eur. J. Biochem. 82, 25-31.

32. Crompton, M., Künzi, M., and Carafoli, E. (1977) Eur. J. Biochem., 79, 549-558.

33. Crompton, M., Heid, I., and Carafoli, E. (1980) FEBS Letters, 115, 257-259.

34. Nicholls, D.G., (1978) Biochem. J., 170, 511-522.

35. Clark, A.F., and Roman, I.J. (1980) J. Biol.Chem. 255, 6556-6558.

36. Haworth, R.A., Hunter, D.R., and Berkoff, H.A. (1980) FEBS Letters, 110, 216-218.

37. Al-Shaikhaly, M.H.M., Nedergaard, J., and Cannon, B. (1979) Proc.Natl.Acad.Sci.(USA) 76, 2350-2353.

38. Affolter, H., and Carafoli, E. (1980) Biochem. Biophys.Res. Commun. 95, 193-196.

39. Mitchell, P., and Moyle, J. (1967) Biochem. J. 109, 1147.

40. Roos, I., Crompton, M., and Carafoli, E. (1978) FEBS Letters, 94, 418-421.

41. Carafoli, E., Crovetti, F., and Ceccarelli, D. (1973) Arch. Biochem. Biophys. 154, 40-46.

42. Malmström, K., and Carafoli, E. (1975) Arch. Biochem. Biophys. 171, 418-423.

43.   Roman, I., Gmaj, P., Nowicka, C., and Angielski,
        S. (1979) Eur. J. Biochem. 102, 615-623.
44.   Hugentobler, G., and Gazzotti, P., unpublished ob-
        servations.
45.   Lehninger, A.L., Vercesi, A., and Bababunmi, E.A.
        (1978) Proc. Natl.Acad.Sci. (USA) 76, 4340-4344.
46.   Lötscher, H.R., Winterhalter, K.H., Carafoli, E.,
        and Richter, C. (1979) Proc.Natl.Acad.Sci. (USA)
        76, 4340-4344.
47.   Lötscher, H.R., Winterhalter, K.H., Carafoli, E.,
        and Richter, C. (1980) J. Biol.Chem. in press.
48.   Nicholls,D.G.,and Brand, M.D. (1980) Biochem. J.
        188, 113-118.
49.   Rossi, C.S., and Lehninger, A.L. (1964) J. Biol.
        Chem. 239, 3971-3980.
50.   Carafoli, E., Rossi, C.S., and Lehninger, A.L.
        (1965) J. Biol. Chem. 240, 2254-2261.
51.   Roos, I., Crompton, M., and Carafoli, E. (1980)
        Eur. J. Biochem., in press.
52.   Lötscher, H.R., Schwerzmann, K., and Carafoli, E.
        (1979) FEBS Letters 99, 194-198.
53.   Hunter, D.R., Haworth, R.A., and Southard, J.H.
        (1976) J. Biol. Chem. 251, 5069-5077.
54.   Hunter, D.R., and Haworth, R.A. (1979) Arch. Biochem.
        Biophys. 195, 468-477.
55.   Nicholls, D.G. (1978) Biochem. J. 176, 463-474.
56.   Denton, R.N., Randle, P.J., and Martin, B.R. (1972)
        Biochem. J. 128, 161-163.
57.   Cooper, R.H., Randle, P.J., and Denton, R.M. (1974)
        Biochem. J. 143, 625-641.
58.   Denton, R.M., Richards, P.A., and Chin, J.G. (1978)
        Biochem. J. 176, 899-906.
59.   McCormack, J.G., and Denton, R.M. (1979) Biochem. J.
        180, 533-544.
60.   Hansford, R., and Chappell, J.B. (1967) Biochem.
        Biophys.Res.Commun. 27, 686-692.
61.   Otto, D.A., and Ontko, J.A. (1978) J. Biol.Chem.
        253, 789-799.
62.   Malmström, K., and Carafoli, E. (1976) Biochem.
        Biophys.Res.Commun. 69, 658-664.
63.   Carafoli, E., Proc.Symp.on Exercise Bioenergetics
        and Gas Exchange, P. Cerretelli, and B.J. Whipp,
        Ed., North-Holland, in press.

INTEGRATING FUNCTIONS OF BIOMEMBRANES AND THE BIOLOGICAL ROLE OF PROTONIC POTENTIAL

Vladimir P. Skulachev

Department of Bioenergetics,
Laboratory of Molecular Biology
and Bioorganic Chemistry,
Moscow State University,
Moscow 117234, USSR

It is absolutely impossible to imagine the history of modern bioenergetics without Lars Ernster. This is primarily due to his outstanding contributions to the creation of a general picture of energy transductions in the living cell. One of these contributions, namely the discovery of energy-linked transhydrogenase (1-3), will be discussed below. Another aspect of Lars Ernster's activity in the international scientific community which deserves high appreciation is his successful attempt to unite bioenergetists of different countries and integrate bioenergetic investigations all over the world. The example of it is the organization of the International Bioenergetic Group and of the First European Bioenergetic Conference. The role of Ernster in both of these recent events is hard to overestimate.

Thus Lars Ernster can be rightfully regarded

C. P. Lee, G. Schatz, G. Dallner (eds.), Mitochondria and Microsomes
in honor of Lars Ernster   ISBN 0-201-04576-1

as one of the founders of the science of bioenergetics.

GENERAL PATTERN OF ENERGY TRANSDUCTIONS IN THE
CELL. A scheme of main energy transductions in
living organisms is shown in Fig. 1. According to
it (4, 5), the energy of light and of respiratory
or glycolytic substrates is utilized by the enzymes
of photosynthetic and respiratory redox chains or
of substrate-level phosphorylations to generate one
of the two convertible forms of energy: (i) ATP or
(ii) protonic potential difference ($\Delta\bar{\mu}H$) postulated
by Mitchell (6, 7). It is also indicated that light
energy can be converted into $\Delta\bar{\mu}H$ in a redox chain-
independent fashion via bacteriorhodopsin, a retinal-containing protein of halophilic bacteria.
$\Delta\bar{\mu}H$ or ATP can be used to support different types
of work when they are transduced into chemical, osmotic, mechanical or electric energy or into heat.

CHEMICAL WORK. For ATP, chemical work means bio-
syntheses driven by ATP or other high-energy compounds equilibrated with ATP. The main reaction of
$\Delta\bar{\mu}H$-dependent chemical work is the formation of ATP
by H^+-ATP-synthetase. In photosynthetic bacteria
such as Rhodospirillum rubrum, there is a reversible
membrane-linked pyrophosphatase, a very active enzyme capable of synthesizing inorganic pyrophosphate
(PP_i) at the expense of $\Delta\bar{\mu}H$ energy. The PP_i formation by bacterial chromatophores in the light was
discovered in Ernster's group by M. Baltscheffsky
(for review, see ref. 8). Membrane potential generation coupled with PP_i hydrolysis was described by

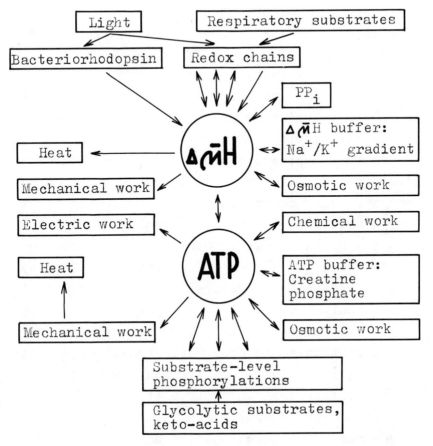

Fig. 1.   Energy transductions in living cells

Isaev et al. (9). Apparently this process is not
specific for bacteria. Recently, respiratory-de-
pendent PP_i synthesis was observed in mitochondria
by Kulaev et al. (10). The rate of this reaction
proved to be much lower than in chromatophores.

Besides ATP and PP_i syntheses, $\Delta\bar{\mu}H$ energy
can be transduced into chemical one when reverse
electron transfer occurs in the respiratory chain.
Such a process can be used by a cell to increase

its reducing power, supporting reductive syntheses
that utilize NAD(P)H. In such bacteria as <u>Thioba-
cillus ferrooxidans</u>, utilizing Fe^{2+} ion oxidation
by O_2 as the only energy source, reverse electron
transfer from Fe^{2+} to NAD(P)H is the only mechanism
capable of regenerating NAD(P)H from $NAD(P)^+$ (for
review, see ref. 11).

In the respiratory chain, there are four re-
actions that are, in principle, competent in $\Delta\bar{\mu}H$
formation. They correspond to the following steps
of the transfer of the reducing equivalents: NADPH
$\longrightarrow NAD^+$, NADH $\longrightarrow$ CoQ, $CoQH_2 \longrightarrow$ cytochrome $\underline{c}^{3+}$
and cytochrome $\underline{c}^{2+} \longrightarrow O_2$. The last reaction (cy-
tochrome oxidase) is irreversible, at least under
physiological conditions, so it generates protonic
potential but never consumes this potential. The
NADH-CoQ reductase and $CoQH_2$-cytochrome $\underline{c}$ reductase
systems are reversible, but are usually used to
form $\Delta\bar{\mu}H$, rather than to utilize it. As to the
first (transhydrogenase) reaction, it is clearly a
$\Delta\bar{\mu}H$ consumer under normal conditions. It is the
energy-requiring reversal of this reaction, discovered by Ernster (1, 2), that was the original indication of the existence of a coupling site in
this region of the respiratory chain. Ernster's
energy-linked transhydrogenase proved to be the
precedent of a system utilizing the substrate-to-
product concentration ratio as the only energy
source to produce $\Delta\bar{\mu}H$. In our group, it was found
that in submitochondrial particles transhydrogenase
forms a membrane potential (12) whose direction de-
pends upon the $[\text{NADPH}] \times [\text{NAD}^+] / [\text{NADP}^+] \times [\text{NADH}]$ ratio

(13). When this ratio was positive, the membrane potential polarity was found to be the same as in the respiring or ATP-hydrolyzing particles. When it was negative, the oppositely directed membrane potential was formed. In these experiments, phenyldicarbaundecaborane (PCB⁻) was used as a membrane potential probe. This observation was recently confirmed by Drachev et al. (14) who directly measured electric potential generation in transhydrogenase proteoliposomes attached to a phospholipid-impregnated Millipore filter.

OSMOTIC WORK. Transport of substances against concentration gradients (osmotic work) is an important function of $\Delta\bar{\mu}H$. It is $\Delta\bar{\mu}H$ or one of its constituents (electric potential difference, $\Delta\Psi$, or ΔpH) that is an energy source for any uphill transport processes in mitochondria. As to the plasma membrane of the animal cell, this role is performed by ATP which is utilized directly by transport systems or indirectly via the formation of ΔpNa by Na^+, K^+-ATPase and the subsequent ΔpNa use by a Na-substrate symporter. In bacteria, there is a great variety of transport systems driven by the total $\Delta\bar{\mu}H$, $\Delta\Psi$, ΔpH, ΔpNa, ATP, phosphoenolpyruvate or acetyl phosphate [for review see (15)]. Some of these mechanisms will be considered later in connection with the $\Delta\bar{\mu}H$ buffering.

MECHANICAL WORK AND HEAT PRODUCTION. The most important example of the mechanical work at the expense of the ATP energy is, certainly, the contractile activity of actomyosin. The same system is

used in the muscles of warm-blooded animals as a
mechanism of urgent heat production in response to
a sudden decrease in the ambient temperature (shi-
vering thermogenesis). In this group, it was found
that there is an alternative mechanism of thermogenesis with no ATP involved, which develops in
muscles of warm-blooded animals during their adapt-
ation to cold. The mechanism in question consists
in uncoupling of respiration and phosphorylation
(direct $\Delta \bar{\mu} H \longrightarrow$ heat transduction) (16, 17). Recently, these data obtained in laboratory experi-
ments with pigeons and mice were confirmed by Grav
and Blix in a study of subcapular muscle mitochondria from fur seals acclimated to cold under natural
conditions (18).

Unlike muscle mitochondria whose thermogenic
function seems to be accessory to the main one
(ATP-synthesis), mitochondria of brown fat were
found to be specialized in heat production [for review, see (19)].

MOTILITY OF FLAGELLAR BACTERIA. The $\Delta \bar{\mu} H \longrightarrow$ mechanical energy transduction was postulated by Mitchell
as a mechanism of rotation of bacterial flagella
(20, 21).

In Adler's group (22), it was found that the
motility of an E.coli mutant deficient in an oxidative phosphorylation enzyme can only be supported
by respiration, but not by glycolysis, differing
in this respect from the wild strain which utilizes
both respiratory and glycolytic energy for swimming.
ATP exhaustion in the mutant cell was without any

effect on motility, whereas the addition of an uncoupler immediately arrested the motility with no
decrease in the ATP level. The authors concluded
that the motility of this E.coli mutant is supported by an "intermediate of oxidative phosphorylation",
rather than by ATP.

When it became clear that (i) the "intermediate of oxidative phosphorylation" is identical to
$\Delta\bar{\mu}H$ and (ii) the reason why the studied mutant
failed to carry out respiratory ATP synthesis is a
defect in H^+-ATP-synthetase, I interpreted the above data as evidence for $\Delta\bar{\mu}H$-supported motility of
bacteria (23).

To verify this suggestion, we carried out a
systematic study of the possible role of $\Delta\bar{\mu}H$ in
bacterial locomotion (23-25). In these experiments,
Drs. A.N. Glagolev and T.N. Glagoleva took part.
The results of the study performed on Rhodospirillum rubrum and E.coli can be summarized as follows:
(i) Motility rate correlates with $\Delta\bar{\mu}H$, rather than
with ATP level.
(ii) Both $\Delta\Psi$ and ΔpH constituents of $\Delta\bar{\mu}H$ can be
used to support motility.
(iii) Not only enzymatically-formed, but also artificially imposed $\Delta\Psi$ and ΔpH of a proper direction
can be used by the flagellar motor.
(iv) Bacterial locomotion does not require any ion
gradient but that of H^+.

The results of our experiments were first
published in 1975 in a preliminary form (23) and in
1976 as a regular paper (24) (see also ref. 25).
In 1977-1978 three papers dealing with other kinds

of bacteria appeared. One of them, from the laboratories of Berg and Harold (26), reported about the motion of a <u>Streptococcus</u> at the expense of artificially formed $\Delta\Psi$ and ΔpH. Independently, a group in Japan observed artificial $\Delta\Psi$ and ΔpH-supported motility of <u>Bacillus subtilis</u> (27) and valinomycin + K^+ inhibition of motility under conditions of enzymatic energization (28).

Thus, <u>R.rubrum</u>, <u>E.coli</u>, <u>Streptococcus</u> and <u>Bacillus subtilis</u>, i.e. all the flagellar bacteria tested for $\Delta\bar{\mu}H$-supported motility, demonstrated this kind of locomotion. This seems to be a convincing proof that such a mechanism is a common feature of bacteria using a flagellar motor.

MOTILITY OF CYANOBACTERIA. The study of the motility of cyanobacteria (blue-green algae) carried out in this laboratory by Drs. A.N. Glagolev and T.N. Glagoleva has shown that unicellular flagellar bacteria are not the only type of biological systems performing $\Delta\bar{\mu}H$-supported mechanical work by means of a protonic motor.

The studied cyanobacteria are multicellular prokaryotes. They glide along the surface of a solid substrate. Instead of flagella, they have fibrils localized between the outer cell membrane and the peptidoglucan layer of the cell wall. The rotation of these fibrils was assumed to induce moving helical waves on the cell wall so that the organism moves along the substrate surface (29).

As shown by our experiments, the motility of cyanobacteria (<u>Oscillatoria</u> and <u>Phormidium</u>) is similar to that of R.rubrum and E.coli. In particular, gliding induced by artificially imposed $\Delta\Psi$ and ΔpH was demonstrated (30).

CHLOROPLAST ROTATION. In 1838 Donne described the rotation of chloroplasts in protoplasm drops squeezed out of a Chara alga cell. During the next century, this phenomenon was rediscovered several times (for review, see ref. 31).

We repeated Donne's experiments to determine the energy source for chloroplast rotation. Drs. A.N. Glagolev and E.H. Motzenok took part in the experiments. Nitella from Lake Baikal was studied.

It was found that the average rate of chloroplast rotation is 1 revolution per 2-3 sec. The rotation is observed for many hours. Sometimes, the direction of rotation changes. The direction of rotation may be constant for as long as an hour. In fact, the behavior of rotating chloroplasts in protoplasm drops resembles that of a motile bacterium attached with its flagellum to glass.

Sometimes this rotation can be observed in intact algal cells. This happens if one or several chloroplasts detach from a motionless chloroplast multilayer inside the cell. Such chloroplasts were found to rotate when carried by the protoplasm stream (see also ref. 31).

We undertook inhibitor analysis of chloroplast rotation in protoplasm drops, comparing this process with the movement of cytoplasm (cyclosis) in intact cells of the same alga. Cyclosis is known to be ATP-dependent. Some results of this study are given in Table 1.

It was found that the chloroplast rotation is resistant to dicyclohexyl carbodiimide (DCCD) in the light but not in the dark. It is suppressed by NH_4^+ which specifically discharges ΔpH, the main constituent of $\Delta\bar{\mu}H$ in chloroplasts. If the NH_4^+ concentration was not too high, its inhibitory action was found to be overcome by an increase in the light intensity. Cytochalasin B, an inhibitor of the ATP-dependent movements of intracellular structures, had no effect upon chloroplast rotation, but abolished the cyclosis completely.

The light-actuated rotation of chloroplasts was studied in detail using an infra-red microscope. It was shown that switching on the actinic light immediately initiates rotation (lag period, if any, is less than 1 sec). Switching off the light stops rotation after several min required, apparently, for the ΔpH formed under illumination to be dissipated.

Table 1. Effect of Different Agents on Rotation of Chloroplasts in Protoplasm Drops and on Cyclosis in Nitella

Additions	Chloroplast rotation	Cyclosis
None	+	+
DCCD	−	+
DCCD + light	+	+
DCCD + NaF		−
light	+	+
light + CCCP	−	−
light + DNP	−	−
light + DNP + ATP	−	+
light + $(NH_4)_2SO_4$	−	+
light + cytochalasin B	+	−

The above data may thus indicate $\Delta\bar{\mu}H$-support-
ed rotation of chloroplasts. Apparently, this is
the first example of (i) $\Delta\bar{\mu}H$-linked mechanical work
in eukaryotic cells and (ii) self-propelled motil-
ity of an intracellular organelle. Certainly, this
study raises questions as to the mechanism of such
motility, its biological function, relation to ATP-
supported movements of intracellular components,
etc.

ELECTRIC WORK. Electric energy production supported by ATP usually includes Na^+, K^+-ATPase if we
consider the outer membrane of animal cells. The
enzyme exchanges three intracellular Na^+ ions for
two extracellular K^+ ions, charging thereby the
cell interior negatively. Na^+, K^+-ATPase seems to
be specific only for animals. At the same time,
the electronegativity of the intracellular space
is the feature all types of living cells have in
common. In fungi, this electric asymmetry is, apparently, created by H^+-ATPase, an enzyme differing
from H^+-ATP-synthetase of mitochondria, chloroplasts
and bacteria (for review, see ref. 5). Probably,
the H^+-ATPase of the plasma membrane of the fungal
cell specialized in ATP $\longrightarrow$ $\Delta\bar{\mu}H$ energy transduction, whereas H^+-ATPase synthetase usually carries
out the opposite process (32). However, further
investigations are needed to identify fungal plasma
membrane ATPase as an enzyme catalyzing electrogenic
H^+ uniport.

In plants, the situation is even more obscure.
Nevertheless, current indirect evidence favors the

notion that the fungal type mechanism of electrogenesis is the most probable one.

In bacteria, the cytoplasmic membrane always
contains $\Delta\bar{\mu}H$-generating enzymes which catalyze H^+
extrusion from the cytosol to the outer medium.
This process, in principle, is sufficient to maintain a membrane potential, the cell interior negative. However, involvement of some additional
electrogenic mechanisms is also possible, especially under certain extremal conditions and in biological niches. Thus, in the extremely halophilic
Halobacterium halobium the electrogenic light-
-driven Na^+ pump is apparently operative (33-36).
The system is retinal-linked, like bacteriorhodopsin, not to be confused with the "classical" H^+-
bacteriorhodopsin. The same bacteria, apparently,
possess yet another retinal-containing protein
which is probably a photosensor responsible for a
photophobic reaction of this microorganism (maximum at 370 nm vs. 570 nm for H^+-bacteriorhodopsin
and 588 nm for Na^+-bacteriorhodopsin) (37-39).
It is not clear whether the 370 nm pigment is competent in the photoelectrogenic activity inherent
in both H^+- and Na^+-bacteriorhodopsin.

VISUAL RHODOPSIN-LINKED ELECTROGENESIS. As to the
animal photosensor, visual rhodopsin, its electrogenic activity was recently demonstrated in this
group.

The first indication of the photoelectric
response of visual rhodopsin was obtained in 1964
(40) when a so-called early receptor potential

(ERP) in the retina was described. ERP was biphasic. The first phase of ERP was found to last
less than 1 μsec, which is too fast to be associated with any postrhodopsin events in the visual process. Recently, a photoelectric signal was observed by Montal and co-worker (41-43) in a model system, namely a Teflon film covered on one side
with rhodopsin. The amplitudes of ERP (as well as
of the photoeffect in the Montal system) were small,
no more than several mV. Therefore ERP was regarded
as an epiphenomenon accompanying a light-induced
conformational change in rhodopsin (44).

Ostrovsky et al. (45) incorporated rhodopsin-
containing photoreceptor discs from the retina rod
cells into the phospholipid-impreganted Millipore
filter using the procedure previously employed in
this group for a study of bacteriorhodopsin proteoliposomes (46). In such a system, illumination induced formation of an electric potential difference
across the filter (up to 25 mV). A systematic comparative study of the photoelectric activities of
animal and bacterial rhodopsins was recently carried out in this group together with Professor M.A.
Ostrovsky's laboratory (47-49). Photoreceptor
discs or bacteriorhodopsin sheets, i.e. H. halobium
membrane fragments containing bacteriorhodopsin,
were incorporated into a collodion film impregnated
with a decane solution of phospholipids.

Under continuous illumination, both animal
and bacterial rhodopsins were found to generate a
potential difference across the film, the rhodop-

sin-free side being positive. Bacteriorhodopsin-
-supported $\Delta\Psi$ persists throughout the illumination
period. With animal rhodopsin, $\Delta\Psi$ reaches ~ 0.05 V
and then instantaneously drops to zero.

The study of the 15 ns laser flash-induced
single turnover of two rhodopsins revealed their
striking similarity. (i) In both cases, the electrogenesis consists of three phases, the 1st being
oppositely directed to the 2nd and 3rd. (ii) The
1st phase τ is faster than 200 ns, that of the 3rd
phase is several ms. The 2nd phase τ is ~ 50 µs
for bacterial and ~ 500 µs for animal rhodopsin.
(iii) The amplitudes of the phases increase in the
order of 1st < 2nd < 3rd. (iv) At low flash energies, $\Delta\Psi$ decay takes seconds. The overall amplitude of $\Delta\Psi$ is 20-35 mV for animal and 30-60 mV for
bacterial rhodopsin.

The major difference between the two rhodopsins was found in responses to repeated flashes.
Animal rhodopsin generates maximal $\Delta\Psi$ at the 1st
flash, the next flash being less electrogenic than
the previous one. Parallel to the amplitude decrease, acceleration of $\Delta\Psi$ decay takes place. The
addition of 11-cis-retinal improves the amplitude
but not the $\Delta\Psi$ decay time. Bacteriorhodopsin showed no dependence of the amplitude and the decay
upon the number of flashes.

It was concluded that not only bacterial, but
also animal rhodopsin is a photoelectric generator,
animal rhodopsin being capable of only a single
turnover.

In connection with the above data, the following facts should be mentioned. (i) The direction, form and kinetics of flash-induced $\Delta\Psi$ in our system closely resemble those of ERP. (ii) Lewis (50) showed that the Raman spectral kinetics of bacterial and animal rhodopsins are very similar. He assumed a light-induced H^+ release inside the disc. (iii) Ostrovsky et al. (45) described a transient light-dependent PCB^- uptake by discs. (iv) There are indications of a light-induced increase in the permeability of the photoreceptor membrane. We found that this effect takes not more than 1 ms. This means that it may be involved in the photoreception.

LONG-DISTANCE LATERAL POWER TRANSMISSION. This phenomenon was recently discovered in this group by Chailakhian et al. (51). Experiments were carried out with filamentous blue-green algae (cyanobacteria) Phormidium uncinatum. These bacteria form multicellular trichomes representing linear sequences of many hundreds of cells. Trichomes can be several millimeters long. The cytoplasmic membrane of cyanobacteria belongs to the category of coupling membranes bearing $\Delta\bar{\mu}H$. If there is an electric conductance between trichome-composing cells, one may hope that the electric potential difference generated across the cytoplasmic membrane near to one end of the trichome can be transmitted along the trichome and utilized, say, in its distal part to perform work.

In this study, the light and motility of

trichomes were used as a source of $\Delta\Psi H$ generation
and $\Delta\bar{\mu}H$-dependent work, respectively. The latter
proved to be a convenient intrinsic probe for $\Delta\bar{\mu}H$
since it was found in this laboratory by Glagolev
et al. (30) that the motility of cyanobacterial
trichomes is directly supported by $\Delta\bar{\mu}H$, like the
motility of flagellar bacteria (see above).

Trichomes of Ph. uncinatum were exposed to
white light under conditions where light was the
only energy source to form $\Delta\bar{\mu}H$ and, hence, to support motility. Illumination was carried out by a
small light beam forming a spot of weak light which
covered only about 5% of the trichome length. It
was shown that such partial illumination can support motility.

The light spot-supported motility was found
to be resistant to DCCD which completely inhibited
illumination-induced increase in the ATP level.
These relationships could be predicted if we assum-
ed $\Delta\bar{\mu}H$ or one of the constituents, $\Delta\Psi$ or ΔpH, to
be transmitted along the trichome from the illumin-
ated to dark part of the organism. This explanation
was confirmed by the measurement of the $\Delta\Psi$ transmission along trichomes.

Ph. uncinatum trichomes are sufficiently
long to be studied with a classical electrophysio-
logical technique using extracellular electrodes.

The trichomes were put into a groove in a
plastic plate. The groove was filled with distil-
led water and four electrodes were placed close to
the trichome: electrodes 1 and 4 in the vicinity
of its ends, and the other two (2 and 3) at a di-

stance of 1/3 and 2/3 of the trichome length, respectively. When a small portion at the end of
the trichome, close, e.g., to electrode 1, was illuminated with a light beam, there appeared a potential difference between the electrodes as if
(i) there was a positive charge extrusion from the
illuminated part of trichome and (ii) the trichome
was a unitary electric system with high electric
conductance between the cells.

$\Delta\Psi$ proved to be higher, and increased faster
when measured between electrodes 1 and 4 placed
near the trichome ends than between a middle electrode and electrode 4. Computer simulation of
these data carried out in this group by Dr. M.
Kara-Ivanov revealed good agreement between (i) the
measured amplitudes and the kinetics of $\Delta\Psi$ and (ii)
those calculated if we assume that electric cable
properties are inherent in the cyanobacterial trichome (Fig. 2).

Damaging the trichome between illuminated and
dark parts completely prevented the electric transmission effect. For example, there was no light
spot-generated potential difference between electrodes 3 and 4 when the spot was situated near
electrode 1 and the middle part of the trichome
(between electrodes 2 and 3) was damaged.

Thus, the data of the experiments on filamentous cyanobacteria can be considered as a precedent of intercellular power transmission in the
form of membrane potential. It is possible that
other coupling membrane structures can also be

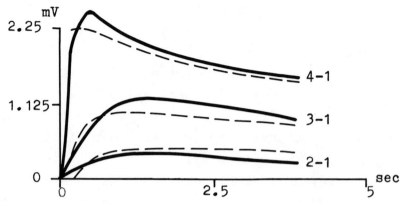

Fig. 2. Lateral $\Delta\Psi$ transmission along cyanobacterial trichomes. The figure shows re-
sults of measurements with four extracellular electrodes (solid lines) and of
computer simulation assuming electric
cable properties of trichomes (dashed
lines) (51). For explanations, see the
text.

used as electric cables for power transmission.
For example, the inner membranes of giant mitochondria were suggested to play the role of intracellular "electric wires" (for discussion, see
refs. 5, 52-54).

THE $\Delta\bar{\mu}H$ BUFFERING. A ROLE OF Na^+/K^+ GRADIENT.
Assuming that $\Delta\bar{\mu}H$ is a convertible and transport-
able form of energy for the cell, we are faced
with the problem of how the living cell stabilizes
a $\Delta\bar{\mu}H$ level. Under conditions of varying activities of energy producing and consuming mechanisms,
$\Delta\bar{\mu}H$ should be buffered by some system(s) performing
this function, like, e.g., creatine phosphate,
which specializes in buffering the level of ATP,
another convertible form of energy. In fact, the

problem of buffering should be more important for
$\Delta\bar{\mu}H$ than for ATP if we take into account the quantity of energy equivalents stored in the form of
ATP and $\Delta\bar{\mu}H$. The energy produced by a $\Delta\bar{\mu}H$ generator must be **pri**marily accumulated as $\Delta\Psi$ since the
electric capacitance of the membrane ($\approx 1 \ \mu F/cm^2$)
is too low for the transported H^+ ions, required
for capacitance charging, to shift pH in the solution on either side of the membrane. The necessary amount is as low as 1 μmol H^+ ions/g protein,
which is in the same order as the amount of enzymes
in the coupling membrane and much lower than that
of ATP (55-57).

Analyzing the probable ways of stabilization
of $\Delta\bar{\mu}H$ in bacteria I suggested that Na^+/K^+ gradient
can function as a specialized $\Delta\bar{\mu}H$ buffer (57).

The hypothesis of the $\Delta\bar{\mu}H$ buffering by means
of Na^+/K^+ gradients consists of the following.
There is a mechanism of K^+ uniport responsible for
electrophoretic K^+ influx into the cell. This
process is accompanied by transduction of $\Delta\Psi$ to
ΔpH and ΔpH. The latter is formed due to discharge of electric capacitance of the membrane by
K^+ electrophoresis which greatly increases the
amount of H^+ ions to be transported across the
membrane to form $\Delta\bar{\mu}H$. It is also postulated that
ΔpH (created when K^+ is imported into the cell) is
used to extrude Na^+ ions by means of the Na^+/H^+
antiport.

If this were the case, any decrease in the
activity of $\Delta\bar{\mu}H$ generators (or increase in the
activity of $\Delta\bar{\mu}H$ consumers), resulting in a certain

$\Delta \bar{\mu} H$ decrease, should cause a downhill K^+ efflux and a Na^+ influx producing $\Delta \Psi$ and ΔpH, respectively. These fluxes must stabilize $\Delta \bar{\mu} H$, preventing its decrease (Fig. 3).

This laboratory undertook a study to verify the above hypothesis (5, 58, 59). In the experiments, an E. coli strain deficient in H^+-ATP-synthetase was used to avoid ATP-linked side-effects.

First, the motility of the bacteria was studied since this $\Delta \bar{\mu} H$-dependent type of work can be easily and directly measured. In this study Drs I.I. Broun and A.N. Glagolev participated.

In H^+-ATP-synthetase-deficient E. coli cells, respiration is the only mechanism competent in the de novo $\Delta \bar{\mu} H$ production. $\Delta \bar{\mu} H$ buffer must prolong the motility for some time after transition to anaerobiosis.

The cells were kept with K^+ and without Na^+ in an open vessel. Then they were diluted with an incubation medium containing KCl, NaCl or Tris-HCl; the mixture was placed into a special closed chamber to measure both respiration and motility rates. After several minutes, oxygen consumption by bacteria resulted in anaerobiosis.

The data of a typical experiment are given in Fig. 4. One can see that incubation of E. coli in the medium with K^+ and without Na^+ results in a rapid decrease in the motility rate which falls to zero 1.5 min after O_2 exhaustion. Under these conditions, high $[K^+]_{in}$ is counterbalanced by high $[K^+]_{out}$, whereas Na^+ is present neither inside

A

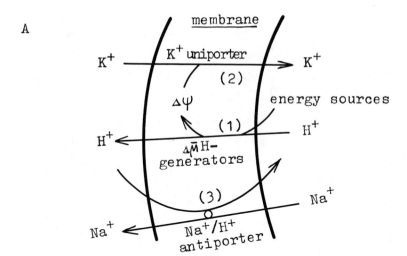

B

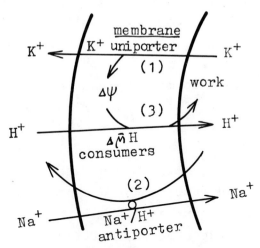

Fig. 3. A $\Delta\bar{\mu}$H buffering by means of the Na^+/K^+ gradient across bacterial membrane. A and B, formation and utilization of the Na^+/K^+ gradient, respectively.

nor outside the cell. Thus, there is no Na^+/K^+ gradient, and, according to the above hypothesis, no $\Delta\bar{\mu}$H-buffering.

In a medium containing $Tris^+$ as a univalent cation, the motility ceases after 11 min. In this case, there is a K^+ gradient across the bacterial membrane because $[K^+]_{in}$ is high and $[K^+]_{out}$ is very low.

The maximal duration of motility (up to 14 min) is observed in a Na^+ medium. In the latter case, there are gradients of both K^+ and Na^+ of opposite directions.

In all the three cases, aeration resulted in immediate activation of motility.

In another experiment, the bacteria were incubated in an anaerobic $Tris^+$ medium. After cessation of motility, NaCl was added to create ΔpNa across the bacterial membrane. This addition was found to initiate motility which could be observed for several minutes.

The simultaneous measurements of K^+ and O_2 showed that the addition of bacteria to a medium supplemented with a small amount of K^+ immediately resulted in a $[K^+]_{out}$ decrease which brought about a K^+ release after complete O_2 exhaustion. It should be stressed that no K^+ ionophore was added in either of our experiments.

In further experiments, anilinonaphthalene sulfonate (ANS^-) was used as a probe for membrane energization. The cells were added to an anaerobic incubation medium containing either K^+ or Na^+. In the K^+-containing medium, a biphasic increase in the ANS^- fluorescence took place, the first phase being smaller than the second. Subsequent

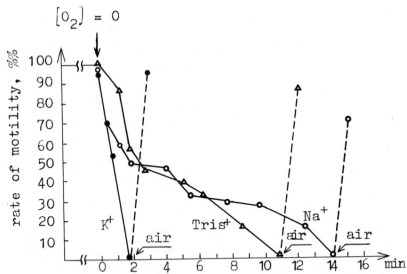

Fig. 4.   Anaerobic motility of E. coli cells supported by the Na^+/K^+ gradient. Incubation mixture contained: (1) K^+ mixture:
0.05 M KCl, 1% glycerol, 10 mM MOPS-Tris,
pH 7.0, 1×10^{-4}M EDTA; (2) Na^+ mixture: as
(1) but 0.05 M NaCl instead of KCl; (3)
$Tris^+$ mixture: as (1) but KCl was omitted
and MOPS-Tris concentration was increased
up to 0.05 M.

aeration resulted in a strong fluorescence decrease,
which indicated membrane energization. The effect
of air was inhibited by uncouplers.

In the Na^+-containing medium, the first phase
(ANS^- binding to non-energized bacterial membranes)
was of the same magnitude as in the K^+ medium.
However, the development of the second phase was
greatly delayed indicating that the anaerobiosis-
-induced membrane deenergization proceeded much
more slowly.

Similar data showing K^+ efflux-linked $\Delta\Psi$
stabilization were obtained with E. coli cells

preloaded with K^+ and then incubated in an anaerobic K^+-free medium, when dis-$C_3(5)$ fluorescence
was used as a probe for $\Delta\Psi$ (Drs. Yu. Kim and A.
Zinchenko took part in the experiments).

Data indicative of the existence of a Na^+/H^+
antiporter in bacteria have been reported by several laboratories (for review, see 15, 59). There-
fore, the key postulate of the $\Delta\bar{\mu}H$ buffer hypothesis is the assumption concerning the electrogenic
K^+ uniporter which is actuated in response to a
$\Delta\Psi$ decrease so that ΔpK formed under energization
conditions proves to be a source of energy for $\Delta\Psi$
generation.

Of great importance is the question of regulation of K^+ accumulation in the bacterial cell.
It is hardly possible that a K^+ channel (or K^+
uniporter) is always operative. Indeed, it is necessary to prevent (i) inhibition of other $\Delta\Psi$-de-
pendent functions by K^+ electrophoresis and (ii)
increase in the intracellular K^+ concentration
above some optimal level. Thus K^+ channels should
be strongly controlled when there is an excess of
energy sources. When the energy is exhausted,
$\Delta pK \rightarrow \Delta\Psi$ energy transition must be actuated via
opening of the K^+ channels. This process should
also be controlled as soon as the K^+ conductance
increase is sufficiently large to support $\Delta\bar{\mu}H$-de-
pendent functions, but not so large as to induce
an immediate loss of all the stored K^+ ions.

Sometimes $\Delta\Psi$ is, apparently, too small to
maintain $\left[K^+\right]_{in}$ at a sufficiently high level.
This may take place, e.g., under conditions when

the outer K^+ concentration is very low or $\Delta\bar{\mu}H$ is
mainly in the form of ΔpH. In these cases, a
$\leftarrow \Delta\Psi$-independent mechanism of K^+ accumulation
should be actuated. This seems to be K^+-ATPase
(60-62).

The downhill K^+ efflux both through the K^+
channel and K^+-ATPase might be used to form ATP,
in the former case indirectly, via $\Delta\Psi$ utilized by
H^+-ATP-synthetase, and in the latter directly, if
K^+-ATPase can serve also as K^+-ATP-synthetase.
In fact, Wagner et al. (63, 64) observed that the
K^+ gradient formed across the Halobacterium halobium membrane in the light stabilized the ATP level in the dark.

Apparently, there are auxiliary mechanisms
of Na^+ extrusion other than Na^+/H^+ antiport. In
Halobacterium halobium, a ΔpH-independent, light-
driven system of Na^+ efflux has been described
(Na-bacteriorhodopsin, see above). The Na-bacteriorhodopsin activity is usually much weaker than
that of the Na^+/H^+ antiporter localized in the same
membrane (35, 36). However, under conditions when
ΔpH is too small, Na-bacteriorhodopsin might be
significant for maintaining low $\left[Na^+\right]_{in}$. Another
way to overcome limitations of the ΔpH magnitude
is to organize the Na^+ efflux as, say, an antiport
of two Na^+ against one H^+. Such a system proves
to be electrogenic. Hence it can use not only ΔpH
but also $\Delta\Psi$. According to the data of Schuldiner
and Fishkes (65), the electroneutral Na^+/H^+ antiport becomes electrogenic when external pH is
shifted from neutral to alkaline, i.e. if the con-

ditions prove to be unfavorable for maintaining ΔpH.

An unusual system of Na^+ transport was recently described by Harold et al. in <u>Streptococcus fae-</u><u>calis</u> (66, 67). The system in question is apparently, Na^+-ATPase which catalyzes Na^+ extrusion from the cell in a $\Delta\bar{\mu}$H-independent manner. Again, this system is activated when ΔpH of the medium becomes alkaline (> 7.4). At acidic pH it is not operative and is replaced, most probably, by a Na^+/H^+ antiporter.

There is a specific requirement of any energy buffer: a buffering component should be formed when energy is abundant enough to prevent a competition for energy between the buffer formation process and other energy-linked reactions. For the ATP buffering compound, creatine phosphate, this is the case due to chemical properties of creatine phosphate and ATP. The standard free energy of creatine phosphate is higher than that of ATP, so creatine phosphorylation starts only when ATP/ADP ratio becomes high, i.e. when the overall rate of ATP-producing reactions is higher than that of ATP-consuming reactions. In the case of Na^+/K^+ gradient, the situation is more complicated. If the activities of K^+ uniport and Na^+/H^+ antiport mechanisms were always high, they would compete with H^+-ATP-synthetase and other $\Delta\bar{\mu}$H-diven processes. On the other hand, if they were always low, the Na^+/K^+ gradient would be ineffective as a $\Delta\bar{\mu}$H buffer.

Apparently, the solution to this problem is

allosteric regulation of the K^+ and Na^+ porters by
ATP, the level of which reflects the state of cellular energetics. It was found that the main $\Delta\Psi$-
driven K^+ transport system in E. coli is operative
only when the ATP level is high (60-62). Similar
relationships are, apparently, true for the Na^+/H^+
antiporter in S. faecalis (66, 67).

CONCLUSION. In the living cells various types of
work (chemical, osmotic, mechanical, electrical and
regulatory heat production) are driven by protonic
potential ($\Delta\bar{\mu}H$) with no ATP involved. Thus, $\Delta\bar{\mu}H$
is not only a transient intermediate of membrane-
linked ATP syntheses, but also a convertible energy currency for the cell.

Among the $\Delta\bar{\mu}H$-supported activities, mechanical work was recently demonstrated. It can be exemplified by the motility systems of (i) flagellar
bacteria and (ii) blue-green algae. The suggestion
is made that the rotatory movement of chloroplasts
in Nitella also utilizes the energy of $\Delta\bar{\mu}H$, rather
than that of ATP.

In multicellular cyanobacteria it was found
that $\Delta\bar{\mu}H$ can be used for power transmission over
distances of millimeters. It is suggested that in
large eukaryotic cells, e.g., in muscle fibers,
giant mitochondria may serve as power-transmitting
structures.

The Na^+/K^+ gradient can be used to stabilize
$\Delta\bar{\mu}H$ in bacteria. One can speculate that the primary function of unequal distribution of these
cations between the microbial cell and the medium
is $\Delta\bar{\mu}H$ buffering.

REFERENCES

1. Danielson, L. and Ernster, L. (1963) Biochem. Biophys. Res. Communs., 10, 91-96.
2. Danielson, L. and Ernster, L. (1963) Biochem. Z., 338, 188-205.
3. Ernster, L. and Lee, C.P. (1964) 6th Intern. Biochem. Congr. Abstr., 10, 729-730.
4. Skulachev, V.P. (1977) FEBS Lett., 74, 1-9.
5. Skulachev, V.P. (1980) Can. J. Biochem., 58, 161-175.
6. Mitchell, P. (1961) Nature, 191, 144-148.
7. Mitchell, P. (1966) Chemiosmotic Coupling in Oxidative and Photosynthetic Phsophorylation, Glynn Research, Bodmin.
8. Baltscheffsky, H., Baltscheffsky, M. and Thore, A. (1971) Curr. Top. Bioenerg., 4, 273-325.
9. Isaev, P.I., Liberman, E.A., Samuilov, V.D., Skulachev, V.P. and Tsofina, L.M. (1970) Biochim. Biophys. Acta, 216, 22-29.
10. Mansurova, S.E., Shakhov, Yu.A., Belyakova, T.N. and Kulaev, I.S. (1975) FEBS Lett., 55, 94-98.
11. Skulachev, V.P. (1969) Energy Accumulation in the Cell, pp. 45-48, Nauka, Moscow.
12. Grinius, L.L., Jasaitis, A.A., Kadziauskas, J.P., Liberman, E.A., Skulachev, V.P., Topali, V.P., Tsofina, L.M. and Vladimirova, M.A. (1970) Biochim. Biophys. Acta, 216, 1-12.
13. Dontsov, A.E., Grinius, L.L., Jasaitis, A.A., Severina, I.I. and Skulachev, V.P. (1972) J. Bioenerg., 3, 277-303.
14. Drachev, L.A., Kondrashin, A.A., Semenov, A. Yu., Skulachev, V.P. (1980) Europ. J. Biochem. (accepted).
15. Harold, F.M. (1980) Curr. Top. Membr. Transport (accepted).
16. Skulachev, V.P. (1963) Proc. 5th Intern. Congr. Biochem., 5, 365-375.
17. Skulachev, V.P. and Maslov, S.P. (1960) Biokhimiya, 25, 1058-1064.
18. Grav, H.J. and Blix, A.S. (1979) Science, 204, 87-89.
19. Smith, R.E. and Horwitz, B.A. (1969) Physiol. Rev., 49, 330-425.

20.   Mitchell, P. (1956) Proc. Roy. Phys. Soc.
      (Edinburgh) 25, 32-34.
21.   Mitchell, P. (1972) FEBS Lett. 28, 1-5.
22.   Larsen, S.H., Adler, J., Gargus, J.J. and
      Hogg, R.W. (1974) Proc. Natl. Acad. Sci. USA
      71, 1239-1243.
23.   Skulachev, V.P. (1975) Proc. FEBS Meet. 10,
      225-238.
24.   Belyakova, T.N., Glagolev, A.N. and Skula-
      chev, V.P. (1976) Biokhimiya USSR, 41, 1478-
      1483.
25.   Glagolev, A.N. and Skulachev, V.P. (1978)
      Nature 272, 280-282.
26.   Manson, M.D., Fedesei, P., Berg, H.C., Ha-
      rold, F.M. and Van der Drift, C. (1977) Proc.
      Natl. Acad. Sci USA 74, 3060-3064.
27.   Matsuura, S., Shioi, J. and Imae, Y. (1977)
      FEBS Lett. 82, 187-190.
28.   Shioi, J., Imae, Y. and Oosawa, F. (1978) J.
      Bacteriol., 133, 1163-1165.
29.   Halfen, L.N. and Castenholz, R.W. (1970)
      Nature, 225, 1163-1165.
30.   Glagoleva, T.N., Glagolev, A.N. Gusev, M.V.
      and Nikitina, K.A. (1980) FEBS Lett., 117,
      49-53.
31.   Jarosch, R. (1956) Oesterreich. Akad. Wissen-
      schaften, Mathem. Naturwissenschaft, Klasse
      N 6, 58-60.
32.   Skulachev, V.P. (1979) in Membrane Bioener-
      getics, (C.P. Lee, G. Schatz and L. Ernster,
      eds.), pp. 373-392, Addison-Wesley Publishing
      Co., London - Amsterdam - Don Mills - Ontario
      - Sydney - Tokyo.
33.   Lindley, E.V. and MacDonald, R.E. (1979) Bio-
      chem. Biophys. Res. Communs. 88, 491-499.
34.   MacDonald, R.E., Greene, R.V., Clark, R.D.
      and Lindley, E.V. (1979) J. Biol. Chem. 254,
      11831-11838.
35.   Greene, R.V. and Lanyi, J.K. (1979) J. Biol.
      Chem. 254, 10986-10994.
36.   Lanyi, J.K. and Weber, H.J. (1980) J. Biol.
      Chem. 255, 243-250.
37.   Hildebrand, E. and Dencher, N.A. (1975)
      Nature 257, 46-48.
38.   Sperling, W. and Schmiz, A. (1980) Biophys.
      Struct. Mech. 6, 165-169.

39. Dencher, N.A. and Hildebrand, E. (1979) Z.
     Naturforsch. 34, 841-847.
40. Brown, K.T. and Murakami, M. (1964) Nature,
     201, 626-628.
41. Trissl, H.W., Darson, A. and Montal, M.
     (1977) Proc. Natl. Acad. Sci USA 74, 207-210.
42. Trissl. H.W. (1979) Photochem. Photobiol. 29,
     579-588.
43. Montal, M. (1979) Biochim. Biophys. Acta 559,
     231-257.
44. Cone, R.A. and Pak, W.Z. (1972) in Handbook
     of Sensory Physiology (ed. W.R. Loewenstein),
     Springer Verlag, Heidelberg, v. 1, pp. 345-
     367.
45. Bolshakov, V.I., Kalamkarov, G.R. and Ostrov-
     sky, M.A. (1979) Dokl. Akad. Nauk USSR 248,
     1485-1488.
46. Drachev, L.A., Kaulen, A.D. and Skulachev,
     V.P. (1978) FEBS Lett. 87, 161-167.
47. Bolshakov, V.I., Drachev, A.L., Drachev, L.A.,
     Kalamkarov, G.R., Kaulen, A.D., Ostrovsky,
     M.A. and Skulachev, V.P. (1979) Dokl. Akad.
     Nauk USSR 248, 1462-1466.
48. Skulachev, V.P., Drachev, L.A., Kalamkarov,
     G.R., Kaulen, A.D. and Ostrovsky, M.A.
     (1980) in 1st Europ. Bioenerg. Conference
     Abstr., pp. 433-434, Patron Editore, Bologna.
49. Drachev, L.A., Kalamkarov, G.R., Kaulen, A.D.,
     Ostrovksy, M.A. and Skulachev, V.P. (1980)
     FEBS Lett., 119, 125-131.
50. Lewis, A. (1976) Feder. Proc. 35, 51-53.
51. Chailakhian, L.M., Glagolev, A.N. Glagoleva,
     T.N., Murvanidze, G.A., Potapova, T.V. and
     Skulachev, V.P. Biochim. Biophys. Acta (sub-
     mitted).
52. Bakeeva, L.E., Chentsov, Yu.A. and Skulachev,
     V.P. (1978) Biochim. Biophys. Acta 501, 349-
     369.
53. Davidson, M.T. and Garland, P.B. (1977) J.
     Gen. Microbiol., 98, 147-153.
54. Skulachev, V.P. (1980) Biochim. Biophys.
     Acta (accepted).
55. Mitchell, P. (1968) Chemiosmotic Coupling
     and Energy Transduction (Glynn Research,
     Bodmin).
56. Mitchell, P. (1977) FEBS Lett. 78, 1-20.

57.   Skulachev, V.P. (1978) FEBS Lett. 87, 171-
       179.
58.   Broun, I.I., Glagolev, A.N. Grinius, L.L.,
       Skulachev, V.P. and Chetkauskayte, A.V.
       (1979) Dokl. Acad. Nauk USSR 247, 971-974.
59.   Skulachev, V.P. (1979) in Cation Flux Across
       Biomembranes (Y. Mukohata and L. Packer,
       eds.), pp. 303-319. Academic Press, New York
       - San Francisco - London.
60.   Rhoads, D.B. and Epstein, W. (1977) J. Biol.
       Chem. 252, 1394-1401.
61.   Epstein, W., Whitelaw, V. and Hesse, J.
       (1978) J. Biol. Chem. 253, 6666-6668.
62.   Epstein, W. and Laimins, L. (1980) Trends in
       Biochem. Sci. 5, 21-23.
63.   Wagner, G. and Oesterhelt, D. (1976) Ber.
       Deutsch. Bot. Ges. 89, 289-292.
64.   Wagner, G., Hartman, R. and Oesterhelt, D.
       (1978) Europ. J. Biochem. 89, 169-179.
65.   Schuldiner, S. and Fishkes, H. (1978) Bio-
       chemistry 17, 706-710.
66.   Heefner, D.L. and Harold, F.M. (1980) J.
       Biol. Chem. (accepted).
67.   Heefner, D.L., Kobayashi, H. and Harold, F.
       M. (1980) J. Biol. Chem. (accepted).

THE COUPLING OF PROTON TRANSLOCATION TO ATP FORMATION[1]

Paul D. Boyer

Department of Chemistry and Molecular Biology Institute
University of California, Los Angeles, California 90024

INTRODUCTION

A salient observation made some years ago by Lars
Ernster and C.-P. Lee is that a small amount of oligomycin
can help restore coupled phosphorylation by submitochondrial
particles, likely by decreasing nonproductive H^+-translocation
across the coupling membrane (1,2). Their observation impinges
directly on the topic of this essay, namely how proton translocation driven by an electrochemical proton gradient can be
coupled to ATP formation.

Initially, I will discuss experiments and theory that
to me favor the view that proton translocation is coupled to
ATP synthesis by an interacting conformational network of proteins in the ATP synthase complex. This will blend into considerations of how components of the ATP synthase may function
in bringing about energy-linked binding changes of reactants

[1] Preparation of this paper and researches in our laboratory
were supported by U.S. Public Health Service Grant GM 11094,
NSF Grant PCM 75-18884, and Department of Energy DEATO 3-76-
ER70102.

C. P. Lee, G. Schatz, G. Dallner (eds.), Mitochondria and Microsomes
    in honor of Lars Ernster   ISBN 0-201-04576-1

at the catalytic site.   These considerations will include the
interesting question of whether the use of energy-driven conformational changes is in one sense concerted, such that the
energy-utilizing transitions occur essentially in the same
time sequence as the yielding transitions, or whether other
energized states intervene between the conformational machinery
of proton translocation and energy-linked binding changes at
the catalytic site.   The brief essay will close with comments
on some collaborative experiments that may help clarify the
role that components of the ATP synthase complex have in the
conformational interactions

SOME CHARACTERISTICS OF ENERGY-LINKED ION TRANSLOCATION

The transport ATPases - Although how ions are translocated against concentration gradients across membranes
remains far from adequate understanding at the molecular level,
there is emerging recognition of prominent features of how
this is accomplished for the active transport of Na^+, K^+ and
Ca^{++}.   The transport ATPases appear to be molecular machines
in which protein conformational events accompanying ATP binding
and cleavage are used to change the exposure and affinity of
ion binding sites.   The ion binding sites are likely not part
of or intimately associated with the ATP binding sites.   Binding of the transported ions with ATP, ADP or P_i is not part of
the transport process.   Instead, energy originating from ATP
binding and cleavage is transmitted through the three-dimensional structure of the protein to focus on the ion binding
sites, so that the groups forming the binding loci change their
exposure from one side of the membrane to the other.   These
energy-requiring conformational changes are accompanied by
requisite changes in ion affinity.   Transported ions appear to

have free access to the binding sites from the medium--there
is no demonstrated participation of inlet or exit channels,
although these are not ruled out.

Proton translocation across coupling membranes appears
to share some of the salient characteristics of the reversible
Na^+,K^+- and $Ca^{++}-ATPases$. The Na^+,K^+- and $Ca^{++}-transport$
ATPases differ from the mitochondrial and chloroplast ATPases
in that the latter do not form protein acyl phosphate intermediates. But this does not imply basically different mechanisms. In both instances, energy-linked changes in binding and
group exposure occur, likely mediated through conformational
interactions. The similarity is emphasized by the recent
reports from several laboratories of participation of an enzyme
acyl phosphate intermediate in membrane ATPases of E. coli and
Neurospora (3-5). As shown by Dame and Scarborough for the
Neurospora enzyme (5), these ATPases with a phosphorylated
intermediate function in proton translocation. Such results
direct attention to the probability that a structure as complex
as the mitochondrial and chloroplast ATPases is not required
to achieve proton translocation coupled to ATP cleavage. Perhaps control features, including the need to modulate kinetic
characteristics so as to favor either ATP formation or ATP
utilization, underlie the more complex ATP synthases found in
various energy transducing membranes.

An important feature of oxidative phosphorylation and
photophosphorylation, as emphasized by Peter Mitchell, is that
either a transmembrane pH gradient or membrane potential can
readily be used to drive ATP synthesis. That is, an electrochemical proton gradient or protonmotive force provides the
energy. This is not a possibility limited to hydrogen ions.
There is no a priori reason why energy derivable from a

membrane potential must be used by proton translocation.
Mechanisms could be envisaged for coupling of electrochemical
Na^+, K^+ or other ion gradients to ATP synthesis. But there are
advantages to proton systems. An important one is that protonmotive force across a membrane can be generated by vectorial
oxidation-reduction reactions. This includes action of intracellular buffers that can   convert H^+ release to $\Delta\Psi$. A further advantage may be that only a single group (e.g. COO^- or
NH_2) is required to form a proton binding site. This may
allow smaller protein subunits to function in proton translocation than would be plausible for the more complicated
binding sites for the alkali and alkaline earth cations.

Roles for ion channels - An ion channel may be considered as a means of selectively allowing penetration of an
ion into or across an otherwise impermeable membrane. The ion
is regarded as becoming transiently associated with groups
that form the channel or to have exclusive access to aqueous
space in the channel. In some membranes, channels for ion
translocation do have a vital role, as in nerve transmission,
with dissipation of ion gradients and accompanying potential.
In contrast to the conformationally driven ion pumps, such
channels are not designed to use ion gradients to drive ATP
synthesis. To use ion gradients for covalent bond formation,
a barrier to transduce the energy must be interposed between
ion entry on one side and ion release on the other side of
the membrane.

Participation of channels is not a necessary part of
the energy-transducing mechanism but they have frequently been
suggested. Whether a transmembrane migration of a proton
involving only a single group should be considered a channel
is uncertain. It may be preferable to use designations of

channels for mechanisms involving combination of the proton
with more than one group, such as the intriguing suggestions
for proton flow made by Kayalar (6), involving keto-enol shifts
in H-bonded peptide groups, and by Nagle and Morowitz (7),
involving protein side groups in H-bonded chains. Also, proton "wells" as suggested by Mitchell (8) could be regarded as
channels if the "wells" have a gating mechanism at the membrane
surface that somehow prohibits entry of ions other than protons
to an aqueous domain in the membrane.

One attractive type of proton channel might involve
proton flux through part of the F_1-ATPase, in particular the
β-subunit. Such a possibility may be related to observations
from Ernster's laboratory of uncoupler-sensitive inhibition of
F1 by bathophenanthroline complexes (9) and to observations by
Vignais and collaborators of inhibition of the ATPase by
reaction of dicyclohexylcarbodiimide (DCCD) with a specific
locus on the β-subunit (10). Direct involvement of the β-
subunit is in harmony with the recent observation made by
Hackney that this subunit of the F_1-ATPase appears to cross-
link preferentially with phospholipids of the membrane (11).

If proton passage through part of the F_1 occurs, the
structure may be such that only a quite limited number of
groups and a short distance between aqueous phases on both
sides of the membrane are involved. A number of multisubunit
protein complexes appear to have donut-like structures with
central holes that, if present in F_1, could help provide
aqueous access for a proton.

Relatively extensive channels could have some disadvan-
tages for energy-linked proton translocation. Charged groups
or dipoles in a channel might decrease the possibility of
attaining a high activity of protons impinging on an energy

barrier. With a channel blocked by an energy barrier, there
would be no build-up of proton activity by back diffusion from
a possible concentration gradient across the membrane. It
could be structurally difficult to restrict the change in ion
activity or the potential drop to the energy barrier in a long
channel. Such restriction would seem desirable for efficient
coupling. Efficient operation might thus become increasingly
difficult the longer or more complicated the channel, and
particularly difficult if the energy barrier is placed at the
catalytic site where charges abound. Structural complications
could also place limitations in focusing of a channel for
delivery of protons at the catalytic site. Channel leakage
could also pose a problem because protons are hard to retain.
Proton permeability in liposomes in orders of magnitude greater
than for other ions (12), although membranes still impose a
sufficient barrier to allow accumulation and use of protonmotive force.

Channels and the H^+/ATP ratio. Stoichiometry raises
additional questions about channels extending through the
ATPase. One possible way to accomodate translocation of 3
protons for each ATP made would be to have 3 separate channels
in the ATPase. Accommodation of 3 separate channels passing
through a single β-subunit and possibly through interposing
γ- or δ-subunits seems somewhat difficult. Another possible
way would be a single channel that is conformationally modified
by each proton passage in a cumulative manner. Again, this
poses unusual but perhaps not insurmountable structural problems.

At least part and perhaps all of a proton translocation
channel appears to occur in a small hydrophobic membrane protein that reacts readily with DCCD (13). The evidence that
there are present several molecules of the DCCD binding protein

for each ATP synthase, and that these show negative coopera-
tivity of derivitization and inactivation of ATPase when only
one protein in a cluster is modified (14-16), suggests a more
appealing way to handle the stoichiometry problem. Simultaneous, or more likely, successive passage of 3 protons through
3 different proteins in an associated cluster of DCCD binding
proteins could drive conformational changes that are transmitted through protein subunit interactions to the catalytic
site of the associated synthase complex.

A MINIMAL MODEL FOR H^+-TRANSLOCATION

A version of a model for the coupling of proton translocation to ATP formation that I suggested several years ago
(17) is depicted in Fig. 1. The model was based in part on
studies with the DCCD-binding protein (13) and has acquired
some additional credibility from more recent developments
showing that DCCD combines with a single side-chain carboxyl
group flanked by several hydrophobic residues on each side
(18,19). It has gained increased attractiveness with the
accumulating evidence that ATP synthesis is accomplished by
energy-linked binding changes coupled to proton translocation
(20). The model requires at least four different states,
designated A, B, C, and D, with interconversion by steps 1,
2, 3, and 4 and with characteristics as follows:

1. Step 1. Charged group migration from matrix
    to cytoplasmic side. May be driven by mem-
    brane potential.

2. Step 2. Protonation-deprotonation from
    matrix side. This must also be accompanied
    by conformational change so that events at

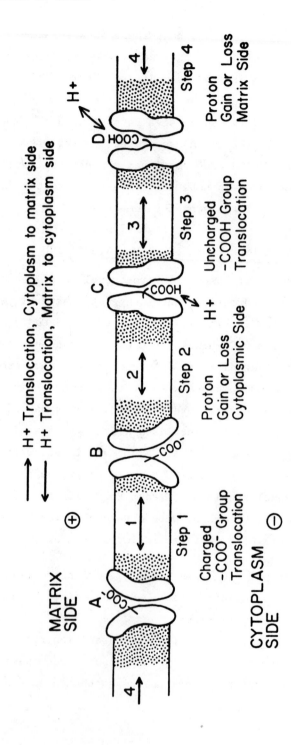

Fig. 1.   A sequence for proton translocation in coupling membranes.

the ATP synthesis site are coordinated with
those at the H^+-translocation site and so
that the subsequent Step 3 will not be a
simple reversal of changes accompanying
Step 1.

3. Step 3.   Uncharged group migration, with
   further conformational shifts coordinated
   with the ATP synthesis site.
4. Step 4.   Protonation-deprotonation from the
   matrix site, with changes akin to those
   listed with Step 2.

It deserves emphasis that all four steps need to be
accompanied by conformational changes interlinked and coordinated with those at the ATP synthesis site, one or more of
which could be associated with major energy changes. Vector-
ality and coupling necessitate that steps of H^+ translocation
cannot procede without associated requisite steps of reactant
binding, interconversion and release.

If this conformational machine were designed to be
driven largely by potential, as may often apply to mitochondria,
a probable preferred state in the absence of a membrane potential would be state A, ready to be driven by negative potential
on the matrix side. If a proton gradient provides the principal driving force, as may often apply to chloroplasts, a probable preferred state in absence of a low pH inside the thylakoids (corresponding to the cytoplasmic side of the mitochondrial membrane) would be state B, ready to be driven by lowered
intrathylakoid pH.

If a prominent component of the use of membrane potential should be to promote migration of the negatively charged
carboxyl group there could be some advantage to the lack of

transient protonation of additional groups in channels.   In
the model of Fig. 1 the potential would drop across a short
distance and could be essentially focused on movement of the
carboxyl group. With an intervening channel, the potential
drop may occur over a longer distance, and thus movement of a
charged group at the energy barrier might be less favored
energetically.

SOME EXPERIMENTAL EVIDENCE FOR CONFORMATIONAL COUPLING

If the translocated proton(s) participate directly in
ATP formation by aiding removal of oxygen from P_i to form
water (8,21) then the reversible oxygen exchange between water
and P_i should depend upon the retention of the effective pro-
ton(s) at the catalytic site. Thus, when ATP is cleaved,
hydrogens from the water oxygen that is incorporated into P_i
would be visualized as entering a proton channel, and to be
essentially retained at the juncture of the channel and the
catalytic site to make possible the reversible removal of
water from P_i, as demonstrated by the $P_i \rightleftharpoons$ HOH exchange. The
initial experimental evidence that this was not the case came
from experiments in the laboratory of Robert Mitchell (22) and
my laboratory (23) showing that the intermediate $P_i \rightleftharpoons$ HOH oxygen exchange accompanying ATP cleavage remained rapid in the
presence of uncoupler concentrations sufficient to stop net
oxidative phosphorylation.

Even more striking evidence came from subsequent experimental observations that with the purified ATPase preparations
extensive reversal of oxygen departure from P_i could readily
be observed during cleavage at low ATP concentrations (24,25).
With submitochondrial particles and oligomycin-sensitive
ATPase preparations reversal of ATP cleavage during net ATP

hydrolysis is observed at somewhat higher ATP concentrations. That these oxygen exchanges reflect a property of the functional ATP synthase complex is shown by a similar behavior of a reconstituted oligomycin-sensitive ATPase (26).

Measurement of the effect of ATP concentration on ^{18}O-isotope distribution in the P_i produced gives strong support to the view that the increased oxygen exchange per P_i released at lower ATP concentration arises from the increased retention of bound substrates at the catalytic site (25). Oxygen exchange with bound reactants continues until ATP binding at an alternate site promotes release of ADP and P_i. This unusual retention of bound reactants, at a substrate concentration well below its K_m value, has subsequently been confirmed by the direct demonstration that each catalytic site retains nearly one bound ADP during steady state cleavage at low ATP concentrations (28).

When F_1-ATPase catalyzes hydrolysis with high ATP concentrations there is very little oxygen exchange accompanying P_i release, demonstrating that if any proton channels exist as part of the mechanism, they are "wide open" in the isolated ATPase. Such a result is as anticipated unless some special closing mechanism accompanied the rupture of attachment of the F_1 to the membrane. This seems unlikely because of the more rapid ATP cleavage by the free ATPase. Removal or modification of limiting channels is not the most attractive reason for rate increase. The more rapid cleavage may be attributable to the removal of the protein-protein interactions that are otherwise required to drive the translocation of protons across the membrane. With the isolated ATPase, the conformational changes initiated at the catalytic site are freed from the constraint of driving the conformational changes required for proton translocation.

These experimental results show that the extensive reversal of phosphate activation, and indeed quite likely of ATP formation*, at the catalytic site of the ATPase occurs without any membrane potential or proton gradient. Such behavior is quite in accord with the developing knowledge of enzyme catalysis. Interconversion of bound reactants and products at enzyme catalytic sites may occur with relative ease-- that is rapidly and more readily than reactant release. Quite independent from experimental results with the ATP synthase, the growing understanding of enzyme catalytic sites could have lead to the suggestion that in oxidative phosphorylation energy should be used to get ADP and P_i appropriately bound and to release the ATP which is formed and which must be tightly bound. But if the protonmotive force drove the conversion of ADP + P_i to ATP at the catalytic site by helping pull water off from P_i, then reactants bound at the catalytic sites of the membrane-bound or the free ATPase would not be expected to be readily interconverted, and further, any ATP that is formed would not be expected to be tightly bound. The energy available from the proton translocation would have been used for covalent bound formation without a driving force remaining to bring about ATP release.

---

* The participation of a phosphorylated intermediate, made unlikely by earlier observations, has been further eliminated by the demonstrated inversion of configuration around the P atom accompanying ATP cleavage (27). However, participation of a species such as a transient metaphosphate remains possible.

EVIDENCE FOR CONFORMATIONAL INTERACTIONS BETWEEN

THE CATALYTIC SITE AND PROTON CHANNEL

The oligomycin and F_1-ATPase inhibition of proton translocation - When F_1-ATPase is removed from mitochondrial membranes the oxidations continue to release or translocate protons but development of protonmotive force is decreased because the remaining F_0 component is more permeable to protons. Oligomycin can help restore protonmotive force by reducing dissipation of protons through F_0. Such behavior was recognized by Ernster as the basis of the observations mentioned earlier (1,2) and the ability of oxidations but not ATP to elicit the analinonaphthalene sulfonate (ANS) response to energization with oligomycin present (29). One possible explanation is that the F_1-ATPase forms part of the proton channel and that when F_1 is removed oligomycin binding physically blocks the uncovered portion of the channel in F_0. But there are indications that the oligomycin site is not carried by the DCCD-binding protein. Enns and Criddle have presented data suggestive of a separate oligomycin-binding protein (30). Mairouch and Godinot have shown that trypsin prevents oligomycin sensitivity by reacting with a protein at the cytoplasmic surface of the inner mitochondrial membrane (31). Observations from several laboratories have indicated that the oligomycin binding site is sensitive to changes at the catalytic site for ATP synthesis (32-34).

At this stage, it appears worthwhile to consider the possibility that the blocking of proton translocation by attachment of the F_1-ATPase to the membrane F_0 or by oligomycin binding to F_0 lacking F_1 results from behavior of the interactive conformational network that drives ATP synthesis. When F_1-ATPase becomes bound its "gears and machinery" need to be

attached to those of the proton translocating F_0 component.
The result is that no proton translocation can occur unless
the catalytic site on F_1 can bind substrates and undergo its
cycle of binding changes. Thus, after F_1 binds to F_0 the key
carboxyl group of the DCCD-binding protein could still have
potential exposure to the aqueous phases on either side of
the membrane, but change of exposure now demands driving of
the conformational machine. This cannot occur without ADP
and P_i binding at the catalytic site. Oligomycin binding
could jam some of the "gears" of the machine even with ADP
and P_i present.

The role of the γ- and δ-subunits - In the binding
change mechanism for ATP synthesis, at any one time there
would be expected to be asymmetry of structure of the α,β-
subunit pairs. Such asymmetry is likely reflected in asymmetry
of interactions of the α,β-subunits with the γ- and δ-subunits
and with any other proteins present as single components in the
ATP synthase complex. If this is the case, then it seems
likely that cyclic changes in properties and orientation of
the smaller, singly represented subunits of the ATP synthase
complex would accompany the synthesis of each ATP molecule.
Some support for this view has come from studies of changes
in the reactivity of the γ-subunits. Of particular interest
is the demonstration from McCarty's laboratory that modifications in the γ-subunit can cause an increased leakage of protons (35,36). This could result from an important role of the
γ-subunit in the energy transductions and a partial blocking
of a conformational state that exists during the transient
exposure of a SH group. When trapped in the conformation, a
restraint on the proton-translocating protein could be released,
and protein leakage favored.

Evidence for conformational restraints from oxygen exchange measurement – Other experimental findings indicate the presence of conformational barriers to ATP cleavage following the initial binding of ATP at the catalytic site. The continuation of the intermediate $P_i \rightleftharpoons$ HOH exchange accompanying ATP cleavage by submitochondrial particles at relatively high ATP concentrations shows that the ATPase on an intact coupling membrane is subject to rate limitation not imposed on the free ATPase. Some or most of this rate limitation

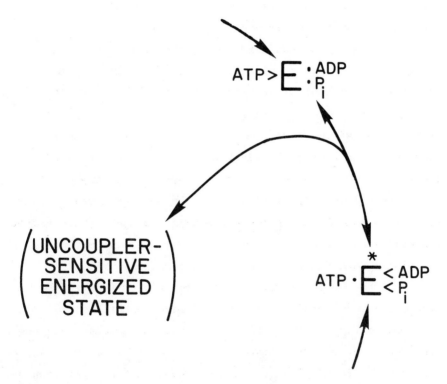

Fig. 2. A portion of a reaction sequence of the binding-change mechanism for ATP synthesis. Reactants that are transitorily tightly bound are indicated by a <. Tightly bound ADP and P_i are readily and reversibly converted to ATP with elimination of HOH.

appears to remain in the presence of sufficient uncoupler to
stop net oxidative phosphorylation as continued $P_i \rightleftharpoons HOH$
exchange accompanies ATP cleavage at high ATP concentration
(22,23). The reaction step likely involved in the rate limitation may be illustrated by a key portion of a schematic
model for the energy-linked binding change mechanism as given
in Fig 2. With the intact synthase where events capable of
proton translocation must accompany ATP cleavage, the transition indicated in Fig. 2 would be expected to be slower than
with the free ATPase. This could allow reversal of the pre-
ceeding step in which ATP hydrolysis occurs.

In Fig. 2 the conformational changes are depicted as
changing binding at one site from "tight" to "loose" and binding at the coordinated catalytic site from "loose" to "tight."
This key conformational change is regarded as linked to an
uncoupler-sensitive energized state. We chose this designation
because of the uncertainty in the nature of the rate limitation
imposed on ATP cleavage with intact membranes in presence of
excess uncoupler. The limitation could be $\Delta\mu_H^+$ itself, a surface or intramembrane proton convertible to $\Delta\mu_H^+$, or a strained
conformation intermediate between the changes accompanying
ATP binding and those of proton translocation.

If the observed rate limitation results from the electrochemical proton gradient, this would mean that the step
depicted in Fig. 2 cannot take place without accompanying proton translocation. Such interaction is attractive as a minimal scheme. It is also attractive because a transient accumulation of a strained conformationally energized structure
would not be required.

To help decide between these alternatives, it would be
useful to have data on how fast the proton channel can operate

to allow flow of protons in both directions. That the translocation can be rapid in one direction (driven by the electrochemical proton gradient formed by oxidations) is shown by the rapid oxidation that can occur when only some of the membrane is without the coupling ATPase. But information on how fast the proton translocation device can be driven in the other direction by ATP cleavage is lacking. The maximal rate of ATP cleavage in a coupled membrane gives only a lower limit.

Some light may be shed on this problem by cooperative experiments we have planned with the encouragement and participation of Lars Ernster. This is to test the effect of various proteins comprising the ATP synthase complex on rate limitation as measured by both ATP cleavage and oxygen exchange in fractionated and reconstituted systems. If protein components of the synthase complex can impose rate limitation without formation of a coupled proton translocation channel, this would be indicative of rate limiting conformational changes intermediate between the catalytic site and the proton channel.

For such experiments, it is necessary to measure oxygen exchange accompanying ATP cleavage at both high and quite low ATP concentrations. For the latter, accumulation of sufficient P_i for analysis can be aided by an ATP regenerating system, such as from $[^{18}O]$phosphoenolpyruvate. Our collaborative experiments have been delayed by some difficulties encountered in preparation of phosphoenolpyruvate highly ^{18}O-labeled in the phosphoryl group to serve for regeneration of low levels of $[^{18}O]$ATP. Unexpected oxygen exchange of phosphoenolpyruvate has been encountered during preparative and assay procedures starting with $[^{18}O]P_i$. We have now obviated the difficulties by defining conditions for these somewhat unexpected exchanges, and are using an H^+-catalyzed exchange with $H^{18}OH$ for preparation of the $[^{18}O]$phosphoenol-pyruvate.

CONCLUSION

Experimental evidence and known properties of proteins support the view that the energy-yielding translocation of protons across a coupling membrane drives ATP synthesis by energy-linked conformational changes in the ATP-synthase complex. As with the transport ATPases, the translocated ion probably is not at any time bound directly to ATP, ADP or P_i. Oligomycin inhibition is readily rationalized as a blocking of required conformational changes. The stoichiometry of 3 protons translocated per ATP made can be accommodated by a required passage of H^+ through 3 of the cluster of the DCCD binding proteins present in the F_o component of the synthase. Cyclic asymmetric interaction of these clusters with the catalytic sites on β subunits of the F_1 component (and likely also involving asymmetric interactions with α, γ and δ subunits) could drive the energy-linked binding changes for ATP synthesis.

There may be distinct advantages to a conformational coupling where a carboxyl group of the DCCD-binding protein is the only group to which the proton becomes attached during its translocation. Conformational transitions accompanying the proton translocation are regarded as driving changes in binding of reactants at the catalytic site. Conversely, conformational changes driven largely or wholly by ATP binding can drive proton translocation. A continued intermediate $P_i \rightleftharpoons HOH$ exchange accompanying net ATP cleavage may be accounted for by rate limitation of this key binding change step. Additional experimental data are needed to ascertain

if this rate limitation results from a required simultaneous
build up of protonmotive force, thus obviating the necessity
of postulating transient energy storage in a strained conformation of the interlocking protein network.

REFERENCES

1. Lee, C. and Ernster, L.   Biochem. Biophys. Res. Commun.
    18, 523 (1965).
2. Lee, C.-P. and Ernster, L.   Europ. J. Biochem. 3, 391
    (1968).
3. Dufour, J.-P., Boutry, M., Goffeau, A.   J. Biol. Chem.
    255, 5735 (1980).
4. Serrano, R. and Malpartida, F. in, "Membrane Bioener-
    getics," C.-P. Lee, G. Schatz and L. Ernster, Eds.,
    Addison Wesley, Reading, Mass., 559-567 (1979).
5. Dame, J. B. and Scarborough, G. A.   Biochemistry 19,
    2931 (1980).
6. Kayalar, C.   J. Membrane Biology 45, 37 (1979).
7. Nagle, J. F., Morowitz, H. J.   Proc. Natl. Acad. Sci.
    USA 75, 298 (1978).
8. Mitchell, P.   Science 206, 1148 (1979).
9. Phelps, D. C., Nordenbrand, K., Nelson, B. D., and
    Ernster, L.   Biochem. Biophys. Res. Commun. 65, 1005
    (1975).
10. Pougeois, R., Satre, M., Vignais, P. V.   Biochemistry
    18, 1408 (1979).
11. Hackney, D. D.   Biochem. Biophys. Res. Commun. 94, 875
    (1980).
12. Nichols, J. W. and Deamer, D. W.   Proc. Natl. Academy
    Sci. 77, 2038 (1980).
13. Beechey, R. B., Holloway, C., Knight, I. G. and Roberton
    A. M.   Biochem. Biophys. Res. Commun. 23, 75 (1966).
14. Sigrist-Nelson, K., Sigrist, H. and Azzi, A.   Eur. J.
    Biochem. 92, 9 (1979).
15. Sebald, W., Graf, T. and Lukins, H. B.   Eur. J. Biochem.
    93, 587 (1979).
16. Kiehl, R. and Hatefi, Y.   Biochem. 19, 541 (1980).
17. Boyer, P. D.   FEBS Letters 58, 1 (1975).
18. Sebald, W., Machleidt, W., and Wachter, E.   Proc. Natl.
    Acad. Sci. USA 77, 785 (1980).
19. Wachter, E., Schmid, R., Deckers, G. and Altendorf, K.
    FEBS Letters 113, 265 (1980).

20. Rosen, G., Gresser, M., Vinkler, C., and Boyer, P. D. J. Biol. Chem. 254, 10654-10661 (1979).
21. Mitchell, P. FEBS Letters 43, 189-194 (1974).
22. Russo, J. A., Lamos, C. M. and Mitchell, R. A. Biochem. 17, 473-480 (1978).
23. Kayalar, C., Rosing, J., and Boyer, P. D. J. Biol. Chem. 252, 2486-2491 (1977).
24. Choate, G. L., Hutton, R. L., and Boyer, P. D. J. Biol. Chem. 254, 286-290 (1979).
25. Hutton, R. L. and Boyer, P. D. J. Biol. Chem. 254, 9990-9993 (1979).
26. Ernster, L., Carlsson, C., and Boyer, P. D. FEBS Letters 84, 283-286 (1977).
27. Webb, M. R., Grubmeyer, C., Penefsky, H. S. and Trentham, D. R. J. Biol. Chem. 255, 11637-11639 (1980).
28. Gresser, M. J. Fed. Proc. 39, 1240 (1979).
29. Ernster, L., Nordenbrand, K., Lee, C.-P., Avi-Dor, Y. and Hundal, T., in Energy Transduction in Respiration and Photosynthesis, eds. E. Quagliariello, S. Papa and C. S. Rossi (Adriatica Editrice, Bari) p. 57 (1971).
30. Enns, R. K. and Criddle, R. S. Arch. Biochem. Biophys. 182, 587-600 (1977).
31. Mairouch, H. and Godinot, C. Proc. Natl. Acad. Sci. 74, 4185-4189 (1977).
32. Azzi, A. and Santato, M. FEBS Letters 7, 135-139 (1970).
33. Bertina, R. M., Stefnstra, I. A. and Slater, E. C. Biochim. Biophys. Acta. 368, 279-297 (1974).
34. Klein, G., Satre, M. and Vignais, P. FEBS Letters 84, 129-134 (1977).
35. McCarty, R. E. and Fagan, J. Biochemistry 12, 1503-1507, (1973).
36. Moroney, J. V. and McCarty, R. E. J. Biol. Chem. 254, 8951-8955 (1979).

BIOCHEMICAL MECHANISM OF PROTONMOTIVATED PHOSPHORYLATION

IN $F_O F_1$ ADENOSINE TRIPHOSPHATASE MOLECULES

Peter Mitchell

Glynn Research Institute, Bodmin, Cornwall, PL30 4AU,
England

The main object of this review is to consider how the fundamental process of protonmotivated phosphorylation catalysed by the reversible protonmotive $F_O F_1$ ATPases of mitochondria, bacteria and chloroplasts may be explained by the topological organisation and precisely articulated diffusional movements of specific ligand-conducting components in the chemiosmotic catalytic domain extending through the $F_O F_1$ osmoenzyme molecules.

The general principle of enzyme-catalysed chemical group translocation (1) was first used to outline a hypothetical OH^--group (or O^{2-}-group) translocation mechanism for proton-motivated phosphorylation by a reversible phosphatase or ATPase at a symposium in September 1960 in Stockholm (2), where Lasse Ernster began to help me to become initiated into the mysteries of mitochondrial biochemistry. I have been much indebted to him for producing with C. P. Lee a masterly review of experimental knowledge of oxidative phosphorylation in 1964 (3), that opened up to the studious outsider a uniquely rich

C. P. Lee, G. Schatz, G. Dallner (eds.), Mitochondria and Microsomes
in honor of Lars Ernster   ISBN 0-201-04576-1

mine of information, and also for his continuing help and friendship, despite the differences in our scientific opinions (4). So, it is a special pleasure to offer my present attempt to make further progress in understanding and describing the biochemistry of protonmotivated phosphorylation as a tribute to Lasse Ernster in celebration of his sixtieth birthday.

Since the reversible protonmotive ATPase was postulated twenty years ago without experimental support for its existence (2,5), and Racker pioneered the identification and purification of the $F_O F_1$ ATPases (6), so much experimental evidence has been published about the ubiquitous occurrence and remarkably uniform structure of these ATPases (6-17) in the coupling membranes of mitochondria (18,19), bacteria (20,21) and chloroplasts (22-24) that it is not possible to review the literature at all comprehensively in an article of this length. Therefore I propose to concentrate attention mainly on conceptual and mechanistic questions about the $F_O F_1$ ATPases that appear to have been comparatively neglected.

STRUCTURES OF F_O AND F_1

Following leads by Racker, research in many laboratories (6-25) has established that the complete protonmotive ATPases of mitochondria, bacteria and chloroplasts consist of a relatively hydrophobic laminar part (F_O), of molecular weight around 100,000, that is physically continuous with the hydro-carbon osmotic-barrier domain B of the primary membrane, and a relatively hydrophilic knob-shaped part (F_1), of molecular weight around 380,000, that is attached to F_O and projects from the surface of the membrane into the protonically neutral or negative aqueous domain N (25,26). The cumulative evidence

strongly supports the notion that F_1 projects about 10 nm into
the aqueous N domain (17,27) and that, contrary to the
suggestion by Kozlov and Skulachev (8), it is not embedded in
the non-polar B domain of the membrane. The F_O component of
the $F_O F_1$ ATPases is tightly associated with lipid, but the F_1
component is not, and it seems very unlikely that the hydrophilic surface of F_1 would readily enter the hydrophobic B
domain of F_O or the neighbouring hydrocarbon B domain of the
primary membrane. Thus, it is probable that F_1 is largely
shielded from the electric field across the B domain that
arises from the electric membrane potential $\Delta\psi$ between the
aqueous P and N domains (28).

Both F_O and F_1 are composed of several polypeptide subunits, and there are only minor variations of composition in
$F_O F_1$ ATPases from different sources (6-25). The main
constituents of F_O are a proton-conducting oligomeric proteolipid that reacts specifically with DCCD, and polypeptides
involved in connecting F_O to F_1, such as OSCP, F_6 and the 28-K
subunit of mitochondrial F_O preparations (17,25), that are
represented as Fixers in Fig. 1. There are five types of
polypeptide subunit in F_1, which are named α, β, γ, δ and ε in
descending order of molecular weight from about 55,000 to
about 10,000. A consensus of opinion has not yet been reached
about the numbers of each subunit in the molecules of F_1. The
majority view appears to be that F_1 from beef heart mito-
chondria and bacteria has the composition $\alpha_3 \beta_3 \gamma \delta \varepsilon$, but that
there are only 2 α and 2 β subunits in F_1 from chloroplasts
(7-24). However, the known tendency of the large subunits to
dissociate from F_1 preparations, pointed out by Kagawa
(10-12) and others (18,19,29), suggests that the subunit
composition of F_1 from chloroplasts, as well as from mito-

chondria and bacteria, may well be $\alpha_3\beta_3\gamma\delta\epsilon$. This would imply that about 85% of the F_1 molecule consists of the α and β subunits.

Chemical tagging, crosslinking and other studies (8-19,30) have shown that γ, δ and ϵ are associated with the region of F_1 that binds components of F_0 through salt linkages (31). The interaction of δ with the rest of F_1 depends on Mg^{2+}, and δ plays a major part in binding F_1 to F_0 (17,32-34). However, the α and β subunits are amphiphilic (35) and anisometric (36-38), and there is evidence that both types of subunit are associated with components of F_0 (39-41). Therefore it seems likely that both α and β extend most if not all the way between the P and N poles of F_1, as indicated crudely by the dashed lines between the equatorial α and β symbols in Fig. 1.

There is much physical evidence for changes of conformation in and between the subunits of F_1 (6-19). The most dramatic correspond to relatively slow transitions of F_1 between catalytically inactive molecular species containing occluded adenine nucleotide molecules, and catalytically

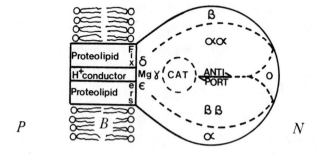

Fig. 1   Diagram of F_0F_1 ATPase. Greek letters show sub-
         units of F_1. Subunits $\alpha\alpha$ and $\beta\beta$ are above and
         below plane of paper. CAT = catalytic site.
         O = entrance to open substrate-binding site.

active species in which the adenine nucleotides are exchange-
able with those in the medium (9,42-48). But, amongst other
observations (6-19), studies of variations of hydrogen
exchange rate (37,49) and of the molecular asymmetry (38) of
the subunits of F_1 in the presence and absence of adenine
nucleotides also imply that F_1 undergoes remarkably extensive
changes of conformation in the course of catalysing the
chemiosmotic process of reversible protonmotive ATP
hydrolysis. As I have pointed out earlier (50), such
extensive conformational changes are predictable in osmo-
enzymes (51) catalysing chemiosmotic reactions that involve
the specifically controlled diffusion of large substrates,
such as adenine nucleotides, into and out of the catalytic
site in the osmotic barrier B domain of the osmoenzyme polypeptide system (Fig. 1). Very closely articulated
conformational movements are presumably required to prevent
non-specific leakage of solutes (such as hydrogen ions)
through the B domain during the chemiosmotic reaction (50).

FUNCTIONS OF F_O AND F_1

Racker's pioneer work and many subsequent studies (6-25)
have shown that the catalytic site for ATP hydrolysis by $F_O F_1$
is contained in F_1; and my suggestion that F_O is a specific
oligomycin-sensitive or DCCD-sensitive conductor of protons
through the B domain of the membrane (52-56) has been widely
confirmed and is now generally accepted (6-25).

The uniport of H^+ ions through the oligomeric proteolipid
component of F_O is normally interrupted by the presence of F_1
on the N side of F_O. This important observation may be interpreted to mean, either that F_1 inhibits proton conduction

through the proteolipid indirectly (for example, by inducing a conformational change in F_O), or that F_1 forms an ion-tight seal against the N side of F_O and thus controls the proton flow by lying directly in the protonic pathway in series with F_O between the P and N aqueous domains, as indicated in Fig. 1. Although there is at present no experimental evidence to rule out the former explanation, I have adopted the latter as a working hypothesis because it is consistent with the direct chemiosmotic type of mechanism that seems to me to stand a much better chance of being true to reality than indirect or proton pump types of chemiosmotic mechanism (54,57).

As the proton conducting region of F_O passes through the non-polar proteolipid of F_O, it would be expected to act as what I have described as a proton well (28), which transforms the electric component $\Delta\psi$ of the total protonic potential difference $\Delta\bar{\mu}H^+$ across the primary membrane into a corresponding pH component $-Z\Delta pH$, where Z is the conventional factor 2.303 RT/F (28,51,55,57). Thus, we would expect that, irrespective of whether the total protonic potential difference $\Delta\bar{\mu}H^+(B)$ across the B domain of the primary membrane were mainly due to electric or pH components, the effect of $\Delta\bar{\mu}H^+(B)$ on F_1 would be equivalent to the acidification of the P pole of F_1, at the bottom of the proton well through F_O, relative to the surface of F_1 in contact with the aqueous domain N, by an amount approaching $-\Delta\bar{\mu}H^+(B)/Z$. Recent work by Maloney and Schattschneider (58), comparing the kinetics of phosphorylation of ADP in bacterial F_OF_1, driven either by $\Delta\psi(B)$ or by $-Z\Delta pH(B)$, extended earlier less quantitative observations on ADP phosphorylation in mitochondria and chloroplasts (cited in 58), and showed that F_1 does not detect any difference between equivalent values of $\Delta\psi(B)$ and

$-Z\Delta pH(B)$. It seems likely, therefore, that we are correct in our expectation that the predominant biochemical action on F_1 of so-called energisation of the membrane arises from the protonic potential difference $\Delta\bar{\mu}H^+(F_1)$, applied via F_O, between the P pole and the N surface of F_1, that corresponds to what can appropriately be called the equivalent pH difference $-\Delta\bar{\mu}H^+(F_1)/Z$. This equivalent pH difference, $Eq\Delta pH(F_1)$, may include an electric component $-\Delta\psi(F_1)/Z$ due to locally trapped protons (see 51) and other fixed charges in the P and N domains of the F_1 molecules. But the value of the electric potential difference $\Delta\psi(F_1)$ across F_1 would be determined by the biochemistry of F_1 (51), and would be virtually independent of the electric potential difference $\Delta\psi(B)$ across the primary membrane at constant $\Delta\bar{\mu}H^+(B)$. At or near protonic equilibrium, the relationship between the total protonic potential difference across the B domain of the primary membrane $\Delta\bar{\mu}H^+(B)$, and the equivalent pH difference $-\Delta\bar{\mu}H^+(F_1)/Z$ between the P pole and N surface of F_1, should be given by:

$$Eq\Delta pH(F_1) \quad = \quad -\Delta\bar{\mu}H^+(F_1)/Z \quad = \quad \Delta pH(F_1) - \Delta\psi(F_1)/Z \qquad (1)$$
$$\leq \quad -\Delta\bar{\mu}H^+(B)/Z$$

where the difference Δ means the value of the variable on the P side minus that on the N side. Thus, for a $\Delta\bar{\mu}H^+(B)$ of 240 mV, and with the aqueous domain N at pH 8, the value of $EqpH_P(F_1)$, the equivalent pH at the P pole of F_1, would be near 4.

Enzymes located in homogeneous aqueous domains generally exhibit optimum catalytic activity at a single pH optimum (near the pH of the natural domain), because the protonation states of groups in the catalytic site and elsewhere influence

the proportion of the enzyme molecules in the most catalytically active states. The $F_O F_1$ ATPases, which are located
between two aqueous phases at different pH values, exhibit
so-called energy-dependent activation phenomena (9,42-48,
59-63) that may, I suggest, be regarded biochemically as
manifestations of a dual pH optimum, corresponding functionally to the single pH optimum of normal enzymes (2,64).
Fig. 2 illustrates a hypothetical example of a dual pH
optimum, given by a plot of the activity of a reversible $F_O F_1$
ATPase against the equivalent pH values, pH_P and pH_N, at the
P pole and N surface of F_1, respectively.

The dual pH optimum for the kinetic catalytic activity of
F_1 in the $F_O F_1$ ATPases corresponds to a value of $\Delta\bar{\mu}H^+(F_1)$ that
may poise the ATPase reaction near thermodynamic equilibrium.
For that reason, the kinetic and thermodynamic or

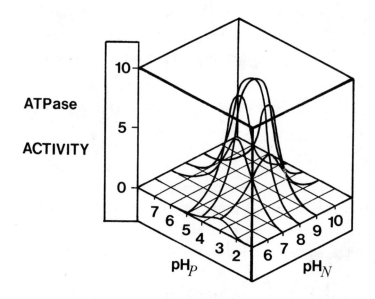

Fig. 2   Hypothetical plot of dual pH optimum of $F_O F_1$
ATPase. pH_P and pH_N represent pH values of
the P pole and N surface of F_1 respectively.
ATPase activity (αV) is in arbitrary units.

stoichiometric functions of the F_OF_1 ATPases have tended to
become confused when explicit measures were not taken to
identify and consider them separately.

All the F_OF_1 ATPases studied exhibit dual pH optima,
although they have never, as far as I know, been derived
systematically from the plot of a pH-activity surface, as
illustrated by Fig. 2. From studies of ATPase activity at
given pH values, we can infer that the pH-activity surface is
strongly dependent on the phosphate and other anion content of
the medium (65,66), as well as on its Mg^{2+} and other cation
content (6-25). Comparatively recent observations on slow
active/inactive state transitions of the ATPases of chloroplasts, mitochondria and bacteria (42-48,67,68) have revealed
that the kinetic activity of the ATPases when saturated with
substrate can only be properly described by an expression of
the form:

$$\Sigma \alpha V \;=\; \alpha_1 V_1 + \alpha_2 V_2 + \alpha_3 V_3 \text{ etc.,} \tag{2}$$

where each α represents the proportion of the enzyme molecules
in a certain activity state, characterised by the corresponding velocity of turnover V (43,68). In practice, most (but
see 68) F_OF_1 ATPases appear to exhibit only one major active
state (42-48,59-63,67), and the kinetic activity under
saturated conditions may be described by αV, in which V
normally represents a turnover number of around 200 sec^{-1}, and
the active/inactive state transitions giving the equilibrium
activity ratio α generally have a turnover number of less than
10 sec^{-1}. The pH-activity surfaces for most F_OF_1 ATPases can
probably be taken to approximate to pH-αV surfaces in which V
is a constant, at least over much of the pH area.

When pH_P and pH_N are about the same, and are in the pH range 5 to 9, the F_OF_1 ATPases of chloroplasts and certain bacteria are virtually inactive, but those of mitochondria retain considerable activity (6-25,42-48). It follows that the pH-activity surface near the dual pH optimum falls towards zero activity (zero α) much more steeply in the F_OF_1 ATPases of chloroplasts and certain bacteria (see 69) than in those of mitochondria. This functional difference can be traced to the obvious physiological difference that $\Delta\bar{\mu}H_P^+(F_1)$ will tend to be subject to relatively wide variations in chloroplasts and bacteria, because of the natural variation of environmental factors such as light intensity or availability of respiratory substrates. A relatively steep dependence of the reversible F_OF_1 ATPase activity (α) on $pH_P(F_1)$ would tend to minimise variation of $\Delta\bar{\mu}H_P^+(F_1)$ by controlling the protic power consumption by the F_OF_1 ATPase molecules under conditions of dim light or low respiratory power availability. It would also minimise ATP hydrolysis by F_OF_1 under conditions of darkness or zero respiratory power availability.

In intact chloroplasts, the physiological control of the activity of the reversible F_OF_1 ATPase probably involves participation of the thioredoxin system, which may modulate the F_OF_1 ATPase activity by reducing thiol groups in one of the subunits of F_1 (64). Recent observations by Petty and Jackson (70) showed that the rate of ADP phosphorylation catalysed by the F_OF_1 of *Rhodopseudomonas capsulata* appeared to depend on the quantity of reducing equivalents traversing the photoredox chain, rather than on the value of $\Delta\bar{\mu}H^+(B)$. This prompts the speculative suggestion that the state of activity α of the reversible F_OF_1 ATPase of *R. capsulata*, may be sensitively controlled by some effector system that signals between the photoredox chain and the reversible ATPase molecules. Perhaps this might be the thioredoxin/ phosphothioredoxin system found recently in *Escherichia coli* (71).

It is generally agreed that the main function of F_1 in the $F_O F_1$ ATPases of mitochondria, bacteria and chloroplasts is to couple the reversible hydrolysis of ATP to the translocation of protons (or their equivalent) from the aqueous domain N to the P pole of F_1, and thence via F_O to the aqueous domain P (6-26). The experimental work in my laboratory giving an $\leftarrow H^+/P$ ratio of 2 for the $F_O F_1$ ATPase of mitochondria has been widely confirmed (see 55-57). Estimates of the $\leftarrow H^+/P$ ratio for the $F_O F_1$ ATPase of certain bacteria (56-58,72) likewise indicate a value of 2. But the $\leftarrow H^+/P$ ratio for the $F_O F_1$ ATPase of chloroplasts has proved to be particularly difficult to measure accurately, and estimates range between 2 and 3 (24,73-75).

The observed fact that $F_O F_1$ functions as a reversible protonmotive ATPase implies that F_1 must have 4 subfunctions, which are described in the following paragraphs.

1. <u>Heterolytic phosphorylium transferring function of F_1.</u> The hydrolytic function of F_1 involves the labilisation of the P-O bond between the γ phosphorus atom and the γ-β bridging oxygen atom of ATP. It also involves labilisation of a P-O bond of inorganic phosphate (76). But it does not involve the formation of any known intermediaries (6-25). For general chemical reasons (77,78), the P-O bond labilisation in the catalytic site of F_1 is expected to occur heterolytically by a nucleophilic displacement on phosphorus —— both valency electrons being retained or supplied by the leaving or attacking oxygen atoms (79-83). Thus, the enzyme-bound $^+PO_3^{2-}$ group, which Lipmann called phosphorylium (78) and Wimmer and Rose

called reactive PO_3^- (81), is enabled to migrate reversibly between an enzyme-bound ADP^{3-} group and an enzyme-bound O^{2-} group, giving alternately enzyme-bound ATP^{4-} and enzyme-bound PO_4^{3-} groups, in specific but unknown protonation states. To avoid previous misunderstandings, let me emphasise that the chemically intrinsic charge-translocating function of this nucleophilic displacement reaction (which turns out to be very significant chemiosmotically) is independent of whether it proceeds by an associative S_N2 mechanism, a dissociative S_N1 mechanism, or by an intermediate mechanism (57).

Photoaffinity labelling and other inhibitor studies have shown that the covalent binding of a single tyrosine, arginine, lysine or carboxyl group in the β subunit of F_1 is sufficient to suppress catalytic activity (6-25,84). Therefore, the conclusion has generally been drawn that the β subunits contain the catalytic site that binds P_i + ADP or ATP, and that F_1 functions by an alternating site mechanism (85). At all events, it seems likely that only one catalytic site is active in the F_1 molecule at a given time, as indicated in Fig. 1. Labelling of α subunits generally occurred in the above studies, but it did not correlate with loss of activity as well as the labelling of β subunits. For that reason, an allosteric, regulatory or non-catalytic function has been attributed to the α subunits (8-14,19,23,24). However, it should not be overlooked that both α and β subunits bind ADP and ATP (37,38,86,87), and Lunardi and Vignais (88) recently showed that mitochondrial F_1 can be completely inactivated by modification of only one α subunit.

Mapping of the F_1 molecule using resonance energy transfer methods (14) indicates that the catalytic site is buried in the B domain of F_1, as represented in Fig. 1. Three other lines of evidence help to confirm this view. The nucleotide specificity of mitochondrial F_1 is decreased by detaching it from F_O, and is increased again by combining F_1 with OSCP (89) —— suggesting that the normally high nucleotide specificity of F_1 may depend on the normal route of access of the nucleotides from the N side of the catalytic site, which is bypassed when the P side becomes accessible by detachment of F_1 from F_O. Certain inhibitors are active against isolated F_1, but apparently fail to gain access to the catalytic site when F_1 is attached to F_O or to a water/octane interface (8,90). When the $F_O F_1$ ATPase undergoes so-called energisation, the F_1 component exhibits a greatly enhanced hydrogen exchange, it reacts much more readily with certain inhibitors, such as NEM, sulphate and permanganate, it releases occluded adenine nucleotides, and it undergoes a conformational transition to a more catalytically active state (9,42–48,59–63). These related phenomena may be most simply attributed to the poising of the catalytic site of F_1 at the dual pH optimum, with its P side at around pH 4 when its N side is at around pH 8, provided that it is situated in the B domain of F_1 (Fig. 1). Further support for this view comes from observations on the induction of proton leakage through F_1 by reagents that disturb the structure of F_1, particularly in the region of the small subunits (91,92), which may function as a proton gate (10,11,13,16,17).

2. Phosphate and adenine nucleotide translocating function of F_1. The widely held view that the catalytic P-O bond labilising site is buried in the B domain of F_1 requires that the part of F_1 on the N side of the catalytic site must be conformationally mobile so that P_i + ADP and ATP may exchange between the aqueous domain N and the catalytic site in F_1. In other words, the N side of F_1 would function as a $(P_i + ADP)/ATP$

antiporter, as originally suggested in the context of the direct chemiosmotic mechanism (53,54).

3. <u>Substrate binding functions of F_1</u>. Both the specific catalysis of phosphorylium group transfer in the catalytic site of F_1, and the specific catalysis of $(P_i + ADP)/ATP$ antiport in the neighbouring ligand-conducting region of F_1, involve the tight binding of the ligands ATP^{4-}, ADP^{3-}, PO_3^-, O^{2-} and PO_4^{3-} in specific states of protonation and/or electrovalent bonding with Mg^{2+} and with functional groups of amino acids in the α or β subunits of F_1. Important work in the laboratories of Boyer (93,94), Robert Mitchell (95) and Wimmer and Rose (96) on the facile exchange of O between H_2O in the medium and PO_4^{3-} in the catalytic site of F_1, when the catalytic site contains the ligands $Mg^{2+} + PO_4^{3-} + ADP^{3-}$, implies that the transition to the ligands $Mg^{2+} + O^{2-} + ATP^{4-}$ occurs in the active site with a significant frequency, even when there is no significant $\Delta\bar{\mu}H^+(F_1)$ to poise the *overall* reaction towards ADP phosphorylation. It follows that the ligand binding energies (and associated conformational mobilities) in the active site facilitate the conduction of the phosphorylium ligand ($^+PO_3^{2-}$ or reactive PO_3^-) between bound ADP^{3-} and bound O^{2-} by lowering the free energy of activation for P-O bond latching and unlatching. They may also increase the frequency of ATP^{4-} formation significantly in the active site by lowering the free energy of bound $Mg^{2+} + O^{2-} + ATP^{4-}$ relative to that of bound $Mg^{2+} + PO_4^{3-} + ADP^{3-}$, compared with the corresponding free energies in aqueous solution. But the degree of stabilisation of bound ATP^{4-} in the catalytic site does not appear to be very great in the absence of a significant $\Delta\bar{\mu}H^+(F_1)$, since the catalysis by mitochondrial F_1 of O exchange between bound PO_4^{3-} and H_2O is much less with $P_i + ADP$ as substrates

than with low concentrations of ATP (93). These binding and ligand conduction functions of F_1 follow classical principles of catalysis by enzymes and porters (50,51,97).

As originally pointed out by Pauling (see 97), enzymic catalysis depends fundamentally on a binding-change mechanism, in which reactions that normally proceed slowly, because they have transition states of high free energy and low probability in aqueous solution, can proceed fast in the catalytic site of an enzyme, because tight binding of the transition state complexes in the catalytic site lowers their free energy and raises their probability. Therefore, the binding functions of F_1 may properly be regarded as normal attributes of the classical ligand-conducting catalytic function of F_1 (26,53-57,97). They are not special attributes that provide discriminatory evidence in favour of the exclusively conformationally coupled or indirect type of chemiosmotic mechanism advocated by the proton pump school (98), as many biochemists seem to have assumed or suggested (e.g., see: 85, and Boyer, Chance, Ernster and Slater in 99).

4. Chemicomotive function of F_1. The fact that the F_0F_1 ATPase is protonmotive implies that the process of hydrolysis of ATP by F_1 has a vectorial (or higher tensorial order) chemicomotive function (see 57,97). That function either includes the translocation of protons (or their equivalent) in a direct chemiosmotic mechanism (26), or it provides a conformationally active or otherwise physically active interface through which a chemically separate proton pump in F_0 or F_1 is driven in accordance with an indirect chemiosmotic mechanism (85).

MECHANISMS OF F_O AND F_1 FUNCTION

The proton conducting and proton well functions of F_O are carried out by the oligomeric proteolipid, probably by a general acid-base relay mechanism involving glutamate or aspartate, arginine and tyrosine residues (10,11,13,100-103). The types of conduction mechanism available have been well discussed by Glasser (104). Now that the amino acid sequence has been determined for the proteolipid of F_O from a number of sources (see 101), it seems likely that the detailed mechanism of proton conduction may soon be worked out. As Glasser has emphasised (104), proton conduction by acid-base relay mechanisms generally involves, not only a cooperative charge-transfer process that can be regarded as the motion of a charged defect along the chain, but also a reorientation process that can be regarded as the sequential motion of an orientational (or Bjerrum) defect along the chain. The successful identification of the mechanism may well be facilitated by concentrating attention on the latter requirement.

As illustrated in Fig. 3, there are two main mechanistic

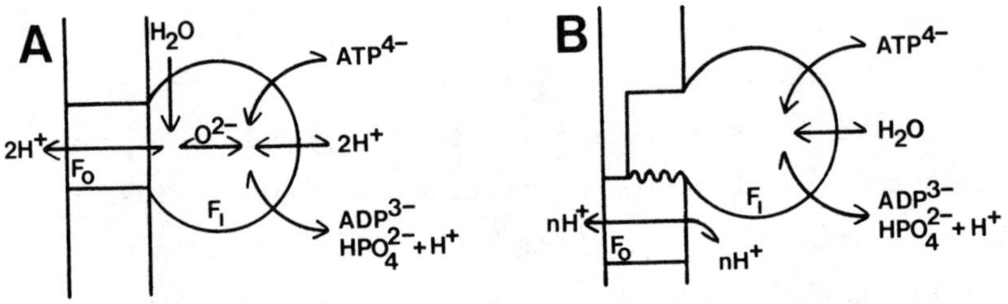

Fig. 3   Diagrams of F_OF_1 ATPase translocating protons by (A) direct and (B) indirect chemiosmotic mechanisms. The ATPase/proton pump interface in B is represented by ∿∿ , and n represents the number of H^+ ions pumped per ATP molecule hydrolysed.

hypotheses of the chemiosmotic function of the protonmotive $F_O F_1$ ATPases: A, the direct chemiosmotic hypothesis according to which the protons (or their equivalent) are conducted directly through the catalytic site of F_1 (1-3 and see 99); and B, the conformationally coupled proton pump hypothesis or indirect chemiosmotic hypothesis, according to which the protons (or their equivalent) are not conducted directly through the catalytic site of F_1 but through a proton pump in F_O that is coupled to the catalytic site by means of a conformationally active or otherwise physically active polypeptide interface (105,106 and see 99). I shall not describe the chemical coupling hypothesis of Griffiths, involving lipoate in the ATPase mechanism, because it is invalidated by recent experimental tests (see 84); nor shall I discuss the non-osmotic protonic anhydride hypothesis of Williams, which concerns an ATPase that is not protonmotive (107,108 and see 57), because it is hardly relevant in the context of the proton-motive chemiosmotic function of the $F_O F_1$ ATPases.

My original proposals for the direct chemiosmotic mechanism of the reversible protonmotive ATPases of mitochondria, bacteria and chloroplasts (1,2,5) were founded on the view that the ATPase reaction was chemically heterolytic (78) as well as being spatially orientated (see 51,57). The reversible dehydration of P_i + ADP to give H_2O + ATP was not, therefore, attributed to the withdrawal of H_2O from P_i + ADP to a microscopic domain at low H_2O potential, as assumed by Williams (107,108). But, as indicated by the reverse of the hydrolytic reaction shown in Fig. 3A, it was conceived as occurring heterolytically by the successive withdrawal of H^+ and of O^{2-} (or OH^-) groups along specific spatially separate ligand-conducting pathways leading from the catalytic site region of the ATPase to aqueous domains at the same water

potential, but at low and high protonic potentials respectively on either side of the ATPase molecules (2,5,57). It is
essential to appreciate that my description of the *net* translocation of O^{2-} groups in diagrams such as Fig. 3A has never
been intended to represent the translocation of free O^{2-} ions.
As I have endeavoured to explain previously (52-57), the O^{2-}
group conduction pathway may correspond partly to the pathway
of *net* O^{2-} conduction via $(PO_4^{3-} + ADP^{3-} + nH^+)/(ATP^{4-} + nH^+)$
antiport on the N side of the catalytic site in F_1, and partly
to the pathway of *net* O^{2-} abstraction from bound
$Mg^{2+} + PO_4^{3-} + ADP^{3-}$ in an unknown state of protonation and
electrovalent binding in the catalytic site of F_1, as discussed earlier in this article. In the direct chemiosmotic
ATPase mechanism giving a $\leftarrow H^+/P$ ratio of 2, the essential
point is that the ligands $PO_4^{3-} + ADP^{3-} + nH^+$ travelling to and
arriving at the catalytic site of F_1 contain one O^{2-} group
more than the ligands $ATP^{4-} + nH^+$ produced at and travelling
from the catalytic site, where n represents the same unknown
degree of protonation of the substrates travelling each way on
the $(P_i + ADP)/ATP$ antiporter system of F_1 (53-57). The
protons shown in Fig. 3 dissociate spontaneously from the substrates on the N side of F_1, but the antiporter system of F_1
is specific for the relative protonation states shown, and the
O^{2-} group leaves through the P side of the catalytic site of
F_1 only when the P-O bond of the phosphate group is labilised
by the binding of $Mg^{2+} + PO_4^{3-} + ADP^{3-}$ in a specific state of
protonation and electrovalent bonding in the catalytic site of
F_1 (26,57). It is also important to appreciate that the
protons assumed to equilibrate with the P side of the
catalytic site in F_1, via the proton-conducting proteolipid
system through F_0, are supposed to be present, not in the free

state, but bound to functional groups in F_1 that are specifically positioned so that the protons are directed to the O^{2-}
groups leaving the phosphorus centre on the P side of the
catalytic site (109).

The in-line nucleophilic displacement reaction on phosphorus can be most simply represented for ADP phosphorylation
as follows:

$$^-O-P \overset{O^-}{\underset{O}{\diagup}}^{\diagup O^-} + {}^-OADP \leftrightarrow {}^-O \overset{O^-}{\underset{O}{\diagup}}^{O^-}_{P} OADP \leftrightarrow O^{2-} + \overset{^-O}{\underset{O}{\diagup}}^{O}_{P}-OADP \qquad (3)$$

Let me again emphasise that the ionic forms shown for simplicity here are not supposed to represent free reacting species.
They will be bound in various ways, and their charges may be
neutralised or modified in an enzyme catalytic site. The
intermediate transition state, which is presumably promoted by
specific binding in the catalytic site, may theoretically lie
between two extremes corresponding, either to the formation of
a bipyramidal pentacoordinate phosphorus centre in a highly
associative (S_N2) mechanism,

$$^-O-P \overset{O^-}{\underset{O}{\diagup}}^{\diagup O^-} + {}^-OADP \leftrightarrow {}^-O-\overset{O^-}{\underset{O^-}{\overset{O}{|}}}_{P}-OADP \leftrightarrow O^{2-} + \overset{^-O}{\underset{O}{\diagup}}^{O}_{P}-OADP \qquad (4)$$

or to the formation of a planar triply-coordinate metaphosphate anion as reactive intermediate in a dissociative (S_N1)
mechanism (79-83),

$$^-O-P\begin{matrix}O^-\\ \nearrow\\ \\ \searrow\\ O\end{matrix} + {}^-OADP \leftrightarrow O^{2-} + P\begin{matrix}O^-\\ \parallel\\ \\ \parallel\\ O\end{matrix} + {}^-OADP \leftrightarrow O^{2-} + \begin{matrix}{}^-O\\ \searrow\\ \\ \nearrow\\ O\end{matrix}P-OADP \qquad (5)$$

According to the direct chemiosmotic hypothesis, the in-line nucleophilic displacement reaction, whether proceeding by an associative or dissociative mechanism, would be promoted: A, by $(PO_4^{3-} + {}^-OADP)/(PO_3^{2-}-OADP)$ antiport or its equivalent (corresponding to O^{2-} uniport) between the catalytic site of F_1 and the aqueous domain N, where net deprotonation occurs at about pH 8 (reverse reaction in Fig. 3); and B, by the withdrawal of O^{2-} by protonation (giving H_2O) at the P side of the catalytic site of F_1 at an equivalent pH of about 4 (53-55). The P and N sides would correspond to the left and right sides of the reacting phosphorus centre shown in equations (3) to (5).

The absolute dependence of O exchange between F_1-bound P_i and medium H_2O on F_1-bound ADP, and other more general considerations, prompted the suggestion (109) that the nucleophilic displacement might proceed in F_1 by an associative mechanism, and that the pulling effect of the protons conducted through F_0 to the P side of the catalytic site, inducing the O^{2-} group to leave the phosphorus centre, might be due to the incipient formation of an oxonium group ($-OH_2^+$) that would accept an electron from the phosphorus centre and leave as H_2O. But, in view of severe criticisms by Boyer (110,111), and although I do not think that my suggestions were invalidated by Boyer's arguments (see 79-84), it is necessary to remark that neither the suggested oxonium group formation, nor the associative type of nucleophilic displacement reaction, are essential features of the proposed direct chemiosmotic mechanism (112).

The O^{2-} group might, for example, accept one proton and leave the phosphorus centre as a complexed OH^- ion (involving Mg^{2+} and other ligands in the catalytic site), before attack and conversion to H_2O by the second proton (26,113). It is the heterolytic charge-translocating property of the nucleophilic displacement reaction, shown by equation (3), and not the intermediate through which it proceeds, that is essential to the direct chemiosmotic mechanism. As a matter of fact, Jencks (see 83) has expressed doubts as to whether meaningful distinctions can generally be made between associative and dissociative mechanisms in the special environment of enzyme catalytic sites. It is relevant, nevertheless, that Webb, Grubmeyer, Penefsky and Trentham (114) have recently observed that the configuration of the γ phosphorus centre of a chiral ATP analogue is inverted during hydrolysis by a mitochondrial F_1 ATPase. Their observations may most readily be explained in terms of an in-line nucleophilic displacement mechanism, possibly involving a pentacoordinate transition state (84).

Until more information is available about the topology of the catalytic site of F_1, it is difficult to make experimentally useful conjectures about details of the catalytic mechanism. It may be worth remarking, however, that the partial stabilisation of ATP in the catalytic site, enabling significant O exchange to occur between F_1-bound P_i and medium H_2O in the absence of a significant $\Delta\bar{\mu}H^+(F_1)$, could be due to a permanent electric dipole across the phosphorus centre, positive on the P side. As can be seen from equation (3), the nucleophilic displacement reaction producing F_1-bound ATP and F_1-bound O^{2-} (presumably in equilibrium with H_2O) in the catalytic site would be promoted by pulling the valency electrons across the phosphorus centre from the N side (right) to the P side (left). The positive aspect of this dipole might, perhaps, be provided by Mg^{2+}, acting as a Lewis acid, and cationic functional groups of amino acids, acting as proton-conducting general acids, in the P region of the catalytic site (113).

The proposition that F_1 has the osmotic function of a $(PO_4^{3-} + ADP^{3-})/ATP^{4-}$ antiporter (or its equivalent) as well as the chemical function of an ATPase (53,54) gave rise to the expectation that, owing to what was described as an energetic push-pull effect, the exit of P_i + ADP from the F_1 molecules during ATP hydrolysis might depend on the availability and entry of ATP; and likewise, the exit of ATP from the F_1 molecules during ADP phosphorylation might depend on the availability and entry of P_i + ADP (55-57). In other words, according to the direct chemiosmotic hypothesis, the F_0F_1 ATPase might be expected to exhibit typical alternating site kinetics (57) ─── as has, in fact, been observed particularly well in recent experimental work from the laboratories of Boyer (85,94) and of Robert Mitchell (95).

The above considerations lead to the conclusion that none of the experimental facts about the chemiosmotic function of the F_0F_1 ATPases are incompatible with the direct chemiosmotic hypothesis. The relative lack of biochemical detail provided by the indirect chemiosmotic or proton pump hypothesis (84,85) seems, at first sight, to allow more flexibility for the protonmotive stoichiometry, which is specified by nothing more informative than the symbol n in Fig. 3B. But, as pointed out previously (109), although it would not be so neat chemically or mechanistically, the direct chemiosmotic mechanism could give a $\leftarrow H^+/P$ ratio of 3 or 4 if the antiporter function of F_1 were specific for the equivalent of $(PO_4^{3-} + ADP^{3-})/ATP^{3-}$ antiport or $(PO_4^{3-} + ADP^{3-})/ATP^{2-}$ antiport, respectively.

The weakest aspect of the proton pump hypothesis is the lack of solid experimental evidence for a conformationally or otherwise physically actuated proton pump in F_O, or for any other physically actuated proton pump of the type advocated by the proton pump school (98,99,105,106,110,111). There is a corresponding vagueness about the physical interface, denoted by 〰 in Fig. 3, through which the catalytic site of F_1 and the proton pump of F_O are supposed to be energetically coupled. Indeed, there may even be some doubt whether the proton pump school still consider the proton pump of the $F_O F_1$ ATPases to be situated in F_O (84).

I suggest that progress in our understanding of the biochemical mechanism of the $F_O F_1$ ATPases may be made by exploiting a crucial distinction between the direct chemiosmotic mechanism and the indirect or proton-pump mechanism. In the former case, but not in the latter case, the application of a protonic potential difference $\Delta \vec{\mu} H^+ (B)$ across the B domain of the membrane containing the $F_O F_1$ ATPase should depress the equivalent pH of the region of the catalytic site nearest F_O, relative to the pH of the aqueous N domain, by an amount approaching $\Delta \vec{\mu} H^+ (B)/Z$. This should be detectable, for example, by the use of endogenous (or specifically located exogenous) pH indicator groups, or by the reactivity of specific tagging agents. Perhaps the known enhancement by $\Delta \vec{\mu} H^+ (B)$ of the reactivity of NEM, sulphate and permanganate with functional groups affecting the catalytic activity of F_1 in the $F_O F_1$ ATPase of chloroplasts (42,44,91) may be relevant.

Assuming a purely proton-conducting and F_1-locating function of F_O, it may be useful to try to outline the type of mechanism by which the chemical and osmotic properties of F_1 could be related to the possible subunit structure $\alpha_3 \beta_3 \gamma \delta \epsilon$, and to the conformational articulations of F_1.

POSSIBLE CYCLIC TERNARY STATE F_OF_1 MECHANISM

My thesis is based on the idea (50,57) that the chemical, catalytic and osmotic translocation processes that occur through F_1 require specifically articulated conformational movements in and between the polypeptide subunits of the molecule so that, while the net conduction of O^{2-} groups through the substrate antiporter and catalytic site regions of F_1 is facilitated, the non-specific leakage of protons and other ions through these regions of the B domain of F_1 is barred. Further, it is assumed that, to minimise dissipation of energy through viscous resistance to changes of the shape of F_1 during the catalytic cycle (115), positional exchange mechanisms will generally be preferred to those involving net

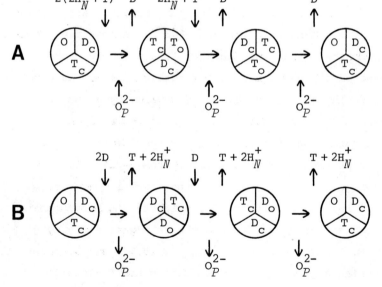

Fig. 4   Diagrams of cyclic ternary state F_OF_1 ATPase mechanism for ATP hydrolysis (A) and ADP phosphorylation (B). In each case, sufficient substrate is added for one complete cycle of three turnovers. Symbols are defined in text.

migration of given components.

Developing previous proposals (57), I suggest that $F_O F_1$ might have a cyclic ternary state mechanism with the following characteristics.

1. There are three sites for binding the substrates P_i + ADP or ATP, corresponding to pairs of α and β subunits, numbered clockwise, 1, 2 and 3, looking at F_1 from the P pole.

2. Each binding site occupied by substrate R in fully active F_1 may have: either a closed potentially-catalytic state, denoted by R_c, in the P region of F_1 that is accessible to the proton conductor through F_O via a proton gate, or an open substrate-exchanging state, denoted by R_o, in a region of F_1 that is accessible to substrate (i.e. P_i + ADP or ATP) in the aqueous N domain. An unoccupied open or closed site is represented by O or C, respectively.

3. The binding of P_i + ADP or ATP by an open site is specific for relative protonation states corresponding to PO_4^{3-} + ADP^{3-} and ATP^{4-} respectively, so that $2H^+$ ions more would be dissociated from H_3PO_4 + ADP than from ATP on binding to an open site. The substrates PO_4^{3-} + ADP^{3-} and ATP^{4-}, or their relative protonation-state equivalents, are abbreviated here to D and T respectively.

4. The positional exchange of occupied closed with occupied open sites corresponds to the substrate antiport reaction, $D_o/T_c \rightleftharpoons T_o/D_c$.

5. Sites in the closed potentially-catalytic states D_c or T_c can change to T_c or D_c, respectively, by donating an O^{2-} group to or accepting an O^{2-} group from the proton conductor through F_O via a proton gate.

6. In the fully active substrate-saturated F_1, there is one open site that may be either D_o or T_o, depending on the

relative concentrations of P_i + ADP and ATP in the aqueous N domain, and two closed sites, one of which is normally D_c while the other is T_c.

7. The transformation of T_c to D_c or of D_c to T_c in a closed site connected to the proton gate may occur at the same time as the antiport reaction, T_o/D_c or D_o/T_c respectively, between the other two sites.

Using the properties specified above, Fig. 4 shows diagrams of the successive states of the F_1 molecule observed from the N side of F_oF_1: A, during ATP hydrolysis; and B, during ADP phosphorylation. It will be seen that the F_1 molecule retains the same ternary state during turnover. But, beginning from the initial configuration that happens to have been chosen in Fig. 4, the state (not the F_1 molecule itself) rotates clockwise during ATP hydrolysis (A), and counter-clockwise during ADP phosphorylation (B). Thus, the F_1 molecule could be subject to the minimum of conformational disturbance and dissipative interaction with its environment during activity. However, a simple transition of the F_1 molecule to a state having all three substrate-binding sites in the closed configuration could readily bring it to a catalytically inactive condition, characterised by tightly bound (or occluded) nucleotide.

Limitations of space do not permit more detailed consideration of this rather speculative cyclic ternary state F_oF_1 ATPase model in the light of the extensive experimental knowledge reviewed in this article. But I hope that, at all events, this conjecture may help to stimulate further conceptual and experimental exploration.

ACKNOWLEDGEMENTS

I thank Jennifer Moyle for helpful discussion and
criticism, and Robert Harper and Stephanie Key for expert
assistance, during the preparation of this paper. I am
grateful to Richard L. Cross for sending me a preprint of
ref. 84, and I thank Glynn Research Ltd. for general financial
support.

REFERENCES

1. Mitchell, P. and Moyle, J. (1958) Nature 182, 372-373.
2. Mitchell, P. (1961) in Biological Structure and Function,
   Proc. First IUB/IUBS Internat. Symp., Stockholm 1960
   (T. W. Goodwin and O. Lindberg, eds.) Vol. 2, pp. 581-
   599, Academic Press, London.
3. Ernster, L. and Lee, C. P. (1964) Ann. Rev. Biochem. 33,
   729-788.
4. Ernster, L. (1977) Ann. Rev. Biochem. 46, 981-995.
5. Mitchell, P. (1961) Nature 191, 144-148.
6. Racker, E. (1976) A New Look at Mechanisms in Bioenergetics, Academic Press, New York.
7. Pedersen, P. L. (1975) J. Bioenerg. 6, 243-275.
8. Kozlov, I. A. and Skulachev, V. P. (1977) Biochim.
   Biophys. Acta 463, 29-89.
9. Harris, D. A. (1978) Biochim. Biophys. Acta 463, 245-273.
10. Kagawa, Y. (1978) Biochim. Biophys. Acta 505, 45-93.
11. Kagawa, Y., Sone, N., Hirata, H. and Yoshida, M. (1979)
    J. Bioenerg. Biomembr. 11, 39-78.
12. Yoshida, M., Sone, N., Hirata, H., Kagawa, Y. and Ui, N.
    (1979) J. Biol. Chem. 254, 9525-9533.
13. Kagawa, Y., Sone, N., Hirata, H., Yoshida, M., Rögner, M.
    and Ohta, S. (1979) in Membrane Bioenergetics (C. P. Lee
    et al., eds.) pp. 177-188, Addison-Wesley, Reading, Mass.
14. Baird, B. A. and Hammes, G. G. (1979) Biochim. Biophys.
    Acta 549, 31-53.
15. Hundal, T. and Ernster, L. (1979) in Membrane Bioener-
    getics (C. P. Lee et al., eds.) pp. 429-445, Addison-
    Wesley, Reading, Mass.
16. De Gomez-Puyou, M. T., Gomez-Puyou, A., Nordenbrand, K.
    and Ernster, L. (1979) in Function and Molecular Aspects
    of Biomembrane Transport (E. Quagliariello et al., eds.)
    pp. 119-133, Elsevier/North-Holland, Amsterdam.

17. Alfonzo, M. and Racker, E. (1979) Can. J. Biochem. 57, 1351-1358.
18. Penefsky, H. S. (1979) Adv. Enzymol. 49, 223-280.
19. Senior, A. E. (1979) in Membrane Proteins in Energy Transduction (R. A. Capaldi, ed.) pp. 233-278, Marcel Dekker, New York.
20. Harold, F. M. (1977) Curr. Top. Bioenerg. 6, 83-149.
21. Downie, J. A., Gibson, F. and Cox, G. B. (1979) Ann. Rev. Biochem. 48, 103-131.
22. Nelson, N. (1976) Biochim. Biophys. Acta 456, 314-338.
23. McCarty, R. E. (1978) Curr. Top. Bioenerg. 7, 245-278.
24. Shavit, N. (1980) Ann. Rev. Biochem. 49, 111-138.
25. Racker, E. (1979) in Membrane Bioenergetics (C. P. Lee et al., eds.) pp. 569-591, Addison-Wesley, Reading, Mass.
26. Mitchell, P. (1979) in Membrane Bioenergetics (C. P. Lee et al., eds.) pp. 361-372, Addison-Wesley, Reading, Mass.
27. Soper, J. W., Decker, G. L. and Pedersen, P. L. (1979) J. Biol. Chem. 254, 11170-11176.
28. Mitchell, P. (1969) Theoret. Exp. Biophys. 2, 159-216.
29. Begusch, H. and Hess, B. (1979) FEBS Lett. 108, 249-252.
30. Ludwig, B., Prochaska, L. and Capaldi, R. A. (1980) Biochemistry 19, 1516-1523.
31. Telfer, A., Barber, J. and Jagendorf, A. T. (1980) Biochim. Biophys. Acta 591, 331-345.
32. Abrams, A., Jensen, C. and Morris, D. H. (1976) Biochem. Biophys. Res. Commun. 69, 804-811.
33. Sternweis, P. C. and Smith, J. B. (1977) Biochemistry 16, 4020-4025.
34. Sternweis, P. C. (1978) J. Biol. Chem. 253, 3123-3128.
35. Andreu, J. M. and Munoz, E. (1979) Biochemistry 18, 1836-1844.
36. Larson, R. J. and Smith, J. B. (1977) Biochemistry 16, 4266-4270.
37. Ohta, S., Tsuboi, M., Yoshida, M. and Kagawa, Y. (1980) Biochemistry 19, 2160-2165.
38. Paradies, H. H. (1980) FEBS Lett. 120, 289-292.
39. Abrams, A., Morris, D. and Jensen, C. (1976) Biochemistry 15, 5560-5566.
40. Leimgruber, R. M., Jensen, C. and Abrams, A. (1978) Biochem. Biophys. Res. Commun. 81, 439-447.
41. Hackney, D. D. (1980) Biochem. Biophys. Res. Commun. 94, 875-880.
42. Jagendorf, A. T. (1975) Fed. Proc. 34, 1718-1722.
43. Moyle, J. and Mitchell, P. (1975) FEBS Lett. 56, 55-61.
44. Grebanier, A. E. and Jagendorf, A. T. (1977) Plant Cell Physiol. Spec. Issue 103-114.
45. Gräber, P., Schlodder, E. and Witt, H. T. (1977) Biochim.

Biophys. Acta 461, 426-440.

46. Wagner, R. and Junge, W. (1980) FEBS Lett. 114, 327-333.

47. Gomez-Puyou, A., De Gomez-Puyou, M. T. and Ernster, L. (1979) Biochim. Biophys. Acta 547, 252-257.

48. Hackney, D. D. (1979) Biochem. Biophys. Res. Commun. 91, 233-238.

49. Naberdryk-Viala, E., Calvet, P., Thiery, J. M., Galmiche, J. M. and Girault, G. (1977) FEBS Lett. 79, 139-143.

50. Mitchell, P. (1963) Biochem. Soc. Symp. 22, 142-168.

51. Mitchell, P. (1977) Symp. Soc. Gen. Microbiol. 27, 383-423.

52. Mitchell, P. (1966) Biol. Rev. 41, 445-502.

53. Mitchell, P. (1972) J. Bioenerg. 3, 5-24.

54. Mitchell, P. (1973) FEBS Lett. 33, 267-274.

55. Mitchell, P. and Moyle, J. (1974) Biochem. Soc. Spec. Publ. 4, 91-111.

56. Mitchell, P. (1976) Biochem. Soc. Trans. 4, 399-430.

57. Mitchell, P. (1977) FEBS Lett. 78, 1-20.

58. Maloney, P. C. and Schattschneider, S. (1980) FEBS Lett. 110, 337-340.

59. Harris, D. A., von Tscharner, V. and Radda, G. K. (1979) Biochim. Biophys. Acta 548, 72-84.

60. Shoshan, V. and Selman, B. R. (1979) J. Biol. Chem. 254, 8801-8807.

61. Bar-Zvi, D. and Shavit, N. (1980) FEBS Lett. 119, 68-72.

62. De Gomez-Puyou, M. T., Nordenbrand, K., Muller, U., Gomez-Puyou, A. and Ernster, L. (1980) Biochim. Biophys. Acta 592, 385-395.

63. De Gomez-Puyou, M. T., Gavilanes, M., Gomez-Puyou, A. and Ernster, L. (1980) Biochim. Biophys. Acta 592, 396-405.

64. Mills, J. D., Mitchell, P. and Schürmann, P. (1980) FEBS Lett. 112, 173-177.

65. Mitchell, P. and Moyle, J. (1971) J. Bioenerg. 2, 1-11.

66. Lambeth, D. O. and Lardy, H. A. (1971) Eur. J. Biochem. 22, 355-363.

67. Rechtenwald, D. and Hess, B. (1977) FEBS Lett. 80, 187-189.

68. Laget, P. P. (1979) Arch. Biochem. Biophys. 192, 474-481.

69. Apel, W. A., Dugan, P. R. and Tuttle, J. H. (1980) J. Bacteriol., 142, 295-301.

70. Petty, K. M. and Jackson, J. B. (1979) Biochim. Biophys. Acta 547, 474-483.

71. Conley, R. R. and Pigiet, V. (1978) J. Biol. Chem. 253, 5568-5572.

72. Petty, K. M. and Jackson, J. B. (1979) Biochim. Biophys. Acta 547, 463-473.

73. Fiolet, J. W. T. and Van de Vlugt, F. C. (1975) FEBS

Lett. 53, 287-291.

74. McCarty, R. E. (1978) in The Proton and Calcium Pumps
    (G. F. Azzone et al., eds.) pp. 65-70, Elsevier/North
    Holland, Amsterdam.
75. Witt, H. T. (1979) Biochim. Biophys. Acta 505, 355-427.
76. Boyer, P. D. (1967) Curr. Top. Bioenerg. 2, 99-149.
77. Pauling, L. (1960) The Nature of the Chemical Bond, 3rd
    Ed., Cornell University Press, Ithaca, New York.
78. Lipmann, F. (1960) in Molecular Biology (D. Nachmansohn,
    ed.) pp. 37-47, Academic Press, New York.
79. Mildvan, A. S. (1974) Ann. Rev. Biochem. 43, 357-399.
80. Mildvan, A. S. (1979) Adv. Enzymol. 49, 103-126.
81. Wimmer, M. J. and Rose, I. A. (1978) Ann. Rev. Biochem.
    47, 1031-1078.
82. Rose, I. A. (1979) Adv. Enzymol. 50, 361-395.
83. Knowles, J. R. (1980) Ann. Rev. Biochem. 49, 877-919.
84. Cross, R. L. (1981) Ann. Rev. Biochem. 50, in press.
85. Boyer, P. D. (1979) in Membrane Bioenergetics (C. P. Lee
    et al., eds.) pp. 461-479, Addison-Wesley, Reading, Mass.
86. Kozlov, I. A. and Milgrom, Y. M. (1980) Eur. J. Biochem.
    106, 457-462.
87. Slater, E. C., Kemp, A., Van der Kraan, I., Muller,
    J. L. M., Roveri, O. A., Verschoor, G. J., Wagenvoord,
    R. J. and Wielders, J. P. M. (1979) FEBS Lett. 103, 7-11.
88. Lunardi, J. and Vignais, P. V. (1979) FEBS Lett. 102,
    23-28.
89. MacLellan, D. H. and Asai, J. (1968) Biochem. Biophys.
    Res. Commun. 33, 441-447.
90. Kozlov, I. A., Shalamberize, M. V., Novikova, I. Yu.,
    Sokolova, N. I. and Shabarova, Z. A. (1979) Biochem. J.
    178, 339-343.
91. Moroney, J. V., Andreo, C. S., Vallejos, R. H. and
    McCarty, R. E. (1980) J. Biol. Chem. 255, 6670-6674.
92. Underwood, C. and Gould, J. M. (1980) Biochim. Biophys.
    Acta 589, 287-298.
93. Choate, G. L., Hutton, R. L. and Boyer, P. D. (1979)
    J. Biol. Chem. 254, 286-290.
94. Hackney, D. D., Rosen, G. and Boyer, P. D. (1979) Proc.
    Nat. Acad. Sci. U.S.A. 76, 3646-3650.
95. Mitchell, R. A., Lamos, C. M. and Russo, J. A. (1980)
    Biochim. Biophys. Acta 592, 406-414.
96. Wimmer, M. J. and Rose, I. A. (1977) J. Biol. Chem. 252,
    6769-6775.
97. Mitchell, P. (1979) Eur. J. Biochem. 95, 1-20.
98. Wikström, M. and Krab, K. (1979) Biochim. Biophys. Acta
    549, 177-222.
99. Boyer, P. D., Chance, B., Ernster, L., Mitchell, P.,

Racker, E. and Slater, E. C. (1977) Ann. Rev. Biochem. 46, 955-1026.

100. Sone, N., Ikeba, K. and Kagawa, Y. (1979) FEBS Lett. 97, 61-64.

101. Hoppe, J., Schairer, H. U. and Sebald, W. (1980) FEBS Lett. 109, 107-111.

102. Friedl, P., Friedl, C. and Schairer, H. U. (1980) FEBS Lett. 119, 254-256.

103. Sebald, W., Machleidt, W. and Wachter, E. (1980) Proc. Nat. Acad. Sci. U.S.A. 77, 785-789.

104. Glasser, L. (1975) Chem. Rev. 75, 21-65.

105. Boyer, P. D. (1974) BBA Library 13, 289-301.

106. Slater, E. C. (1974) BBA Library 13, 1-20.

107. Williams, R. J. P. (1962) J. Theoret. Biol. 3, 209-229.

108. Williams, R. J. P. (1976) Chem. Soc. Spec. Publ. 27, 137-161.

109. Mitchell, P. (1974) FEBS Lett. 43, 189-194.

110. Boyer, P. D. (1975) FEBS Lett. 50, 91-94.

111. Boyer, P. D. (1975) FEBS Lett. 58, 1-6.

112. Mitchell, P. (1975) FEBS Lett. 50, 95-97.

113. Mitchell, P. (1981) Chemistry in Britain 17, 14-23.

114. Webb, M., Grubmeyer, C., Penefsky, H. S. and Trentham, D. R. (1980) J. Biol. Chem. 255, 11637-11639.

115. Beece, D., Eisenstein, L., Frauenfelder, H., Good, D., Marden, M. C., Reinisch, L., Reynolds, A. H., Sorensen, L. B. and Yue, K. T. (1980) Biochemistry 19, 5147-5157.

THE STOICHIOMETRY OF H$^+$ EJECTION COUPLED TO

MITOCHONDRIAL ELECTRON FLOW, MEASURED WITH A

FAST-RESPONDING OXYGEN ELECTRODE[1]

A. L. Lehninger, B. Reynafarje, P. Davies,

A. Alexandre, A. Villalobo, and A. Beavis

Department of Physiological Chemistry,
Johns Hopkins University School of Medicine,
Baltimore, MD 21205

INTRODUCTION

Earlier reports from this laboratory have provided evidence that the maximum H$^+$/site ejection ratio for the three energy-conserving sites in mitochondrial electron transport is close to 4 (reviewed in 1). This conclusion was supported by measurements on sites 1 + 2 + 3 (2-4), sites 2 + 3 (2-5), sites 1 + 2 (1,6), site 2 alone (7), and site 3 alone (1,6,8,9), using different instrumental techniques, different sets of electron donors and acceptors, and mitochondria from different tissues (2-5). Independent studies have also come to the conclusion that the H$^+$/site ratio is close to 4.0 (10). While there is agreement among all reporting laboratories that the H$^+$/site ratio is 4 for

[1]Supported by NIH Grant GM05919, NCI Grant CA25360, and NSF Grant PCM78-18190.

C. P. Lee, G. Schatz, G. Dallner (eds.), Mitochondria and Microsomes
in honor of Lars Ernster   ISBN 0-201-04576-1

459

site 2 (reviewed in 7,11), there is significant disagreement regarding the H^+/site ratio for site 3, the cytochrome
oxidase reaction, when measured separately. Earlier
measurements with ferrocyanide as electron donor in this
(8) and another laboratory (10) yielded H^+/O ratios for
site 3 approaching 4. However, some groups have reported
values near 2 (13-15) and others conclude that no H^+ ejection
is directly coupled to the cytochrome oxidase reaction (16-
19).

It has been proposed that H^+/O ejection ratios exceeding 2 for the cytochrome oxidase reaction, such as we have
reported, are greatly overestimated owing to the slow response of the Clark oxygen electrode to changes in oxygen
concentration, and consequent underestimation of the rate
of oxygen uptake. However, the response time of Clark-
type oxygen electrodes can be made quite adequate for H^+/O
rate ratio measurements with stretched ultrasensitive membranes and optimal adjustment of the operating electrical
characteristics. Appropriate tests established that under
our conditions the oxygen electrode yielded theoretical
H^+/O rate ratios for the scalar reaction $2H^+ + \frac{1}{2}O_2 \rightarrow H_2O$
promoted by the cytochrome oxidase reaction in the presence
of protonophores, as well as theoretical H^+/O rate ratios
for the oxidation of acetaldehyde to acetate by xanthine
oxidase-catalase systems. But in any case, we have reported
(8) that the vectorial H^+/site ejection ratio for the cytochrome oxidase reaction in rat liver mitochondria approaches
4.0 and the H^+/site uptake ratio in the presence of protonophores is close to the theoretical 2.0, when use of the
oxygen electrode was avoided, i.e. when kinetically-com-
patible measurements of both H^+ ejection and electron flow

were carried out by dual-wavelength spectrophotometry, which
has essentially no instrumental dead time. Nevertheless,
the possible inadequacy of oxygen consumption measurements
cannot be disregarded, since it has come to our attention
that commercially available oxygen electrodes as often used
in other laboratories are inadequate for H^+/O ratio measurements.

There is, however, a related difficulty that has not
been adequately recognized. Determination of H^+/site ratios
from initial rate measurements require a very short mixing
time, since the measurements must be obtained as early as
possible, since H^+ ejection is accompanied by an increasing
ΔpH gradient across the membrane, which causes an increasing
rate of H^+ back-flow into the matrix and consequent underestimation of the H^+/O ejection ratio. Since the initial
rates of H^+ ejection and electron flow are quasi-linear for
only very short periods, serious errors in estimating their
true initial rates from recorder traces can occur.

We have been able to eliminate these technical problems
by two measures (1) use of a fast-responding membrane-less
oxygen electrode, and (2) detailed kinetic analysis of the
initial rates of electron flow and H^+ ejection to determine
their precise time course. The new electrode, a modification
of that described earlier by Davies and Grenell (20,21),
consists of a 32 guage platinum wire fused into and made
flush with the end of a 4 cm length of 3 mm flint glass
tubing. Its tip is coated with successive microscopic
layers of finely-powdered glass sintered at 750° and slowly
annealed. The number of sintered layers was optimized to
yield maximal speed of response and minimal interference
from stirring and convection noise. The 90% full-scale

response times of such electrodes, tested by an oxygen jet
and recorded with an oscillograph, was <50 ms and could be
made less than 20 ms. Details of preparation and testing
of the electrodes will be given elsewhere. H^+ traces were
obtained with a Beckman combination glass electrode selected
for fast response; the reference cell of the latter was
common to both the O_2 and H^+ electrodes. The reactions were
carried out in a closed glass cell at 25° with a gas phase
of minimum volume. Very rapid stirring of the cell contents
was provided by a magnetic bar at a speed of about 2,000
RPM, with the cell geometry carefully arranged to avoid
vortex formation. Additions were made through a permanently-
fixed port leading into an empirically determined optimal
position in the chamber. The geometry of the cell and the
electrodes, as well as the stirring parameters, were empirically adjusted to achieve a combined mixing and response
time of 0.2 s or less.

Interference in the response of the pH electrode because of the opposite current generated by the oxygen electrode was made insignificant by keeping changes in oxygen
concentration quite small or by precise compensation of the
oxygen electrode current with an external electronic circuit,
to be described elsewhere. The signals from the H^+ and O_2
electrodes were suitably preamplified and fed into a multi-
channel Soltec Model 330 strip-chart recorder run at 120 cm
per min or into a Soltec Model L direct-recording oscillograph, depending upon the rate of the reaction to be measured. All the experiments described here were carried out
with rat liver mitoplasts, which made possible the use of
added ferrocytochrome c as direct electron donor in site 3
experiments, thus avoiding the permeability barrier to

to cytochrome c imposed by the outer membrane.

RESULTS

THE H^+/O EJECTION RATIO DURING OXIDATION OF β-HYDROXY-
BUTYRATE. H^+ ejection and oxygen consumption measurements
with the new electrode were carried out using the NAD-linked
substrate β-hydroxybutyrate for estimation of the H^+/O ratio
for sites 1 + 2 + 3. The mitoplasts were suspended in a
medium containing K^+ + valinomycin, to allow entry of K^+
in electroneutral exchange for the ejected H^+, oligomycin
to prevent ATP hydrolysis or synthesis, and N-ethylmaleimide
(NEM) to prevent reentry of H^+ by H^+/$H_2PO_4^-$ symport (5),
conditions similar to those described in earlier reports
(1-5). Mitoplasts were first preincubated aerobically with
all the components plus 0.2 mM Zn^{2+}, to prevent electron
flow through site 2 from NAD-linked substrates (22). The
H^+/O ejection ratio for NAD-linked substrates can be greatly
underestimated with rat liver mitochondria because of their
considerable endogenous respiration during the aerobic preincubation in the presence of valinomycin, with the consequence that a large part of the matrix buffering power is
consumed before the NAD-linked test substrate is added, a
complication not recognized in some studies (23). Rat heart
(4) and Ehrlich ascites tumor mitochondria (3), which are
readily depleted of endogenous substrates, have yielded
average H^+/site ratios close to 4 for NAD-linked substrates.
We have eliminated the problem of endogenous electron flow
in rat liver mitochondria and mitoplasts based on the fact
that Zn^{2+} inhibits electron transport at a point between

ubiquinone and cytochrome c (22). Unlike inhibition of this
site by antimycin A, inhibition by Zn^{2+} can be instantly
removed, by chelation with excess EGTA. After preincubation
of the mitoplasts aerobically in the presence of β-hydroxybutyrate and Zn^{2+}, to start electron flow excess EGTA is
added. Although chelation of Zn^{2+} is accompanied by in-
stantaneous release of H^+ from EGTA, this is easily distinguished, due to its high rate and small extent, from the
slower but more extensive vectorial H^+ ejection coupled to
oxygen uptake. The traces in Figure 1 show that the rates
of both O_2 uptake and H^+ ejection decline continuously with
time, neither showing a sufficiently linear early rate for
simple determination of their slopes. However, kinetic
analysis of the traces showed that the initial rates could
be evaluated in a simple manner. When the H^+ ejection and
oxygen uptake rates obtained at 0.5 s intervals from the
traces of this experiment were plotted on a semilog basis
(Figure 2), regression analysis of the points gave straight
lines in both cases, with regression coefficients exceeding
0.98, indicating almost exact adherence to a simple exponential rate law. It will be noted that the rate of H^+
ejection declined faster than the rate of O_2 uptake, as
expected, since the ejection of H^+ generates an increasing
ΔpH gradient across the membrane, thus causing H^+ to flow
back into the matrix at an increasing rate (24). To obtain
true initial rates of both O_2 uptake and H^+ ejection the
plots were extrapolated back to zero time, at which ΔpH
is zero. The interceptors gave true initial rates of 1370
ng-ions H^+ ejected and 114 ng-atoms oxygen consumed per min
per mg. The H^+/O rate ratio was thus 1370/114 = 12.0,
equivalent to an average of 4.0 H^+ ejected per site for the

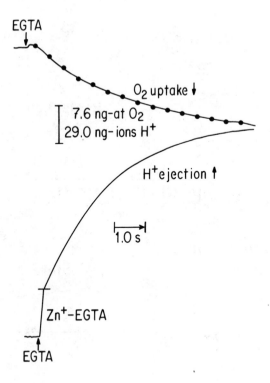

Fig. 1 Rates of H⁺ ejection and O₂ consumption during the
        oxidation of β-hydroxybutyrate. The medium (1.6 ml,
        25°C) contained 250 mM sucrose, 40 mM KCl, 1.5 mM
        Hepes pH 7.05, 6 mM β-hydroxybutyrate, 2 μM valino-
        mycin, 80 μM N-ethylmaleimide, 9.2 μM $ZnCl_2$, and 2
        mg RL mitoplast protein per ml. Electron flow was
        initiated by addition of 0.5 mM EGTA.

3 sites traversed, in essential agreement with earlier
measurements (1-4). It is also seen that the H^+/O ratio
declines with time; by the 3rd second the H^+/O rate ratio
had already declined to about 9.0.

    The addition of rotenone, antimycin A, or cyanide
almost completely abolished both H^+ ejection and oxygen consumption. In the presence of FCCP the rate of oxygen consumption was linear with time rather than exponential; under

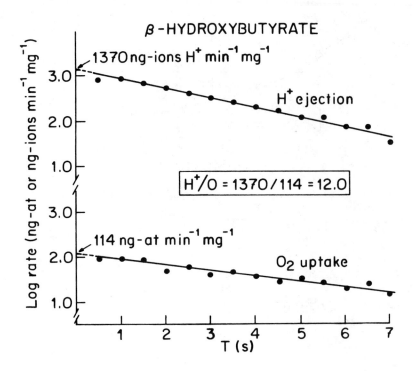

Fig. 2   Semilog plots of the rates of H^+ ejection and O_2
consumption during β-hydroxybutyrate oxidation.
The logarithms of the rates of corresponding points
at 0.5 s intervals over a 7 s period, obtained from
the trace in Figure 1, were plotted against time
and the lines fitted by regression. The rates at
zero time, at which ΔpH is zero, were 1370 ng-ions
H^+ ejected and 114 ng-atoms oxygen consumed per min
per mg, giving an H^+/O ejection ratio of 12.0 or
an average H^+/site of 4.0.

these circumstances no net H^+ ejection took place. This
observation suggests that the major cause of the constantly
declining rate of O_2 uptake in the presence of K^+ + valinomycin is the buildup of the ΔpH gradient across the membrane
and the resulting alkalinization of the matrix.

H$^+$/O ratios very close to 12 for NAD-linked oxidation could also be observed when the reaction rate was slowed down by low temperatures or by use of lower concentrations of valinomycin. Under these conditions the mixing time is not an important factor in relation to the relatively long reaction periods over which the rates remained precisely exponential.

THE H$^+$/O RATIO WITH SUCCINATE AS SUBSTRATE. Traces were also obtained from oxygen and H$^+$ electrodes when electron flow was initiated by the addition of succinate to de-energized aerobic rat liver mitoplasts preincubated in a medium containing rotenone to prevent electron flow from endogenous NAD-linked substrates, K$^+$ + valinomycin, oligomycin, EGTA to prevent Ca^{2+} cycling, and N-ethylmaleimide (in this case Zn^{2+} was not present). After a short mixing period of about 0.1 s, oxygen consumption and H$^+$ ejection took place at rates that declined continuously with time from the beginning, as in the case described above. Kinetic analysis of the traces showed that the initial rates could be evaluated by the same procedure used with β-hydroxybutyrate above. Figure 3 shows semilog plots of the rates of the observed oxygen consumption and H$^+$ ejection vs time taken from a typical experiment; the rates were determined from the traces at successive intervals. The points were fitted by regression analysis and gave rectilinear plots with regression coefficients exceeding 0.99, again indicating near-exact adherence to a simple exponential rate. Extrapolation of the plots to the zero time intercepts on the ordinate, at which point the H$^+$ back-decay rate is zero,

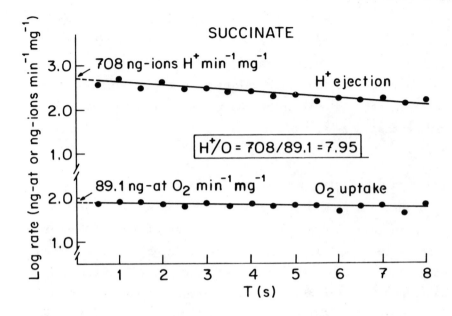

Fig. 3 The H^+/O ejection ratio during electron flow from succinate to oxygen. The system (1.6 ml) contained sucrose, KCl, Hepes, valinomycin, and N-ethylmalei-mide as given in Figure 1, supplemented with 2 μM rotenone and rat liver mitoplasts (1.0 mg protein per ml). After aerobic preincubation 1.0 mM succinate was added to initiate electron flow. From the traces (not shown) the rates of H^+ ejection and oxygen uptake at successive 0.5 s were obtained and log rates plotted vs time. The regression coefficient of the plots exceeded 0.99. Extrapolation of the plots to zero time gave an H^+/O ejection ratio of 7.95.

yielded initial rates of 708 ng-ions H^+ ejected min^{-1} and 89.1 ng-atoms oxygen consumed min^{-1}. The H^+/O ejection ratio from the initial rates was thus 708/89.1 = 7.95, equivalent to 3.98 H^+ ejected per $2e^-$ per energy-conserving site for electron flow from succinate to oxygen, in agreement with earlier reports from this laboratory (1-6). The

rate of H^+ ejection again declined more rapidly with time
than the rate of oxygen uptake. As a result the observed
H^+/O rate ratio declined to about 5.0 by the 5th second,
due to the increasing rate of H^+ back-decay as ΔpH increased.

Control experiments with succinate showed that cyanide
and antimycin A inhibited both oxygen consumption and H^+
ejection at least 96-97 percent, as did 2-heptyl-4-hydroxy-
quinoline N-oxide (HQNO). When excess protonophore (FCCP,
10 μM) was added to the system instead of valinomycin, H^+
ejection was completely abolished but oxygen consumption
took place at a rate about the same as the initial rate
with valinomycin, but instead of declining exponentially
with time it was linear almost to the point of oxygen exhaustion. This observation confirmed that the major cause
of the decline in the rate of oxygen consumption and H^+
ejection in the presence of K^+ + valinomycin is the build-
up of the ΔpH gradient across the membrane.

THE H^+/O EJECTION RATIO OF THE CYTOCHROME OXIDASE REACTION
MEASURED DURING DECAY OF AN OXYGEN PULSE. The great power
of the new oxygen electrode is best shown by its use in
resolving the kinetics of oxygen uptake and H^+ ejection in
the cytochrome oxidase reaction, particularly in view of
the disagreement among different laboratories with respect
to the H^+/O ratio. When anaerobic mitoplasts supplemented
with 60 μM ferrocytochrome c are given a pulse of dissolved
oxygen the added oxygen is consumed so fast that not even
a blip corresponding to the addition of oxygen and its
reduction can be seen in the trace when a standard Clark
oxygen electrode is employed. However, the fast response

of the membrane-less oxygen electrode used here made possible monitoring of the very rapid O_2 concentration changes
during addition of oxygen and its rapid reduction, as
shown in Figure 4. Injection of the oxygen pulse gave a

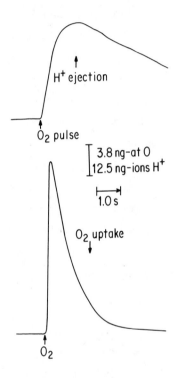

Fig. 4 H^+ ejection and O_2 uptake during the oxidation of
ferrocytochrome c. The test system contained 250 mM
sucrose, 40 mM KCl, 1.5 mM Hepes pH 7.05, 2 μM
rotenone, 4 μM antimycin A, 40 nmoles N-ethylmaleimide per mg, 1 μg oligomycin per mg, 0.5 mM EGTA
valinomycin per mg, and mitoplasts (2.0 mg per ml).
The system was made anaerobic and ferrocytochrome c
(60 μM) was added. The oxygen pulse was injected
to start the reaction. The total amount of oxygen
added is indicated by the horizontal bar.

very fast rise of oxygen concentration, followed by a high
rate of O_2 consumption that obviously began before mixing
was complete. Accompanying the uptake of oxygen was very
fast H^+ ejection which attained a peak in 1.5 s, followed
by reuptake of H^+. The rates of oxygen consumption taken
from the trace at successive 0.1 s intervals between 0.3
and 1.7 s could be fitted with a rectilinear semilog plot
of rate vs time, with a regression coefficient of 0.99.
The rate of H^+ ejection also gave a linear semilog plot,
but showed a much more rapid decline with time than the
rate of O_2 uptake (Fig. 5). The extrapolated initial rate
of O_2 uptake at zero ΔpH was 692 ng-at min^{-1} mg^{-1} and of
H^+ ejection was 2850 ng-ions min^{-1} mg^{-1}, to give an initial
H^+/O ejection ratio of 4.12, in agreement with values approaching 4 reported earlier from this laboratory with less
accurate methods (1,8,9,25,26). The H^+/O ratio declined
very rapidly with time to about 1.7 at about 0.5 s.

The traces in Figure 4 were allowed to run for an
additional 3 min after the added ferrocytochrome c was completely oxidized and oxygen consumption had come to a halt.
After the peak of net H^+ ejection occurred, reuptake of H^+
followed due to the increased permeability of the membrane
to H^+. H^+ uptake from the medium finally came to a halt
at a point corresponding to a net overall uptake of 59.5 ng-
ions H^+ starting from the time of addition of oxygen. Since
the total amount of oxygen consumed during ferrocytochrome
oxidation was 29.8 ng-at, the final net H^+/O uptake ratio
was 1.99, in close agreement with the expected value 2.0
for the scalar equation $2H^+ + \frac{1}{2}O_2 \rightarrow H_2O$ of the cytochrome
oxidase reaction.

Appropriate control experiments showed that both H^+

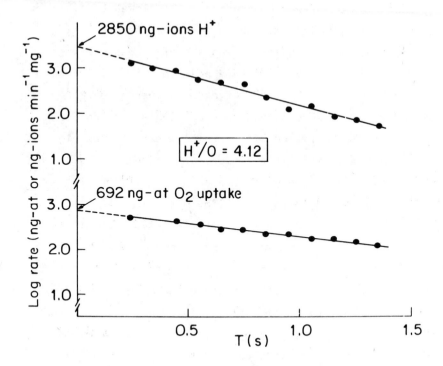

Fig. 5   The H^+/O ratio for the oxidation of ferrocytochrome
c. Semilog plots of the rates of H^+ ejection and O_2
uptake vs time for the period 0.25 to 1.5 s after
injection of O_2 are shown. The initial rates at
zero time (zero ΔpH) are used to calculate the H^+/O
ratio for the cytochrome oxidase reaction.

ejection and oxygen consumption were abolished by addition

of cyanide. When FCCP was present, H^+ uptake occurred as

oxygen consumption proceeded. Again, as in the case of the

other substrates described above, the rate of oxygen con-

sumption in the presence of FCCP was linear with time,

rather than decreasing exponentially. The H^+/O uptake ratio

in this case was close to the theoretical value of 2.0 for

the scalar equation $2H^+ + \frac{1}{2}O_2 \rightarrow H_2O$. Most important, the

addition of 2-heptyl-4-hydroxyquinoline-N-oxide (HQNO) in
concentrations sufficient to block the oxidation of succinate
had no effect on the rates of O_2 uptake or H^+ ejection,
coupled to oxidation of ferrocytochrome c, or on the H^+/O
ratio. This observation, in concurrence with other observations (see 11,12), renders unlikely the view of Papa et al.
(19) that H^+ ejection observed during electron transfer from
cytochrome c to oxygen is actually associated with electron
flow through an HQNO-inhibited step of site 2 on the oxygen
side of the antimycin block.

It will be noted in Figure 4 that the O_2 uptake curve
was rather steep, even with the chart speed at its maximum
rate of 120 cm/min. In order to obtain more accurate rate
measurements, the electrode signals in such experiments have
also been recorded at much higher chart speeds with a direct-
recording cathode ray oscillograph.

DISCUSSION

The experiments briefly outlined in this paper, to be
described in detail elsewhere, eliminate some important
sources of possible errors in the determination of the H^+/O
ejection ratio of mitochondrial electron transport. First,
the extremely responsive membrane-less oxygen electrode used
here makes possible much more accurate time-resolution of
the kinetics of oxygen consumption; moreover, it is more
convenient and direct than other methods for measurement of
oxygen consumption, such as the cumbersome hemoglobin oxygenation method, which requires a number of corrections and
assumptions and has yielded low H^+/O ratios (27).

Secondly, rapid mixing and accurate time resolution of the rates of O_2 uptake and ejection established that both decrease exponentially with time in the presence of K^+ + valinomycin, regardless of the nature of the substrate. Since the rates of oxygen uptake with all these substrates in the presence of the uncoupler FCCP were linear, it is clear that the exponential decrease in the rate of oxygen consumption in the presence of K^+ + valinomycin is due to the increasing pH gradient formed across the inner membrane as H^+ is extruded and K^+ is taken up in exchange. Recognition of the exponential decrease in the rate of oxygen consumption makes possible direct and accurate extrapolation of the rates of O_2 uptake and H^+ extrusion to zero time, at which no pH gradient is present. This procedure, made possible by the new oxygen electrode, supplants the less accurate estimation of initial rates from linear tangents drawn to the early portions of essentially non-linear traces, as used in earlier reports from this (2-5,8) and other (10-12) laboratories. While the latter procedure can give reliable results, it requires very short mixing times and use of the earliest possible portions of the traces, conditions not met in many reports giving low H^+/O ratios.

The third conclusion is that the measured H^+/O rate ratio declines with time because the ΔpH gradient increases with time, causing an increasing rate of back-flow of H^+. This is particularly true in the site 3 measurements. H^+ back-flow is obviously the result of the large increase in ΔpH, accompanied by depletion of the buffering capacity of the matrix and its consequent alkalinization, which leads to inhibition of electron flow and an increase in the permeability of the membrane to H^+. It must be emphasized that

the H^+ ejection accompanying oxidation of each of the three
substrates in the presence of K^+ + valinomycin is entirely
vectorial in nature, since it conforms to two important
criteria (1) it is abolished in the presence of the protonophore FCCP and (2) it is accompanied by uptake of K^+ ions,
which has been measured by a new method. The K^+/O uptake
ratio for NAD-linked electron transport has been found to be
close to 12, for succinate oxidation close to 8.0, and for
oxidation of ferrocyanide close to 6.0 (28).

The technical improvements in these "second generation"
measurements yield H^+/O stoichiometric ratios that fully
verify earlier measurements reported from this (1-8) and
another (10,29) laboratory carried out with less accurate
methods. Also implicit in the data described here are the
possible causes of the failure of certain other laboratories
to observe maximum H^+/O ejection ratios, particularly for
site 3 of the chain. These causes may be briefly listed
(1) mixing times that are relatively long in relation to
the rate of oxygen consumption, (2) failure to begin recording the changes in O_2 and H^+ concentration soon enough
to obtain initial rates, (3) failure to recognize the intrin-
sically non-linear nature of respiration under the special
conditions of such experiments, (4) use of oxygen electrodes
with sluggish response and (5) use of intact mitochondria,
rather than mitoplasts, in site 3 measurements with added
ferrocytochrome c as substrate (cf. 30), and (6) failure to
recognize that oxidation of endogenous substrates via NAD
can severely deplete matrix buffering power prior to addition
of the specific electron donor to be tested (cf. 31). It
may also be pointed out that in those reports in which it is
concluded that the H^+/O ejection ratio is 2 or less for the

cytochrome oxidase reaction, the rates of oxygen consumption
were extremely low, only a few percent of V_{max} for cytochrome oxidase of mitoplasts or mitochondria. The rate of
O_2 consumption in the experiments described here were much
higher and near maximal, from 600 to 1000 ng-at oxygen per
min per mg protein. It appears possible that cytochrome
oxidase exhibits its maximal efficiency in its H^+ translocating function only under conditions in which its two (or
more) cytochrome c binding sites are fully occupied.

Finally, it may be pointed out that the H^+/site ratio
of 4 confirmed in the experiments outlined here are fully
consistent with other types of stoichiometric measurements
carried out in this laboratory. We have already reported
(8) that the H^+/ATP_{in} synthesis ratio is very close to 3;
more refined recent measurements are being prepared for
publication. Since the ADP_{out}/ATP_{in} translocation process
is electrophoretic and inward transport of P_i requires support with 1 H^+ during oxidative phosphorylation, we have
concluded (8) that 3 H^+ equivalents per site are required
for synthesis of ATP_{in} from ADP_{in} and $P_{i_{in}}$ and that 1 H^+ is
required for the associated membrane transport steps, for a
total of 4 H^+ utilized per site. This conclusion necessarily
implies that the P/O ratio for NAD-linked electron transport
to oxygen is a maximum of 3.0, contrary to recent conclusions
that the intrinsic P/O ratio is less than 3.0 (31,32). We
have recently reexamined the P/O stoichiometry of oxidative
phosphorylation by an approach combining kinetic and non-
equilibrium thermodynamic analyses which made it possible
to establish the magnitude of energy "leak" in state 3
respiration. Under conditions of zero energy leak the intrinsic or mechanistic P/O ratio of NAD-linked respiration

is very close to 3.0 and that for succinate oxidation is very close to 2.0 (33, further details to be published). Because of the internal consistency of the several methods and approaches we have employed, the H^+/site, K^+/site, and ATP/site ratios developed in these studies are supported by an interlocking web of evidence.

The maximum H^+/site ratio of 4 for the 3 energy-conserving segments of the mitochondrial respiratory chain has some fundamental implications with respect to the molecular mechanisms by which H^+ ejection and charge separation occur. Some of these have been developed elsewhere (34,35).

CONCLUSION

With the use of a fast-responding membrane-less oxygen electrode the kinetics of oxygen uptake and vectorial H^+ ejection by liver mitoplasts supplemented with β-hydroxy-butyrate, succinate, and ferrocytochrome c as electron donors have been examined. The rates of both oxygen consumption and H^+ ejection in the presence of K^+-valinomycin declined in an exponential manner with time, due to the increasing pH gradient across the inner membrane. Extrapolation of the rates to zero time, at which ΔH^+ back-flow is zero, yields true initial rates of O_2 uptake and vectorial H^+ ejection, from which H^+/O ratios can be calculated. Oxidation of β-hydroxybutyrate gave an H^+/O ratio of close to 12, succinate close to 8, and ferrocytochrome c close to 4.0. Possible causes of the low values for the H^+/O ejection ratios reported by other laboratories are discussed.

REFERENCES

1. Lehninger, A. L., Reynafarje, B., Alexandre, A. and
    Villalobo, A. (1979) in Membrane Bioenergetics (C. P.
    Lee, et al., eds.) pp. 393-404, Addison-Wesley.
2. Vercesi, A., Reynafarje, B. and Lehninger, A. L. (1978)
    J. Biol. Chem. 253, 6379-6385.
3. Villalobo, A. and Lehninger, A. L. (1979) J. Biol. Chem.
    254, 4352-4358.
4. Reynafarje, B. and Lehninger, A. L. (1978) J. Biol.
    Chem. 253, 6331-6334.
5. Reynafarje, B., Brand, M. D. and Lehninger, A. L.
    (1976) J. Biol. Chem. 251, 7442-7451.
6. Lehninger, A. L., Reynafarje, B., Alexandre, A., and
    Villalobo, A. (1980) Ann. N.Y. Acad. Sci. 341, 585-592.
7. Alexandre, A. and Lehninger, A. L. (1979) J. Biol. Chem.
    254, 11555-11560.
8. Alexandre, A., Reynafarje, B. and Lehninger, A. L.
    (1978) Proc. Natl. Acad. Sci. USA 75, 5296-5300.
9. Alexandre, A. and Lehninger, A. L. (1981) J. Biol. Chem.
    Submitted.
10. Pozzan, T., DiVirgilio, F., Bragadin, M., Miconi, V. and
    Azzone, G. F. (1979) Proc. Natl. Acad. Sci. USA 76,
    2123-2127.
11. Wikström, M. and Krab, K. (1979) Biochim. Biophys. Acta
    549, 177-222.
12. Wikström, M. and Krab, K. (1980) Current Topics in
    Bioenergetics 10, 51-101.
13. Sigel, E. and Carafoli, E. (1978) Eur. J. Biochem. 89,
    119-123.
14. Sigel, E. and Carafoli, E. (1979) J. Biol. Chem. 254,
    10572-10574.
15. Casey, R. P., Chappell, J. B. and Azzi, A. (1979)
    Biochem. J. 182, 149-156.
16. Mitchell, P. and Moyle, J. (1967) in Biochemistry of
    Mitochondria (E. C. Slater, et al., eds.) pp. 53-74,
    Academic Press and Polish Sci. Publ., London and Warsaw.
17. Moyle, J. and Mitchell, P. (1978) FEBS Letts. 88,
    268-272.
18. Papa, S., LoRusso, M., Guerrieri, F., Izzo, G. and
    Capuano, F. (1978) in The Proton and Calcium Pumps
    (G. F. Azzone, et al., eds.) pp. 367-374, Academic Press.
19. LoRusso, M., Capuano, F., Boffoli, D., Stefanelli, R.
    and Papa, S. (1980) Biochem. J. 182, 133-147.

20. Davies, P. W. and Grenell, R. G. (1962) J. Neurophysiol. 25, 651–683.
21. Davies, P. W. (1962) in Physical Techniques in Biological Research (W. L. Nastuk, ed.) Vol. 2, Chapter 3, Academic Press.
22. Kleiner, D. and von Jagow, G. (1972) FEBS Letts. 20, 229–232.
23. Papa, S., Capuano, F., Markert, M., and Altamura, N. (1980) FEBS Letts. 111, 243–248.
24. Nicholls, D. (1974) Eur. J. Biochem. 50, 305–315.
25. Lehninger, A. L., Reynafarje, B. and Alexandre, A. (1978) in Frontiers of Biological Energetics (P. L. Dutton, et al., eds.) Vol. 1, pp. 384–392, Academic Press.
26. Alexandre, A. (1980) Fed. Proc. 39, 1706.
27. Capuano, F., Izzo, G., Altamura, N. and Papa, S. (1980) FEBS Letts. 111, 249–254.
28. Beavis, A. (1981) Fed. Proc. In Press.
29. Azzone, G. F., Pozzan, T. and DiVirgilio, F. (1979) J. Biol. Chem. 254, 10206–10212.
30. Krab, K. and Wikström, M. (1979) Biochim. Biophys. Acta 548, 1–15.
31. Brand, M.D., Harper, W.G., Nicholls, D.E., and Ingledew, W.J. (1978) FEBS Letts. 95, 125–129.
32. Hinkle, P. C. and Yu, M. L. (1979) J. Biol. Chem. 254, 2450–2455.
33. Beavis, A. (1980) Fed. Proc. 39, 2056.
34. Lehninger, A. L., Reynafarje, B. and Alexandre, A. (1977) in Structure and Function of Energy-Transducing Membranes (K. van Dam and B. V. van Gelder, eds.) pp. 95–106, North-Holland Publishing Co., Amsterdam.
35. Lehninger, A. L. (1978) in Protons and Ions Involved in Fast Dynamic Phenomena (P. Laszlo, ed.) Elsevier, Amsterdam.

THE GENERATION AND ROLE OF THE ELECTRIC FIELD

IN $\Delta\tilde{\mu}_H{}^+$ FORMATION AND ATP SYNTHESIS

T.E. Conover[a] and G.F. Azzone[b]

[a]Department of Biological Chemistry
Hahnemann Medical College
Philadelphia, Pa. 19102

[b]N.C.R. Unit for the Study of Physiology of Mitochondria and Institute
of General Pathology, University
of Padova, Padova, Italy

INTRODUCTION

The importance of charge separation and the electrogenic movement of electrons and protons across membranes of
high capacitance in electron transfer-coupled energy transduction was suggested twenty years ago by Mitchell (1) and by
Williams (2) and has now become almost universally accepted.
The model proposed by Mitchell (3,4,5,6,7) has received the

Abbreviations: ANS, 8-anilino-1-naphthalene sulfonate; DCCD,
dicyclohexylcarbodiimide; DCMU, 3-(3,4-dichlorophenyl)-1,
1-dimethylurea; DMO, 5,5-dimethyl-2,4,-oxazolidinedione; FCCP,
carboxyl cyanide p-trifluoromethoxyphenylhydrazone; TPMP$^+$,
triphenylmethylphosphonium ion; ΔG_p, steady state phosphorylation potential or the Gibbs free energy of ATP synthesis.

C. P. Lee, G. Schatz, G. Dallner (eds.), Mitochondria and Microsomes
   in honor of Lars Ernster   ISBN 0-201-04576-1

most attention perhaps for reasons of its precise formulation
and experimental approachability. In this hypothesis H^+
translocation during electron transport and the reversible
H^+ pump of the ATPase complex are coupled through an electrochemical H^+ gradient across their common membrane. Such systems have been found in mitochondria, bacteria, and chloroplasts and seem to represent a common basis of coupling electron transport to the processes of ATP synthesis, ion movement and possibly flagellar rotation (8).

The fundamental role of the electrochemical H^+ gradient in electron transfer-coupled ATP synthesis is supported
by three basic observations: a) the existence of redox- and
ATP-linked H^+ pumps which can generate large electrochemical
gradients in energized systems (6); b) the uncoupling of energized systems by protonophoric agents (6); and c) the electrochemical gradient-driven synthesis of ATP by extracted
F_0-F_1 ATPase in reconstituted proteoliposomes (9). The acceptance of these three facts, however, still leaves room
for considerable debate as to the link between electron
transport and generation of an electrochemical H^+ gradient
and how this serves as an intermediate in ATP synthesis.

It is the purpose of this chapter to review some properties of the electrochemical H^+ gradient and its generation
with an emphasis on the role of the electric field. Other
relevant aspects of the problem of energy-coupling such as
the questions of H^+ stoichiometry, energy utilization, and
mechanisms of the ATP-linked H^+ pump function will be considered in other chapters of this volume. The important considerations of methodology used in determination of the electrochemical H^+ gradient will likewise not be dealt with except
when relevant to the interpretation of data. Several recent

reviews of these problems are available (10,11, see also 12).

DEFINITION OF THE ELECTROCHEMICAL PROTON GRADIENT, $\Delta\tilde{\mu}_H{}^+$

Operation of the redox-and ATP-linked H^+ pumps in the membrane of competent systems is linked to the generation of an electrochemical H^+ gradient, $\Delta\tilde{\mu}_H{}^+$, or proton motive force, Δp.

$$\Delta\tilde{\mu}_H{}^+ = F\Delta\Psi - 2.3RT \, \Delta pH \qquad (1)$$

$$\Delta p = \frac{\Delta\tilde{\mu}_H{}^+}{F} = \Delta\Psi - \frac{2.3RT}{F} \, \Delta pH \qquad (2)$$

Equation (2) indicates that the frequently used term Δp is in fact $\Delta\tilde{\mu}_H{}^+/F$. In this review the terminology of $\Delta\tilde{\mu}_H{}^+$ will be reserved for use as an expression of the electrochemical potential as a free energy function and Δp will be used for these values expressed in millivolts. The magnitude of $\Delta\tilde{\mu}_H{}^+$ reflects the difference in electrochemical potential of the H^+ between two aqueous phases separated by the membrane. The expression $\Delta\Psi$ refers to the membrane potential as determined between the two aqueous phases and ΔpH is the difference in pH of the two aqueous phases. This commonly used equation is conveniently simple but does involve numerous assumptions which have been discussed elsewhere (13,14).

THERMODYNAMIC COMPETENCE OF $\Delta\tilde{\mu}_H{}^+$ UNDER STEADY STATE CONDITIONS

There has been considerable effort recently to determine both $\Delta\tilde{\mu}_H{}^+$ and ΔG_p under the steady state situation where elec-

Table I.   Magnitude of Chemiosmotic Parameters in Steady

Input	$\Delta\Psi$ (mV)	$-59\Delta pH$ (mV)
Rat liver mitochondria		
β-hydroxy-butyrate	-160^a	-70^h
succinate	-170^b	-20^h
glutamate-malate	-163^b	$-44^{h,i,k}$
succinate	-150^b	-20^n
Blowfly flight-muscle mitochondria		
pyruvate	$-125^{a,c}$	-55^h
α-glycerol phosphate	$-130^{a,c}$	-40^h
Beef heart submitochondrial particles		
succinate	100^d	100^j
NADH	145^c	$<30^k$
succinate	140^c	$<30^k$
Rhodospirillum rubrum chromatophores		
1.6×10^5 ergs/cm^2/s	60^e	134^m
0.2×10^5 ergs/cm^2/s	51^e	115^m
	100^c	$<20^k$
	258^g	212^o
Rhodopseudomonas capsulata chromatophores		
	$227(ave)^f$	$190(ave)^m$
1.2×10^5 ergs/cm^2/s	211^f	203^m
0.04×10^5 ergs/cm^2/s	222^f	164^m
Spinach chloroplasts		
2.0×10^5 ergs/cm^2/s	$<10^o$	230^m
0.2×10^5 ergs/cm^2/s	$<10^o$	183^m
	0^o	198^l
Halobacterium halobium	-85^b	-60^i

a	Rb^+ or K^+ distribution with valinomycin
b	$TPMP^+$ distribution
c	SCN^- distribution
d	NO_3^- distribution
e	ANS fluorescence enhancement
f	Carotenoid band shift
g	Oxonol VI band shift
h	Acetate distribution

State Conditions

$\Delta\tilde{\mu}_{H^+}$ (kcal/H$^+$)	ΔG_p (kcal/mol)	$\Delta G_p / \Delta\tilde{\mu}_{H^+}$	Reference
5.30	13.47	2.54	15
4.38	12.45	2.85	16
4.77	15.9	3.33	17
3.6 ~ 4.6	9 – 11	2.40	18
4.15	12.17	2.93	147
3.92	13.19	3.36	147
4.60	11.53	2.51	19
4.04	10.3	2.55	20
3.92	11.1	2.83	20
4.47	11.75	2.63	21
3.60	11.11	3.08	21
2.77	14.1	5.09	22
6.23	13.0	2.09	23
9.62	15.36	1.60	24
9.55	14.3	1.50	25
8.90	12.2	1.37	25
5.30	13.4	2.53	26
4.22	11.9	2.82	26
4.57	14.3	3.13	27
3.34	9.74	2.92	28

i DMO distribution
j NH_4^+ distribution
k Methylamine distribution
l Hexylamine distribution
m 9-aminoacridine fluorescence quenching
n P^{31} nuclear magnetic resonance
o Calculated or estimated to be negligible

tron transfer and ATP synthesis are in static head condition, i.e., where the output flow is zero and the output force is maximal. Under this condition the output force of electron transfer should approach that of ATP hydrolysis, and $\Delta\tilde{\mu}_H+$, if this is the energy conserving intermediate, should bear a stoichiometric relation to these values. Table I lists the results of such studies in a variety of systems involving both oxidative phosphorylation and photophosphorylation. The values for $\Delta\Psi$ and 2.3RT/F ΔpH are given as well as for $\Delta\tilde{\mu}_H+$ and ΔG_p.

For mitochondria Δp ranges between -170 and -230 mV. Consistent with earlier determination of $\Delta\Psi$ and ΔpH (29), the $\Delta\Psi$ constitutes the major part of the $\Delta\tilde{\mu}_H+$ in this organelle.

The ratio of ΔG_p to $\Delta\tilde{\mu}_H+$ should give the minimal number of H^+ required by the reversible ATP-linked H^+ pump in order to synthesize an ATP molecule at static head or state 4 condition. With rat liver mitochondria the values vary between 2.4 and somewhat above 3. This may be compared with the H^+/ATP ratio of 3 determined by direct measurement (30), although values of 2 have also been reported (31,32). The H^+/ATP ratio predicted in the original proposal of Mitchell was two; however it has been suggested that an additional H^+ may be needed for exporting ATP from the matrix of the mitochondria to the medium (33). The $\Delta G_p/\Delta\tilde{\mu}_H+$ values obtained with intact mitochondria are in reasonable agreement with this requirement. However, values greater than two for the $\Delta G_p/\Delta\tilde{\mu}_H+$ ratio are also obtained with beef heart submitochondrial particles where there should be no necessity for energy utilization in the movement of ATP as the particles are presumably inverted in orientation. Similar values are obtained with photosynthetic bacteria and chloroplasts during photophosphorylation (21,26,27) as well as with Halobacterium halobium (28) which utilizes bacteriorhodopsin for $\Delta\tilde{\mu}_H{}^+$ generation. Recent determination of H^+ fluxes during photophosphorylation in chloroplasts also indicate a requirement for 3 H^+ per ATP (34).

Much higher values of both $\Delta\Psi$ and ΔpH were observed by Baccarini-Melandri and co-workers (24,25) with chromatophores from Rhodopseudomonas capsulata resulting in $\Delta G_p/\Delta\tilde{\mu}_H{}^+$ ratios below two. The values of $\Delta\Psi$ were determined from the electrochromic shift of the carotenoid absorption band calibrated by valinomycin-induced K^+ diffusion potentials. It has been suggested that such a calibration may lead to an overestimation of $\Delta\Psi$ if the carotenoids are responding to localized charge in contrast to delocalized membrane potential (10,35,36). The same reservations may apply to higher values of $\Delta\Psi$ reported by Bashford and co-workers (23) in Rhodospirillum rubrum since the carotenoid band shift was used to calibrate the changes observed in the extrinsic probe employed, oxonol VI. Baccarini-Melandri and co-workers also utilized 9-aminoacridine fluorescence quenching for ΔpH determination and this too has been suggested to overestimate the true values (37,38).

Kell and co-workers (22) have observed with Rhodospirillum rubrum chromatophores that light excitation gave a $\Delta\Psi$ of 100 mV with negligible ΔpH but still generated a high ΔG_p. The resulting $\Delta G_p/\Delta\tilde{\mu}_H{}^+$ ratio is therefore high ($\sim$5). Such high values are also found with mitochondria under various conditions of uncoupling (16,17). For instance, with the uncoupler, FCCP, a significantly greater decrease of $\Delta\tilde{\mu}_H{}^+$ than of ΔG_p is observed (16,25). A similar lack of consistent correlation between $\Delta\tilde{\mu}_H{}^+$ and rate of respiration in the

presence of uncoupler as compared with ADP has also been reported (39,40, see also 148).

It must be concluded that a rather consistent rela-
tionship of $\Delta\tilde{\mu}_H+$ to ΔG_p is observed in most systems investigated. The apparent stoichiometry, however, is 3 which is
higher than that predicted by Mitchell. It is difficult, in
addition, to explain the inconsistencies of the observations
made in partially uncoupled systems with a simple chemios-
matic model.

RELATIVE COMPETENCE OF $\Delta\Psi$ AND ΔpH IN ATP SYNTHESIS

An implication of Mitchell's original proposal is the
thermodynamic equality and interconvertibility of the $\Delta\Psi$ and
ΔpH terms in the expression of $\Delta\tilde{\mu}_H+$. Support of this conclusion is found in the effects of ionophoric antibiotics
which dissipate one component at the expense of the other.
Therefore a neutral ionophore such as valinomycin in the
presence of K^+ dissipates $\Delta\Psi$ by allowing K^+ to distribute
in response to potential and ΔpH is increased in compensation. Lipophilic ions behave in much the same manner. On
the other hand the carboxylic ionophore, nigericin, facilitates an electroneutral exchange of K^+ for H^+ allowing the
ΔpH to be converted into $\Delta\Psi$. The relative insensitivity
of steady state phosphorylation in mitochondria and photosynthetic bacteria to these lipophilic ions and ionophores
under conditions where osmotic swelling does not become a
predominant factor has long been recognized (15,16,20,21,25,
41). This insensitivity would appear to indicate that steady
state phosphorylation in these systems may be equally well

driven by $\Delta\Psi$ or ΔpH.

The ability of a diffusion potential to drive ATP synthesis in mitochondria has been well demonstrated (42,43,44). However, the most direct support of the competence of both ΔpH and $\Delta\Psi$ in ATP synthesis has been obtained with chloroplast suspensions. The demonstration of ATP synthesis driven by an artificially-generated pH gradient by Jagendorf and Uribe (45) was one of the earliest confirmations of the chemiosmotic hypothesis. More recently Witt and co-workers (46, 47) have been able to demonstrate the ability of $\Delta\Psi$ to support ATP synthesis by the use of an external field. ATP synthesis in this case was independent of ΔpH or electron transport and the amount of ATP synthesized per external voltage pulse was comparable to that synthesized with a saturating light pulse.

GENERATION OF $\Delta\tilde{\mu}_H^+$ IN PHOTOPHOSPHORYLATION

Rise of an electric field as the primary event

The initial events in the development of $\Delta\tilde{\mu}_H^+$ are far better understood in the case of bacterial chromatophores and plant chloroplasts than in that of mitochondria. This is due to the ability to initiate single turnovers of the light-driven electron transfer system by short saturating light flashes and to the presence of an apparent intrinsic monitor of $\Delta\Psi$, the electrochromic shift of the carotenoid absorption band (48,49).

In chromatophores from Rhodopseudomonas sphaeroides, the carotenoid absorption band shift shows a polyphasic re-

sponse on a single flash excitation (50). These phases have
been distinguished and associated with specific electron tran-
fer events (51). Phase I of the band shift ($t_{\frac{1}{2}}$ = 1 μs) is
associated with the primary light reaction of the reaction
center and the formation of P^+X^- where P is the reaction center bacteriochlorophyll and X is the primary electron accept-
or thought to be a special form of ubiquinone associated with
iron. Consistent with this is the insensitivity of this fast
phase to low temperatures (52) and to the oxidation state of
the other electron carriers or presence of electron transport
inhibitors (51). Phase II, which has a $t_{\frac{1}{2}}$ = 150 μs, appears
to be associated with the reduction of the reaction center (P^+)
by ferrocytochrome c_2 and is also insensitive to antimycin A.
A much slower response (phase III) with a $t_{\frac{1}{2}}$ >1 ms is inhibited by antimycin A and is thought to be associated with the reduction of ferricytochrome c_2 by an electron donor, ZH_2, with
a E_h = 155 mV which is thought to be a reduced ubiquinone
molecule. This reaction completes the cyclic flow of electrons through the reaction center-ubiquinone-cytochrome bc_2
complex (51).

Since the photoexcitation of the reaction center ap-
pears to generate a $\Delta\Psi$ negative to the outside and since phase
I and II responses appear to be additive, it is suggested
that the primary acceptor of the reaction center is located
near the outer membrane-aqueous interface with the cytochrome
c_2 at the inner surface of the membrane and the bacteriochlorophyll dimer of the reaction center located centrally to these
reactants. Photoactivation would therefore generate a two-
step charge transfer across the membrane as one electron
moves from the reaction center towards the outer boundary
and then another electron is transferred from the inner sur-

face to the reaction center (51).

The nature of the electrogenic reaction involved in phase III of the carotenoid shift is not understood at this time. The overall electron transfer would appear to be opposite to that indicated by the carotenoid shift, i.e., transfer from outer surface back to cytochrome c_2 at the inner surface. Various modifications of the Q cycle proposed by Mitchell (53,54) involving cytochrome b_{50} in an electrogenic transfer of an electron from the inner face toward the outer boundary have been suggested (55,56).

In the light-driven turnover of a single electron the uptake of two H^+ is observed at the outer surface of the chromatophore and appears to involve only the slower electrogenic reactions (57). The first proton (H^+_I) is taken up with a $t_{\frac{1}{2}}$ of 150 µs and is thought to involve the formation of a ubisemiquinone during the oxidation of the primary acceptor (X^-) of the reaction center (58). While there appears to be no significant physical barrier to the H^+ uptake, the primary acceptor apparently turns over too rapidly to allow diffusion-limited H^+ uptake to occur during its reduction. The uptake of H^+_I is insensitive to antimycin A during a single turnover. The second proton (H^+_{II}) is bound with a $t_{\frac{1}{2}}$ = 1.5 ms which correlates well with the rate of reduction of ferricytochrome c_2 by the reductant ZH_2. This H^+ uptake is abolished by antimycin A (55,59).

Proton release in bacterial chromatophores has been less studied then the H^+ binding reactions primarily because it occurs at the inside surface of chromatophore vesicles. It has been suggested that a H^+ should be released on the oxidation of either reduced ubiquinone or ferrocytochrome b_{50} dependent on the internal pH. By the use of cresol red

distributed internally in chromatophores it was determined
that a light-driven internal acidification occurs at pH 6.0
with a $t_{\frac{1}{2}}$ = 20-30 ms (60).

Carotenoid spectral shifts in association with light
excitation in green plant chloroplast preparations were observed even before the studies on chromatophores (61). The
correlation of a rapid band shift ($t_{\frac{1}{2}}$ < 20 ns) with the single flash excitation of the reaction centers has been established by Witt and co-workers (62). It was also shown that
selective inhibition of photosystem I (PSI) or photosystem
II (PSII) results in dimunition of the carotenoid band shift
by 50% indicating that the electrical field rise was due to
equal contributions of the two reaction centers of chloroplast photosynthetic electron transport. The rapidity of
the spectral shift and its insensitivity to low temperature
suggest that the response is due to the primary light-driv-
en charge separation occurring at the reaction centers. The
polarity of the potential generated indicates an orientation
of both reaction centers with the primary electron acceptor
near the outer membrane surface and the primary electron
donor to the inside as was observed with chromatophore vesicles (47,62).

A slower phase in the carotenoid band shift was observed in intact Chorella (63,64), but the significance of this
has been questioned (47). However, Velthuys (65,66) has reported a slow carotenoid shift ($t_{\frac{1}{2}}$ > 1 ms) in isolated chloroplasts under conditions where both reaction centers were
functioning as well as under conditions where the ubiquinone
pool was pre-reduced and only the reaction center of PSI
was excited. It was therefore suggested that the oxidation
of the reduced ubiquinone pool by PSI is also involved in

the generation of an electric field. The $t_{\frac{1}{2}}$ of the slow
phase of the carotenoid band shift would appear to be in reasonable agreement with electron transfers occurring in this
region.

As with the bacterial chromatophores the protolytic
events occur on a slower time scale than the fast phase of
the carotenoid band shift. There is an uptake of H^+ at the
external surface of the chloroplast associated with both
photosystems. This is attributed to the reduction of plasto-
quinone by PSII in one case and to the reduction of the terminal acceptor by PSI in the other (62). The observation that
the H^+ uptake is considerably slower ($t_{\frac{1}{2}}$ = 60 ms) than the
electron transfers involved ($t_{\frac{1}{2}}$ < 2 ms) has led to the suggestion that a diffusion barrier exists between the site of
acceptor reduction and the external aqueous phase. Mechanical and chemical treatments were shown to accelerate the
rate of protonation to the levels of the rates of reduction
(67). The electron stoichiometry is 1 H^+/e^- for PSII, but
is variable for PSI depending on the terminal acceptor employed (68).

Two sites of H^+ release into the internal space have
been identified using neutral red as internal pH indicator
(69,70). These are attributed to the oxidation of water by
PSII and oxidation of plastohydroquinone by PSI. The time
constants for H^+ release agree well with the rate of the respective redox reactions, namely $t_{\frac{1}{2}}$ = 300 μs for the H^+ from
H_2O and $t_{\frac{1}{2}}$ = 20 ms for that from plastohydroquinone.

Although the H^+/e^- stoichiometry of H^+ translocation
observed by the above workers is two overall, or one for each
photosystem, others have reported higher stoichiometries
(65,71). Velthuys (65) has reported the uptake of H^+ by

chloroplasts during the oxidation of plastohydroquinone by
PSI under conditions which also gave rise to the slower, second phase of the carotenoid band shift. Evidence for additional H^+ release in the chloroplast interior has also been
reported (66). An electrogenic reaction and H^+ translocation
during the oxidation of plastohydroquinone would appear to
be analogous to the observation of the antimycin A-sensitive
phase III of the carotenoid band shift and uptake of H^+_{II}
observed in bacterial chromatophores.

It is clear that in any case in both photosynthetic
systems there are electrogenic events which precede the translocation of H^+ and which, indeed, are probably independent
of H^+ movements. In chromatophores of Rhodopseudomonas sphaeroides at pH 10.0, light-induced electron turnover is accompanied by no H^+ uptake during electron transfer due to the
fact that the pK's of the reduced H^+ acceptors are 8.5 or
lower. There is no significant effect of pH on the carotenoid
band shifts, however, suggesting that electrogenic electron
transfer occurs without electroneutral transmembrane hydrogen
transfer (59). It therefore seems that the electrical field
rise as reflected in the carotenoid band shift is independent
of the protolytic reactions, a conclusion which may also be
reached on the basis of the rapid, H^+-independent phase I
band shift

Properties of the membrane potential

The theoretical calculation of the $\Delta\Psi$ generated by
a single turnover flash in chloroplast preparations is approximately 50 mV. This value is close to the mean value determined by a variety of other approaches (47) including the
use of the extrinsic probe, oxonol VI (72). Approximately

four to five single turnovers of the reaction centers are
required to give the maximal response of the carotenoid band
shift when excited by flashes in short succession (47,73).
The maximal $\Delta\Psi$ generated would therefore appear to be approximately 200 mV.

Most of the estimations of $\Delta\Psi$ during single turnovers
in bacterial chromatophores have utilized carotenoid band
shifts calibrated by use of K^+ diffusion potentials. Values
estimated for $\Delta\Psi$ after a single saturating flash vary from
100 mV (50) to 140 mV (74). The value obtained during a single flash in the presence of antimycin A ($H^+/e^- = 1$) was between 48 mV and 135 mV depending on the method of calibration
(75). As with chloroplasts 4 to 6 flashes are required to
develop maximal carotenoid response (74,76). The value for
the maximal $\Delta\Psi$ was estimated to be 430 mV (48) in agreement
with the number of turnovers required.

The values of $\Delta\Psi$ in chromatophores appear to be consistently at least twice the values in chloroplasts. This
is unexpected since the electrogenic events and H^+ translocated per turnover is approximately the same and the number of
flashes required to give maximal $\Delta\Psi$ is likewise similar. Assuming the capacitance of the membranes is roughly equivalent
one would expect the potential obtained should also be equivalent. An explanation of this discrepancy may lie in the
method used for the calibration of the carotenoid band shift
in the studies on chromatophores. As mentioned previously,
the use of K^+ diffusion potentials to calibrate the carotenoid band shift assumes that the response of carotenoids to
delocalized diffusion potential is equivalent to the response
to light-induced charge separation. If the carotenoids are
indeed responding to more than the bulk phase membrane poten-

tial during light excitation of reaction centers, the values reported for the bacterial chromatophores may be overestimated (10). This would be in agreement with the difference
in the steady state $\Delta\Psi$ also reported in chromatophores where
anion distribution was used for estimation instead of the
carotenoid band shift. (35, see also Table I).

The rise of the electric field clearly reflects a
localized event, so the potential must be generated as a localized charge separation. It has been suggested that this
is rapidly delocalized by ion movements at the membrane surface. The evidence that this occurs is two fold. In chloroplasts it has been shown that gramicidin D, added at the low
level of one gramicidin D molecule to every 10^5 chlorophyll
molecules or essentially one gramicidin D molecule per thylakoid, gives a rapid total decay of electrical field reflected by the carotenoid shift (77). This suggests that a single
ion-permeable pore per thylakoid is sufficient to decay $\Delta\Psi$
and the carotenoid shift. Similar determinations with bacterial chromatophores using valinomycin have led to similar
conclusions (78,79).

A further line of evidence for the rapid delocalization resulted from the use of electrostatic induction techniques (80,81,82). If a chloroplast suspension is irradiat-
ed by a single unidirectional source the thylakoid will assume a slightly asymmetric charge due a greater light intensity on one side than the other and this asymmetry may be
measured electrostatically. A comparison of the decay rate
of electrical measurement with that shown by the carotenoid
band shift indicated that electrical asymmetry decays 10^4
times more rapidly than does the carotenoid band shift. This
was interpreted to mean that delocalization of asymmetrical

membrane charge occurs with a $t_{\frac{1}{2}}$ of 0.1 μs in comparison
to the much slower decay of membrane potential measured by
the carotenoid band shift. However it should be pointed
out that the loss of asymmetrical charge distribution in
this case may have been produced by either ion redistribution
at the membrane surface or by randomization of the thylakoid
orientation which would not necessarily involve "de-localiza-
tion."

Although the above evidence indicates that the electrical field generated by light reactions is delocalized,
the possibility still remains that there is a significant
difference in the carotenoid response to bulk phase membrane
potential as generated by K^+ diffusion and membrane surface
potential as probably generated in the light reactions.
The important differences in the nature of membrane surface
potential and bulk phase membrane potential have been discussed in several reviews. (47,83).

Formation of steady state $\Delta\tilde{\mu}_H{}^+$

Steady state $\Delta\Psi$ observed in both chloroplasts and
chromatophores during continuous illumination is about 50%
of the maximal value observed after the first four or five
turnovers of photoactivated electron transfer (48,84,85,86).
Even lower values (10-30 mV) have also been reported for
chloroplasts in continuous light (87,88). The carotenoid
band shift observed in chromatophores after saturating flash
excitation decays in a biphasic manner. The fast component
with a $t_{\frac{1}{2}} = 1s$ is attributed to a back transfer of electrons
to the reaction center bacteriochlorophyll while the slow
component (t$_{\frac{1}{2}} = 10s$) is due to the electrophoretic movement

of ions in response to the electric field (89,90). In chromatophores this electrophoretic movement appears to be primarily a slow efflux of H^+ (91). In chloroplasts, however, ions
such as K^+, Mg^{++}, and Cl^- move readily in response to $\Delta\Psi$
and allow the acidification of the internal space. With the
decrease in pH, H^+ efflux may become more significant (47).
The increased ion movements plus a possible decrease in electron transfer rates due to the lower pH may account for the
fall in the $\Delta\Psi$ during steady state illumination.

GENERATION OF $\Delta\tilde{\mu}_{H}+$ IN OXIDATIVE PHOSPHORYLATION

Absence of a primary proton translocation

The early events in energy-coupling during oxidation-
reduction have been more difficult to follow in mitochondria
due particularly to the lack of a satisfactory means of following the rise of $\Delta\Psi$. Indeed the question as to whether
there is any significant membrane potential generated in mitochondria during respiration has been raised by some (12).
Most determinations of $\Delta\Psi$ have depended on the distribution
of radioactive cations, or anions in the case of submitochondrial particles, a procedure which is not readily adapted
to kinetic studies. Extrinsic probes such as cationic dyes
which exhibit spectral band shifts or quenching of fluorescence on uptake into the mitochondria have been utilized with
calibration by K^+ diffusion potentials (92,93). However the
reaction time for these so-called redistribution signals is
generally in the range of hundreds of milliseconds, which
is much slower than the carotenoid band shifts observed in

photosynthetic systems.

More work has been done on H^+ translocation presumably due to the availability of more direct means for following H^+ movements. Mitchell and Moyle (94) originally correlated the H^+ extruded during an oxygen pulse to anaerobic mitochondria to the rise in $\Delta\tilde{\mu}_{H}+$, even though the extrusion was in excess of the 1 ng ion H^+/mg protein which was calculated to be necessary to raise Δp to -220 mV. However it was demonstrated with the use of EGTA (95,96) and ruthenium red (97) that when Ca^{++} uptake by mitochondria is prevented, little H^+ extrusion is seen. It was argued by Mitchell (3) that the absence of H^+ movement is to be expected due to the high capacitance of the mitochondrial membrane and the consequent rapid rise of $\Delta\Psi$. Similar explanations of the low H^+ extrusion observed in E. coli in the absence of permeant anions (98) have been criticized by Gould and co-workers (99,100) on the basis of studies indicating that the kinetics and stoichiometry of H^+ extrusion by E.coli during oxygen pulses are inconsistent with the H^+ extrusion as the primary act of $\Delta\tilde{\mu}_{H}+$ generation. Archibald et al (101) quantitated the H^+ movements of rat liver mitochondria in the presence of cation exchange resins to bind cations such as Ca^{++} during oxygen pulses and concluded that the 1 ng ion H^+/mg protein which was expected is not seen.

The problem of determining both $\Delta\Psi$ and H^+ movements under the same conditions has been approached by the use of merocyanine 540 as a probe of the electrical field. Merocyanine 540 because of the presence of a sulfonic acid group is only poorly permeant to membranes (102). The difficulty in the use of penetrant probes such as safranine O or cyanine dyes is that as penetrant cations they will affect both $\Delta\Psi$

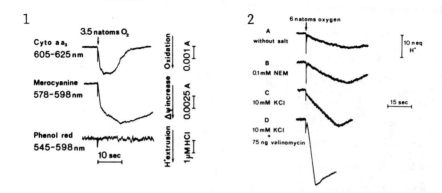

Fig. 1. Response of cytochrome aa₃, merocyanine and
phenol red on oxygen addition to anaerobic
mitochondria

>Cuvette contained 0.21 M sucrose, 0.5 mM MOPS,
pH 7.2, 0.2 mM EGTA, 1.0 mM succinate, 10 μM
merocyanine 540, 10 μM phenol red, 4 μM rotenone, 6 μg oligomycin, 3 μg catalase and 6 mg
rat liver mitochondria. Volume 3 ml. Oxygen,
3.5 natoms, added as H_2O_2. Temperature 15°.

Fig. 2. Proton translocation on oxygen addition to
anaerobic mitochondria

>Cuvette contained 0.21 M sucrose, 0.5 mM TES,
pH 7.2, 0.2 mM EGTA, 1.0 mM succinate, 20 μM
phenol red, 4 μM rotenone, 7.5 μg oligomycin,
3 μg catalase and 7.5 mg rat liver mitochondria. Volume 3 ml. Oxygen, 6.0 natoms, added as H_2O_2. Temperature 15°. 578 - 620 nm.

and H^+ extrusion during the process of redistribution on an
oxygen pulse. Figure 1 shows the oxidation of cytochrome
aa₃, the response of merocyanine 540 and H^+ extrusion during
sequential oxygen pulses on the same incubation mixture.
Cytochrome aa₃ shows a rapid oxidation leveling off when state
4 steady state is reached. This should mark the attainment
of maximal $\Delta\tilde{\mu}_{H^+}$. Merocyanine 540 shows a biphasic response
with a fast phase similar to cytochrome aa₃ oxidation and
a slow phase which continues after apparently reaching max-

imal $\Delta\tilde{\mu}_H+$. This is presumably due to a slow diffusion limited electophoretic uptake of the dye into or across the mitochondrial membrane as opposed to the rapid redistribution or reorientation which reflects the rise of an electrical field. Most importantly, however, during this rise in potential little if any extrusion of H^+ is observed.

Zanotti and Azzone (103) have reported that the oxygen required to give maximal $\Delta\Psi$ is 0.5 natoms/mg protein using the rate of safranine O uptake as a measure of the dimensions of $\Delta\Psi$. Similar values have been obtained using merocyanine 540.[1] Based on stoichiometries of $4H^+$ to $8H^+$ per oxygen it may be calculated from the requirement of 0.5 natoms oxygen/mg protein to raise $\Delta\tilde{\mu}_H+$ from zero to maximal state 4 values that the extrusion of 2 to 4 ng ion H^+/mg protein should be seen. In the presence of Ca^{++} or K^+ and valinomycin the merocyanine 540 response is largely suppressed and the rapid extrusion of 4 to 6 H^+ per oxygen may be measured.

Using phenol red in the absence of merocyanine 540 as a monitor of pH change under more sensitive conditions, a slow H^+ release was observed in the absence of permeant ions. The H^+ extrusion under various conditions is shown in Figure 2. For the most part, the H^+ extrusion is proportional to the size of the oxygen pulse over a wide range and is dependent on the permeability of the cations present. The amount of rapid H^+ extrusion during the development of $\Delta\tilde{\mu}_H+$ appears to be negligible.

Papa and co-workers (104,105) have reported the rapid extrusion of 2 ng ion H^+/mg protein on the introduction of oxygen to anaerobic beef heart mitochondria. The differences

---

[1] T.E. Conover and G.F. Azzone, unpublished experiments.

in these observations may be related to differences in the
mitochondrial preparations used and to the absence of precau-
tions to eliminate Ca^{++} from the reaction mixture in the studies with beef heart mitochondria. These workers have also
studied rapid H^+ uptake occurring in submitochondrial particles during anaerobic-aerobic transitions (105,106,107).
While in the presence of valinomycin and K^+, three phases
of H^+ uptake were observed and correlated with various electron transfer steps, only the antimycin-insensitive rapid
phase ($t_{\frac{1}{2}}$ = 33 ms) was observed in the absence of valinomycin.
The extent of this H^+ uptake is about 2 ng ion/mg protein
and is presumed to be associated with the cytochrome c oxidase
reaction. Again the extent of ion conductance in these preparations is not known.

It seems probable that the release of H^+ from mitochondria is primarily the result of compensating electrophoretic
movement of ions occurring under the influence of $\Delta\Psi$ with
the rate of release related to the permeability of the ions.
Consequently, it appears that the $\Delta\tilde{\mu}_{H}^+$ may be generated by
a primary charge separation or electrogenic electron transfer
across the membrane, or if there is a H^+ translocation, that
the H^+ does not equilibrate readily with the bulk phase. This
might be due to a physical shielding of the H^+ extruded, similar to that reported for H^+ uptake in chloroplasts (67),
or to the H^+ residing primarily in the regions of the membrane-aqueous interfaces such as the Stern-Graham layer as
suggested by Kell (83).

Support of an apparent secondary role for H^+ translocation is also found in the simultaneous observation of the
$\Delta\Psi$ and H^+ translocation in submitochondrial particles (108).
Initiation of electron transport by addition of NADH produced

a rapid response of the membrane potential probe, oxonol VI, $(t_{\frac{1}{2}} < 1s)$ and a slow response of the pH probe, 9-aminoacridine. The acidification of the vesicles appears to coincide with a slow decay of the oxonol VI response, suggesting that, as in chromatophores and chloroplasts a rapid primary generation of a $\Delta\Psi$ is followed by slower compensating ion movements to give the change in ΔpH.

In a study of the effects of ionophores on the rate of oxidation of the respiratory carriers in submitochondrial particles treated with oligomycin during anaerobic-aerobic transitions, it was observed that valinomycin in the presence of K^+ stimulated the oxidation of flavoprotein, cytochrome b, and ubiquinone to rates comparable to those seen in the presence of uncoupler (109). Nigercin was without effect. Stimulation by valinomycin and K^+ of cytochrome c oxidation in submitochondrial particles during an oxygen pulse has also been reported (104). As neither valinomycin nor nigercin affect steady state oxidation rates in these systems, it must be concluded that during the onset of electron tranport, $\Delta\Psi$ is generated initially and is the major factor in the restriction of electron flow.

Generation and properties of the membrane potential

The rate of $\Delta\Psi$ formation has not been accurately determined in either mitochondria or submitochondrial particles, due to the lack of rapidly responding probes. Oxonol VI has been frequently used in submitochondrial particles, and a response to an anaerobic-aerobic transition in the order of 50 ms has been observed (110). A comparison of the probe response with the carotenoid band shift in chromatophores

suggests that this value is limited by the response time of
the probe (111).

It may be calculated from the data of Zanotti and
Azzone (103) that in mitochondria with succinate as substrate
4 to 5 single electron turnovers of the cytochrome $\underline{aa}_3$ is
required for the generation of maximal $\Delta\Psi$. This is similar
to the number of single flash turnovers required for maximal
carotenoid response in chromatophores and chloroplasts. The
maximal $\Delta\Psi$ generated during an optimal oxygen pulse to mitochondria has been determined to be -160 to -170 mV which is
a similar value to that observed during steady state.[2] In
contrast to chloroplasts and chromatophores, with mitochondria
there appears to be little decay of $\Delta\Psi$ in the attainment of
steady state, which is in agreement with the minor contribution of ΔpH to $\Delta\tilde{\mu}_H+$ in this organelle.

RELATION OF THE ELECTRIC FIELD TO ATP SYNTHESIS

ATP synthesis and the decay of potential

The role of $\Delta\Psi$ in ATP synthesis is indicated in
both the effect of ADP and P_i on the decay of $\Delta\Psi$ as well as
the dependence of ATP synthesis on the rise of $\Delta\Psi$. In chromatophores following saturating light flashes, ADP and P_i
produce a rapid decay phase in the carotenoid band shift which
is not seen in the absence of ADP. This fast decay, which
has a $t_{\frac{1}{2}}$ in the range of 10 to 100 µs, accounts for only 10%
or 20% of the total decay observed (90,91, 112). In mitochon-

---

[2] J.N. Pile and T.E. Conover, unpublished experiments.

dria following inhibition of respiration, addition of phosphate acceptor gives a similar accelerated decay of $\Delta\Psi$ as
followed by the extrinsic probe, safranine O. This rapid decay is about 35% of the eventual total decay (113). The accelerated decay of $\Delta\Psi$ is quantitatively similar to the 20%
to 30% decrease observed in steady state $\Delta\tilde{\mu}_H^+$ under state 3
conditions relative to state 4 values in both photosynthetic
systems (21,22,114) as well as mitochondria (15,29,39,147) and
is consistent with the function of $\Delta\Psi$ in the synthesis of
ATP.

Rise of potential and the initiation of ATP synthesis

There is considerable evidence that ATP synthesis
commences very shortly after the generation of $\Delta\Psi$. In saturating continuous light or with saturating single turnover
flashes, ATP synthesis starts with little or no lag in both
chloroplasts (73,115,116) and chromatophores (78,117,118).
Lags under these conditions are 5 ms or less. However during
single turnover flashes in chromatophores, the first flash
does show a lower efficiency of ATP synthesis than do subsequent flashes (76,118). Since negligible ΔpH is generated
in the first seconds of illumination, ATP synthesis must be
driven at least initially by $\Delta\Psi$. ATP synthesis occurs also
in chromatophores during the first few turnovers in the presence of antimycin A when H^+ translocating reactions are inhibited (118).

The importance of $\Delta\Psi$ in the initiation of ATP synthesis in photophosphorylation has been verified by the use of
valinomycin and nigericin. Nigericin shows little effect
on ATP synthesis during illumination; however, valinomycin

produces marked lags in the initiation of ATP synthesis in
both chloroplasts (73,115,119) and chromatophores (117).

ATP synthesis during NADH oxidation in beef heart submitochondrial particles starts immediately after an oxygen
pulse but at a several fold lower rate than steady state synthesis. Initiation of ATP synthesis with the addition of ADP
to respiring particles and ATP synthesis driven by artificially imposed electrochemical gradients gives rates comparable
to steady state conditions (120). In rat liver submitochondrial particles ATP synthesis is estimated to begin within
20 ms of giving an oxygen pulse (121). Addition of valinomycin and K^+ to beef heart submitochondrial particles causes a
delay of more than 100 msec in ATP synthesis although the
steady state phosphorylation rate is not affected. These
findings indicate once more the role of $\Delta\Psi$ in the initiation
of ATP synthesis (120).

Critical threshold potential and the activation of the H^+ Pump

The existence of a critical threshold potential necessary for ATP synthesis was suggested by Junge and co-workers
(122) to explain the observation that 80% to 90% of the carotenoid response was resistant to ADP-accelerated decay. A minimal ΔpH requirement for photophosphorylation has also been indicated for chloroplasts (123,124,125). Indeed, when the energy input into chloroplasts suspensions is limited or reduced
by the use of low intensity light, decreased frequency of
light pulse, or the use of DCMU to block PSII function, a significant lag in the initiation of ATP synthesis is observed,
suggesting the time-dependent generation of the required ΔpH
(73,115). In mitochondrial systems, also, a minimal $\Delta\tilde{\mu}_H^+$ for

ATP synthesis has been reported (120,126).

In attempting to explain the presence of a threshold potential it should be kept in mind that if one assumes a reversible ATP-linked H^+ pump which is responsible for the generation of ATP in the presence of a H^+ gradient, $\Delta\tilde{\mu}_H{}^+$ and ΔG_p should be at equilibrium. Therefore any increase in $\Delta\tilde{\mu}_H{}^+$ should generate an increase in ΔG_p. Pick et al (123) have reported observing no influence of ΔG_p on the threshold ΔpH required for phosphorylation in chloroplasts. However it has been recently reported in chromatophores, at least, that the lag in ATP synthesis is related to the ΔG_p present, longer energization time being required when the ΔG_p is high (76).

A number of explanations have been put forth for the existence of a critical threshold potential. It has been suggested that the apparent threshold in chloroplasts is a result of two parallel pathways of H^+ efflux or ΔpH decay, a basal non-energy linked efflux with a linear dependence on the internal H^+ concentration and the efflux through the reversible ATP-linked H^+ pump which is dependent on the square (125,127) or cube (124) of the internal H^+ concentration. More recently an obligatory activation by $\Delta\Psi$ of the chloroplast H^+ pump ATPase system was proposed as an explanation for the observation of critical threshold potential (47,128). Harris and Crofts (73) suggested that the activation involves the displacement of an inhibitory protein subunit off the CF_1-ATPase complex of the H^+ pump.

These latter conclusions on chloroplasts are supported by the studies of Gomez-Puyou et al (129) and Harris et al (130) on submitochondrial particles. A lag in ATP synthesis following initiation of electron transport is seen in the presence of bound F_1-ATPase inhibitor protein (131). The lag is accounted for by the slow dissociation of the inhibitory protein which occurs on membrane energization and reverses on deenergization. It appears that $\Delta\Psi$ is the primary component involved, but ATP concentration may also have an influence (129). The rate of ATP synthesis may depend on the number of ATPase complexes activated by dissociation of the inhibitor protein (130).

It seems possible therefore that $\Delta\Psi$ functions in two capacities during ATP synthesis, both as a possible input force and as the activator of the H^+ pump ATPase complex. Any study quantitating the relation of $\Delta\tilde{\mu}_H^+$ to ATP synthesis needs to consider both functional roles for $\Delta\Psi$.

RELATION OF $\Delta\tilde{\mu}_H^+$ TO THE RATE OF ATP SYNTHESIS

In submitochondrial particles Sorgato and co-workers (20, see Table I) reported that both NADH and succinate generate a similar $\Delta\tilde{\mu}_H^+$ and ΔG_p although the rate of oxidation of succinate is 70% less than that of NADH. Inhibition of succinate oxidation by 50 or 60% with malonate does not significantly alter the $\Delta\tilde{\mu}_H^+$. Nevertheless the rate of ATP synthesis is inhibited to a degree comparable to the rate of respiration indicating that ATP synthesis reflects the percent inhibition of respiration and not the $\Delta\tilde{\mu}_H^+$ (132).

Similar observations have been made in Rhodopseudomonas capsulata chromophores by Baccarini-Melandri and co-workers (25) who reported that increasing concentrations of antimycin A or decreasing light intensity reduce the rate of photophosphorylation to negligible values with only a 20% decrease in $\Delta\tilde{\mu}_H^+$. The dependency of ATP synthesis on actual electron

transfer was confirmed in the use of single turnover flashes
of saturating intensity. ATP synthesis was observed to stop
within 800 ms after a flash even though $\Delta\tilde{\mu}_H^+$ appeared to still
exceed critical threshold values and could be initiated again
only on a new turnover of the electron transport chain (76).

Portis and McCarty (124) have reported that in chloroplasts the relationship of the rate of ATP synthesis to ΔpH is
logarithmic, i.e., the log of the phosphorylation rate varies
linearly with ΔpH. This is true whether the ΔpH is varied
with light intensity or with the use of uncouplers. The relationship of phosphorylation rate and ΔpH in chloroplasts
shown by Avron and co-workers (123,133) may indicate a similar
situation. While it has not been studied, it may well be that
the $\Delta\tilde{\mu}_H^+$ in organelles where $\Delta\Psi$ is the major component may
also reflect this relationship to the rate of phosphorylation.
This would imply that small changes in $\Delta\tilde{\mu}_H^+$ near maximal values would be seen in large changes in phosphorylation rate.
While it is possible that this may explain some of the preceding observations, the reports of Sorgato and co-workers (132)
and Melandri and co-workers (76) still imply that in addition
to the generation of a sufficient $\Delta\tilde{\mu}_H^+$, ATP synthesis may also
require the transfer of electrons in some more direct or concerted manner. Conceivably this might involve the maintenance
of the H^+ pump ATPase in an activated, inhibitor-dissociated
state as discussed in the previous section.

CAVEATS AND CONCLUSIONS

Much evidence suggests that the initial event in the
energy-generating electron transfers of oxidative phosphoryl-

ation as well as photophosphorylation is the production of an
electric field by a vectoral electrogenic transfer. This may
be detected by electrochromic probes and by its influence on
the distribution of ions in the bulk aqueous phases, but perhaps not by microelectrode measurements (134,135). Furthermore, under conditions of high membrane conductance, this electric field will be manifested by the movement of H^+ into or
out of the bulk phases. However, the answer to the question of
whether the $\Delta\tilde{\mu}_H+$ as measured in these bulk phases is an intermediate in the coupling of the charge separation of electron
transport and the reversible, ATP-linked H^+ pump is less as-
sured. Various inconsistencies in the relationship of $\Delta\tilde{\mu}_H+$
determined in the bulk phases and ATP synthesis or respiratory
control have suggested that this $\Delta\tilde{\mu}_H+$ is only indirectly related to the coupling events in ATP synthesis and that generation of H^+ associated with localized regions of the membrane
or in unstirred layers of the membrane interfaces may be the
true link between oxidation-reduction and ATP synthesis (16,
19,39,40,83,136,137). This is certainly not to deny that under certain conditions simple $\Delta\tilde{\mu}_H+$ is totally competent for
ATP synthesis (138,139,140,141,142).

Perhaps the most direct support for the existence of
coupling via something other than $\Delta\tilde{\mu}_H+$ comes from studies on
bacteria. Using an uncoupler-resistant mutant of Bacillus
megaterium, Decker and Lang (143) have found that this mutant
can maintain normal respiration-linked ATP synthesis in the
presence of uncoupler concentrations which dissipate $\Delta\tilde{\mu}_H+$ in
both the mutant as well as the wild type. However, ATP
synthesis driven by acid pulses, as well as the energy-linked
uptake of amino acids in the mutuant are both inhibited by uncoupler at the same concentrations as required in the wild

type. The mutant shows no ATPase activity even in the presence of uncoupler implying that the ATP synthesizing complex
is irreversible and suggesting that dissipation of $\Delta\tilde{\mu}_H{}^+$ can affect coupling only when the H^+ pump is reversible.

Another anomalous situation is the case of the extreme
alkalophilic bacterium, Bacillus alcalophilus, which shows
optimal growth at pH 10.5. At high pH, the bacteria generate
a large $\Delta\Psi$, but have a Δp which is depressed (-15 to -80 mV)
because of the highly unfavorable ΔpH. ATP levels are nevertheless maintained by a DCCD-and uncoupler-sensitive process
(144). Vesicles prepared from this bacteria with the same orientation as the intact cells are able to maintain a ΔG_p
of 10.65 kcal/mol at pH 10.5 when the Δp is only -40 mV
($\Delta\tilde{\mu}_H{}^+$ <1 kcal/mol). Lowering the external pH to 9.0 which
brought ΔpH to zero and Δp to -125 mV, raised the ΔG_p slightly to 11.58 kcal/mol[3]. Clearly there are apparent thermody-
namic discrepancies between $\Delta\tilde{\mu}_H{}^+$ as measured and the ΔG_p
which is generated by the organism.

Finally mention must be made of the studies of Storey
and co-workers (145,146) on submitochondrial particles prepared from rabbit skeletal muscle which appear to be open membrane fragments with no indication of any vesicular permeability barriers. These preparations which have no capacity for
the generation of a $\Delta\tilde{\mu}_H{}^+$ nevertheless show uncoupler-sensitive
respiratory control and respiration-linked quinacrine fluores-
ence responses in the same manner as do vesicular beef heart
submitochondrial particles indicating that electrogenic and
possibly protolytic events during electron transport can occur in the absence of the formation of $\Delta\tilde{\mu}_H{}^+$.

---

[3]T.A. Krulwich, personal communication

The suggestion that energy coupling events in the electron transfer-driven synthesis of ATP may be localized in the membrane or membrane-aqueous interface layers, does not do violence to the concept of chemiosmotic coupling. The original formulation by Mitchell of the chemiosmotic hypothesis has proven invaluable as a stimulus for experimental design and, indeed, as applied to the problems of membrane transport, the evidence overwhelmingly supports its predictions of the importance of $\Delta\tilde{\mu}_H^+$ in these processes. However the closer one moves towards the molecular events of electron transfer-coupled ATP synthesis, the more modifications of the original formulation seem to be required. Whether this shall ultimately mean that the less easily tested postulates of Williams were the more nearly correct will only be judged in time.

ACKNOWLEDGEMENTS

The authors wish to express their gratitude to Dr. Lars Ernster for years of guidance, inspiration and friendship. It is to him that we dedicate this review of a field to which he has contributed so greatly.

REFERENCES

1. Mitchell, P. (1961) Nature 191, 144-148.
2. Williams, R.J.P. (1961) J. Theoret. Biol. 1, 1-13.
3. Mitchell, P. (1966) Biol. Rev. 41, 445-502.
4. Mitchell, P. (1966) "Chemiosmotic Coupling in Oxidative Phosphorylation and Photosynthetic Phosphorylation." Glynn Research Ltd. Bodmin, U.K.
5. Mitchell (1968) "Chemiosmotic Coupling and Energy Transduction." Glynn Research Ltd. Bodmin, U.K.

6. Mitchell, P. (1976) Biochem. Soc. Trans. 4, 399-430.
7. Mitchell, P. (1979) Eur. J. Biochem. 95, 1-20.
8. Skulachev, V.P. (1980) Can. J. Biochem. 58, 161-175.
9. Kagawa, Y. (1978) Biochim. Biophys. Acta 505, 45-93.
10. Rottenberg, H. (1975) J. Bioenerg. 7, 61-75.
11. Rottenberg, H. (1979) Methods in Enzymology 55, 680-688.
12. Tedeschi, H. (1980) Biol. Rev. 55, 171-206.
13. Morowitz, H.J. (1978) Amer. J. Physiol. 235, R99-R114.
14. Walz, D. (1978) Biochim. Biophys. Acta 505, 279-353.
15. Nicholls, D.G. (1974) Eur. J. Biochem. 50, 305-315.
16. Azzone, G.F., Pozzan, T. and Massari, S. (1978) Biochim. Biophys. Acta 501, 307-316.
17. Holian, A. and Wilson, D.F. (1980) Biochemistry 19, 4213-4221.
18. Shen, C., Boens, C. and Ogawa, S. (1980) Biochem. Biophys. Res. Comm. 93, 243-249.
19. Azzone, G.F., Pozzan, T., Viola, E. and Arslan, P. (1978) Biochim. Biophys. Acta 501, 317-329.
20. Sorgato, C., Ferguson, S.J., Kell, D.B. and John, P. (1978) Biochem. J. 174, 237-256.
21. Leiser, M. and Gromet-Elhanan, Z. (1977) Arch. Biochem. Biophys. 178, 79-88.
22. Kell, D.B., Ferguson, S.J. and John, P. (1978) Biochim. Biophys. Acta 502, 111-126.
23. Bashford, C.L., Baltscheffsky, M., and Prince, R.C. (1979) FEBS Lett. 97, 55-60.
24. Casadio, B., Baccarini-Melandri, A., Zannoni, D. and Melandri, B.A. (1974) FEBS Lett. 49, 203-207.
25. Baccarini-Melandri, A., Casadio, B., and Melandri, B.A. (1977) Eur. J. Biochem. 78, 389-402.
26. Avron, M. (1976) in The Structural Basis of Membrane Function (Hatefi, Y. and Djavadi-Ohaniance, L., eds.) pp. 227-238, Academic Press, New York.
27. McCarty, R.E. (1978) in The Proton and Calcium Pumps (Azzone, G.F., Avron, M., Metcalfe, J.C., Quagliariello, E. and Siliprandi, N., eds.) pp. 81-91, North-Holland, Amsterdam.
28. Michel, H. and Oesterhelt, D. (1980) Biochemistry 19, 4607-4614.
29. Mitchell, P. and Moyle, J. (1969) Eur. J. Biochem. 7, 471-484.
30. Alexandre, A., Reynafarje, B. and Lehninger, A.L. (1978) Proc. Natl. Acad. Sci., U.S.A., 75, 5296-5230.
31. Mitchell, P. and Moyle, J. (1968) Eur. J. Biochem. 4, 530-539.

32. Thayer, W.S. and Hinkle, P.C. (1973) J. Biol. Chem. 248, 5395-5402.
33. Klingenberg, M. and Rottenberg, H. (1977) Eur. J. Biochem. 73, 125-130.
34. Rumberg, B. and Rothenow, M. (1980) Hoppe-Zeyler's J. Phys. Chem. 361, 1462.
35. Michels, A.M. and Konings, W.N. (1978) Eur. J. Biochem. 85, 147-155.
36. Ferguson, S.J., Jones, O.T.G., Kell, D.B. and Sorgato, M.C. (1979) Biochem J. 180, 75-85.
37. Gräber, P. and Witt, H.T. (1976) Biochim. Biophys. Acta 423, 141-163.
38. Elema, R.P., Michels, P.A.M. and Konings, W.N. (1978) Eur. J. Biochem. 92, 381-387.
39. Padan, E. and Rottenberg, H. (1973) Eur. J. Biochem. 40, 431-437.
40. Azzone, G.F., Pozzan, T., Massari, S. and Bragadin, M. (1978) Biochim. Biophys. Acta 501, 296-306.
41. Bashford, C.L. and Thayer, W.S. (1977) J. Biol. Chem. 252, 8459-8463.
42. Cockrell, R.S., Harris, E.J. and Pressman, B. (1967) Nature 215, 1487-1488.
43. Rossi, E. and Azzone, G.F. (1970) Eur. J. Biochem. 12, 319-327.
44. Azzone, G.F. and Massari, S. (1971) Eur. J. Biochem. 19, 97-107.
45. Jagendorf, A.T. and Uribe, E. (1966) Proc. Natl. Acad. Sci., U.S.A., 55, 170-177.
46. Witt, H.T., Schlodder, E. and Gräber, P. (1976) FEBS Lett. 69, 272-276.
47. Witt, H.T. (1979) Biochim. Biophys. Acta 505, 355-427.
48. Jackson, J.B. and Crofts, A.R. (1969) FEBS Lett. 4, 185-189.
49. Witt, H.T. (1971) Q. Rev. Biophys. 4, 365-477.
50. Jackson, J.B. and Crofts, A.R. (1971) Eur. J. Biochem. 18, 120-130.
51. Jackson, J.B. and Dutton, P.L. (1973) Biochim. Biophys. Acta 325, 102-113.
52. Dutton, P.L. (1971) Biochim. Biophys. Acta 226, 63-80.
53. Mitchell, P. (1975) FEBS Lett. 56, 1-6.
54. Mitchell, P. (1976) J. Theor. Biol. 62, 327-367.
55. Petty, K.M., Jackson, J.B. and Dutton, P.L. (1977) FEBS Lett. 84, 299-303.
56. Prince, R.C. and Dutton, P.L. (1977) Biochim. Biophys. Acta 462, 731-747.
57. Cogdell, R.J., Jackson, J.B. and Crofts, A.R. (1972) J. Bioenerg. 4, 413-429.

58.   Petty, K.M. and Dutton, P.L. (1976) Arch. Biochem.
      Biophys. 172, 335-345.
59.   Petty, K.M., Jackson, J.B. and Dutton, P.L. (1979)
      Biochim. Biophys. Acta 546, 17-42.
60.   Petty, K.M. and Dutton, P.L. (1976) Arch. Biochem.
      Biophys. 172, 346-353.
61.   Duysens, L.N.M. (1954) Science 120, 353-354.
62.   Schliephake, W., Junge, W. and Witt, H.T. (1968) Z.
      Naturforsch. 23B, 1571-1578.
63.   Joliot, P., Delosme, R. and Joliot, A. (1977) Biochim.
      Biophys. Acta 459, 47-57.
64.   Bouges-Bocquet, B. (1977) Biochim. Biophys. Acta 462,
      371-379.
65.   Velthuys, B.R. (1978) Proc. Natl. Acad. Sci., U.S.A.,
      75, 6031-6034.
66.   Velthuys, B.R. (1980) FEBS Lett. 115, 167-170.
67.   Auslander, W. and Junge, W. (1974) Biochim. Biophys.
      Acta 357, 285-298.
68.   Junge, W. and Auslander, W. (1974) Biochim. Biophys.
      Acta 333, 59-70.
69.   Auslander, W. and Junge,W. (1975) FEBS Lett. 59, 310-
      315.
70.   Graber, P. and Witt,H.T.(1975) FEBS Lett. 59, 310-315.
71.   Fowler, C.F. and Kok, B. (1976) Biochim. Biophys. Acta
      423, 510-523.
72.   Schuurmans, J.J., Casey, R.P. and Kraayenhof, R.
      (1978) FEBS Lett. 94, 405-409.
73.   Harris, D.A. and Crofts, A.R. (1978) Biochim. Biophys.
      Acta 502, 87-102.
74.   Takamiya, K. and Dutton, P.L. (1977) FEBS Lett. 80,
      279-284.
75.   Packham, N.K., Berriman, J.A. and Jackson, J.B. (1978)
      FEBS Lett. 89, 205-210.
76.   Melandri, B.A., Venturoli, G., De Santis, A. and
      Baccarini-Melandri, A. (1980) Biochim. Biophys. Acta
      592, 38-52.
77.   Junge, W. and Witt, H.T. (1968) Z. Naturforsch. 23B,
      244-254.
78.   Saphon, S., Jackson, J.B., Zerbs, V. and Witt, H.T.
      (1975) Biochim. Biophys. Acta 408, 58-66.
79.   Packham, N.K., Greenrod, J.A. and Jackson, J.B. (1980)
      Biochim. Biophys. Acta 592, 130-142.
80.   Fowler, C.F. and Kok, B. (1974) Biochim. Biophys. Acta
      357, 308-318.
81.   Witt, H.T. and Zickler, A. (1973) FEBS Lett. 37, 307-
      310.
82.   Witt, H.T. and Zickler, A. (1974) FEBS Lett. 39, 205-

208.

83.   Kell, D.B. (1979) Biochim. Biophys. Acta 549, 55-99.
84.   Barber, J. (1972) FEBS Lett. 20, 251-254.
85.   Evans, E.H. and Crofts, A.R. (1974) Biochim. Biophys.
      Acta 333, 44-51.
86.   Gräber, P. and Witt, H.T. (1974) Biochim. Biophys.
      Acta 333, 389-392.
87.   Rottenberg, H., Grunwald, T. and Avron, M. (1972) Eur.
      J. Biochem. 25, 54-63.
88.   Strichartz, G.R. and Chance, B. (1972) Biochim. Bio-
      phys. Acta 256, 71-84.
89.   Jackson, J.B., Greenrod, J.A., Packham, N.K. and
      Petty, K.M. (1978) in Frontiers in Bioenergetics
      (Dutton, P.L., Leigh, J.S. and Scarpa, A., eds.)
      pp. 316-325, Academic Press, N.Y.
90.   Petty, K.M. and Jackson, J.B. (1979) Biochim. Biophys.
      Acta 547, 463-473.
91.   Saphon, S., Jackson, J.B. and Witt, H.T. (1975) Bio-
      chim. Biophys. Acta 408, 67-82.
92.   Laris, P.C., Bahr, P.D. and Chaffee, R.R.J. (1975)
      Biochim. Biophys. Acta 376, 1415-1425.
93.   Åkerman, K.E.O. and Wikstrom, M.K.F. (1976) FEBS Lett.
      68, 191-197.
94.   Mitchell, P. and Moyle, J. (1965) Nature 208, 1205-
      1209.
95.   Chance, B. and Mela, L. (1966) Nature 212, 372-376.
96.   Chappell, J.B. and Haarhoff, K. (1967) in Biochem-
      istry of Mitochondria (Slater, E.C., Kaniuga, Z. and
      Wojtczak, L., eds.) pp. 75-91, Academic Press, London.
97.   Pozzan, T. and Azzone, G.F. (1976) FEBS Lett. 71, 62-
      66.
98.   Scholes, P. and Mitchell, P. (1970) J. Bioenerg. 1,
      309-323.
99.   Gould, J.M. and Cramer, W.A. (1977) J. Biol. Chem.
      252, 5875-5882.
100. Gould, J.M. (1979) J. Bacteriol. 138, 176-184.
101. Archibald, G.P.R., Farrington, C.L., Lappin, S.A.,
      McKay, A.M. and Malpress, F.H. (1979) Biochem. J.
      180, 161-174.
102. Waggoner, A.S. (1979) Ann. Rev. Biophys. Bioeng. 8,
      47-68.
103. Zanotti, A. and Azzone, G.F. (1980) Arch. Biochem.
      Biophys. 201, 255-265.
104. Papa, S., Guerrieri, F. and Lorusso, M. (1974) in Mem-
      brane Proteins in Transport and Phosphorylation
      (Azzone, G.F., Klingenberg, M.E., Quagliariello, E.
      and Siliprandi, N., eds.), pp. 177-186, North-Holland,

Amsterdam.
105. Papa, S., Guerrieri, F. and Lorusso, M. (1974) Biochim. Biophys. Acta 357, 181-192.
106. Papa, S., Lorusso, M. and Guerrieri, F. (1975) Biochim. Biophys. Acta 387, 425-440.
107. Papa, S. (1976) Biochim. Biophys. Acta 456, 39-84.
108. Bashford, C.L. and Thayer, W.S. (1977) J. Biol. Chem. 252, 8459-8463.
109. Papa, S., Scarpa, A., Lee, C.P. and Chance, B. (1972) Biochemistry 11, 3091-3097.
110. Wikström, M. and Krab, K. (1979) Biochim. Biophys. Acta 549, 177-222.
111. Bashford, C.L., Chance, B. and Prince, R.C. (1979) Biochim. Biophys. Acta 545, 46-57.
112. Petty, K.M. and Jackson, J.B. (1979) FEBS Lett. 97, 367-372.
113. Lemasters, J.J. and Hackenbrock, C.R. (1980) J. Biol. Chem. 255, 5674-5680.
114. Pick, U., Rottenberg, H. and Avron, M. (1973) FEBS Lett. 32, 91-94.
115. Ort, D.R. and Dilley, R.A. (1976) Biochim. Biophys. Acta 449, 95-107.
116. Davenport, J.W. and McCarty, R.E. (1980) Biochim. Biophys. Acta 589, 353-357.
117. Melandri, B.A., De Santis, A., Venturoli, G. and Baccarini-Melandri, A. (1978) FEBS Lett. 95, 130-134.
118. Petty, K.M. and Jackson, J.B. (1979) Biochim. Biophys. Acta 547, 474-483.
119. Vinkler, C., Avron, M. and Boyer, P.D. (1978) FEBS Lett. 96, 129-134.
120. Thayer, W.S. and Hinkle, P.C. (1975) J. Biol. Chem. 250, 5336-5342.
121. Lemasters, J.J. and Hackenbrock, C.R. (1976) Eur. J. Biochem. 67, 1-10.
122. Junge, W., Rumberg, B. and Schröder, H. (1970) Eur. J. Biochem. 14, 575-581.
123. Pick, U., Rottenberg, H. and Avron, M. (1974) FEBS Lett. 48, 32-36.
124. Portis, Jr., A.R. and McCarty, R.E. (1974) J. Biol. Chem. 249, 6250-6254.
125. Gräber, P. and Witt, H.T. (1976) Biochim. Biophys. Acta 423, 141-163.
126. Massari, S. and Azzone, G.F. (1970) Eur. J. Biochem. 12, 310-318.
127. Schroder, H., Siggel, U. and Rumberg, B. (1975) in Proc. 3rd Internat. Cong. on Photosynthesis, Rehovat,

1974 (Avron, M., ed.), pp. 1031-1039, Elsevier, Amsterdam.

128.   Junge, W. (1970) Eur. J. Biochem. 14, 582-592.
129.   Gómez-Puyou, A., Tuena de Gómez-Puyou, M. and Ernster, L. (1979) Biochim. Biophys. Acta 547, 252-257.
130.   Harris, D.A., von Tscharner, V. and Radda, G.K. (1979) Biochim. Biophys. Acta 548, 72-84.
131.   Pullman, M.E. and Monroy, G.C. (1963) J. Biol. Chem. 238, 3762-3769.
132.   Sorgato, M.C., Branca, D. and Ferguson, S.J. (1980) Biochem. J. 188, 945-948.
133.   Avron, M. (1978) FEBS Lett. 96, 225-232.
134.   Maloff, B.L., Scordilis, S.P. and Tedeschi, H. (1977) Science 195, 898-900.
135.   Maloff, B.L., Scordilis, S.P., Reynolds, C. and Tedeschi, H. (1978) J. Cell Biol. 78, 199-213.
136.   Williams, R.J.P. (1978) FEBS Lett. 85, 9-18.
137.   Williams, R.J.P. (1978) Biochim. Biophys. Acta 505, 1-44.
138.   Racker, E. and Stoeckenius, W. (1974) J. Biol. Chem. 249, 662-663.
139.   Ryrie, I.J. and Blackmore, P. F. (1976) Arch. Biochem. Biophys. 176, 127-135.
140.   Wilson, D.M., Alderette, J.F., Maloney, P.C. and Wilson, T.H. (1976) J. Bacteriol. 126, 327-337.
141.   Sone, N., Yoshida, M., Hirata, H. and Kagawa, Y. (1977) J. Biol. Chem. 252, 2956-2960.
142.   Pick, U. and Racker, E. (1979) J. Biol. Chem. 254, 2793-2799.
143.   Decker, S.J. and Lang, D.R. (1978) J. Biol. Chem. 253, 6738-6743.
144.   Guffanti, A.A., Susman, P., Blanco, R. and Krulwich, T.A. (1978) J. Biol. Chem. 253, 708-715.
145.   Scott, D.M., Storey, B.T. and Lee, C.P. (1979) Biochem. Biophys. Res. Commun. 87, 1058-1065.
146.   Storey, B.T., Scott, D.M. and Lee, C.P. (1980) J. Biol. Chem. 255, 5224-5229.
147.   Johnson, R.N. and Hansford, R.G. (1977) Biochem. J. 164, 305-322.
148.   Nicholls, D.G. and Bernson, V.S.M. (1977) Eur. J. Biochem. 75, 601-612.

MITOCHONDRIAL ANCESTOR MODELS: PARACOCCUS, RHIZOBIUM AND RHODOSPIRILLUM

Herrick Baltscheffsky and Margareta Baltscheffsky

Department of Biochemistry, Arrhenius Laboratory, University of Stockholm, S-106 91 Stockholm, Sweden

INTRODUCTION

Early ideas on a bacterial origin of mitochondria (1, 2) may well have fallen into disrepute because of uncritical over-statements, as was pointed out by Lederberg in 1952 (3), and because of being too far ahead of any possibilities for incisive experimental testing. They were revived again ten years later, in 1962, when mitochondrial DNA was observed by Nass and Nass (4) and Ris (5).

The hypothesis that the eukaryotic cell is the result of ancient symbioses, was vigorously put forward (6) and elaborated in great detail (7) by Margulis (Sagan), who more recently defended her "serial endosymbiosis theory" (8) against several critics (9-11). Support for the hereditary endosymbiosis concept and thus a bacterial origin of mitochondria has appeared from numerous comparative studies on, for example, ribosome size, protein synthesis inhibition patterns, and enzyme structural properties.

C. P. Lee, G. Schatz, G. Dallner (eds.), Mitochondria and Microsomes
in honor of Lars Ernster   ISBN 0-201-04576-1

519

Recent views on which of the presently living bacteria, if any, can serve as living models for the suggested prokaryotic ancestors of mitochondria, have essentially been based on metabolic and structural characteristics. The very recent findings of somewhat different genetic codes in mitochondria from yeast, Neurospora, and man (12-14) as well as of unusual nucleotide sequences in human mitochondria (15) have led to reconsideration of the question whether they indeed originate from recognizable ancestors of present day bacteria (15). One may predict by extrapolation that we will witness, in the eighties, further intensification of research and debate about identifiable prokaryotic relatives of the ancestors of mitochondria in aerobic eukaryotes, chloroplasts of photosynthetic eukaryotes, and hydrogenosomes (16,17) of anaerobic trichomonads.

In this mini-review, in the volume appropriately honoring Lars Ernster for his long and successful dedication to mitochondrial and microsomal structure and function, we shall focus our attention on some aspects of mitochondrial origins and evolution. An attempt will be made to briefly describe and summarize pros and cons regarding the concept that mitochondria are more or less recent products of bacterial colonization of eukaryote hosts. We also discuss the apparent significance in this respect of the current "revolution" in bacterial taxonomy (18), based on the determination of 16S ribosomal RNA nucleotide sequences. And finally, we single out and describe three examples from present day bacteria, which in our opinion all may serve as models for mitochondrial ancestry: Paracoccus, Rhizobium and Rhodospirillum.

EARLIER AND MORE RECENT CONSIDERATIONS

Which bacteria of this day and age show close resemblance to the mitochondrion? The well-known choice of Paracoccus denitrificans, by John and Whatley (19), was based on
an impressive array of demonstrated similarities and evolutionary considerations (19-21). These authors emphasized in
1975 (19) that P. denitrificans in the future probably will
be viewed as a representative of a small group of aerobic
bacteria with obvious affinity to the mitochondrion. They
also pointed out (19) that the purple non-sulfur photosynthetic bacterium Rhodopseudomonas spheroides has a mitochondrial
type of respiratory chain, when grown aerobically in the dark
(22), and thus probably should be included in the group.

Based on determinations of amino acid sequences and
three-dimensional structure of cytochrome c from many photosynthetic and non-photosynthetic organisms, as well as other
parameters, both mitochondria and P. denitrificans were recently suggested to have emerged from photosynthetic bacterial ancestors (23-25).

In 1979, Ambler and coworkers (26,27) issued warnings
that interspecific gene transfer among photosynthetic bacteria may contribute to the evolutionary process and thus
blur the record of ancestry. However, this would not at the
current stage of knowledge appear to be a serious problem,
at least in the case of cytochrome c (28,29). On the other
hand, the very recent finding that the ATPase subunit 9 gene
of yeast mitochondria has complementarity to one segment of
Neurospora mitochondrial DNA (30) indicates gene transfer between eukaryote nuclear and mitochondrial DNA. As was earlier
shown, the gene for subunit 9 in Neurospora (in contrast to

yeast) is located in the nuclear DNA (31), and thus possibly
a silent gene or only part of the gene has been conserved
in the mitochondrial genome of Neurospora.

In this connection one may recall that D.C. Smith
recently drew attention to the observation that a colonizing
microbe (due to the fact that invaded cells manifest dramatic
reactions to colonization by symbiotic microbes), "is subject
to greatly increased selection pressure, perhaps analogous
to that originally noted by Darwin for plants and animals
under domestication" (32). The speed of symbiotic selection
is well illustrated by the findings of Jeon and Jeon (33)
that the harmful effects of a naturally occurring bacterial
infection of a laboratory strain of Amoeba proteus dis-
appeared after only 1000 generations and that, meanwhile,
the amoeba had become obligatorily dependent upon the bacteria.

The transfer of segments of DNA to or from the main
genetic store of the host cell nucleus clearly can obscure
the origin of the cytoplasmic DNA. Another problem is whether fundamental differences between nuclear and cytoplasmic DNA can be expected to arise as a result of strong selection pressure after the symbiosis was established, for
example as a defense mechanism in a competitive element of
symbiotic collaboration? As has recently been pointed out
by Heckman et al. (13), the data available do not permit a
choice between: 1) the unusual codon-reading patterns in
mitochondria representing a simple, and hence perhaps primitive form of the genetic code (34); and 2) the genetic code
and codon-reading patterns in mitochondria being due to unique selection pressures operating within the mitochondrion.
The fact that CUA is translated as threonine in yeast mito-

chondria and as leucine in <u>Neurospora crassa</u> mitochondria
would appear to support the latter alternative, by indicating that the use of CUA as a codon for threonine in yeast
mitochondria arose well after the evolutionary divergence
of these two organisms (13). Notably, mammalian mitochondria
use CUA as a codon for leucine.

As is illustrated in Table I, current knowledge about
coding variability in mitochondria from different sources
appears to be in line with the idea of increased selection
pressure on apparent descendants of colonizing microbes.

Table I. Examples of coding variability in mitochondria
          (compiled after Sanger (35))

Codon	Normal	Mitochondria		
		Mammals	N. crassa	Yeast
UGA	Term	Trp	Trp	Trp
CUN	Leu	Leu	Leu	Thr
AUA	Ile	Met	?	Ile

N = nucleotide (C, U, G, A); Term = termination.

Although there still would seem to remain the possibility that mitochondria do not have an endosymbiotic origin,
we consider it to be small, in view of the collected evidence supporting the endosymbiotic theory. So, there is hope
that a fully satisfactory, logical endosymbiotic picture of
mitochondrial origins and evolution can be obtained in a not
too distant future - a picture which has reconciled the seemingly conflicting traits of uniformity in metabolic and
structural properties and diversity in the processing of in-

formation. Let us thus continue the search for plausible mitochondrial ancestor models among the bacteria living today.

SOME IMPLICATIONS OF THE CURRENT "REVOLUTION" IN BACTERIAL
TAXONOMY

As has been discussed above, mitochondria and some
bacteria show great similarities in metabolic and protein
structural properties but pronounced differences at the informational nucleic acid level. Can the current "revolution"
in bacterial taxonomy contribute towards a solution of the
problems concerning the possible origins of mitochondria
from more or less recent relatives of present-day bacteria?
The data obtained by Woese et al. (18,36-38) from comparative
analysis of oligonucleotide sequences in ribosomal 16S RNA
from numerous bacterial species have led to several new conclusions, some of which appear to be of more than passing
significance for the elucidation of mitochondrial origins:

1. The eukaryotic condition evolved from the prokaryotic one,
   which in its turn evolved from a series of less complex
   entities, such as "progenotes", "protocells", or "metabo-
   lons", which, we would like to suggest, should all be termed preprokaryotes. The lines of descent leading to the
   eukaryote host, as well as to some endosymbiotic ancestors
   of mitochondria, may well have diverged from the line leading to the various known prokaryotes already at the pre-
   prokaryote stage (36,37).

2. The question may well be what types of endosymbiosis occurred at each stage in evolution (37) rather than whether and
   when endosymbiosis occurred. Thus, photosynthesis of the
   bacterial type, with no splitting of water, may well have

evolved not only in the free-living cell, but also in the
endosymbiont stage to a) water splitting photosynthesis of
algal and higher plant type, and b) respiration, when
the environment became aerobic:

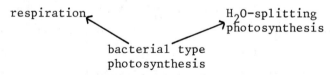

respiration H_2O-splitting
                                         photosynthesis

                    bacterial type
                    photosynthesis

The indications that early photosynthetic bacteria were
ancestors of mitochondria of currently living eukaryotes
(11,23-25,37) can well be reconciled with both these environmental alternatives.

3. Three phylogenetic lines of descent appear to have diverged before the level of complexity found in the prokaryotic cell was reached (18): the eubacteria (true bacteria),
   the archaebacteria, and the line represented in the 18S
   ribosomal RNA of the eukaryotic cytoplasm (the cytoplasmic
   aspect of the eukaryotic cell). Among the eubacteria, photosynthetic bacteria hold a prominent place, according to
   Fox et al. (18): "Four of the eight major groups are in
   fact defined by photosynthetic bacteria. Moreover, the traditional tendency to separate the photosynthetic from the
   nonphotosynthetic bacteria phylogenetically does not hold.
   Photosynthetic lines of descent tend to be intermixed with
   the lines of nonphotosynthetic species".

4. The purple photosynthetic bacteria and their closer relatives comprise three distinct sublines. The first of these,
   "purple nonsulfur-1" (PNS-1) is defined by the majority of
   Rhodopseudomonas species together with Rhodospirillum
   and Rhodomicrobium species. It contains, in addition Paracoccus and Rhizobium (and Aquaspirillum) (18). This position of Rhizobium was unexpected and is of particular sig-

nificance for the discussion below. Fig. 1 gives some of
these relationships, as they have been compiled from Fig:s
1 and 4 in ref. 18.

5. Aerobic respiration has arisen many times (18,39), in the
   purple bacterial group in particular (18), and although
   "the evolution of aerobic metabolism is not always associated with photosynthetic phenotypes initially, it often
   is" (18). The close relationships between the membrane-
   -bound photosynthetic and respiratory electron transport
   chains, and the apparent evolutionary relation between
   them has long been recognized (39,40). The ancestral bacterial phenotype, long generally held to be an anaerobic
   heterotroph, may well rather be "a photosynthetic one,
   perhaps autotrophic as well" (18).

   Based on the information given in the points 1-5
above, it is possible to discuss the evolutionary picture
given in Fig. 1 in more detail. It is seen that the nonphotosynthetic bacteria, Paracoccus and Rhizobium appear to have
evolved from purple photosynthetic ancestors, and that they
share with Rhodospirillum rubrum a common line of descent,
emerging from a "common ancestral state" (18). Mitochondrial
ancestors may, in principle, have branched off from this early ancestral state and/or from any of the branches leading to
the bacteria mentioned above. The first alternative may accomodate the differences between bacteria and mitochondria at
the informational levels, whereas the latter alternatives
would appear to be in line with all the similarities at the
metabolic and structural levels.

   Realizing that other plausible bacterial relatives of
mitochondria may well exist, we will nevertheless focus attention on the three candidates, Paracoccus denitrificans (ear-

lier discussed in detail by John and Whatley (19-21), Rhizobium trifolii, and Rhodospirillum rubrum.

MITOCHONDRIAL ANCESTOR MODELS

Paracoccus denitrificans

Since the early work of John and Whatley (19-21),
Paracoccus has appeared to be, on all accounts, the bacterium
showing the greatest number of known similarities with mitochondria (23,41) and thus the best mitochondrial ancestor model. Additional similarities were summarized by Whatley et al.
in 1979 (42), who in a broader context discussed various respiratory, phosphorylation, and membrane structural characteristics. In 1980 Whatley (43) on the basis of an incisive
comparison of respiratory chain and respiratory control properties confirmed and extended the great similarities of the
electron transport chains of Paracoccus denitrificans (and
Rhodopseudomonas sphaeroides) and of mitochondria. The properties compared included: cytochrome and quinone characteristics, antimycin, rotenone, and other inhibitor sensitivities,
electron transport control, membrane phosphatidyl choline content, and phosphorylation and ATPase reactions.

While there appears to be little, if any, reason for
doubt that Paracoccus is a very plausible model for a bacterial ancestor of mitochondria, it is a fascinating task to
discuss also our two alternative bacterial candidates for such
a model.

Rhizobium trifolii

The particular reason for selecting Rhizobium trifolii

among the Rhizobia is found in the rather detailed knowledge
of its respiratory chain (44). On the other hand, the new
phylogenetic location of Rhizobium (38) is based on oligonucleotide sequence analysis of 16S rRNA from Rhizobium legu-
minosarum, and Rhizobium japonicum was used in the analysis
of details in the symbiotic interaction, during leghaemoglobin synthesis, between leguminous plant host and Rhizobium
in the root nodule (45).

Fig. 1 gives the fundamental general background for
our novel suggestion that Rhizobium should be considered as
a mitochondrial ancestor model. Its new phylogenetic position
as a descendant from purple nonsulfur photosynthetic bacteria
in combination with its well known capability to enter into
an endosymbiotic relationship with a leguminous plant root
cell are of particular significance in this context.

The synthesis of leghaemoglobin, which in the symbiotic nodules protects nitrogen fixation from oxygen and which
is found only in nodules and not in uninfected legume tissues
or free-living bacteria, is a striking example of the coope-
ration of the symbiotic partners, i.e. the root cells and the
bacteria, which differentiate into bacteroids. Whereas the
globin of leghaemoglobin is the product of plant genes (45),
the haem is synthesized by the bacteria, which seems to contain both δ-aminolevulinic acid synthetase (46) and ferro-
chelatase (47), two enzymes in the haem biosynthetic pathway.

As has recently been pointed out by Beringer et al.
(48), this symbiotic interaction is not unlike the situation
in animal haemoglobin-synthesizing cells, in which globin
synthesis occurs on the cytoplasmic polysomes while the two
above-mentioned steps in haem biosynthesis take place in mitochondria (49). The mitochondrial δ-aminolevulinic acid syn-

thetase has been purified from rat liver (50) and found to
be loosely associated with the inner mitochondrial membrane
(51).

The electron transport chain of Rhizobium trifolii
appears to be, with its cytochrome oxidase and other components (44), basically about as similar as that of Paracoccus
to a mitochondrial one. Neither the metabolic analysis of
Rhizobia nor the structural determination of proteins involved in their respiration and phosphorylation have, however,
yet proceeded far enough to allow very detailed comparisons.
The genetic analysis of Rhizobium species has advanced in
recent years and linkage maps of standard genetic markers
are available (48). The stage seems to be set for obtaining
soon more knowledge about such metabolic and nucleic acid
characteristics which would allow closer evaluation of the
possible evolutionary relationships of these endosymbiotic
invaders of leguminous plant root cells to mitochondria.

Rhodospirillum rubrum

The photosynthetic purple nonsulfur bacterium Rhodospirillum rubrum, which can grow anaerobically in the light
and aerobically in the dark, shows many remarkable similarities to mitochondria, for example in some of the key reactions of energy conversion (52-54). Certain of the similarities to the mitochondrial electron transport system are found
in the photosynthetic, rather than the aerobic, electron
transport of R. rubrum.

Interacting with the photosynthetic, cyclic electron
transport system of R. rubrum are the two membrane-bound dehydrogenases oxidizing succinate and NADH. The former has
been solubilized from the bacterial chromatophore membrane,

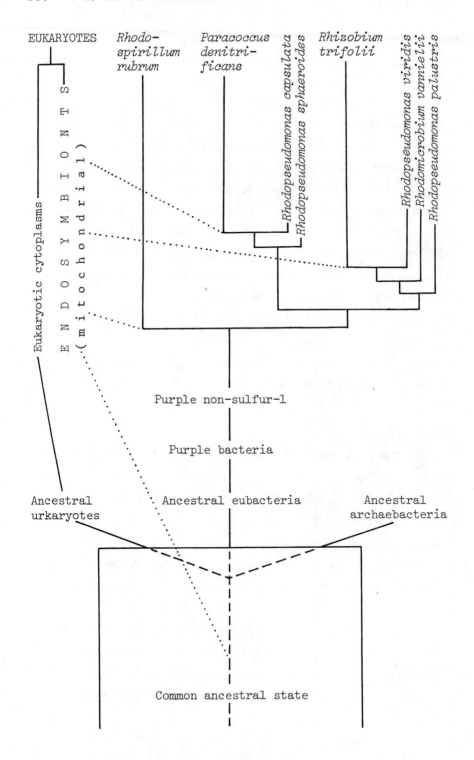

Fig. 1. Scheme for some possible endosymbiotic origins of
         mitochondria. The scheme was compiled from appropri-
         ate parts of Figures 1 and 4 in Ref. 18, which gives
         the experimental background for those evolutionary
         relationships which are included as well as for ma-
         ny others which have been left out.

         The suggested evolutionary links between ancestors
         of the three selected mitochondrial ancestor models
         among bacteria (Paracoccus denitrificans, Rhizobium
         trifolii, Rhodospirillum rubrum) and mitochondria are
         indicated by dotted lines. The fourth dotted line
         emerges from the "common ancestral state" before its
         branching off along the three phylogenetic lines of
         descent leading to ancestral urkaryotes, ancestral
         eubacteria, and ancestral archaebacteria, and repre-
         sents one of the possible ways to account for the ob-
         served differences between mitochondria and bacteria
         at the levels of molecular information.

         It should be emphasized that the "Purple bacteria"
         in the scheme are only one of several branches emer-
         ging from the "Ancestral eubacteria", and so is "Purp-
         le non-sulfur-1" from "Purple bacteria". Ref. 18 gi-
         ves more information on this and on urkaryotes and
         archaebacteria.

purified, characterized, and compared with the mitochondrial
one (55). As was found by Hatefi et al. (55,56) there are
great similarities between the succinate dehydrogenases from
R. rubrum and bovine heart mitochondria. They extend from
molecular characteristics to identical solubilization properties and, perhaps most striking, complete exchangeability in
the sense that the R. rubrum enzyme when reconstituted with
depleted mitochondrial membranes gave the same rate of succinate oxidation as the mitochondrial enzyme (55).

The membrane-bound NADH-dehydrogenase has not yet
been isolated. Rotenone partly inhibits the NADH-oxidase activity (57) and completely inhibits the reversed electron
transport from succinate to NAD, both when driven with light
and with ATP in the dark (58).

Many of the mitochondrial features shared by Paracoccus (19,42) are also shared by R. rubrum, as is seen in Table II. An exception is that cytochrome aa_3 as oxidase has not
been shown to exist in dark-aerobically grown R. rubrum. However, this could well be due to the very limited investigations carried out on this problem.

Additional properties of mitochondria, which have
been found to exist in R. rubrum, are listed in Table III.

The lists of similarities given in Table II and III
are about as extensive as has been given for Paracoccus (42).
This is perhaps no surprise in view of the evidence that Paracoccus has emerged from photosynthetic purple nonsulfur
bacteria and clearly shows evolutionary relationship to R.
rubrum (18,38). It should be pointed out that these similarities to mitochondria in our view make R. rubrum about as plausible a candidate for a mitochondrial ancestor model as Paracoccus.

Table II.   Some mitochondrial features shared by <u>Paracoccus</u>
            (from Whatley et al., ref. 42) compared with
            <u>R. rubrum</u>

Mitochondria and <u>Paracoccus</u>	<u>R. rubrum</u>
2 b-type and 2 c-type cyto-chromes readily distinguish-able by difference spectro-photometry	2 b-type cytochromes, probably 2 c-type cytochromes as well (59-62)
cytochrome aa_3 as oxidase	cytochrome aa_3 not found
ubiquinone 10 sole quinone	ubiquinone 10 main quinone, also rhodoquinone (63)
sensitive to low concentra-tions of antimycin	sensitive to low concentra-tions of antimycin (64)
NADH respiration inhibited by rotenone	NADH respiration partly inhibited by rotenone (57)
respiratory control released by ADP or uncouplers	electron transport control re-leased by ADP or uncouplers (65,66)
a rapid oxygen-induced, anti-mycin-dependent reduction of b-type cytochrome	a rapid oxidant (ferricyanide) induced antimycin-dependent reduction of b-type cytochro-me[a]

[a]M. Baltscheffsky, unpublished observation.

        The cytochrome composition of the photosynthetic elect-
ron transport system in <u>R. rubrum</u> includes at least one b-type
and one c-type cytochrome, but two recent publications and an
old observation suggest that we will soon have a more complete
picture of the entire b-c segment, showing great similarity
to the same segment in mitochondria. The b-type cytochrome

Table III. Additional mitochondrial features shared by
            R. rubrum

"cross-over" between b and c type cytochromes (65,66)

energy-linked transhydrogenase (67)

succinate dehydrogenase with similar absorption and EPR
spectra (55)

energy linked, membrane-bound inorganic pyrophosphatase (68)

oligomycin sensitivity of ATP synthesis (69) and hydrolysis
(70)

---

earlier established in R. rubrum has a midpoint potential of
-105 mV (59), and recently a new cytochrome b with a midpoint
potential of +50 mV was described (60). The exact location of
this cytochrome in the electron transport, however, still remains to be shown. The other recent finding is the isolation
of a membrane bound c-type cytochrome from Rps. spheroides
with an α absorption maximum of 552 nm (61). This cytochrome
would be analogous to cytochrome c_1 in mitochondria. The
existence of two c-type cytochromes in R. rubrum has been indicated in a spectrum of c-type cytochromes oxidized in reversed electron transport, where, in addition to cytochrome c_2
with a maximum at 552 nm, another absorption maximum at 554 nm
appeared (62).

Remarkable similarities also exist in the organization
of the enzyme complex catalyzing the final steps of energy
conversion in mitochondria and in R. rubrum, the ATPase, as
well as in the membrane-bound inorganic pyrophosphatase. The
coupling factor F_1 from both sources has five subunits with
a similar distribution of molecular weights (54) and both the

membrane-bound ATPase activity and that of the $F_o F_1$ complexes
are sensitive to oligomycin (71). Less is known about the alternative enzymes, the membrane-bound inorganic pyrophosphata-
ses. In R. rubrum this enzyme functions as coupling factor for
photophosphorylation of Pi to PPi (72,52). In beef heart mitochondria it has been claimed to similarly couple respiration
to pyrophosphate synthesis (73).

In order to determine more stringently the possible
evolutionary relationship between R. rubrum and mitochondria,
more attention should be given to the details of its dark,
aerobic electron transport and phosphorylation, as well as to
structural and functional properties of selected tRNA molecules and pertinent parts of the bacterial genome.

DISCUSSION AND OUTLOOK

Time and evolution never proceed backwards. Thus,
there often is an element of speculation and extrapolation in
studies of molecular evolution. This certainly is true about
the particular area discussed here. However, the new experimental methods for determining nucleotide sequences in DNA,
which already have made possible a complete linear description of the human mitochondrial genome (74), lead to expecta-
tions of rapid progress towards a clearer perception of mitochondrial ancestry and relations to existing bacterial species.
And so do studies on RNA (74) and protein structure and function.

There is an urgent need for linkage maps of the genomes in the bacteria which may show evolutionary relationships to mitochondria. Taking into account the rapidity of
current mapping techniques one may expect to have soon, if a

concentrated effort can be made, linkage maps of Paracoccus, Rhizobium and/or Rhodospirillum which at least may be compared with that already exists for Escherichia coli and Salmonella typhimurium (75). Such studies, in combination with DNA hybridization and nucleotide sequencing, may be expected to be most useful in the near future. On the other hand, the apparent natural occurrence of genetic transfer between genomes in various kinds of endo-relationship (30,76), and in general, gives reason for a note of caution, notwithstanding the observations indicating that such transfers are not very frequent.

It has been a recurring theme in our discussion that endosymbiosis leading to mitochondria may have happened many times, at widely differing stages of biological evolution. Thus mitochondria in different groups of eukaryotes may have different origins. Endosymbionts developing to mitochondria may perhaps in earlier periods have had, as Rhodospirillum rubrum, the capacities for both photosynthesis and nitrogen fixation or, as Rhizobium, the capacity for nitrogen fixation. Adaptations from an anaerobic to an aerobic environment, with development of respiration, may have occurred among both free-living bacteria and endosymbionts. Among both there should have been appearances and disappearances even of fundamental functions.

But, how was the exquisite symphony of interactions between host and invader orchestrated?

ACKNOWLEDGEMENT. The authors are supported by grants from The Swedish Natural Science Research Council.

REFERENCES

1. Portier, P. (1918) "Les Symbiotes", Masson, Paris.
2. Wallin, I.J. (1927) "Symbionticism and the Origin of Species", Williams & Williams, Baltimore.
3. Lederberg, J. (1952) Physiol. Rev. 32, 403-430.
4. Nass, M.M.K. and Nass, S. (1962) Exptl. Cell Res. 26, 424-427.
5. Ris, H. (1962) 5th Intern. Congr. Electron Microscopy, Philadelphia 2, xx-1.
6. Sagan, L. (1967) J. Theor. Biol. 14, 225-274.
7. Margulis, L., Origin of Eukaryotic Cells (1970) Yale University Press, New Haven and London, pp. 349.
8. Margulis, L. (1975) BioSystems 7, 266-292.
9. Raff, R.A. and Mahler, H.R. (1972) Science 177, 575-582.
10. Avers, C. (1974) Evolution, Harper & Row, New York.
11. Uzzell, T. and Spolsky, C. (1974) Amer. Sci. 62, 334-343.
12. Macino, G., Coruzzi, G., Nobrega, F.G., Li, M. and Tzagoloff, A. (1979) Proc. Natl. Acad. Sci. USA 76, 3784-3785.
13. Heckman, J.E., Sarnoff, J., Alzner-DeWeerd, B., Yin, S. and RajBhandary, U.L. (1980) Proc. Natl. Acad. Sci. USA 77, 3159-3163.
14. Barrell, B.G., Bankier, A.T. and Drouin, J. (1979) Nature 282, 189-194.
15. Eperon, I.C., Anderson, S. and Nierlich, D.P. (1980) Nature 286, 460-467.
16. Lindmark, D.G. and Müller, M. (1973) J. Biol. Chem. 248, 7724-7728.
17. Whatley, J.M., John, P. and Whatley, F.R. (1979) Proc. R. Soc. London, Ser. B 204, 165-187.
18. Fox, G.E. et al. (1980) Science 209, 457-463.
19. John, P. and Whatley, F.R. (1975) Nature 254, 495-498.
20. John, P. and Whatley, F.R. (1977) Biochim. Biophys. Acta 463, 129-153.
21. John, P. and Whatley, F.R. (1977) Adv. Bot. Res. 4, 51-115.
22. Dutton, P.L. and Wilson, D.F. (1974) Biochim. Biophys. Acta 346, 165-212.
23. Dickerson, R.E., Timkovich, R. and Almassy, R.J. (1976) J. Mol. Biol. 100, 473-491.
24. Schwartz, R.M. and Dayhoff, M.O. (1978) Science 199, 395-403.
25. Dickerson, R.E. (1981) in Evolution of Protein Structure and Function (D.S. Sigman and M.A. Brazier, eds.) pp. 173-202, Academic, New York.
26. Ambler, R.P. et al. (1979) Nature 278, 659-660.

27. Ambler, R.P., Meyer, T.E. and Kamen, M.D. (1979) Nature 278, 661-662.
28. Dickerson, R.E. (1980) Nature 283, 210-212.
29. Woese, C.R., Gibson, J. and Fox, G.E. (1980) Nature 283, 212-214.
30. Agsteribbe, D., Samallo, J., de Vries, H., Hemsgens, L.A.M. and Grivell, L.A. (1980) in Organization and Expression of the Mitochondrial Genome (A. M. Kroon and C. Saccone, eds.), pp. 51-60, North Holland, Amsterdam.
31. Sebald, W., Sebald-Althaus, M. and Wachter, E. (1977) in Mitochondria 1977, Genetics and Biogenesis of Mitochondria (W. Bandlow, et al., eds.) pp. 433-440, de Gruyter, Berlin.
32. Smith, D.C. (1979) Proc. R. Soc. London, Ser. B 204, 115-130.
33. Jeon, K.W. and Jeon, M.S. (1976) J. Cell Physiol. 89, 337-344.
34. Jukes, T.H. (1973) Nature 246, 22-26.
35. Sanger, F. (1981) Biosci. Reports 1, 3-18.
36. Woese, C. and Fox, G.E. (1977) J. Mol. Evol. 10, 1-6.
37. Woese, C. (1977) J. Mol. Evol. 10, 93-96.
38. Gibson, J., Stackebrandt, E., Zahlen, L.B., Gupta, R. and Woese, C.R. (1979) Curr. Microbiol. 3, 59-64.
39. Broda, E. (1970) Progr. Biophys. Mol. Biol. 21, 143-208.
40. Baltscheffsky, H. (1963) Sven. Kem. Tidskr. 75:1, 1-8.
41. Taylor, F.J. (1979) Proc. R. Soc. London, Ser. B 204, 267-286.
42. Whatley, J.M., John, P. and Whatley, F.R. (1979) Proc. Roy, Soc. London, Ser. B 204, 165-187.
43. Whatley, F.R. (1980) Vierteljahrsschrift Naturforsch. Ges. Zürich 125:1, 61-72.
44. De Hollander, J.A. and Stouthamer, A.H. (1980) Eur. J. Biochem. 111, 473-478.
45. Sidloi-Lumbroso, R., Kleiman, L. and Schulman, H.M. (1978) Nature 273, 558-560.
46. Nadler, K.D. and Avissar, Y.J. (1977) Plant Physiol. 60, 433-436.
47. Porra, R.J. (1975) Anal. Biochem. 68, 289-298.
48. Beringer, J.E., Brewin, N., Johnston, A.W.B., Schulman, H.M. and Hopwood, D.A. (1979) Proc. Roy. Soc. London, Ser. B 204, 219-233.
49. Gravick, S. and Sassa, S. (1971) In Metabolic regulation - metabolic pathways (H.J. Vogel, ed.) vol. 5, pp. 79-95, Academic, New York.
50. Paterniti, J.R. and Beattie, D.S. (1979) J. Biol. Chem. 254, 6112-6118.
51. Patton, G.M. and Beattie, D.S. (1973) J. Biol. Chem. 248, 4467-4474.

52. Baltscheffsky, M. (1978) In The Photosynthetic Bacteria (R.K. Clayton and W.R. Sistrom, eds.) pp. 595-613, Plenum, New York.
53. Baccarini-Melandri, A. and Melandri, B.A. (1978) ibid. pp. 615-628.
54. Johansson, B.C. and Baltscheffsky, M. (1975) FEBS Letters 53, 221-224.
55. Hatefi, Y., Davis, K.A., Baltscheffsky, H., Baltscheffsky, M. and Johansson, B.C. (1972) Arch. Biochem. Biophys. 152, 613-618.
56. Davis, K.A., Hatefi, Y., Crawford, I.P. and Baltscheffsky, H. (1977) Arch. Biochem. Biophys. 180, 459-464.
57. Thore, A., Keister, D.L. and San Pietro, A. (1969) Arch. Microbiol. 67, 378-396.
58. Keister, D.L. and Yike, N.J. (1967) Arch. Biochem. Biophys. 121, 415-422.
59. Dutton, P.L. and Baltscheffsky, M. (1972) Biochim. Biophys. Acta 267, 172-178.
60. Niederman, R.A. et al. (1980) Biochem. J. 186, 453-459.
61. Wood, P.M. (1980) Biochem. J. 189, 385-391.
62. Baltscheffsky, M. (1969) Arch. Biochem. Biophys. 133, 46-53.
63. Kakuno, T., Bartsch, R.G., Nishikawa, K. and Horio, T. (1971) J. Biochem. (Japan) 70, 79-94.
64. Baltscheffsky, H. and Baltscheffsky, M. (1958) Acta Chem. Scand. 12, 1333-1334.
65. Baltscheffsky, M. (1967) Biochem. Biophys. Res. Commun. 28, 270-276.
66. Baltscheffsky, M. and Baltscheffsky, H. (1972) in Structure and Function of Oxidation-Reduction Enzymes (Å. Åkeson and A. Ehrenberg, eds.) pp. 257-262, Pergamon Press, Oxford.
67. Keister, D.L. and Yike, N.J. (1966) Biochem. Biophys. Res. Comm. 24, 519-525.
68. Baltscheffsky, M. (1968) in Regulatory Functions of Biological Membranes (J. Järnefelt, ed.) pp. 277-286, Elsevier, Amsterdam.
69. Baltscheffsky, H. and Baltscheffsky, M. (1960) Acta Chem. Scand. 14, 257-263.
70. Bose, S.K. and Gest, H. (1965) Biochim. Biophys. Acta 96, 159-162.
71. Oren, R. and Gromet-Elhanan, Z. (1977) FEBS Lett. 79, 147-150.
72. Baltscheffsky, H., von Stedingk, L.-V., Heldt, H.-W. and Klingenberg, M. (1966) Science 153, 1120-1122.
73. Mansurova, S.E., Shakov, Yu.A. and Kulaev, I.S. (1977) FEBS Lett. 74, 31-34.

74. Anderson, S. et al. (1981) Nature 290, 457–465.
75. Sanderson, K.E. and Hartman, P.E. (1978) Bacteriol. Rev. 42, 471–519.
76. Schell, J. et al. (1979) Proc. Roy. Soc. London, Ser. B 204, 251–266.

Part II
Microsomes

A Limited History of the
Biochemistry of Microsomal Membranes
Philip Siekevitz
Rockefeller University, New York, N.Y. 10021

A subtitle of this chapter would be "With Particular
Reference to the Work of Lars Ernster," for my old and
dear friend, "Lasse" Ernster, has been involved with
research on microsomes for the last quarter-century, ever
since the review by Lindberg and Ernster (1) in 1954.

I think that the history of microsomes should begin
with the isolation in 1938 by Claude of a high-speed pel-
let (2), and the later (3) naming of it "microsomes" in
1943, when Claude found that it contained most of the RNA
of the cytoplasm and was therefore a new, singular subcellular fraction. However, the morphological nature of
this fraction was not elucidated until 1956 (4) when it
could be equated with fragmented endoplasmic reticulum,
and that the microsomal RNA was associated with what was
called at that time ribonucleoprotein particles attached
to the membrane. All that was known, up to 1949, about microsomes, mostly cell biology, is reviewed by Claude in
two publications in 1948 (5) and 1949 (6). However, soon
after came the first biochemical characterization of this

fraction, when it was found by Hogeboom and Schneider that
both the NADH- and the NADPH-cytochrome c reductases were
in this fraction, with the former enzyme being concentrated there (7-9). At about the same time, it was discovered
that this fraction had esterase activity (10), that it
could reductively cleave p-dimethylaminoazobenzene (11),
and that it was the sole depository of glucose-6-phos-
phatase activity in liver and kidney (12, 13). The excel-
lence of these early observations is attested to by the
fact that histochemical work a decade later confirmed the
findings in the case of glucose-6-phosphatase (14, 15) and
in the case of esterase (16, 17), so that we have no need
today, after much work, to alter these early conclusions.

Because in the early 1950's it was known that cytochrome c was exclusively a mitochondrial hemoprotein,
research was started to look for the "natural" electron
acceptor for pyridine nucleotide oxidation in microsomes.
And it was found, in 1954-55, in a variety of tissues, initially by several laboratories (18-20), later by many
other laboratories, and named cytochrome b_5, with a new
peak appearing, upon reduction, at 555-557 nm. That this
cytochrome was actually a participant in the oxidations by
NADH-cytochrome c reductase was indicated soon thereafter
in 1956 (21, 22). A couple of years later came the first
indication, (23-25) that another cytochrome, initially
named CO-binding pigment, existed in microsomes, for upon
reduction and gassing with CO, a new spectral peak appeared at 450 nm. More extensive work was done later on
the properties of this cytochrome by Omura and Sato and
named by them cytochrome P-450 (26, 27). Thus, by the
middle 1960's, cytochrome b_5 had been purified (21), as

well as its reductase (28), and direct evidence had been given for the interactions of these two with NADH (29). Also, it was found that cytochrome P-450 acts as an oxygen-activating enzyme (30) as well as interacting with its substrates (31). These studies, as well as a detailed study of the NADPH-cytochrome c reductase (32), gave impetus to the scheme of two electron transport chains in microsomes, one oxidizing NADH via cytochrome b_5, and the other oxidizing NADPH via cytochrome P-450, but having possible electron-flow interconnections between them (33, 34).

By this time, direct evidence was being obtained linking these electron transport chains with the results of studies which had been done a decade before. Based on experiments on the metabolism of carcinogenic dyes (35), Brodie et al. (cf. 36) postulated in 1958 that the NADPH system of microsomes is the common pathway for the hydroxylation of many drugs and aromatic compounds. Other work, linking the ω-oxidation of fatty acids (37) and the hydroxylation of steroids (38) to the NADPH-cytochrome P-450 microsomal system, gave forth the concept of the mixed function oxidase for the microsomal complex (cf. 39). Concomitant with this work, and bearing directly upon it, was the discovery (40, 41) of the phenomenon of the drug-induced synthesis of the liver microsomal hydoxylating system, and the subsequent finding (42-44), initially by Ernster's group, that the injection of phenobarbital increases the concentrations of only two microsomal components, the NADPH-cytochrome c reductase and the cytochrome P-450. Hardly a year went by before four different laboratories (45-48) reported on the occurrence of at

least one other form of cytochrome P-450, differing in
spectral properties, and appearing when 3-
methylcholanthrene, rather than phenobarbital, was injected. By the time 1968 rolled around, there was more than
enough work to have a symposium, "Microsomes and Drug
Oxidations", the result of which was a publication containing papers by almost all the leaders in the field
(49).

All during this time, Ernster and his group have been
involved with many aspects of these electron transport
chains in microsomes. The phenobarbital-induction work
has already been mentioned (43). They carried on the
findings of Gillette et al. (50) on the NADPH oxidase
system, and showed (51-55) that this system was involved
in a newly-discovered lipid peroxidation by microsomes and
that this peroxidation was activated by ADP and that iron
was involved. Further, they showed (56) that this peroxidation and the oxidative demethylation competed for a common NADPH oxidizing enzyme, the first indication of linkage in the microsomal electron transport chains.
Ernster's later findings on the stereospecificity of the
NADPH oxidation (57) first indicated that all the reactions involving NADPH, as lipid peroxidation, hydrocarbon
demethylation, steroid hydroxylation, ω-oxidation of fatty
acids, were via a common flavo enzyme, the so-called
NADPH-cytochrome c reductase. Particularly interesting
were the studies (58-62) on the structural relationship of
the electron-transport chain enzymes to the microsomal
membrane, including the first clear demonstration (60) of
the occurrence of the electron transport chains on the
membranes of the microsomes, and the binding of the elec-

tron transport enzymes to the membrane, which showed that the NADH-linked flavo enzyme and the glucose-6-phosphatase were firmly membrane-bound while the NADPH-linked flavo enzyme and cytochrome b_5 were loosely bound. The early work with G. Dallner (63, 64) on subfractionation of microsomal membranes led to the further, and still continuing, investigation by Dallner's group of subclasses of these membranes containing differences in enzymatic composition. During this period, reviews by Ernster and his co-workers appeared on all the above topics (63, 65-69).

At about the same time that work was ongoing on the electron transport chains, many investigators were finding (1953-1962) that microsomes are the predominant sites for phospholipid synthesis in animal cells. Thus it was first found that acylation of phosphatidic acid occurs there (70), then that neutral glyceride formation occurs there (71), as well as the acylation of lysolecithin (72), and the methylation of phosphatidylaminoethanol (73). Later the intervening role of cytidine diphosphate choline in phosphatidylcholine synthesis in microsomes was shown (74, 75), and that the synthesis of palmityl glyceride (76) and of phosphatidic acid (77) also occur in microsomes. The earlier discovery (78) of a phosphatidic acid phosphatase

in microsomes now began to make sense. During the four years of 1956-1960 the role of microsomes in steroid synthesis and degradation became apparent. First, it was shown that many of the enzymes synthesizing (79) and degrading cholesterol (80) are microsomal. Some of the specific individual synthetic (81-85) and degradative (86-88) enzymes were soon thereafter also shown to be microsomal in origin. By the early 1960's it had become

clear that liver microsomes, with the aid of some soluble
enzymes, were the major site of steroid metabolism in the
body. What was not clear at that time was whether microsomes had a role to play in fatty acid metabolism, outside
of the control of of these pathways by the redox states of
the pyridine nucleotides. A review (89) cites more
specifically the localization of the various steroid-
involving enzymes known to be in microsomes at the time
(1963).

As it happened countless times in other instances,
and early observation, this one by Marsh in 1951 (90),
that a Ca^{2+}-dependent factor in muscle was instrumental in
muscle relaxation, was not further explicated until much
later, when in 1958, work by Ebashi (91-94) and then by
others (95-98) showed that this "relaxing factor" was microsomal, that it bound Ca^{2+} but only in the presence of
ATP, and that this binding correlated with its muscle-
relaxing ability; a foundation had been laid for future
work on the role of the sarcoplasmic reticulum to seques-
ter Ca^{2+} and thus control muscle contractility. By the
middle of the 1960's a large amount of literature had accumulated on microsomal enzymes (cf. 89 for review), in-

cluding, in addition to the above, esterases, ATPases,
glucoronide metabolism, various miscellaneous enzymes, and
(by Ernster's group) nucleotide diphosphatases (99, 100).
Thus, by this time, all the fundamentals of microsomal
membrane biochemistry had been laid.

Work continued in the late 1960's and early 1970's on
the involvement of microsomes in phospholipid metabolism,
particularly on the acylation of glycerol phosphate (101)
and lysophosphatides (102); Van Deenen's group and Land's

group were particulary active in elucidating the specificity of fatty acid attachment to the 1, or 2 position of glycerol (cf. Rev 103). The specificity of microsomes as the site for phosphatidylcholine synthesis was firmly established (104-107). At this time, a novel exchange reaction was discovered, that of phosphatidylcholine between microsomes and mitochondria, by two groups independently (107, 108), giving rise to the idea that the major phosphatides of the mitochondria are synthesized by microsomes and then transferred to mitochondria. More evidence quickly accumulated in keeping with this hypothesis (cf. rev. 109). Thus, in the CDP-choline and CDP-ethanolamine pathways, microsomes, mitochondria and a soluble fraction are necessary (107, 110) with the microsomes supplying the enzymes, the mitochondria the ATP, and the soluble fraction the cytidine nucleotides (111). Just a few years before, the synthesis of C_{18} and C_{20} fatty acids by the malonyl CoA plus NADPH pathway was first shown for microsomes (112,113).

At about this time, in the middle 1960's, work from five laboratories (114-118) showed, almost simultaneously, that the microsomal fraction was the site of various glycosyl transferases, which shunted sugars onto covalent attachment to proteins, to synthesize the glycoproteins. Later, it was indicated (119) that probably only the inner core of sugars on the glycoproteins were attached by microsomal enzymes, and that the rest of the sugars were probably attached in the Golgi, a finding corroborated at the same time by electron microscopic histochemistry (120, 121). An interesting finding, later, was that dolichol phosphate seems to be involved in glycoprotein synthesis

as the sugar carrier (122) and that the microsomes are the
site of this involvement (123).

During this period work was progressing on the enzymatic hydroxylation system of microsomes and the involvement of cytochrome P-450. The main thrust was the confir-
mation of earlier beliefs that one hydroxylase was responsible for the attack on such diverse compounds as drugs
and steroids. The earlier indication of the Conney group
(124-126) that the steroid hormones may be the physiological substrates of the system was confirmed and extended by
Ernster's group (127). Enough work on the inducers of the
drug metabolizing system was being done that a review of
the topic could be written (128). During the mid-1970's
came the first indication of a phospholipid involvement in
the regulation of the system (129) and also the first success, from Coon and from Sato (130, 131), at purification
of cytochrome P-450, this after phenobarbital induction,
with the same groups (132,133) purifying the NADPH-
cytochrome P-450 reductase. These studies were based on
an earlier one, again from Coon's group (134), of the

reconstitution of the system from three soluble microsomal
components and the finding that the NADPH-specific flavoprotein reduces cytochrome P-450 directly without the
intervention of an iron-sulfur protein. More evidence accumulated at this time (135-138) for the existence of at
least three and probable four forms of cytochrome P-450.
Early work on cytochrome b_5 (139, 140) showed, by a comparison of the preparations purified by proteolytic digestion or by detergent extraction, that the cytochrome was
attached to the membrane by a hydrophobic region which
could be split off by proteolysis, solubilizing a still

intact and active heme-protein fragment. Later, similar
work on the purification of both the NADPH- (141, 142) and
NADH- (143) cytochrome c reductases after solubilization
by either proteolytic digestion or detergent treatment of
the microsomal membrane showed that proteolysis cleaved
off a portion of the flavoprotein (143, 144), rendering it
inactive, and this portion was hydrophobic, attaching the
protein to the microsomal membrane (143). By this time,
evidence was accumulating for the initial suggestion (145)
that the second reducing equivalent to cytochrome P-450
comes from reduced cytochrome b_5 by way of NADH oxidation
(146-151). By the mid 1970's more than enough new information had accumulated on the hydroxylations and cytochrome P-450 that two reviews could appear (141, 152), one
by Orrenius and Ernster (141).

In the early 1970's work proceeded on the sarcoplasmic reticulum Ca^{2+} pump. ATP could be generated from
ADP and P_i by this pump (153, 154) and a phosphoenzyme was
formed (155). The first purification of the enzyme was

accomplished by MacLennan in 1970 (156), and a hydrolysis
of this showed that the polypeptide contained an aspartyl
phosphate group (157) presumably the phosphate intermediate between P_i and ATP. Numerous groups confirmed these
findings, including the isolation of a 100,000 M_r protein
(cf. rev. 158-160). The phospholipid requirement was elucidated by reconstitution experiments (161, 162), as well
as the activity of tryptic peptides (163, 164).

At the same time (1960's) that work was ongoing on
the identification and characterization of microsomal membrane proteins, experiments on the turnover of these in
the membranes were being performed. Using radioactive la-

belling techniques it was found that the proteins and
phospholipids by and large turnover independently (165).
A difference was found among the proteins, as bands on SDS
gels (166), and when two individual purified proteins,
cytochrome b_5 and NADPH-cytochrome c reductase, were examined, different turnover times were found for them (165).
Even the heme and apoprotein moities of cytochrome b_5 had
different half-lives (167). When it was found that another membrane protein, the NAD glycohydrolase, gave still
another turnover rate (168), it was clear that the microsomal membrane was not being degraded and re-synthesized
as a unitary whole, but its different components were in-
dividually being turned over. These results considering
turnover rates were paralled by results showing that in
the developing and differentiating liver microsomal system, various enzymes, even including those of the same
electron transfer chains, were synthesized and inserted
into the membrane at different periods (169, 170). In the

case of phenobarbital induction of NADPH-cytochrome c
reductase it was found that the effect was due to an increased rate of synthesis (171, 172) plus a decreased rate
of degradation (171, 173). As an adjunct to these
studies, the first demonstration of the in vitro synthesis
by microsomes and the transfer through the microsomal
membrane of a secretory enzyme, pancreatic amylase, was
shown (174), a finding which led to the current work on
the synthesis and transfer of secretory proteins in cells.
The role of membranes in this regard, including microsomal
membranes, has been noted in a recent review (175).

    We now come to the problem of the topology of the mi-
crosomal membranes, of how macromolecules are related to

one another in the lateral and transverse planes of the membrane. The work on the transverse topology of the phospholipids was started in the early 1970's, by using phospholipases (176) or non-penetrating reagents (177), or the phospholipid exchange protein (178, 179). Specifically in the case of microsomal membranes, the cytoplasmic side of these is the locale of the enzyme synthesizing the phospholipids (180-182). However, there are discrepancies in the literature regarding the localization of the phospholipids themselves in the transverse plane of the microsomal membranes, though all the investigators (183-187) used the phospholipase method (cf. review (188) for a critique). The situation regarding microsomal membrane proteins seems to be more exact, in that by the late 1970's, using EM histochemistry (189), or protease digestion of membranes, or release from membranes (cf. 183) or inhibition by antibodies (190), and other methods, the cyto-

plasmic or luminal surface localization of some dozen enzymes had been made known. In the case of the lateral plane, it was known that while various enzymes were qualitatively distributed all over the surface of the membrane (cf. 189, 190) there were regions of this membrane surface which were quantitatively different in their composition. Using fractionation methods of microsomal vesicles differing in size, density, and surface charge, it has been possible to find subfractions differing in their concentrations of the cytochromes b_5 and P-450, of the NADH- and NADPH-cytochrome c reductases, of the glucose-6-phosphatase and nucleoside diphosphatase, as well as the glycoproteins. A good deal of this work was done in the late 1960's and early 1970's by Dallner's and Ernster's

group (191-197) and by others (198-200). Further work in
the past decade also further strengthened the earlier evidence for the binding of the membrane antipathic proteins
by their hydrophobic side-chains (cf. rev 201), and for
the incorporation of cytochrome b_5 (202) and cytochrome
P-450 (203) in microsomes in an enzymatically normal way.
By this time it had become well established that the cytochrome P-450 and the NADPH-cytochrome P-450 reductase complex is used in the oxidation of xenobiotics, steroids,
and fatty acids, and that the cytochrome b_5 and NADH-
cytochrome b_5 reductase is complexed with a fatty acyl-CoA
desaturase in the desaturation of fatty acids (204, 205).
Further, the earlier evidence (33, 34) that the two membrane complexes intersect in electron flow was amply confirmed in the mid-1970's (206-208). A comprehensive review of enzyme topology in intracellular membranes, including that of the microsomes, is that of DePierre and
Ernster in 1977 (201).

I would like to close by referring to the recent work
of Ernster and his associates, the examination of the aryl
hydrocarbon monooxygenase system in liver and lung and its
relationship to xenobiotic tumor production (cf. 209).
Not only have experimental papers been produced, but a
very comprehensive survey of the whole field, "Metabolism
of Polycyclic Hydrocarbons and its relationship to Cancer", has recently appeared (209). Thus, Ernster's love
affair with microsomes has progressed throughout these
over twenty years, from simple enzymatic surveys to an examination of the complex relationship of these organized
microsomal enzymes to the world in which they exist.

REFERENCE

1. Lindberg, O. and Ernster, L. (1954) Protoplasmatologia, III, A4, 1-136, Springer-Verlag, Vienna.
2. Claude, A. (1938) Proc. Soc. Exp. Biol. Med. 39, 398-403
3. Claude, A. (1943) Science 97, 451-456.
4. Palade, G.E. and Siekevitz, P. (1956). J. Biophys and Biochem. Cytol. 2, 171-200; 671-690
5. Claude, A. (1948) Harvey Lecture 43, 121-164.
6. Claude, A. (1949) Adv. Protein Chem. 5, 423-440
7. Hogeboom, G.H. (1949) J. Biol. Chem. 177, 847-858.
8. Hogeboom, G.H. and Schneider, W.C. (1950) J. Nat. Cancer Inst. 10, 983-987
9. Hogeboom, G.H. and Schneider, W.C. (1950) J. Biol. Chem. 186, 417-428.
10. Omachi, A., Barnum, C.P. and Glick, D. (1948) Proc. Soc. Exp. Biol. Med. 67, 133-136.
11. Mueller, G.C. and Miller, J.A. (1949) J. Biol. Chem. 180, 1125-1136
12. Hers, H.G. and de Duve, C. (1950). Bull. Soc. Chim. Biol. 32, 20-29
13. Hers, H.G., Berthet, J., Berthet, L. and de Duve. C. (1951). Bull. Soc. Chim. Biol. 33, 21-41
14. Tice, W. and Barrnett, R. J. (1962). J. Histo. and Cytochem. 10, 754-762
15. Goldfischer, S., Essner, E. and Novikoff, A.B. (1964). J. Histo.and Cytochem. 12, 72-95
16. Wachstein, M., Meisel, E. and Falcon, C. (1961) J. Histo. and Cytochem. 9, 325-339
17. Holt, S.J. and Hicks, R.M. (1966). J. Cell Biol. 29, 361-366
18. Strittmatter, C.F. and Ball, E. G. (1954) J. Cell Comp. Physiol. 43, 57-78
19. Bailie, M.J. and Morton, R.K. (1955) Nature 176, 111-113
20. Chance, B. and Williams, G.R. (1954) J. Biol. Chem. 209, 945-951
21. Strittmatter, C.F. and Velick, S.F. (1956) J. Biol. Chem. 221, 253-264; 277-286
22. Garfinkel, D. (1956) Biochim. Biophy. Acta. 21, 199:
23. Garfinkel, D. (1957) Arch. Biochem. Biophys. 77, 493-509
24. Ryan, K.J. and Engel, L. (1957) J. Biol Chem. 225, 103-114

25. Klingenberg, M. (1957) Arch. Biochem, Biophys. 75, 376-386

26. Omura, T. and Sato, R. (1962) J. Biol. Chem. 237, PC 1375

27. Omura, T. and Sato, R. (1964) J. Biol. Chem. 239, 2370-2378; 2379-2385

28. Strittmatter, P. and Velick, S. F. (1957) J. Biol. Chem. 228, 785-799

29. Strittmatter, P. (1961) J. Biol. Chem. 236, 2336-2341

30. Estabrook, R.W., Cooper, D.Y. and Rosenthal, O. (1963) Biochem. Zeit. 338, 741-755

31. Imai, Y. and Sato, R. (1966). Biochem. Biophys. Res. Comm. 22, 620-626

32. Masters, B.S. , Kamin, H. Gibson, O. H. and Williams, C.H. (1965) J. Biol. Chem. 240, 921-931

33. Sato, R., Omura, T. and Nishibashi, H. (1965) in Oxidases and Related Redox System, Eds., King, T. E., Mason, J.S. and Morrison, M., John Wiley, New York, 2, p.861-878

34. Estabrook, R. W. and Cohen, B. (1968) in Microsomes and Drug Oxidations. Eds. Gillette, J. R., Conney, D.H., Cosmides, G.J., Estabrook, R. W., Fouts, J.R. and Mannering, G.J. Academic Press, New York. p. 95-110

35. Brown, R. R., Miller, J. A. and Miller E.C. (1954) J. Biol Chem. 209, 211-222

36. Brodie, B. B., Gillette, J. R. and La Du, B.N. (1958) Ann. Rev. Biochem. 27, 427-454

37. Lu, A.Y.W. and Coon, M.J. (1968) J. Biol. Chem. 243, 1331-1332

38. Kuntzman, R., Johnson,M., Schneidman, K. and Conney, A.H. (1964) J. Pharmacol. Exp. Therap. 146, 280-285

39. Mason, H.S. (1965). Ann. Rev. Biochem. 34, 595-634

40. Conney, A.H. and Burns, J.J. (1959) Nature 184, 363-364

41. Remmer, H. (1959) Arch. Exp. Path. Pharmakol. 235, 279-285

42. Orrenius, S. and Ernster, L. (1964) Biochem. Biophys. Res. Comm. 16, 60-65

43. Orrenius, S., Ericsson, J.L., and Ernster, L. (1965) J. Cell. Biol. 25, 627-639

44. Remmer, H. and Merker, H.J. (1965) Ann. N.Y. Acad. Sci. 123, 79-97

45. Sladek, N.E. and Mannering, J.G. (1966) Biochem. Biophys. Res. Comm. 24, 668-674

46. Remmer, H., Schenkman, J.B., Estabrook, R.W., Sasame,

H., Gillette, J. R., Narasimhulu, S., Cooper, D.Y., and Rosenthal, O. (1966) Mol. Pharmacol. 2, 187-190

47. Imai, Y. and Sato, R. (1966) Biochem. Biophys. Res. Comm. 22, 620-626

48. Kuntzman, R., Levin, W., Schilling, G. and Alvares, A (1968) in Microsomes and Drug Oxidations. Eds. Gillette, J.R., Conney, A.H., Cosmides, G.J., Estabrook, R.A., Fouts, J.R. and Mannering, G.J. Academic Press. N.Y. p.349-363

49. Gillette, J.R., Conney, A.H. Cosmides, G.J.,Estabrook, R.W., Fouts, J.R., and Mannering, G.J., Eds. Microsomes and Drug Oxidations, Academic Press. N.Y.

50. Gillette, J.R., Brodie, B.B., and La Du, B.N. (1957) J. Pharm. Exp. Therap. 119, 532-540

51. Hochstein, P. and Ernster, L. (1963) Biochem. Biophys. Res. Comm. 12, 388-394

52. Hochstein, P. and Ernster, L. (1964) in CIBA Found. Symp., Cell Injury. Eds. A.V.S. DeReuck, and J. Knight, p.123-134. J. and A. Churchill. Ltd. London.

53. Hochstein, P., Nordenbrand, K. and Ernster, L. (1964) Biochem. Biophys. Res. Comm. 14, 323-328

54. Orrenius, S., Dallner, G. and Ernster, L. (1964) Biochem. Biophys. Res. Comm. 14, 329-334

55. Nilsson, R., Orrenius, S. and Ernster, L. (1964) Biochem. Biophys. Res. Comm. 17, 303-309

56. Orrenius, S., Dallner, G. and Ernster, L. (1964) Biochem. Biophys. Res. Comm. 14, 329-334.

57. Das, M., Orrenius, S., Ernster, L. and Gnosspelius, Y., (1968) FEBS Letters. 1, 89-92

58. Ernster, L. (1958) Acta. Chem. Scand. 12, 600-602

59. Danielson, L., Ernster, L. and Ljungren, M. (1960) Acta. Chem. Scand. 14, 1837-1838

60. Ernster, L., Siekevitz, P. and Palade, G. E. (1962) J. Cell Biol. 15, 541-562

61. Orrenius, S., Berg, A. and Ernster, L. (1969) Eur. J. Biochem. 11, 193-200

62. Kuylenstierna, B., Nicholls, D. G., Hovmöller, S. and Ernster, L. (1970) Eur. J. Biochem. 12, 419-426

63. Dallner, G. and Ernster, L. (1968) J. Histo. Cytochem. 16, 611-632

64. Dallman, P. R., Dallner, G., Bergstrand, A. and Ernster, L. (1969) J. Cell Biol. 41, 357-377

65. Orrenius, S. and Ernster, L. (1971) in Cell Membranes: Biol. and Pathol. Aspects, p. 38-53

66. Dallner, G., Svensson, H. and Ernster, L. (1972) in Structure and Function of Oxidation Reduction Enzymes, Pergamon Press, Oxford, p.567-574

67. Ernster, L.and Orrenius, S. (1973) Drug Metabolism
     and Distribution 1, 66-73
68. Svensson, H., Dallner, G. and Ernster, L. (1972)
     Biochem. Biophys. Acta 274, 447-461
69. Ernster, L. (1973) in Biochem. of Gene Expression in
     Higher Organisms. Eds. J.K. Pollak and J.W. Lee.
     p.393-409. Australia New Zealand Book Co., Sydney.
70. Kornberg, A. and Price, W.E., Jr. (1953) J. Biol.
     Chem. 204, 345-357
71. Stein, Y. and Shapiro. B. (1958) Biochem. Biophys.
     Acta 30, 271-277
72. Lands, W.E.M. (1960) J. Biol. Chem. 235, 2233-2237
73. Bremer, J. and Greenberg, D.M. (1960) Biochem.
     Biophys. Acta 37 173-175
74. Dils, R.R. and Hubscher, G. (1961) Biochim. Biophys.
     Acta 46, 505-513
75. Wilgram, G. and Kennedy, E. P. (1963) J. Biol Chem.
     248, 2615-2619
76. Senior, J. R. and Isselbacher, K. J. (1962) J. Biol.
     Chem. 237, 1454-1459
77. Pierenger, R. A. and Hokin, L. E. (1962) J. Biol.
     Chem. 237, 653-658
78. Smith, S. W., Weiss, S.B. and Kennedy, E.P. (1957)
     J. Biol. Chem. 228, 915-923
79. Bucher, N.L.R. and McGarrahan, K. (1956) J. Biol.
     Chem. 222, 1-15
80. Forchielli, E. and Dorfman, R. I., (1956) J. Biol.
     Chem. 223, 443-448
81. Popjak, G., Gosselin, L, Gore, I. Y. and Gould, R.
     (1958)   Biochem. J. 69, 238-248
82. Knauss, H.J., Porter, J. W., and Wasson, G. (1959)
     J. Biol. Chem. 234, 2835-2840
83. Bucher, N.L.R., Overath, P. and Lynen, F. (1960)
     Biochim. Biophys. Acta 40, 491-501
84. Goodman, D.S. and Popjak, G. (1960) J. Lipid Res. 1,
     286-300
85. Anderson, D.G., Rice, M.S.and Porter,J.W. (1960)
     Biochem. Biophys. Res. Comm. 3, 591-595
86. Beyer, K.F. and Samuels, L.T. (1956) J. Biol. Chem.
     219, 69-76
87. McGuire, J.S., Jr. and Tomkins, G.M. (1959) Arch.
     Biochem. Biophys. 82, 476-477
88. Endahl, G.L., Kochakian, C.D., and Hamm, D. (1960)
     J. Biol.Chem. 235, 2792-2796
89. Siekevitz P. (1963) Ann. Rev. Physiol. 25, 15-40
90. Marsh, B.B. (1951) Nature 167, 1065-1066

91. Ebashi, S. (1958) Arch. Biochem. Biophys. 76, 410-423

92. Ebashi, S. (1960) J. Biochem. (Japan) 48, 150-151

93. Ebashi, S. (1961) J. Biochem.(Japan) 50, 236-244

94. Ebashi, S. and Lipmann, F. (1962) J. Cell Biol. 14, 389-400

95. Portzehl, A. (1957) Biochim. Biophys. Acta 24, 474-482

96. Nagai, T., Makinose, M. and Hasselbach, W. (1960) Biochem. Biophys. Acta 43, 223-238

97. Berne R.M. (1962) Biochem.J. 83, 364-368

98. Stam, A.C.J. and Honig, C.L. (1962) Biochem. Biophys. Acta 60, 259-264

99. Jones, L.C. and Ernster, L. (1960) Acta Chem. Scand. 14, 1839-1840

100. Ernster, L. and Jones, L. C. (1962) J. Cell Biol. 15, 563-578

101. Possmager, F., Scherphof, G.L., Dubbelman, T.M.A.R. and Van Deenen, L.L.M. (1969) Biochem. Biophys. Acta 176, 95-110

102. Eibe, H., Hill, E.E., Lands, W.E.M. (1969) Eur. J. Biochem. 9, 250-258

103. Lennarz, W.J. (1970) Ann. Rev. Biochem. 39, 359-388

104. Stein, O. and Stein, Y. (1969) J. Cell Biol 40, 461-483

105. Bygrave, F.L. and Buecher, T. (1968) FEBS Lett 1, 42-45

106. Nagley, P.and Hallinan, T. (1968) Biochim. Biophys. Acta 163, 218-225

107. McMurray, W.C. and Dawson, R. M.C. (1969) Biochem. 112, 91-108

108. Wirtz, K.W. A. and Zilversmit, D.B. (1968) J. Biol. Chem. 243, 3596-3602

109. McMurray, W.C. and Magel, W.L. (1972) Ann. Rev. Biochem. 41, 129-160

110. William, M.L. and Bygrave, F.L. (1970) Eur. J. Biochem. 17, 32-38

111. Schneider, W.C. (1963) J. Biol. Chem. 238, 3572-3578

112. Guchhait, R.B., Putz, G.R. and Porter, J.W. (1966) Arch. Biochem. Biophys. 117, 541-549

113. Smith, S. and Dils, R. (1966) Biochim. Biophys. Acta 125, 435-444.

114. Molnar, J., Robinson, G.B. and Winzler, R.J. (1964) J. Biol. Chem. 239, 3157-3162.

115. Sarcione, E.J., Bohne, M. and Leahy, M. (1964) Biochem. 3, 1973-1976.

116. Lawford, G.R. and Schachter, H. (1966) J. Biol. Chem.
     241, 5408-5418.
117. Helgeland, L. (1965) Biochim. Biophys. Acta 101,
     106-112.
118. Cook, G.M.W., Laico, M.T. and Eylar, E.H. (1965)
     Proc. Natl. Acad. Sci. USA 54, 247-252.
119. Schachter, H., Jabbal, I., Hudgin, R.L., Pinteric, L.,
     McGuire, E.J., and Roseman, S. (1970) J. Biol. Chem.
     245, 1090-1100.
120. Peterson, M. and Leblond, C.P. (1964) J. Cell Biol.
     21, 143-148.
121. Neutra, M. and Leblond, C.P. (1966) J. Cell Biol. 30,
     119-136; ibid, 137-150.
122. Behrens, N.H. and Leloir, L.F. (1970) Proc. Natl.
     Acad. Sci. USA 66, 153-159.
123. Dallner, G., Behrens, N.H., Parodi, A.J. and Leloir,
     L.F. (1972) FEBS Lett. 24, 315-317.
124. Conney, A.H. Davison, C, Gastel, R, and Burns, J.J.
     (1960) J. Pharm. Exp. Therap. 130, 1-8
125. Conney, A. H. and Klutch, A (1963) J. Biol. Chem.
     238, 1611-1617
126. Kuntzman, R., Lawrence. D. and Conney, A. H. (1965)
     Mol. Pharmacol. 1, 163-167
127. Das, M.L. Orrenius, S. and Ernster, L. (1968) Eur. J.
     Biochem. 4, 519-523
128. Conney, A.H. (1967) Pharmacol. Rev. 19, 317-366
129. Strobel, H. W., Lu, A.Y.H. , Heidema, J. and Coon,
     M.J. (1970) J.Biol. Chem. 245, 4851-4854
130. Van Der Hoeven, T. A., Haugen, D.A., and Coon, M.J.
     (1974) Biochem. Biophys. Res. Comm. 60, 569-575
131. Imai, Y. and Sato, R. (1974) Biochem. Biophys. Res.
     Comm. 60, 8-14
132. Satake, A., Imai, Y. and Sato, R. (1972) Saikagaku
     44, 765
133. Vermilion, J.L. and Coon, M.J. (1974) Biochem.
     Biophys. Res. Comm. 60, 1315-1322.
134. Lu, A.Y.H. and Coon, M.J. (1968) J. Biol Chem. 243,
     1331-1332.
135. Nebert, D. W., Heidema, J. K., Strobel, H. W., and
     Coon, M.J. (1973) J. Biol. Chem. 248, 7631-7636.
136. Alvares, A. P. and Siekevitz, P. (1973) Biochem.
     Biophys. Res. Comm. 54, 923-929.
137. Welton, A. F. and Aust, S. D. (1974) Biochem.
     Biophys. Res. Comm. 56, 898-906.
138. Van der Hoeven, T. A., Haugen, D. A. and Coon, M.J.
     (1974) Pharmacol. 6, 321a.

139. Ito, A. and Sato, R. (1968) J. Biol. Chem. 243, 4922-4923.
140. Spatz, L. and Strittmatter, P. (1971) Proc. Natl. Acad. Sci. USA 68, 1042-1046.
141. Orrenius, S. and Ernster, L. (1974) in Mol. Mech. of Oxygen Activation, ed. O. Hayaishi, Academic Press, N.Y. pp.215-224.
142. Coon, M.J., Strobel, H.W. and Boyer, R.F. (1973) Drug Meta. Distr. 1, 92-97.
143. Spatz, L. and Strittmatter, P. (1973) J. Biol. Chem. 248, 793-799.
144. Welton, A. F., Pederson, T. C., Buege, J. H., and Aust, S.D. (1973) Biochem. Biophys. Res. Comm. 54, 161-167.
145. Hildebrandt, A. G. and Estabrook, R.W. (1971) Arch. Biochem. Biophys. 143, 66-79.
146. Correia, M.A. and Mannering, G.J. (1972) Drug Meta. Distr. 1, 139-149.
147. Sasame, H. A., Mitchell, J.R., Thorgeirsson, S. and Gillette, J.R. (1972) Drug Meta. Distr. 1, 150-155.
148. Mannering, G.J., Kuwahara, S., and Omura, T. (1974) Biochem. Biophys. Res. Commm. 57, 476-481.
149. Sasame, H. A., Thorgeirsson, S., Mitchell, J.R. and Gillette, J.R. (1974) Life Sci., 14, 35-46.
150. Hrycay, E. and Estabrook, R.W. (1974) Biochem. Biophys. Res. Comm. 60, 771-778.
151. West, S.B., Levin, W., Ryan, D., Vore, M. and Lu, A.Y.H. (1974) Biochem. Biophys. Res. Comm. 58, 516-522.
152. Gunsalus, I.C., Pederson, T.C. and Sligar, S.G. (1974) Ann. Rev. Biochem. 44, 377-407.
153. Makinose, M. and Hasselbach, W. (1971) FEBS Lett. 12, 271-272.
154. Panet, R. and Selinger, Z. (1972) Biochem. Biophys. Acta 255, 34-42.
155. Masuda, H. and DeMeis, L. (1973) Biochem. 12, 4581-4585.
156. MacLennan, D. (1970) J. Biol. Chem. 245, 4508-4518.
157. Bastide, F., Meissner, G., Fleischer, S. and Post, R.L. (1973) J. Biol. Chem. 248, 8385-8391.
158. MacLennan, D.H. and Holland, P.C. (1976) in Enzymes of Biological Membranes, ed. A. Martonosi, Plenum Press, N.Y. 3, 221-259.
159. Hasselbach, W. and Beil, F.U. (1977) in Biochemistry of Membrane Transport, ed. G. Semenza and E. Carafoli, Springer, N.Y. p.461-528.
160. Wilson, D. B. (1978) Ann. Rev. Biochem. 47, 933-965.

161.   Racker, E. and Eytan, E. (1975) J. Biol. Chem. 250, 7533-7534.
162.   Warren, G.B., Houslay, M.D., Metcalfe, J.C. and Birdsall, N.J. (1975) Nature 255, 684-687.
163.   Shamoo, A.E. and Goldstein, D.A. (1977) Biochim. Biophys. Acta 472, 13-53.
164.   Madeira, V.M.C. (1977) Biochim. Biophys. Acta 464, 583-588.
165.   Omura, T., Siekevitz, P. and Palade, G.E. (1967) J. Biol. Chem. 242, 2389-2396.
166.   Arias, I.M., Doyle, D. and Schimke, R.T. (1969) J. Biol. Chem. 244, 3303-3315.
167.   Bock, K.W. and Siekevitz, P. (1970) Biochem. Biophys. Res. Comm. 41, 374-380.
168.   Bock, K.W., Siekevitz, P. and Palade, G.E. (1971) J. Biol. Chem. 246, 188-195.
169.   Dallner, G., Siekevitz, P., and Palade, G.E. (1966) J. Cell. Biol. 30, 73-96.
170.   Dallner, G., Siekevitz, P. and Palade, G.E. (1966) J. Cell Biol. 30, 97-117.
171.   Kuriyama, Y., Omura, T., Siekevitz, P., and Palade, G.E. (1969) J. Biol. Chem. 244, 2017-2026.
172.   Schimke, T., Ganschow, R., Doyle, D., and Arias, I.M. (1968) Fed. Proc. 27, 1223-1230.
173.   Jick, H. and Schuster, L. (1966) J. Biol. Chem. 241, 5366-5369.
174.   Redman, C.M., Siekevitz, P. and Palade, C.E. (1966) J. Biol. Chem. 241, 1150-1158.
175.   Davis, B.D. and Tai, P.-C. (1980) Nature 283, 433-438.
176.   Verkleij, A.J. Zwaal, R.F.A., Roelofsen, B., Comfurius, P., Kastelijn, D. and Van Deenen, L.L.M. (1973) Biochem. Biohys. Acta 323, 178-193.
177.   Bretscher, M. (1972) Nature New Biology. 236, 11-12.
178.   Bloj, B. and Zilversmit, D.B. (1976) Biochem. 15, 1277-1283.
179.   Rothman, J.E., Tsai, D.K., Davidowicz, E.A. and Lenard, J. (1976) Biochem. 15, 2361-2371.
180.   Vance, D.E., Choy, P.C., Farren, S.B., Lim, P.H. and Schneider, W.J. (1977) Nature 270, 268-269.
181.   Brophy, D.J.,Burbach, P., Nelemans, A.S., Westerman,J., Wirtz, K.W.A. and Van Deenen, L.L.M. (1978) Biochem. J. 174, 413-420.
182.   Coleman, R., and Bell, R.M. (1978) J. Cell Biol. 76, 245-253.
183.   De Pierre, J.W. and Dallner, G. (1975) Biochem. Biophys. Acta 415, 411-472.

184. Nilsson, O.S. and Dallner, G. (1977) Biochem. Biophys. Acta 464, 453-458.

185. Nilsson, O.S. and Dallner, G. (1977) J. Cell Biol. 72, 568-583.

186. Sundler, R., Sarcione, S.L., Alberts, A.W. and Vagelos, P.R. (1977) Proc. Natl. Acad. Sci. 74, 3350-3354.

187. Higgins, J.A. and Dawson, R.M.C. (1977) Biochem. Biophys. Acta 470, 342-356.

188. Op den Kamp, J.A.F. (1979) Ann. Rev. Biochem. 48, 47-71.

189. Leskes, A., Siekevitz, P. and Palade, G.E. (1971) J. Cell Biol. 49, 264-287; 288-302.

190. Morimoto, T., Matsubara, S., Sasaki,S., Tashiro, Y. and Omura, T. (1976) J. Cell Biol. 68, 189-201.

191. Dallner, G., Bergstrand, A. and Nilsson, R. (1968) J. Cell Biol. 38, 257-276.

192. Dallman, P.R., Dallner, G., Bergstrand, A. and Ernster, L. (1969) J. Cell Biol. 41, 357-377.

193. Glaumann, H. and Dallner, G. (1970) J. Cell Biol. 47, 34-48.

194. Svensson, H., Dallner. G, and Ernster, L. (1972) Biochem. Biophys. Acta 274, 447-461.

195. Bergman, A. and Dallner, G. (1976) Biochem. Biophys. Acta 433, 496-508.

196. Winquist, L., Eriksson, L. Dallner, G. and Ersson, B. (1976) Biochem. Biophys. Res. Comm. 68, 1020-1026.

197. Winquist, L. and Dallner, G. (1976) Biochim. Biohys. Acta 436, 399-412.

198. Murphy, P.J., Van Frank, R.M. and Williams, T.L. (1969) Biochem. Biophys. Res. Comm. 37, 697-704.

199. Schulze, H.U. and Staudinger, H. (1971) Hoppe-Seyler's Z. Physiol. Chem. 352, 1659-1674.

200. Beaufay, H., Amar-Costesec, A., Thines-Sempoux, D. Wibo, M., Robbi, M. and Berthet, J. (1974) J. Cell Biol. 213-231.

201. De Pierre, J.W. and Ernster, L. (1977) Ann. Rev. Biochem. 46, 201-262.

202. Rogers, M., and Strittmatter, P. (1974) 249, 895-900.

203. Yang, C.S. and Strickhart, F.S. (1975) J. Biol. Chem. 250, 7968-7972.

204. Oshino, N., Imai, Y., and Sato, R. (1971) J. Biochem. (Japan) 69, 155-168; 169-180.

205. Oshino, N. and Omura, T. (1973) Arch. Biochem. Biophys. 157, 395-404.

206. Lu, A.Y.H., and Levin, W. (1974) Biochim. Biophys. Acta 344, 205-240.
207. Mannering, G.J., Kuwahara, S.I. and Omura, T. (1974) Biochem. Biophys.Res. Comm. 57, 476-481.
208. Archakov, A.I., Devichensky, V.M., Karuzina, I.I. and Karjakin, A.V. (1975) Arch. Biochem. Biophys. 166, 295-307; 308-312; 313-317.
209. De Pierre, J.W. and Ernster, L. (1978) Biochim. Biophys. Acta 473, 149-186.

MEMBRANE AND ORGANELLE BIOGENESIS: A BRIEF SYNOPSIS

OF CURRENT CONCEPTS

David D. Sabatini, Gert Kreibich, Takashi Morimoto,

and Milton Adesnik

Department of Cell Biology, New York University
School of Medicine, 550 First Avenue
New York, New York 10016

INTRODUCTION

Studies on organellar and membrane biogenesis have
revealed the existence of a variety of cellular mechanisms
which serve to direct newly synthesized polypeptides to
specific membranes and, when necessary, assist them in
crossing the hydrophobic barriers which lie within the membranes. From a biogenetic standpoint and in attempting to
make a broad conceptual generalization, two main classes of
proteins destined to be incorporated into organelles may
now be recognized. One class consists of polypeptides
which upon discharge from free ribosomes spontaneously adopt
the configuration which is required for their specific

C. P. Lee, G. Schatz, G. Dallner (eds.), Mitochondria and Microsomes
in honor of Lars Ernster   ISBN 0-201-04576-1

interaction with the receiving membrane. The incorporation of these polypeptides is therefore a posttranslational event, the specificity of which must depend on the recognition by specific membrane components of structural features which are expressed in the newly synthesized polypeptides when these are released in the cell sap. Most mitochondrial (c.f. 1), chloroplast (c.f. 2) and probably glyoxysomal (3) and peroxisomal (4,5) proteins, as well as some integral membrane proteins exposed on the cytoplasmic face of membranes (6-9) utilize posttranslational mechanisms of this kind to reach their functional destination.

The second major class of proteins destined to membranes and organelles consists of polypeptides which are synthesized in ribosomes bound to ER membranes. This class includes secretory (c.f. 10,11) and lysosomal proteins (12, 13), as well as integral membrane proteins and glycoproteins which have a transmembrane disposition (c.f. 10,11, 14-18) or are exposed on the luminal or extracellular faces of membranes. When synthesized in vitro in the absence of microsomal membranes these polypeptides adopt, upon release from the ribosome, a configuration which does not allow their posttranslational insertion into the ER membrane. Direct cotranslational insertion into the hydrophobic interior of the ER membrane which occurs in vivo, therefore prevents tertiary folding of the nascent chains from taking place in the aqueous cell sap medium.

POSTTRANSLATIONAL TRANSFER OF POLYPEPTIDES TO THEIR

SITE OF FUNCTION

The majority of proteins within mitochondria and
chloroplasts are coded for by nuclear genes and are synthesized on cytoplasmic polyribosomes. For several such proteins large molecular weight precursors have been detected
(19-31), which are proteolytically processed upon entrance
into the organelle. Most likely the segments which are
removed proteolytically from these precursors serve to
insure that the polypeptides achieve configurations in
which addressing signals, which are not necessarily part of
the transient segments, can be recognized by the organelle
surface receptors.

Our studies on the biosynthesis of cytochrome c
have led to the identification of a peptide segment within
the polypeptide which appears to contain the primary addressing signal which determines the uptake of the apoprotein
into mitochondria (32). This segment must be exposed in the
newly synthesized molecule, but seems to be masked in the
mature protein (holocytochrome c). We found that rat liver
cytochrome c synthesized in vitro in translation mixtures
programmed with mRNA from animals treated with triodothyron-
ine, in which cytochrome c messenger levels are increased
substantially, was incorporated into mitochondria when
these were added to the cell free system after translation
was completed. The specific conformation of the newly
synthesized product that is recognized by the mechanism

which mediates the posttranslational uptake appears to be equivalent to that of the apoprotein which can be prepared from the holocytochrome by chemical removal of heme, since the apoprotein, but not the native holocytochrome, could compete with in vitro product for its incorporation into mitochondria. Competition experiments using separated cyanogen bromide fragments of apocytochrome c demonstrated that a discrete segment of the molecule, which extends from residue 65 to the carboxy terminal end and includes the cytochrome sequence most highly conserved during evolution (33,34), contains the addressing elements exposed in the newly synthesized product. These studies suggest that folding of the polypeptide and acquisition of the heme occur within the mitochondria and that these processes lead to masking of the adressing signal within the mature apo-cytochrome.

The case of cytochrome c demonstrates that the presence of a transient peptide is not always required to maintain a newly synthesized polypeptide in a conformation capable of interacting with membrane receptors. The notion that newly synthesized polypeptides may differ from mature molecules in conformational features which are recognized by the receiving membranes resembles aspects of the "mem-brane trigger folding hypothesis" of Wickner (35), who emphasized that transient peptide segments may serve to alter the folding pathway of newly synthesized polypeptides. Primarily on the basis of his work with the M13 coat pro-tein (36,37) Wickner has also proposed that posttranslation-al asymmetric insertion into the phospholipid bilayer may be a major mechanism for the incorporation of polypeptides into membranes.

Our own studies with the Na^+, K^+ ATPase of the plasma membrane (38), a transmembrane protein composed of two different polypeptides, have shown that the large nonglycosylated catalytic subunit of this enzyme is synthesized in free polysomes and therefore must be posttranslationally incorporated into the plasma membrane. The glycosylated small subunit of the ATPase is, however, synthesized in the rough endoplasmic reticulum and interaction between the two subunits may be an important event in the posttranslational assembly of the mature protein within the membrane.

Work from several laboratories, including our own, has also shown that posttranslational insertion may be an important mechanism for the incorporation of proteins into ER membranes, particularly when the polypeptides do not adopt a transmembrane disposition. Several proteins which are exposed on the cytoplasmic face of the ER, such as cytochrome b_5 and NADH cytochrome b_5 reductase have been shown to be synthesized in free polysomes (6-9) and therefore must be inserted posttranslationally into their membrane sites of function. Hydrophobic segments which are located near the carboxy terminal end of the polypeptides (39) are likely to serve as posttranslational insertion signals, which are responsible for the permanent association of the polypeptide with the ER membrane.

Posttranslational mechanisms also appear to be involved in the biogenesis of peroxisomes. The finding that the peroxisomal enzymes urate oxidase and catalase are synthesized in free polysomes (4,5) indicates that the completed polypeptides must be able to cross the membranes limiting these organelles. It has not yet been shown,

however, whether the newly synthesized polypeptides are
incorporated directly into mature peroxisomes, as is the
case with mitochondrial and chloroplast proteins, or if
incorporation occurs only into developing forms of peroxisomes, which appear to bud from the ER.

COTRANSLATIONAL INSERTION OF PROTEINS INTO ER

MEMBRANES

Studies on the biosynthesis of secretory proteins
first revealed the existence of a mechanism which effects
the "vectorial discharge of nascent polypeptides" (11,40-42),
across the cisternal lumen of the endoplasmic reticulum. It
is now evident, however, (c.f. 10,11), that not all polypeptides released into the ER lumen are destined for
secretion. Some such polypeptides may remain as permanent
residents of the ER (43), bound as peripheral components to
the luminal face of the membranes, while others may be
diverted from the secretory pathway for segregation within
limited organelles. Calsequestrin, a Ca^{++} binding protein
within the sarcoplasmic reticulum of muscle cells (44)
appears to be an example of the first class, while lysosomal
enzymes are examples of products which are discharged from
bound polysomes but later depart from the secretory route
(12,13).

It is also apparent that an important variation of
the basic mechanism of vectorial discharge which translo-
cates proteins into the ER lumen can lead to the direct
incorporation of other specific nascent polypeptides into
the membrane. Several integral membrane proteins of the ER

and plasma membrane have been shown to be synthesized in
bound polysomes and to be first inserted into the rough ER
membranes before being transferred to their ultimate destination. These include the Ca^{++} ATPase of the sarcoplasmic
reticulum (44,45), cytochrome P-450 (46-48) and NADPH
cytochrome P-450 reductase (5,6,9), integral proteins of
the ER which are present in both rough and smooth portions
of this organelle. Integral membrane proteins of the plasma
membrane which have been shown to be synthesized in bound
polysomes include not only the well studied envelope glycoproteins of Vesicular Stomatitis (VSV) (49,50), Sindbis and
Semliki Forest viruses (51-54), but also cellular plasma
membrane proteins such as the hepatocyte 5-nucleotidase
(55), band 3, the major glycoprotein of the erythrocyte
plasma membrane (56) and the glycoprotein subunit of the
Na^+, K^+ ATPase (38). Indirect evidence also indicates that
the heavy chains of the histocompatability antigen (57,58)
and of the IgM immunoglobulin (59) are products of bound
polysomes. Indeed, given the mechanism for core glycosylation of glycoproteins (c.f. 60) it can be expected that all
glycoproteins containing asparagine-linked carbohydrates
are synthesized in bound polysomes.

    Much progress has been made towards an elucidation
of the mechanism which determines the binding of polysomes
translating specific classes of messenger RNA's to ER membranes. Current evidence supports a hypothesis, postulating
(61-63) that amino-terminal segments of nascent polypeptide
chains serve as signals which determine the polysome membrane interaction and in some way assist in the cotranslational translocation of the polypeptides across the membrane (see, for example, (14,64). Many authors have

observed (e.g. 57,58,65-69) that primary translation products synthesized in vitro in mRNA dependent systems (presecretory proteins) contain amino-terminal segments (referred to as signal segments) which are not present in the final secretory products or in the polypeptides found within the microsomal lumen (pro-secretory and secretory proteins). When translation occurs in the presence of microsomal membranes, the signal segment is removed before synthesis of the polypeptide is completed (57,70) and, after polypeptide termination, the finished product is sequestered within the microsomal lumen, where it is inaccessible to exogenous proteases (58).

The existence of transient amino terminal signal segments has been established for all eukaryotic secretory proteins so far examined, with the exception of ovalbumin (71) and for many prokaryotic periplasmic and membrane proteins (c.f. 14). Direct evidence for a role of signal segments in the translocation process has been provided by genetic experiments in bacteria showing that mutations affecting specific residues within the signals prevent passage of the polypeptides across the membrane (72,73).

Although signal segments of different presecretory proteins vary substantially in amino acid composition characteristically all signal segments contain middle regions of highly hydrophobic character, well suited for initiating insertion of the nascent chain into the lipophilic interior of the membrane. In addition, one or two charged residues are generally present near the amino-terminus of the signal. Because such charges may be expected to remain exposed on the cytoplasmic side of the membrane during the vectorial discharge of the nascent polypeptide

(74-76) it is likely that the signal adopts a loop configuration within the membrane.

It is important to note that such configuration of
the insertion signal could be attained even if the signals
were located in the interior of the growing polypeptide, as
may be the case for ovalbumin (77,78) and for certain
membrane proteins in which the amino-terminal region remains
exposed on the cytoplasmic side of the membrane (see below).

An important question concerning membrane biogenesis
relates to the mechanisms which account for the precise
asymmetrical disposition of proteins within membranes. Many
cellular and viral transmembrane proteins, such as glyco-
phorin in red blood cells (79-81), heavy chains of membrane
associated IgM (82,83) and histocompatability antigens
(84,85) as well as the envelope glycoprotein (G) of
Vesicular Stomatitis Virus (VSV) (49), have their amino-
terminal ends exposed on the luminal or extracellular face
of plasma membranes and the C terminal portions on the
cytoplasmic side. The simple transmembrane disposition of
these proteins can be easily explained as resulting from an
interruption of the vectorial discharge which takes place
in the ER membrane, followed by transfer of the polypeptide
to the cell surface by a process of membrane flow which
maintains the transverse orientation of the polypeptide.
This presupposes that the nascent membrane polypeptides
contain transient amino-terminal signals analogous to those
in secretory proteins, which initiate their insertion in
the ER membrane, as well as signals which may be referred
to as "halt" or "stop" transfer signals (16,45,48,86) that
cause the interruption of the vectorial discharge and lead
to retention of the protein in the phospholipid bilayer.

The existence of amino-terminal signal-like segments has been demonstrated for several membrane proteins (86,87), but it is clear that cleavage of the amino-terminal segment is not an obligatory step of the insertion process (48,88). In the case of retinal opsin (89) for example, the amino-terminal portion of the polypeptide is transferred to the cisternal lumen without cleavage, while in cytochrome P-450 it appears to remain associated with the membrane in the mature product.

It should be apparent that the nature and location of "halt transfer signals" should be important factors in determining the transmembrane disposition of polypeptides synthesized in bound polysomes. Transmembrane polypeptides which, like the G protein of VSV cross the membrane only once and have their N-terminal segment on the extracellular or luminal side, need have one halt transfer signal which may simply be represented by the same structural elements of the polypeptide which position the mature membrane protein in the membrane and maintain a stable interaction with the phospholipid bilayer. Sequence information on several polypeptides suggests that halt transfer signals consist of the membrane-embedded hydrophobic segments, which interact strongly with the interior of the membrane, and of adjacent charged residues, which are found on the cytoplasmic side of the membrane and cannot easily penetrate the hydrophobic barrier. It is striking that in the simple membrane polypeptides which have been studied putative halt transfer signals are located very close to the carboxy terminal end of the molecules and therefore only short polypeptide segments remain exposed on the cytoplasmic face of the membranes. Recent work shows that in one interesting

case, that of the μ chain of IgM, replacement of a carboxy-
terminal portion of the molecule containing the putative
halt transfer signal by another segment leads to secretion
of the protein rather than its retention in the membrane.
Most interestingly the two different forms of μ chains
(secretory and membrane bound) found in lymphocytes are the
products of mRNA's which differ only at their 3' ends
(59), and are generated from a single arrangement of coding
(exons) and noncoding (introns) DNA regions which are
utilized differentially as a result of alternate RNA processing pathways (90).

     In spite of the examples just mentioned, there is
no logical reason to postulate that halt transfer signals
must always be located so close to the site of termination
of polypeptide growth. Some integral membrane proteins
synthesized in bound polysomes, such as NADPH cytochrome P-
450 reductase (6) and cytochrome P-450 (48) have large
functional segments exposed on the cytoplasmic side of the
membrane. It is also well known that not all transmembrane
proteins have the simple transmembrane disposition illustrated by the examples previously mentioned. Several well
characterized proteins such as retinal opsin (89) and band
3 of the erythrocyte plasma membrane (91) cross the membrane more than once and some membrane proteins may even
have their amino terminal ends exposed on the cytoplasmic
side of the membrane, as is the case with band 3 (c.f. 91),
a polypeptide which is synthesized in bound polysomes but
does not appear to be co-or posttranslationally cleaved
(56). As we previously mentioned, exposure of the amino
terminus of a membrane protein on the cyotplasmic side of
the membrane may be attained without cleavage of the

primary translation product, if the first insertion signal
is located in the interior of the nascent chain. The
presence of a sequence of insertion and halt transfer
signals within a single polypeptide chain could, therefore,
explain the transmembrane disposition of any polypeptide,
including the existence of multiple crossings of the phospholipid bilayer. Whether the C terminal end is found on
the cytoplasmic or luminal side of the membrane would
depend on whether the last signal encoded in the messenger
before the termination codon was an insertion or a halt
transfer one.

Studies on the biosynthesis of the envelope glycoproteins of the closely related Sindbis and Semliki Forest
Virus appear to explain the different location and orientation of several viral polypeptides in terms of the biosynthetic process which utilizes several insertion and halt
transfer signals within a single polypeptide (53,92,93).
For these viruses one cytoplasmic capsid protein and two
simple transmembrane polypeptides are generated by the
cotranslational processing of a nascent polyprotein which
begins to be synthesized in free polysomes and following
proteolytic cleavage and exposure of an insertion signal
becomes associated with the ER membrane. Hydrophobic segments which anchor the two transmembrane polypeptides in
the envelope of these viruses must represent stop transfer
signals, one of which was far removed from the carboxy
terminal end of the polyprotein. Following stop transfer
by the first signal an interior insertion signal, which may
be first exposed by a second cleavage, appears to re-
establish passage of the nascent chain through the membrane
(94,95).

An important property of halt transfer signals must be that after interrupting the vectorial discharge and perhaps detaching the ribosome from its binding site, they should allow the reutilization of the vectorial discharge apparatus for new rounds of protein synthesis. This function of halt transfer signal is emphasized by their designation as "dissociation sequences" by Silhavy and his associates (96). By genetic manipulations these investigators have produced chimeric molecules containing amino-terminal portions of various lengths derived from either a secretory (maltose binding protein) or a membrane protein (lambda receptor) and carboxy terminal portions of constant length derived from a cytoplasmic protein (β-galactosidase). When linked to the secretory protein or to a short amino-terminal segment of the membrane protein, sequences within the segment derived from the cytoplasmic protein acted as "pseudo halt transfer signals" and prevented passage of distal portions of the polypeptide across the membrane. Insertion via the pseudo halt transfer signal present in the cytoplasmic portion of the chimeric protein was, however, lethal for the cell because it appeared to jam the vectorial transfer apparatus. On the other hand, chimeric protein molecules could be constructed which caused interruption of the vectorial discharge by a natural "dissociation sequence" present in a segment of the membrane protein. In these cases the chimeric protein with its large portion derived from a cytoplasmic protein also became associated with the membrane but cell viability was not altered, presumably because the vectorial transfer apparatus could be reutilized for the synthesis of other important cell proteins.

MEMBRANE GLYCOPROTEINS CHARACTERISTIC OF ROUGH
MICROSOMAL FRACTIONS

Transfer of a polypeptide across the ER membrane
during the elongation phase of protein synthesis involves
the binding of ribosomes to specific sites on membranes of
the rough endoplasmic reticulum (ER), sites which may be
regarded as ribosome receptors (c.f. 97-101). Comparative
biochemical and electrophoretic analyses carried out
initially with rat liver cell fractions, but later extended to other sources (76,102-104), have revealed the
existence of protein compositional differences between
membranes derived from the rough and smooth ER which appear
to be related to the function of the rough portions of the
ER in protein synthesis (99-101, 105,106). In protein
composition, purified rough (RM) and smooth microsomal
membranes (SM) differ mainly by the exclusive presence in
RM of two transmembrane glycoproteins (MW 65,000 and 63,000
daltons) which have been named Ribophorins I and II. It
was found that although these proteins are solubilized when
RM microsomes are treated with DOC and other ionic detergents
routinely used to purify ribosomes, they are recovered in
association with sedimentable polysomes when neutral detergents, such as Kyro EOB and Emulgen 913 are employed, which
solubilize most other membrane components. Treatment of
intact microsomes with low concentrations of glutaraldehyde
or with reversible bifunctional reagents led to the crosslinking of the ribophorins to the ribosomes (99-100),

suggesting a possible role for these proteins in mediating
the attachment of ribosomes to the RM membranes. This
suggestion was reinforced by the finding that in RM of
different isopycnic densities ribophorin content is stoich-
iometrically related to the amounts of ribosomes.

The presence of ribophorin-like protein appears to
be characteristic of rough microsomal membranes from different tissues and different species (76,101-104) and immunoprecipitation with specific antibodies raised against each
of the rat liver ribophorins has shown that there is
extensive immunological crossreactivity between ribophorins
of different species.

Ribophorins have a strong tendency to interact with
each other, and electron microscopic analysis of membrane
remnants obtained after treatment of RM with nonionic detergents has shown that upon removal of other membrane components the proteins tend to accumulate within curved membrane
remnants which bear ribosomes attached to their convex side
(99,101). Our observations suggest that an intramembranous
network of ribophorin polypeptide exists within the rough
ER, which may be an important determinant of the characteristic morphological feature of the rough ER cisternae, and
may control the distribution of ribosome binding sites in
the plane of the membrane. The close proximity and per-
sistant association of ribophorins and ribosomes substan-
tiates the presumption that these proteins play an important
functional role in the binding of polysomes to the ER and/or
the transfer of nascent polypeptides across the ER membranes.

It is clear that our understanding of these
important processes will require a detailed elucidation of
the molecular interactions between ribosomes, nascent

polypeptides and membrane components at the ribosome-membrane junction.

REFERENCES

1. Schatz, G. and Mason, T.L. (1974) Ann. Rev. of Biochem. 43, 51-87.
2. Chua, N.-H. and Schmidt, G.W. (1979) J. Cell Biol. 81, 461-483.
3. Kindl, H., Koller, W. and Frevert, J. (1980) Hoppe Seyler's Z. Physiol. Chem. 361, 465-467.
4. Goldman, B.M. and Blobel, G. (1978) Proc. Natl. Acad. Sci. U.S.A. 75, 5066-5070
5. Kreibich, G., Bar-Nun, S., Czako-Graham, M., Mok, W., Nack, E., Okada, Y., Rosenfeld, M.G. and Sabatini,D.D. (1980a) in Biological Chemistry of Organelle Formation (Ed. Th. Bucher, et al. eds.) pp. 147-163, Proceedings of Mosbach P., Colloquia, Springer Verlag, Berlin.
6. Okada, Y., Sabatini, D.D. and Kreibich, G. (1979) J. Cell Biol. 83, 437a.
7. Rachubinski, R.A., Verma, D.P.S. and Bergeron, J.J.M. (1980) J. Cell Biol. 84, 705-716.
8. Borgese, N. and Gaetani, S. (1980) FEBS Letts. 112, 216-220.
9. Gonzalez, F.J. and Kasper, C.B. (1980) Biochemistry 19, 1790-1796.
10. Palade, G.E. (1975) Science 189, 347-358.
11. Sabatini, D.D. and Kreibich, G. (1976) in The Enzymes of Biological Membranes (A. Martonosi, ed.) pp. 531-579, Plenum Publishing Co., New York, N.Y.
12. Erickson, A.H. and Blobel, G. (1979) J. Biol. Chem. 254, 11771-11774.
13. Rosenfeld, M., Sabatini, D.D., Sabban, E., Kato, K. and Kreibich, G. (1980) J. Cell Biol. 87, 308a.
14. Davis, B.D. and Tai, P.-C. (1980) Nature 283, 433-438.
15. Lodish, H.F., Braell, W.A., Schwartz, A.L., Strous, G.J.A.M. and Zilberstein, A. (1980) International Review of Cytology.
16. Blobel, G. (1978) FEBS 43, pp. 99-108

17. Blobel, G. (1980) Proc. Natl. Acad. Sci. U.S.A.
     77, 1496-1500.
18. Emr, S.D., Hall, M.N. and Silhavy, T.J. (1980)
     J. Cell Biol. 86,701-711.
19. Cote, C., Solioz, M. and Schatz, G. (1979) J. Biol.
     Chem. 254, 1437-1439.
20. Maccecchini, M.-L., Rudin, Y. and Schatz, G. (1979a)
     J. Biol. Chem. 254, 7468-7471.
21. Maccecchini, M.L., Rudin, Y., Blobel, G. and Schatz,G.
     (1979b) Proc. Natl. Acad. Sci. U.S.A. 76, 343-347.
22. Harmey, M. and Neupert, W. (1979) FEBS Letts. 108,
     385-389.
23. Ries, G., Hundt, E. and Kadenbach, B. (1978) Eur. J.
     Biochem. 91, 179-191.
24. Schatz, G. (1979) FEBS Letts. 103, 201-211.
25. Lewin, A.S., Gregor, I., Mason, T.L., Nelson, N. and
     Schatz, G. (1980) Proc. Natl. Acad. Sci. U.S.A. 77,
     3998-4002.
26. Shore, G.C., Carignan, P. and Raymond, Y. (1979)
     J. Biol. Chem. 254, 3141-3144.
27. Mori, M., Miura, S., Tatibana, M. and Cohen, P.P.
     (1979) Proc. Natl. Acad. Sci. U.S.A. 76, 5071-5075.
28. Conboy, J.G., Kalousek, F. and Rosenberg, L.E. (1979)
     Proc. Natl. Acad. Sci. 76, 5724-5727.
29. Dobberstein, B., Blobel, G. and Chua, N.-H. (1977)
     Proc. Natl. Acad. Sci. U.S.A. 74, 1082-1085.
30. Highfield, P.E. and Ellis, R.J. (1978) Nature 271,
     420-424.
31. Schmidt, G.W., Devillers-Thiery, A., Dearuisseaux, H.,
     Blobel, G. and Chua, N.H. (1979) J. Cell Biol. 83,
     615-622.
32. Matsuura, S., Arpin, M., Margoliash, E., Sabatini,D.D.,
     and Morimoto, T. (1979) J. Cell Biol. 83, 437a.
33. Margoliash, E., Fitch, W.M. and Dickerson, R.E. (1971)
     in Molecular Evolution II:Biochemical Evolution and
     the Origin of Life.
34. Urbanski, G.J. and Margoliash, E. (1977) in
     Immunocytochemistry of Enzymes and their Antibodies
     (M.R.J. Salton,ed.) pp. 203-225, John Wiley & Sons,
     New York.
35. Wickner, W. (1979) Ann. Rev. Biochem. 48, 23-45.
36. Ito, K., Mandel, G. and Wickner, W. (1979) Proc. Natl.
     Acad. Sci. 76, 1199-1203.
37. Ito, K., Date, T. and Wickner, W. (1980) J. Biol.
     Chem. 255, 2123-2130.

38. Sherman, J.M., Sabatini, D.D. and Morimoto, T. (1930) J. Cell Biol. 87, 307a.

39. Ozols, J. and Gerard, C. (1977) Proc. Natl. Acad. Sci. U.S.A. 74, 3725-3729.

40. Redman, C.M. and Sabatini, D.D. (1966) Proc. Natl. Acad. Sci. U.S.A. 56, 608-615.

41. Sabatini, D.D. and Blobel, G. (1970) J. Cell Biol. 45, 146-157.

42. Negishi, M., Sawamura, T., Morimoto, T. and Tashiro, Y. (1975) Biochim. Biophys. Acta. 301, 215-220.

43. Kreibich, G. and Sabatini, D.D. (1974) J. Cell Biol. 61, 789-807.

44. Greenway, D.C. and MacLennan, D.H. (1978) Canadian J. Biochem. 56, 452-456.

45. Chyn, T.L., Martonosi, A.N., Morimoto, T. and Sabatini, D.D. (1979) Proc. Natl. Acad. Sci. U.S.A. 76, 1241-1245.

46. Negishi, M., Fujii-Kuriyama, Y., Tashiro, Y. and Imai, Y. (1976) Biochem. Biophys. Res. Comm. 71, 1153-1160.

47. Fujii-Kuriyama, Y., Negishi, M., Mikawa, R. and Tashiro, Y. (1979) J. Cell Biol. 81, 510-519.

48. Bar-Nun, S., Kreibich, G., Adesnik, M., Alterman, L., Negishi, M. and Sabatini, D.D. (1980) Proc. Natl. Acad. Sci. 77, 965-969.

49. Katz, F.N., Rothman, J.E., Knipe, D.M. and Lodish, H.F. (1977) J. Supramol. Structure 7, 353-370.

50. Katz, F.N., Rothman, J.E., Lingappa, V.R., Blobel, G. and Lodish, H.F. (1977) Proc. Natl.Acad. Sci. U.S.A. 74, 3278-3282.

51. Garoff, H., Simons, K. and Dobberstein, B. (1973) J. Mol. Biol. 124, 587-600.

52. Bonatti, S., Cancedda, R. and Blobel, G. (1979) J. Cell Biol. 80, 219-224.

53. Wirth, D.F., Lodish, H.F. and Robbins, P.W. (1979) J. Cell Biol. 81, 154-162.

54. Rothman, J.E. and Lodish, H.F. (1977) Nature 269, 775-780.

55. Bergeron, J.J.M., Berridge, M.V. and Evans, W.H. (1975) Biochim. Biophys. Acta. 407, 325-337.

56. Sabban, E., Sabatini, D.D., Adesnik, M. and Marchesi, B. (1979) J. Cell Biol. 83, 437a.

57. Dobberstein, B., Garoff, H., Warren, G. and Robinson, P.J. (1979) Cell 17, 759-769.

58. Ploegh, H.L., Cannon, L.W. and Strominger, J.L. (1979) Proc. Natl. Acad. Sci. U.S.A. 76, 2273-2277.
59. Rogers, J., Early, P., Carter, C., Calame, K., Bond, M., Hood, L. and Wall, R. (1980) Cell 22, 303-312.
60. Struck, D.K. and Lennarz, W.J. (1980) in The Biochemistry of Glycoproteins and Proteoglycans (W.J. Lennarz, ed.) pp. 35-84, Plenum Press, New York.
61. Blobel, G. and Sabatini, D.D. (1971) in Biomembranes (Manson, L.A., ed.), pp. 193-195, Plenum Publishing Corp. New York.
62. Blobel, G. and Dobberstein, B. (1975a) J. Cell Biol. 67, 835-851.
63. Blobel, G. and Dobberstein, B. (1975b) J. Cell Biol. 67, 852-862.
64. Sabatini, D.D., Kreibich, G., Morimoto, T. and Adesnik, M. (1980) Submitted to Cell.
65. Milstein, C., Brownlee, G.G., Harrison, T.M. and Mathews, M.B. (1972) Nature New Biol. 239, 117-120.
66. Swan, D., Aviv, H. and Leder, P. (1972) Proc. Natl. Acad. Sci. 69, 1967-1971.
67. Kemper, B., Habener, J.F., Mulligan, R.C., Potts, J.T. Jr. and Rich, A. (1974) Proc. Natl. Acad. Sci. U.S.A. 71, 3731-3735.
68. Schechter, I. (1973) Proc. Natl. Acad. Sci. U.S.A. 70, 2256-2260.
69. Devillers-Thiery, A., Kindt, T., Scheele, G. and Blobel, G. (1975) Proc. Natl. Acad. Sci. U.S.A. 72, 5016-5020.
70. Palmiter, R.D., Gagnon, J., Ericsson, L.H. and Walsh, K.A. (1977) J. Biol. Chem. 252, 6386-6393.
71. Palmiter, R.D., Gagnon, J. and Walsh, K.A. (1978) Proc. Natl. Acad. Sci. U.S.A. 75, 94-98.
72. Bassford, P. and Beckwith, J. (1979) Nature 277, 538-541.
73. Bedouelle, H., Bassford, P.J.Jr., Fowler, A.V., Zabin, I., Beckwith, J. and Hofnung, M. (1980) Nature 285, 78-81.
74. Inouye, M. and Halegoua, S. (1979) in Escherichia coli, CRC Crit. Rev. Biochem.
75. Steiner, D.F., Quinn, P.S., Chan, S.J., Marsh, J. and Tager, H.S (1980) Ann. N.Y. Acad. Sci. 343, 1-16.
76. Kreibich, G., Czako-Graham, M., Grebenau, R.C. and Sabatini, D.D. (1980b) Ann. N.Y. Acad. Sci. 343, 17-33.
77. Lingappa, V.R., Lingappa, J.R. and Blobel, G. (1979) Nature 281, 117-121.

78. Meek, R.L., Walsh, K.A. and Palmiter, R. (1980) Fed. Proc. $\underline{39}$, 1367.
79. Bretscher, M.S. (1971) J. Mol. Biol. $\underline{59}$, 351-357.
80. Tomita, M. and Marchesi, V.T. (1975) Proc. Natl. Acad. Sci. U.S.A. $\underline{72}$, 2964-2968.
81. Cotmore, S.F., Furthmayr, H. and Marchesi, V.T. (1977) J. Mol. Biol. $\underline{113}$, 539-553.
82. Vassalli, P., Tedghi, R., Lisowska-Bernstein, B., Tartakoff, A. and Jaton, J.-C. (1979) Proc. Natl. Acad. Sci. U.S.A. $\underline{76}$, 5515-5519.
83. Kehry, M., Ewald, S., Douglas, R., Sibley, C., Raschke, W., Fambrough, D. and Hood, L. (1980) Cell $\underline{21}$, 393-406.
84. Yamane, K. and Nathenson, S.G. (1970) Biochemistry $\underline{9}$, 4743-4750.
85. Peterson, P.A., Rask, L. and Lindblom, J.B. (1974) Proc. Natl. Acad. Sci. U.S.A. $\underline{71}$, 35-39.
86. Lingappa, V.R., Katz, F.N., Lodish, H.F. and Blobel, G. (1978) J. Biol. Chem. $\underline{253}$, 8667-8670.
87. Irving, R.A., Toneguzzo, F., Rhee, S.H., Hofmann, T. and Ghosh, H.P. (1979) Proc. Natl. Acad. Sci. U.S.A. $\underline{76}$, 570-574.
88. Bonatti, S. and Blobel, G. (1979) J. Biol. Chem. $\underline{254}$, 12261-12264.
89. Schechter, I., Burnstein, Y., Zemell, R., Ziv, E., Kantor, F. and Papermaster, D. (1979) Proc. Natl. Acad. Sci. U.S.A. $\underline{76}$, 2654-2658.
90. Early, P., Rogers, J., Davis, M., Calame, K., Bond, M., Wall, R. and Hood, L. (1980) Cell $\underline{20}$, 313-230.
91. Steck, T.L. (1978) J. Supramol. Struct. $\underline{8}$, 311-324.
92. Garoff, H. and Soderlund, H. (1978) J. Mol. Biol. $\underline{124}$, 535-549.
93. Garoff, H., Frischauf, A.-M., Simons, K., Lehrach, H. and Delius, H. (1980) Nature (London) $\underline{288}$, 236-241.
94. Welch, W.J. and Sefton, B.M. (1979) J. of Virol. $\underline{29}$, 1186-1195.
95. Welch, W.J. and Sefton, B.M. (1980) J. of Virol. $\underline{33}$, 230-237.
96. Emr, S.D., Hedgpeth, J., Clement, J.-M., Silhavy, T.J. and Hofnung, M. (1980) Nature $\underline{285}$, 82-85.
97. Borgese, N., Mok, W., Kreibich, G. and Sabatini, D.D. (1974) J. Mol. Biol. $\underline{88}$, 559-580.
98. Sabatini, D.D., Borgese, D., Adelman, M., Kreibich, G. and Blobel, G. (1972) in RNA Viruses/Ribosomes (H. Bloemendal et al. eds.) pp. 147-172 FEBS Symposium North Holland Publishing Co., Amsterdam.

99. Kreibich, G., Ulrich, B.L. and Sabatini, D.D. (1978)
J. Cell Biol. 77, 464-487.
100. Kreibich, G., Freienstein, C.M., Pereyra, B.N.,
Ulrich, B.L. and Sabatini, D.D. (1978b) J. Cell
Biol. 77, 488-506.
101. Kreibich, G., Czako-Graham, M., Grebenau, R., Mok, W.,
Rodriguez-Boulan, E. and Sabatini, D.D. (1978a).
J. Supramolec. Struct. 8, 279-302.
102. Grebenau, R., Sabatini, D.D. and Kreibich, G. (1977)
J. Cell Biol. 75, 234a.
103. Marcantonio, E.E., Rosenfeld, M.G., Sabatini, D.D.
and Kreibich, G. (1980) J. Cell Biol. 87, 208a.
104. Kreibich, G., Bar-Nun, S., Czako-Graham, M., Czichi,
U., Marcantonio, E., Rosenfeld, M.G. and Sabatini,D.D.
(1981) International Cell Biology, 1980-1981,
Springer Verlag, Berlin, pp. 579-589.
105. Kreibich, G., Ulrich, B. and Sabatini, D.D. (1975)
J. Cell Biol. 67, 225a.
106. Rodriguez-Boulan, E., Sabatini, D.D., Pereyra, B.N.
and Kreibich, G. (1978) J. Cell Biol. 78, 894-909.

INDUCTION OF DRUG-METABOLIZING ENZYMES: A STATUS REPORT

Joseph W. DePierre, Janeric Seidegård, Ralf Morgenstern,
Lennart Balk, Johan Meijer and Anders Åström

Arrhenius Laboratory, Department of Biochemistry, University
of Stockholm, 106 91 Stockholm, Sweden

INTRODUCTION

The process of enzyme induction allows an organism
to adjust its metabolism to the challenges and opportunities
provided by its environment. Enzyme induction has been studied
for more than two decades in bacteria and a number of important principles have emerged. However, it is not yet clear
in what ways these principles must be modified to explain induction in mammalian systems.

The fact that the cytochrome P-450 system, which
plays a central role in the metabolism and detoxication of
xenobiotics, is inducible in mammalian cells was first discovered more than 15 years ago. Lars Ernster and his cowork-
er Sten Orrenius performed the initial characterization of a
number of important aspects of this induction (see, for example, ref. 1). Eventually, it also became clear that a number
of other drug-metabolizing enzymes -- including, in particular, epoxide hydrolase, glutathione S-transferase, and UDP-
-glucuronyl transferase -- are also induced when experimental

C. P. Lee, G. Schatz, G. Dallner (eds.), Mitochondria and Microsomes
in honor of Lars Ernster   ISBN 0-201-04576-1

animals are exposed to certain xenobiotics.

In its simplest terms the goal of the drug-metabolizing systems in mammalian cells can be said to be the transformation of hydrophobic xenobiotics to hydrophilic products
which can be excreted in the urine and in the bile. During
this process a number of reactive intermediates -- e.g.,
epoxides, free radicals, and carbonium ions -- are produced,
especially via the catalytic activity of the cytochrome P-450
system. Attack of such reactive intermediates on cellular
proteins, RNA, and DNA is now thought to be responsible for
a large proportion of the toxic and carcinogenic effects of
xenobiotics.

Thus, induction of different drug-metabolizing enzymes
can be expected to have far-reaching consequences for the
reaction of the human organism to various xenobiotics. For
instance, the dose of a given medicine required to obtain the
desired effect, the side-effects of this medicine, and the
potential toxicity and carcinogenicity of chemicals present
in our work environment and in our food are all influenced
by the activity levels of our drug-metabolizing systems.

For the past decade and a half Lars Ernster has continued to investigate mechanistic and functional aspects of
the induction of drug-metabolizing enzymes. A number of over-
simplified principles concerning this induction have arisen
and become generally accepted during this time. In the present
article we would like to question three of these principles
-- viz, that phenobarbital and 3-methylcholanthrene are representative inducers of drug-metabolizing enzymes; that the
induction is substrate induction by a poor substrate; and that
only enzymes directly involved in detoxication are selectively
induced. Much of the data discussed here has been gathered as

a part of our close, continuing, and highly stimulating
collaboration with Professor Ernster.

Needless to say, there are a great many aspects of the
induction of drug-metabolizing systems which will not be
discussed here. These include the following: the different
isoenzymes of various drug-metabolizing enzymes induced by
different xenobiotics (2,3); the molecular mechanism of induction (4,5); genetic aspects of this induction (6); extrahepatic induction (7); and the consequences of induction of
drug-metabolizing enzymes in man (8). (For general reviews,
see also references 9-11).

Are phenobarbital and 3-methylcholanthrene representative
inducers of drug-metabolizing enzymes?

We find it convenient in discussing drug metabolism to
divide this process into three sequential stages (see illustration).

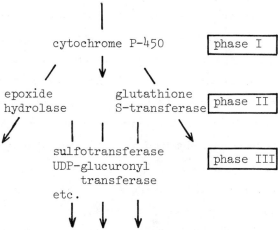

Generally speaking, these three different phases have somewhat different functions. The cytochrome P-450 system intro-

duces oxygen into the substrate molecule and provides thereby a "handle" for the other enzymes to work with. As mentioned
above, the product formed is often reactive and thus potentially toxic and/or carcinogenic. The phase II enzymes epoxide
hydrolase and glutathione S-transferases detoxify these reactive intermediates by converting them to relatively inert dihydrodiols and glutathione conjugates, respectively. Subsequently, the phase III enzymes catalyze the formation of
highly water-soluble metabolites which are then excreted
-- chiefly glucuronides, sulfates and mercapturic acids.
Obviously, this is a rather simplified picture of drug metabolism, but may, as we shall see below, provide a convenient
means of classifying different inducers of drug-metabolizing
enzymes.

When the early induction studies were performed, the
only drug-metabolizing system which had been characterized
to any extent was the cytochrome P-450 system. Indeed, epoxide hydrolase, for instance, had not yet been discovered! Consequently, most of the induction studies which have appeared
in the literature are concerned solely with the effect of
exposure to a given xenobiotic on the levels of the components of the cytochrome P-450 system -- i.e., NADPH-cytochrome P-450 reductase and cytochrome P-450 itself -- and on
the rate of catalysis of phase I reactions. Even today, the
effect of exposure to a given xenobiotic on the phase II enzymes is seldom determined, a situation which, for reasons to
be explained, we would like to see changed.

It became apparent relatively early that a great many
different xenobiotics, if not most, induce the cytochrome
P-450 system, at least to some extent. Phenobarbital and
3-methylcholanthrene (see Figure 1) were found to induce very

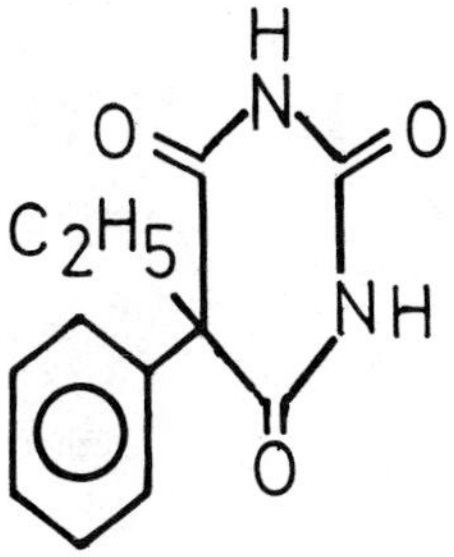

phenobarbital

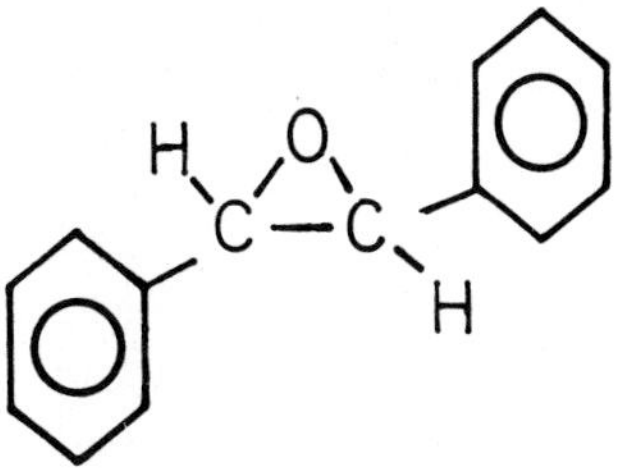

trans-stilbene
oxide

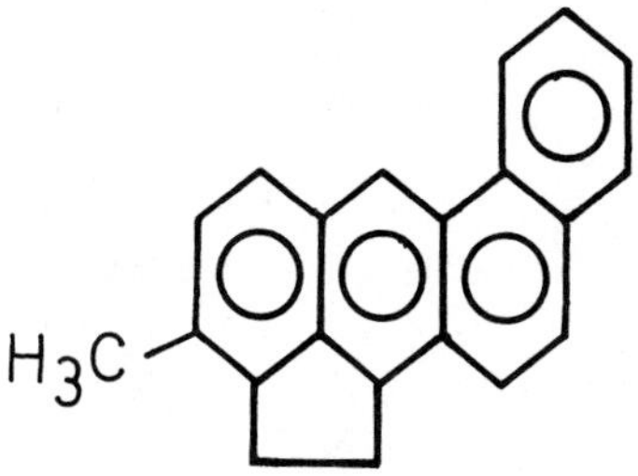

3-methylcholanthrene

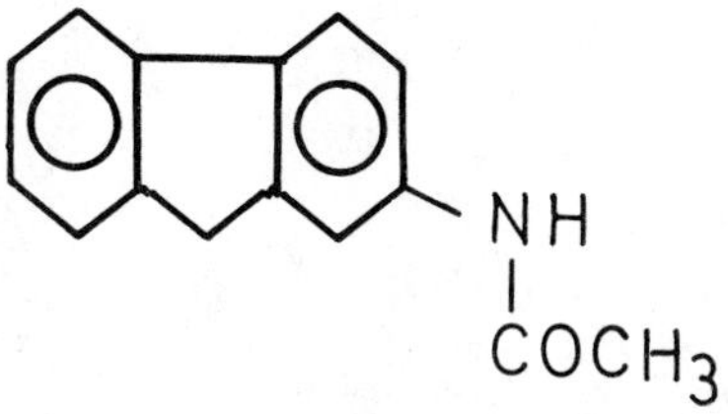

2-acetylaminofluorene

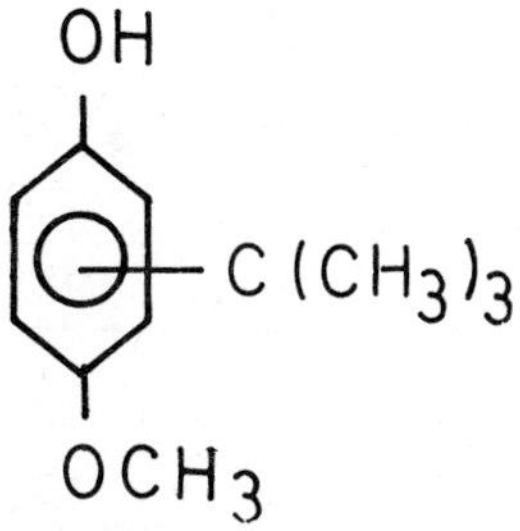

2(3)-tert-butyl-
4-hydroxyanisole

Fig. 1.Five inducers whose effects both on phase I and
phase II enzymes have been investigated

Table 1. Comparison between Five Different Inducers

Parameter	Inducer phenobarbital[a]	3-methylcholanthrene[b]
Dose	80 mg/kg (5 daily i.p. injections)	20 mg/kg (3-5 daily i.p. injections)
NADPH-cytochrome[f] c reductase	increases 300-400%	no change
Cytochrome[f] P-450	increases 300-400%, no spectral shift	increases 100-200%, spectral shift
Cytochrome[f] P-450-catalyzed reaction	many increase 300-400%	benzpyrene mono-oxygenase increases selectively 1000-2000%
Epoxide[f] hydrolase	increases 100-200%	no change
Glutathione[g] S-transferase	increases 40-60%	increases 40-60%

[a] Experiments with rats. See refs. 1, 13, and 14.
[b] Experiments with rats. See refs. 12 and 15.
[c] Experiments with rats. See ref. 16.
[d] Experiments with rats. See ref. 17.
[e] Experiments with mice. See, for example, refs. 18-20.
[f] Specific activity or content per mg microsomal protein.

of Hepatic Drug—Metabolizing Enzymes

trans-stilbene[c] oxide	2-acetylamino-[d] fluorene	2(3)-tert-butyl-[e] 4-hydroxyanisole
400 mg/kg (5 daily i.p. injections)	50 mg/kg (5 daily i.p.injections)	0.75% in the diet
increases 120%	increases 69%	increases 130%
increases 120%, no spectral shift	increases 43%, no spectral shift	no change
many increase 100-200%	N-hydroxylation increases 8-10-fold; other activities?	no change in aryl-hydrocarbon mono-oxygenase, amino-pyrine demethylase; aniline hydroxylase increases 180%
increases 620%	increases 660%	increases 850%
increases 200-300%	increases 100-150%	increases 400-900%

[g]Specific activity per mg supernatant protein. The large
spread of values for glutathione S-transferase activity reflects the fact that this activity with different substrates is induced to different extents.

large increases in the rates of phase I reactions (Table I)
(see, for examples, ref. 1 and 12). Presumably because of
this fact, as well as because they induce different isoenzymes
of cytochrome P-450, induction by phenobarbital and 3-methyl-
cholanthrene has been very extensively investigated. Indeed,
the effects of newly discovered inducers are routinely compared to those of these two classical, supposedly representative inducers.

As is shown in Table I, phenobarbital and 3-methyl-
cholanthrene induce primarily the cytochrome P-450 system and
phase I reactions. Phenobarbital also causes a doubling or
tripling of microsomal epoxide hydrolase activity, while intraperitoneal injection of 3-methylcholanthrene does not
affect this activity at all. Neither of these inducers causes
any change in cytosolic epoxide hydrolase activity (B. Hammock,
personal communication). Both phenobarbital and 3-methylchol-
anthrene induce cytosolic glutathione S-transferase activity
about 50-100%, while neither of these xenobiotics affects the
corresponding microsomal activity (21).

Recently, the effects of at least three other inducers
of drug-metabolizing enzymes -- namely, trans-stilbene oxide,
2-acetylaminofluorene, and 2(3)-tert-butylhydroxyanisole
(Figure 1) -- on both phase I and phase II activities have
been thoroughly investigated. As is also shown by the data
in Table I, all three of these substances bring about a much
greater increase in microsomal epoxide hydrolase and soluble
glutathione S-transferase activities than in the level and
general activities of the cytochrome P-450 system. Trans-
-stilbene oxide has no effect on either the cytosolic epoxide
hydrolase (B. Hammock, personal communication) or microsomal
glutathione S-transferase activities (21). To our knowledge,

the effect of 2-acetylaminofluorene and of 2(3)-tert-butyl-
-hydroxyanisole on the latter two activities has not yet been
tested.

The finding that trans-stilbene oxide, 2-acetylamino-
fluorene, and 2(3)-tert-butyl-hydroxyanisole all induce
microsomal epoxide hydrolase activity to a considerably larger
extent than they induce cytosolic glutathione S-transferase
activity may reflect the relative abundance of the proteins
involved in untreated animals. In the liver of control rats
epoxide hydrolase constitutes approximately 2% of the protein
of the endoplasmic reticulum (e.g., 22,23) and this organelle
contains about 19% of the total cellular protein (24), i.e.,
the microsomal epoxide hydrolase(s) constitutes roughly 0.4%
of the total cellular protein. After induction with trans-
-stilbene oxide or 2-acetylaminofluorene, this figure apparently increases to around 2.5%.

In the same control livers approximately 4.5% of the
total cytosolic protein consists of glutathione S-transferases
A,B, and C (25). Since the cytosolic protein accounts for
roughly 30% of the total cellular protein (calculations based
on our own data), around 1.35% of the total cellular protein
is glutathione S-transferase in control rat liver. After induction with trans-stilbene oxide, this value increases to
3.9% (25). Thus, the total increase in glutathione S-transferase protein seems to be somewhat greater than the increase
in epoxide hydrolase protein.

It can also be calculated on the basis of the purification factor observed upon isolation of cytochrome P-450
(e.g., 2,3 and references therein) that roughly 2% of the
total cellular protein in the livers of phenobarbital-treated
rats and, thus, about 0.5% of the total cellular protein in

control livers consists of this cytochrome. The relatively
large amounts of cellular protein devoted to these three enzymes indicates their central importance in the detoxication
of xenobiotics and, possibly, in other cellular functions
as well.

The data presented in Table 1 suggest that there are
two different kinds of inducers of drug-metabolizing enzymes -- those which induce mainly phase I reactions, as do
phenobarbital and 3-methylcholanthrene, and those which induce chiefly phase II reactions, as do trans-stilbene oxide,
2-acetylaminofluorene, and 2(3)-tert-butyl-hydroxyanisole.
To the latter group can also be added a number of metabolites
and structural analogues of trans-stilbene oxide (26), as well
as several other substances which, like 2(3)-tert-butyl-hydr-
oxyanisole, are antioxidants (e.g., 18 and 27). If this conclusion is correct, then phenobarbital and 3-methylcholan-
threne are representative of only one class of inducers and
the relative numbers of inducers which fall into these two
different classes remains to be seen. Furthermore, future
determination of the effect of various inducers on phase III
enzymes may reveal an additional class or classes of inducers.

Of course, these conclusions are relatively speculative. It is well-known that different inducers cause increases in different isozymes of cytochrome P-450 (e.g., 2;
see also Table 1), and the same is true for UDP-glucuronyl
transferase (e.g., 28,29). On the other hand, trans-stilbene
oxide induces glutathione S-transferases A,B, and C to
approximately the same extent (25). Future studies on the induction of individual proteins by phenobarbital, 3-methyl-
cholanthrene, trans-stilbene oxide, 2-acetylaminofluorene,

2(3)-tert-butyl-hydroxyanisole, and other xenobiotics may
reveal more complicated patterns than those discernible at
present.

All this is of practical interest because substances
which selectively induce phase I reactions would be expected
to increase the steady state level of reactive intermediates,
whereas selective inducers of phase II activities would be
expected to reduce this level. Thus, different inducers
would be expected to affect the toxic and carcinogenic potenc-
ies of various xenobiotics in very different ways. A number
of observations are of interest in this connection. Nebert
and Felton (30) have reported that 3-methylcholanthrene
is less toxic and more tumorigenic in mice strains whose
hepatic aryl hydrocarbon monooxygenase activity (catalyzed
by the cytochrome P-450 system) cannot be induced than in
responsive mice. 2-Acetylaminofluorene is a well-known liver
carcinogen in experimental animals (e.g., 31), whereas 2(3)-
-tert-butyl-hydroxyanisole has been found to reduce the number
of tumors caused in a number of different organs of laboratory animals by chemical carcinogens (see 18 and references
therein). Finally, when the S-9 fraction from rats treated
with trans-stilbene oxide is used in the Ames' test, fewer
revertants are obtained than with control S-9 (32). The relevance of the pattern of induction by these different substances to these observations remains to be determined.

Is the process of induction of drug-metabolizing enzymes simply substrate induction?

It has long been loosely assumed that the induction of

drug-metabolizing enzymes is substrate induction, i.e., that
xenobiotics induce the enzymes by which they are metabolized.
This seems to be a reasonable assumption in light of the fact
that substrate induction is a common phenomenon in bacteria.
In addition, such substrate induction would serve a useful
function, allowing the animal to detoxify and excrete the
xenobiotic to which it is chronically exposed more rapidly.

However, in connection with recent findings on the induction of phase II reactions, the assumption of substrate
induction must be reconsidered. For example, phenobarbital,
3-methylcholanthrene, trans-stilbene oxide, 2-acetylamino-
fluorene, and 2(3)-tert-butyl-hydroxyanisole have all been
found to induce glutathione S-transferase levels significantly (Table I). All of these substances, with the exception of
3-methylcholanthrene, also induce epoxide hydrolase (Table I).
However, to date only trans-stilbene oxide has been found to
be a substrate for epoxide hydrolase (33) and glutathione S-
-transferase (25). Nor is there any reason to believe, judging from their molecular structures, that phenobarbital, 3-
-methylcholanthrene, 2-acetylaminofluorene, or 2(3)-tert-
-butyl-hydroxyanisole themselves will prove to be substrates
for the phase II enzymes.

Other examples can also be added to this list. We have
found that benzil and benzoin (see Figure 2), metabolites of
trans-stilbene oxide in mammalian cells, are also potent inducers of microsomal epoxide hydrolase and cytosolic glutathione S-transferase activities in rat liver (26). There is
no evidence or reason to believe that either of these substances are substrates for the enzymes which they induce.

Thus, either these substances are not the true inducers of the phase II enzymes, or else the process involved

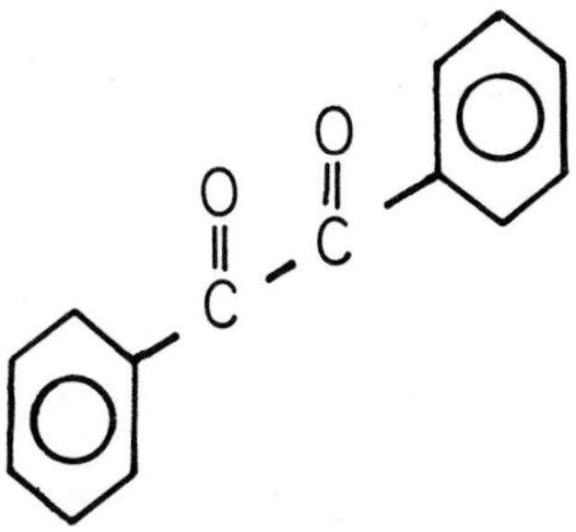

benzil 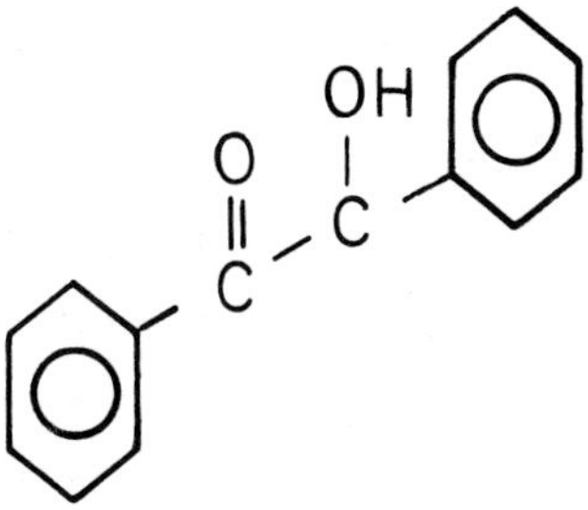   benzoin

Fig. 2 Inducers of drug-metabolizing enzymes which are appa-
       rently not substrates for the enzymes they induce

is not substrate induction. One possibility is that the xenobiotic to which the animal is exposed is metabolized in vivo
to yield a reactive intermediate which is both a substrate for
epoxide hydrolase and/or glutathione S-transferase and the
true inducer. This hypothesis is very difficult to test rigo-
rously, mainly because the relative amounts of different metabolites produced in vivo under the conditions used for induction are seldom known and because metabolites which have not
yet been identified may also be produced.

Despite these difficulties, we have attempted to determine whether trans-stilbene oxide is the true inducer of the
cytochrome P-450 system and the phase II enzymes by comparing
the relative abilities of this substance and of its known metabolites to induce (26). We came to the conclusion that it is
the trans-stilbene oxide molecule itself which induces. Using
a similar approach, Kano and coworkers (34) concluded that it
is 3-methylcholanthrene and benzpyrene themselves, and not
metabolites of these compounds, which induce aryl hydrocarbon
monooxygenase activity in rat Reuber hepatoma H-4-II-E

established cell line. However, in both of these cases the
xenobiotic administered is itself known to be a substrate
for the drug-metabolizing enzymes which are affected to the
greatest extent.

Thus, the question as to whether induction of drug-
-metabolizing enzymes is substrate induction remains, especially in the case of the phase II enzymes. Of course, it
is not impossible that the phase I and II enzymes are induced by different compounds even when a single xenobiotic
is administered, e.g., by the xenobiotic itself and by a
metabolite, or by two different metabolites. There might also
be some type of "coupling" between the genetic control of the
phase I and II system. There is obviously much we still do
not know about the genetic control of drug-metabolizing enzymes -- and of other enzymes as well, for that matter -- in
mammalian systems.

How specific is the induction of drug-metabolizing enzymes?

Anyone who has opened the peritoneal cavity of a rat
which has been induced with phenobarbital, for instance, is
well aware that induction of drug-metabolizing enzymes is
often accompanied by dramatic changes in the general morphology of the liver. In the first place this organ often hypertrophies upon exposure to an inducer. In the case of phenobarbital this increase in weight is due both to an increased
number of hepatocytes, as well as to an enlargment of the
individual cells (35,36). On the other hand, only the number
of hepatocytes, and not their size, is increased upon administration of trans-stilbene oxide (37).

In the case of phenobarbital the cellular content of endoplasmic reticulum also increases, i.e., this organelle also hypertrophies (e.g., 1,38). Most other inducers have a much smaller effect on the amount of endoplasmic reticulum in liver cells. Certain inducers -- including phenobarbital (e.g., 39-42) and trans-stilbene oxide (37) -- also cause small changes in the phospholipid composition of this organelle.

Increases in the specific activity of drug-metabolizing enzymes and the hypertrophy of the liver and of the endoplasmic reticulum are apparently complementary mechanisms for increasing the total drug-metabolizing capacity of the liver. Thus, after i.p. injection of phenobarbital, the specific content and activity of the cytochrome P-450 system in liver microsomes is increased 4- to 5-fold, about 2-2.5 times as much microsomal protein is recovered per gram liver, and the liver weighs 50% more than control livers. Taken together, these changes result in a 12-20-fold increase in the capacity of the liver of rats to metabolize a wide variety of xenobiotics via the cytochrome P-450 system, an increase which is reflected, among other ways, in a dramatically decreased sleeping time after exposure to phenobarbital.

Obviously, such growth in the liver and of the endoplasmic reticulum requires changes in the replication of DNA and in the synthesis and/or degradation of cellular RNA, proteins, and phospholipids (see, e.g., refs. 35,36,43-45). Induction of specific drug-metabolizing enzymes also requires, of course, changes in the synthesis and/or degradation of mRNA and, perhaps, other forms of RNA and/or protein (e.g., 46-50). In all cases to date where the question has been investigated, the increased activity of drug-metabolizing enzymes after

exposure to inducers has been found to reflect a corresponding increase in enzyme protein rather than enzyme modification.

The types of changes which have been reported after
exposure of animals to inducers of drug-metabolizing enzymes
are summarized in Table 2. It should be noted that not all
inducers produce all of these different changes. Indeed, to
date only phenobarbital has been shown to alter all of the
hepatic parameters listed in Table 2.

Table 2. Some hepatic parameters found to be effected by inducers of drug-metabolizing enzymes

> liver weight
>
> number of hepatocytes
>
> cellular content of
> endoplasmic reticulum
>
> DNA synthesis
>
> RNA synthesis
>
> protein synthesis
>
> heme synthesis
>
> phospholipid composition of the
> endoplasmic reticulum
>
> enzyme levels
>
> substrate binding
>
> drug metabolism

See, for example, refs. 35-50.

Of course, in addition to these general changes in hepatic morphology and metabolism, specific increases in the
activity of various drug-metabolizing enzymes also occur after
exposure to inducers. Cytochrome P-450, epoxide hydrolase,
the glutathione S-transferases, and the UDP-glucuronyl transferases have all been shown to be inducible by various xeno-

biotics. To our knowledge, induction of sulfotransferase activity has not yet been reported, but this is probably only
a question of time.

The activities of several other enzymes have also been
found to be increased by inducers of drug-metabolizing systems
and this increase subsequently used as an indication that the
enzyme in question is indeed involved in drug-metabolism. One
example of this is cytochrome b_5. As shown in Figure 3, treatment of rats with 3-methylcholanthrene increases the specific
content of cytochrome b_5 in rat liver microsomes about 80%,
while the specific content of cytochrome P-450 increases 190%.
A number of other inducers have also been shown to increase
the specific microsomal content of cytochrome b_5 -- for instance, trans-stilbene oxide increases this content to 168% of
the control (16), while the increase after administration of
2-acetylaminofluorene is to 171% of the control (17). Indeed,
there is now a great deal of evidence that cytochrome b_5 is
involved in the metabolism of xenobiotics, perhaps as a donor
of the second electron to cytochrome P-450 (see, for example,
ref. 51 and references therein).

The specific activity of DT-diaphorase in rat liver
cytosol has also been shown to increase after treatment with
different inducers. This was first observed with 3-methylchol-
anthrene, which led to the interesting hypothesis that DT-diaphorase functions as a cytochrome P-448 reductase (e.g., 52).
However, this does not seem to be the case (e.g., 53). DT-diaphorase is also induced by trans-stilbene oxide (to 500% of
the control) (54), and by 2-acetylaminofluorene (to 229% of
the control) (17). Figure 4 illustrates the time course of

the induction of DT-diaphorase activity brought about by
treatment of rats with trans-stilbene oxide, as well as the

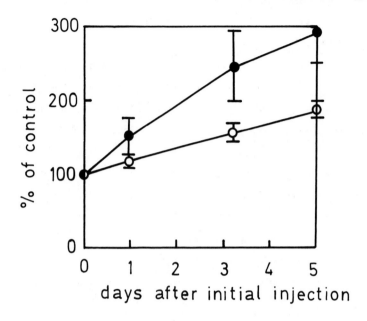

Fig. 3. <u>Time course of the induction of cytochromes P-450</u>
<u>(●) and b</u>$_5$ <u>(o) by 3-methylcholanthrene</u>

Rats were injected once daily for 1-5 days with 20 mg
3-methylcholanthrene per kg body weight in corn oil
or with corn oil alone (controls) and killed 24 hours
after the last injection. Each point represents the
mean (± SD) of determinations from 3 different rats.
The control level of cytochrome P-450 was 0.322±
0.037 nmoles/mg microsomal protein and the control
level of cytochrome b$_5$ was 0.181+0.021 nmoles/mg
microsomal protein. These results were obtained recently in our own laboratory.

time course of return to control levels after cessation of
treatment. We also demonstrated in the same study that the increase in DT-diaphorase activity reflects a corresponding increase in the amount of enzyme protein (54). There is presently great interest in the potential role of DT-diaphorase in
drug metabolism, possibly as a quinone reductase (see, for
example, ref. 53).

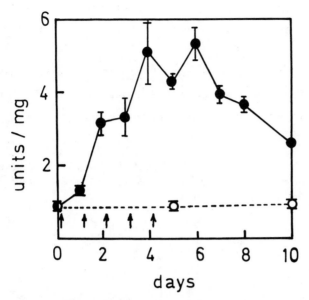

Fig. 4. <u>Time course of the changes in cytosolic DT-diaphorase</u>
<u>brought about by treatment of rats with trans-stil-</u>
<u>bene oxide</u>

Rats were injected once daily for 1-5 days, as indicated by the arrows, with 400 mg <u>trans</u>-stilbene
oxide/kg body weight and killed 24 hours after the
final injection (days 1-5). Other rats were injected
once daily for 5 days and killed 2, 3, and 5 days
after the final injection (days 7, 8, and 10, respectively). Each point represents the mean ($\pm$SD)
of the individual hepatic cytosolic DT-diaphorase
activities for 3 different rats.
•---• = rats injected with <u>trans</u>-stilbene oxide
o---o = rats injected with the carrier, sunflower oil
For further details, see ref. 54.

On the other hand, it has been apparent for some time

that inducers of drug-metabolizing enzymes may also increase

the cellular activity of other enzymes which are very unlikely

to be directly involved in the metabolism of xenobiotics.

For instance, phenobarbital has been found to induce the ac-

tivity of δ-aminolevulinic acid synthetase, the rate-limiting enzyme in heme synthesis (55). This change probably reflects an increased requirement for heme in connection with the induction of cytochrome P-450.

We have also observed that trans-stilbene oxide causes highly significant increases in the hepatic cytosolic activities of glucose-6-phosphate dehydrogenase and of UDP-glucose dehydrogenase (Table 3) (26,56).

Table 3. Induction of Glucose-6-Phosphate Dehydrogenase and of UDP-Glucose Dehydrogenase by trans-Stilbene Oxide, Benzoin, and Benzil

% of control

Inducer	G6PDH	UDPGDH
trans-stilbene oxide	188 ± 35 (12)	193 ± 37 (11)
benzoin	218 ± 31 (6)	182 ± 37 (6)
benzil	214 ± 42 (6)	191 ± 35 (6)

1 mmole of each compound/kg body weight was injected i.p. into rats once daily for 5 days and the animals were killed and the liver removed 24 hours after the final injection. G6PDH = glucose-6-phosphate dehydrogenase; the control activity was 21.6 $\pm$ 1.7 (34) nmol $NADP^+$ reduced/min-mg cytoplasmic protein. UDPGDH = UDP-glucose dehydrogenase; the control activity was 7.20 $\pm$ 0.88 (16) nmol NAD^+ reduced/min-mg cytoplasmic protein. The values given are the means and standard deviations of the number of animals shown in parenthesis. All values shown are significantly different than the controls (P < 0.001). For further details, see refs. 26 and 56.

Phenobarbital has also been demonstrated to induce UDP-glucose dehydrogenase (57). Again, there appear to be relatively straightforward explanations for these effects. Glucose-6--phosphate dehydrogenase is the first and rate-limiting enzyme of the pentose monophosphate shunt, which supplies the cytoplasm with NADPH, the cofactor utilized by the cytochrome

P-450 system. At the same time, UDP-glucose dehydrogenase is
involved in the synthesis of UDP-glucuronic acid, the co-
factor utilized by UDP-glucuronyl transferase.

Upon reflection, it is not at all surprising that enzyme systems supplying prosthetic groups present in drug-
-metabolizing enzymes or cofactors used in drug metabolism
should be induced together with the phase I, II, and III enzymes. It will be interesting to see if inducers of NADPH-
-cytochrome P-450 reductase and of DT-diaphorase also bring
about changes in flavin metabolism, since both of these are
flavoproteins. Can the synthesis of glutathione be induced?
Preliminary results in our laboratory indicate that trans-
-stilbene oxide induces glutathione reductase in rat liver,
thereby presumably providing for the maintenance of glutathione in its reduced state, that is to say, the state in
which it can be conjugated to reactive metabolites of xenobiotics.

In conclusion, it is quite clear that the effects of
inducers of drug-metabolizing enzymes on rat liver are much
more extensive than simply increases in the activities of phase I, II, and III enzymes. Indeed, considering the inter-
relatedness of all the different systems in a biological organism, our original ideas about the specificity of drug induction seem naive.

OUTLOOK

A large number of questions have been raised in these
pages and the answers will be exciting and, probably, very
useful in our efforts to explain the etiology of various diseases and toxic effects and to prevent the same. More thorough characterizations of the effects of inducers on hepatocytes must be performed. How are the activities of the phase

III enzymes affected? Which enzymes not directly involved in drug metabolism are also affected? In addition, induction in organs other than the liver must be studied.

Such detailed characterizations should allow us to distinguish between different classes of inducers. It may turn out that certain inducers -- those which preferentially induce the phase I reactions -- increase our ability to detoxify xenobiotics and excrete their metabolites, but with an increased risk for mutation of a normal cell to a cancer cell as the price tag. Other inducers -- those which preferentially induce the phase II enzymes -- may not carry this price tag.

Of course, the relative activities of phase I and II enzymes is only one of the metabolic factors which affect the toxicity and carcinogenicity of xenobiotics. Recycling, the reactivity of products formed via the cytochrome P-450 system, the ability of these reactive intermediates to serve as substrates for the phase II enzymes, and the ease with which adducts formed between these intermediates and DNA can be repaired are just a few examples of other important factors. Also critical may be the relative levels of different isozymes of cytochrome P-450.

Studies designed to elucidate the molecular mechanism of enzyme induction will also be continued. It is to be expected that such investigations will yield valuable information about genetic control in mammalian systems and may even reveal new principles of gene regulation.

Inducers, in particular phenobarbital, can also be used to initiate the biosynthesis of membrane proteins and phospholipids and the formation of new membranes of the endoplasmic reticulum. After exposure to the inducer is terminated, these membrane components and membranes are degraded, so that the control situation is re-established. Thus, induction of drug-metabolizing enzymes can also be used as a model system for characterizing membrane biogenesis and degradation.

Recently, we have been engaged in the study of mechanisms for rapidly activating drug-metabolizing enzymes. For instance, we have observed that benzil activates epoxide hydrolase activity 5-6-fold when incubated with rat liver microsomes for a few minutes at $37^{\circ}C$ (58). Furthermore, treatment of microsomal glutathione S-transferase with sulfhydryl reagents in vitro results in a 5-6-fold increase in this activity within seconds (59). We are presently trying to determine whether there are also similar mechanisms for activating these enzymes in vivo. Activation of drug-metabolizing enzymes within seconds or minutes would provide a very important complement to the process of enzyme induction, which requires several days for completion.

Finally, much remains to be discovered about the consequences of induction in man. To what extent does it occur and how important is it in our defenses against environmental pollutants? Is the induction of aryl hydrocarbon monooxygenase in lung cells by cigarette smoke (60) involved in the etiology of lung cancer? How does induction of the cytochrome P-450 system affect the metabolism of steroid hormones?

Even though we have been studying the induction of drug-metabolizing enzymes for more than 15 years, we are still only at the beginning.

ACKNOWLEDGEMENTS
The studies from our own laboratory discussed in this mini-review, were supported by Grant No. 1 RO 1CA 26261-02 from the National Cancer Institute, Department of Health, Education, and Welfare, U.S.A., and by grants from the Swedish Natural Science Research Council, the Swedish Medical Research Council, and the National Swedish Environment Protection Board.

REFERENCES

1. Ernster, L. and Orrenius, S. (1965) Fed. Proc. 24, 1190-1199.
2. Lu, A.Y.H. (1979) Drug Metabolism Reviews 10, 187-208.
3. Guengerich, F.P. (1979) Pharmacol. Therap. 6, 99-121.
4. Poland, A. and Glover, E. (1976) J. Biol. Chem. 251, 4936-4946.
5. Greenlee, W.F. and Poland, A. (1979) J. Biol. Chem. 254, 9814-9821.
6. Nebert, D.W. and Jensen, N.M. (1979) Critical Reviews in Biochemistry 6, 401-437.
7. Extrahepatic Metabolism of Drugs and Other Foreign Compounds (Gram, T.E., ed.), SP Medical and Scientific Books: New York, 1980, 601 pp.
8. Goldberg, D.M. (1980) Clin.-Chem. (Winston-Salem, N.C.) 26, 691-699.
9. Symposia Medica Hoechst 14: The Induction of Drug Metabolism (Estabrook, R.W. and Lindenlaub, E., eds.), F.K. Schattauer Verlag: Stuttgart, 1979, 645 pp.
10. Gilette, J.R. (1979) Drug Metabolism Review 10, 59-87.
11. Snyder, R. and Remmer, H. (1979) Pharmacol. Therap. 7, 203-244.
12. Conney, A.H. (1967) Pharmacol. Rev. 19, 317-366.
13. Oesch, F. (1973) Xenobiotica 3, 305-340.
14. Jakoby, W.B. (1977) Adv. Enzymol. 46, 381-412.
15. DePierre, J.W. and Ernster, L. (1978) Biochim. Biophys. Acta 473, 149-186.
16. Seidegård, J., Morgenstern, R., DePierre, J.W. and Ernster, L. (1979) Biochim. Biophys. Acta 586, 10-21.
17. Åström, A. and DePierre, J.W. Biochim. Biophys. Acta, in press.
18. Benson, A.M., Batzinger, R.P., Ou, S.-Y.L., Bueding, E., Cha, Y.-N. and Talalay, P. (1978) Cancer Res. 38, 4486-4495.
19. Cha, Y.-N., Martz, F. and Bueding, E. (1978) Cancer Res. 38, 4496-4498.
20. Cha, Y.-N. and Bueding, E. (1979) Biochem. Pharmacol. 28, 1917-1921.
21. Morgenstern, R., Meijer, J., DePierre, J.W. and Ernster, L. (1980) Europ. J. Biochem. 104, 167-174.

22. Bentley, P. and Oesch, F. (1975) FEBS Lett. 59, 291-295.
23. Lu, A.Y.H., Ryan, D., Jerina, D.M., Daly, J.W. and Levin, W. (1975) J. Biol. Chem. 250, 8283-8288.
24. DePierre, J.W. and Dallner, G. (1975) Biochim. Biophys. Acta 415, 411-472.
25. Guthenberg, C., Morgenstern, R., DePierre, J.W. and Mannervik, B. (1980) Biochim. Biophys. Acta 631, 1-10.
26. Seidegård, J., DePierre, J.W., Morgenstern, R., Pilotti, Å. and Ernster, L. Biochim. Biophys Acta, in press.
27. Kahl, R. and Wulff, U. (1979) Toxicol. Appl. Pharmacol. 47, 217-227.
28. Bock, K.W., Fröhling, W., Remmer, H. and Rexer, B. (1973) Biochim. Biophys. Acta 327, 46-56.
29. Wishart, G.J. (1978) Biochem. J. 174, 671-672.
30. Nebert, D.W. and Felton, J.S. (1976) Fed. Proc. 35, 1133-1141.
31. Solt, D.B., Medline, A. and Farber, E. (1977). Am. J. Pathol. 88, 595-618.
32. Bücker, M., Golan, M., Schmassmann, H.U., Glatt, H.R., Stasiecki, P. and Oesch, F. (1979) Mol. Pharmacol. 16, 656-666.
33. Schmassmann, H., Sparrow, A., Platt, K. and Oesch, F. (1978) Biochem. Pharmacol. 27, 2237-2245.
34. Kano, I., Gielen, J.E., Yagi, H., Jerina, D.M. and Nebert, D.W. (1977) Mol. Pharmacol. 13, 1181-1186.
35. Schlicht, I., Koransky, W., Magour, S. and Schulte-Herman, R. (1968) Naunyn-Schmiedebergs Arch. Pharmakol. Exp. Pathol. 261, 26-30.
36. Koransky, W. and Schulte-Herman, R. (1970) in Proceedings of the Fourth International Congress of Pharmacology (Eigenmann, R., ed.), Vol. IV, pp. 277-283, Schwabe and Co., Basel.
37. Suzuki, Y., DePierre, J.W. and Ernster, L. (1980) Biochim. Biophys. Acta 601, 532-543.
38. DePierre, J.W. and Ernster, L. (1976) FEBS Lett. 68, 219-224.
39. Ishidate, K. and Nakazawa, Y. (1976) Biochem. Pharmacol. 25, 1255-1260.
40. Davison, S.C. and Wills, E.D. (1974) Biochem. J. 140, 461-468.
41. Cooper, S.D. and Feuer, G. (1972) Can. J. Physiol. Pharmacol. 50, 568-575.

42. Schulze, H.U. and Staudinger, H. (1970) Z. Physiol. Chem. 351, 184-193.

43. Eriksson, L.C. (1973) Acta Pathol. Microbiol. Scand., Sect. A, Suppl. 239.

44. Gelboin, H.V., Wortham, J.S. and Wilson, R.G. (1967) Nature 214, 281-283.

45. Bresnick, E. (1966) Mol. Pharmacol. 2, 406-410.

46. Nebert, D.W. and Gelboin, H.V. (1968) J. Biol. Chem. 243, 6242-6249.

47. Nebert, D.W. and Gelboin, H.V. (1968) J. Biol. Chem. 243, 6250-6261.

48. Nebert, D.W. and Gelboin, H.V. (1970) J. Biol. Chem. 245, 160-168.

49. Grundin, R., Jakobsson, S. and Cinti, D.L. (1973) Arch. Biochem. Biophys. 158, 544-555.

50. Conney, A.H. and Gilman, A.G. (1963) J. Biol. Chem. 238, 3682-3685.

51. Sugiyama, T., Miki, T. and Yamano, T. (1980) J. Biochem. (Tokyo) 87, 1457-1467.

52. Lind, C. and Ernster, L. (1974) Biochem. Biophys. Res. Commun. 56, 392-400.

53. Lind, C., Hochstein, P. and Ernster, L. (1980) in Oxidases and related redox systems (King, T.E., Mason, H.S. and Morrison, M., eds.), in press.

54. Lind, C., Höjeberg, B., Seidegård, J., DePierre, J.W. and Ernster, L. (1980) FEBS Lett. 116, 289-292.

55. Tephly, T.R., Hasegawa, E. and Baron, J. (1971) Metab. Clin. Exp. 20, 200-214.

56. Seidegård, J., DePierre, J.W. and Ernster, L. (1980) Acta Chem. Scand. B34, 382-384.

57. Zeidenberg, P., Orrenius, S. and Ernster, L. (1967) Ann. N.Y. Acad. Sci. 29, 310-315.

58. Seidegård, J. and DePierre, J.W. Europ. J. Biochem., in press.

59. Morgenstern, R., DePierre, J.W. and Ernster, L. (1979) Biochem. Biophys. Res. Commun. 87, 657-663.

60. Cantrell, E.T., Warr, G.A., Busbee, D.L. and Martin, R.R. (1973) J. Clin. Invest. 52, 1881-1884.

INDUCTION OF INDOLEAMINE 2,3-DIOXYGENASE

BY VIRUS, ENDOTOXIN AND INTERFERON[*]

Osamu Hayaishi and Ryotaro Yoshida

Department of Medical Chemistry
Kyoto University Faculty of Medicine
Kyoto 606, Japan

Molecular oxygen in tissues serves two functions; first, to act as the ultimate hydrogen acceptor in the process of the biological oxidation of food stuff being reduced to either water or hydrogen peroxide, and second, to transform dietary nutrients into cellular constituents and biologically important substances. The enzymes involved in the former processes are usually referred to as oxidases, whereas those involved in the latter processes are termed as oxygenases (1). Since two types of oxygenases, namely di- and mono-oxygenases,

---

[*] Dedicated to Professor Lars Ernster on the occasion of his 60th birthday.

This work was supported in part by research grants from the Naito Foundation, Nippon Shinyaku Co., Ltd., and the Intractable Diseases Division, Public Health Bureau, the Ministry of Health and Welfare, Japan, and by a Grant-in-Aid for Scientific Research from the Ministry of Education, Science and Culture, Japan.

C. P. Lee, G. Schatz, G. Dallner (eds.), Mitochondria and Microsomes
   in honor of Lars Ernster   ISBN 0-201-04576-1

were independently discovered by Hayaishi (2) and Mason (3) in 1955, many oxygenases have been found to be widely distributed in nature and to play physiologically important roles including the biosynthesis, transformation and degradation of phenolic compounds, amino acids, lipids, vitamins, etc., as well as the metabolic disposal of a variety of drugs and foreign compounds. In fact, two most important pathways of tryptophan in mammals, leading to the formation of pyridine

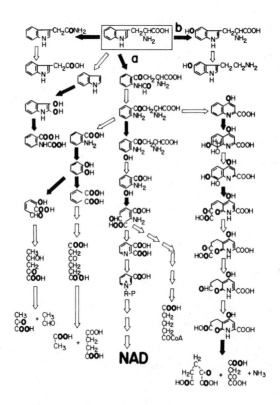

Fig. 1.Metabolism of tryptophan

    ➡, oxygenase reactions;⇨ , reactions catalyzed by enzymes other than oxygenases.
a, tryptophan 2,3-dioxygenase.
b, tryptophan 5-monooxygenase.

nucleotide coenzymes and indoleamines, are catalyzed by the
two well-known oxygenases, (1) tryptophan 2,3-dioxygenase
and (2) tryptophan 5-monooxygenase, respectively (Fig. 1).
The black arrows in this figure denote oxygenase reactions,
and the white arrows indicate reactions catalyzed by enzymes
other than oxygenases.

HISTORY OF INDOLEAMINE 2,3-DIOXYGENASE

    In 1937, Kotake and Ito found that rabbits fed D-
tryptophan excreted D-kynurenine in the urine and suggested
the occurrence of an enzyme in rabbit tissues that catalyzes
the oxidative ring cleavage of D-tryptophan (4). About ten
years ago, an enzyme activity capable of oxidizing D-tryptophan to D-kynurenine was shown in our laboratory by using a
homogenate of rabbit small intestine (5). Then, the enzyme
activity has been found to be ubiquitously distributed in almost all organs except in the liver and the highest specific

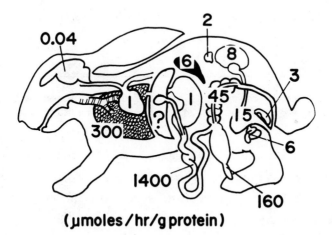

Fig. 2.Distribution of indoleamine 2,3-dioxygenase in rabbit

    The number represents μmoles of product formed per hour
    per gram of protein with L-tryptophan as substrate.

activity was observed in the small intestine, lung and colon, in that order, as shown in Fig. 2. Because the small intestine was the richest source of this enzyme, it was purified from rabbit small intestine (6). The enzyme was purified about 400-fold with a yield of about 8% to apparent homogeneity. Molecular weight was estimated to be about 42,000 and carbohydrate content 4.8%. It contained one protoheme IX per mole of enzyme but other metals including copper were negligible. Molecular activity was 2 katals per mole, namely the turn over number was about 120 (Table I). A highly

Table I. Molecular properties of indoleamine 2,3-dioxygenase in rabbit small intestine

Molecular weight	42,000
Subunit	1
Carbohydrate	4.8%
Protoheme IX	1
Turn over number*	120

*Turn over number represents moles formyl kynurenine formed per min/mole of enzyme at 24° under the standard assay conditions (6).

purified preparation of this enzyme had a broad substrate specificity and was found to catalyze the oxygenative ring cleavage of 5-hydroxytryptophan, serotonin and tryptamine in addition to the D- and L-isomers of tryptophan in the presence of the superoxide anion (7-9). Therefore, the term "indoleamine 2,3-dioxygenase" was proposed to designate this enzyme (10).

INDUCTION OF INDOLEAMINE 2,3-DIOXYGENASE

A. Induction of pulmonary indoleamine 2,3-dioxygenase during
   virus infection

        Because the lung is an aerobic organ and presumably
has an abundance of the superoxide anion and also contain
high levels of indoleamine 2,3-dioxygenase, the lung was selected as an enzyme source for the studies of various changes
of this enzyme activity under a variety of physiological conditions. The indoleamine 2,3-dioxygenase is relatively stable under various physiological conditions, although it
exhibits a daily rhythmic cycle and age-dependent changes
(11).

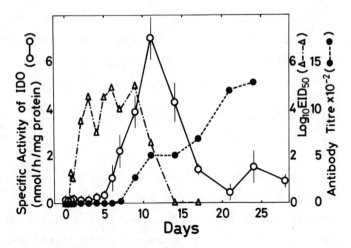

Fig. 3. Indoleamine 2,3-dioxygenase activity and virus titer
        in mouse lung and serum antibody content after influ-
        enza virus infection
        o, pulmonary indoleamine 2,3-dioxygenase activity;
        Δ, virus infectivity titer in the lung; •, antibody
        titer in serum. The enzyme activity is expressed as
        nmole of product formed per hour per mg of protein.
        Each point represents the mean value ± SEM for six
        mice.
        IDO, indoleamine 2,3-dioxygenase.

Recently, we found that when mice were exposed to the influenza virus, the indoleamine 2,3-dioxygenase activity in the lung increased almost linearly from the 5th day after infection (Fig. 3). The enzyme activity was usually at the highest level (100- to 120-fold) around the 11th day and then gradually decreased to normal values in about 3 weeks. On the other hand, the virus replication began within 24 h after the infection, reached a peak by the 3rd day and persisted until the 9th day. Thereafter, the virus was rapidly elimi- nated from the lung, and on the 14th day practically none was found. The serum antibody did not appear until the 9th day after infection, when the virus had began to disappear. The level of serum antibody rapidly increased thereafter, and was detectable in dilutions in 1:1,000 on the 24th day. On the other hand, histological studies indicated that the enzyme induction is closely related to the inflammatory response, particularly infiltration of lymphocytes.

B. Induction of indoleamine 2,3-dioxygenase by bacterial endotoxin (LPS)

Reactions caused by viruses are very complicated and most of them have not been elucidated. To simplify the exper- imental conditions, we then used bacterial endotoxin instead of influenza virus. Bacterial endotoxin, the lipopolysaccha- ride fraction (LPS) of Gram-negative bacteria, is an inflam- matory agent and causes nonspecific immune responses. When LPS derived from Salmonella abortus equi or from Escherichia coli was given intraperitoneally to mice at zero time, the indoleamine 2,3-dioxygenase activity increased almost linear- ly in the lung for about 24 h (Fig. 4). Both preparations of LPS (20 μg/mouse) caused a 30- to 50-fold increase in the

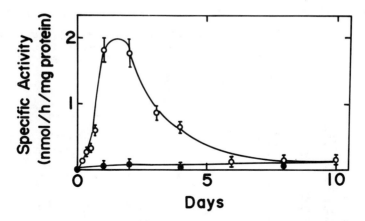

Fig. 4. Induction of indoleamine 2,3-dioxygenase in mouse lung
by LPS (E. coli)

The enzyme activity is expressed as μmole of product
formed per hour per mg of protein. Each point represents the mean ± SEM for five mice. o, + LPS; ●,
control.

specific activity of the high speed supernatant in the lung
at 24 h later. After about 48 h, it gradually decreased and
reached a normal value within 7 days. On the other hand,
there was no increase in the enzyme activity, when saline was
applied.

This increase in the enzyme activity appears to be
specific for the lung, because in all other organs the increment in the enzyme activity was usually less than several fold
at most. The tryptophan cleaving activity in the liver is
known to be due to tryptophan dioxygenase (12), and was essentially unchanged or somewhat decreased. The increase in the
enzyme activity appears to be due to the net synthesis of the
enzyme protein rather than removal of inhibitors, formation
of activators or other reasons. The concomitant administration of actinomycin D or cycloheximide abolished the increase

in the enzyme activity produced by LPS (Fig. 5). Immunological analysis using antibody against indoleamine 2,3-dioxygen-
ase are in line with this conclusion, because the increment
of the enzyme activity was mainly due to an increase in the
amount of immunoreactive protein.

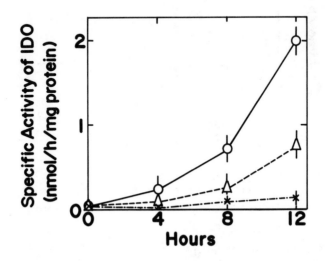

Fig. 5. Effects of actinomycin D and cycloheximide on the
increase in pulmonary indoleamine 2,3-dioxygenase activity induced by LPS treatment

Inhibitors were administered simultaneously with LPS
at zero time and at the maximum three injections each
at 4-hour intervals were given to the animals sacrificed 12 h later. o, LPS (20 µg/mouse) only; Δ, LPS
plus actinomycin D (25 µg/mouse); X, LPS plus cycloheximide (5 mg/mouse).
IDO, indoleamine 2,3-dioxygenase.

The effects of bacterial LPSs derived from Salmonella
and Escherichia S and R mutant strains, and lipid A on the
indoleamine 2,3-dioxygenase activity in mouse lung were examined (Fig. 6). The induction of the indoleamine 2,3-dioxy-
genase activity in the mouse lung appears to be specific for
the lipopolysaccharide fraction, because other inflammatory

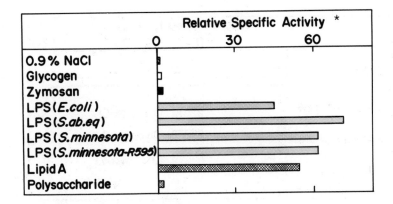

Fig. 6.Effects of LPSs or lipid A on pulmonary indoleamine
2,3-dioxygenase activity

    Mice were sacrificed 24 h after intraperitoneal ad-
ministration of various stimulants; glycogen (2 mg/
mouse), zymosan (50 μg/mouse), LPSs (20 μg/mouse),
lipid A (50 μg/mouse) and polysaccharide fraction
(100 μg/mouse).
* Relative to saline treatment.

agents such as glycogen (2 mg/mouse) or zymosan (50 μg/mouse),

a protein polysaccharide complex of the yeast cell wall, had

no such effect. LPSs, derived from <u>Salmonella</u> <u>minnesota</u>,

were purified by protease treatment and freed from endotoxin

protein. These LPSs (20 μg/mouse) as well as commercial LPSs

produced a 40- to 70-fold increase in the specific activity

of pulmonary indoleamine 2,3-dioxygenase within 24 h. Lipid

A (50 μg/mouse), freed from the polysaccharide fraction of

LPS was also effective. By contrast, polysaccharide fraction

(100 μg/mouse) obtained by acid hydrolysis of LPS, was in-

effective, under the same experimental conditions.

    To determine whether or not the induction was speci-

fic for the indoleamine 2,3-dioxygenase activity, various

other enzyme activities in the lung were determined (Fig. 7).

For example, typical lysosomal enzymes such as β-glucuronidase

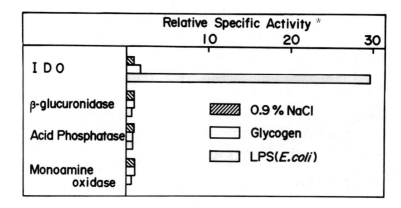

Fig. 7. Induction of various enzymes in the lung
Mice were sacrificed 24 h after intraperitoneal injections of various stimulants. IDO, indoleamine 2,3-
dioxygenase.
* Relative to saline treatment.

and acid phosphatase, as well as monoamine oxidase, another
enzyme involved in the metabolism of biogenic amines, were
not induced to any significant extent. Other enzymes, including superoxide dismutase, prostaglandin synthetase, lip-
oxygenase etc., were not induced either.

C. Induction of indoleamine 2,3-dioxygenase by interferon

These findings raised several interesting possibilities. The LPS fraction of the cell wall of Gram-negative
bacteria is an inflammatory agent and causes various immune
responses. A good correlation was also obtained between
enzyme induction and peribronchial and perivascular infiltra-
tions of lymphocytes. These results, therefore, suggest that
the enzyme induction may be related to a local increase in
the production of substances liberated at the inflammatory
loci, including the superoxide anion, serotonin and interferon. In an attempt to clarify the mechanism of the

induction of indoleamine 2,3-dioxygenase by LPS or during virus infection, in vivo, an in vitro system using mouse lung
slices was developed. When lung slices were incubated in the
presence of LPS (5 µg/ml), indoleamine 2,3-dioxygenase activity in the high speed supernatant fraction was increased for
at least 48 h and gradually decreased. As shown in Fig. 8,

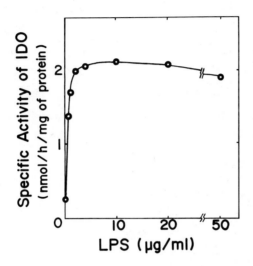

Fig. 8.Effect of LPS (E. coli) on indoleamine 2,3-dioxygenase
activity in mouse lung slices
  LPS (0-50 µg/ml) was added to the culture medium at
  zero time, and incubation carried out for 36 h. Each
  point represents the mean value in duplicate.
  IDO, indoleamine 2,3-dioxygenase.

a dose-dependent increase in the specific activity of indoleamine 2,3-dioxygenase was observed below 1 µg/ml of LPS and a
maximum induction (approximately 10-fold) was obtained at
approximately 5 µg/ml (20 µg/0.2-0.3 g lung slices), a dose
almost identical to that used in vivo (20 µg/mouse, 0.2-0.3
g lung), previously (13).

    In these in vitro systems, the effects of various

other substances, such as the superoxide anion or indoleamines, both of which are utilized by indoleamine 2,3-
dioxygenase (6,8,9), were examined on the enzyme activity.
Methyl viologen (40 µg-1.25 mg/ml), a potent inducer of the
superoxide anion in the lung (14), 5-hydroxy-L-tryptophan
(80 µg-2.25 mg/ml), a precursor of serotonin, or serotonin
(0.352-352 µg/ml) caused no significant effect on the enzyme
activity. However, when slices of mouse lung were incubated
with interferon (10^4 units/ml, 1.4 X 10^6 units/mg protein)
from mouse L cells, a 10- to 15-fold increase in the enzyme
activity was observed within 48 h (Fig. 9). The specific
activity of indoleamine 2,3-dioxygenase in the high-speed
supernatant fraction of lung slices increased almost linearly

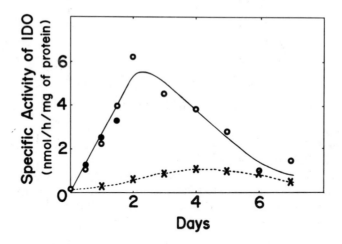

Fig. 9. Effects of mouse L cell interferons on pulmonary
indoleamine 2,3-dioxygenase activity of mouse lung
slices

Mouse interferons (10^4 units/ml) were added to the
culture medium at zero time. o, L cell interferon
(1.4 X 10^6 units/mg protein); •, purified L cell interferon (2 X 10^8 units/mg protein); X, no addition.
IDO, indoleamine 2,3-dioxygenase.

for 48 h, then began to decrease, and reached a normal value
after about 7 days. In the absence of interferon the enzyme
activity increased slowly and reached a plateau (about 4-fold)
after about 4 days. Although this observation was also re-
producible, the exact nature is at present unknown since there
was no increase in interferon titer. Fig. 9 also shows that
a similar effect was observed with highly-purified L cell
interferon (10^4 units/ml, 2 X 10^8 units/mg protein), prepared
by poly (U) affinity column chromatography.

When lung slices were incubated with increasing amounts
of interferon (0-30,000 units/ml), a dose-dependent increase
in the specific activity of indoleamine 2,3-dioxygenase was
observed by either mouse L cell or mouse brain interferon

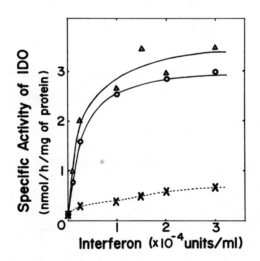

Fig. 10. Dose dependency of interferons on pulmonary indoleamine 2,3-dioxygenase activity

   After addition of interferons (0-3 X 10^4 units/ml),
   lung slices were incubated for 36 h. Each point represents the mean value in duplicate. o, mouse L
   cell interferon; Δ, mouse brain interferon; X, human
   leukocyte interferon.
   IDO, indoleamine 2,3-dioxygenase.

(Fig. 10). Maximum effects were observed with about 10^4 units
per ml. However, human leukocyte interferon showed little
effect on the enzyme activity, since interferon is species-
specific.

Using mouse L-cell interferon, the effect of heat,
α-chymotrypsin or anti-interferon serum treatment was
examined (Fig. 11). When interferon was heated at 60° for
30 minutes, the activity as enzyme inducer decreased by

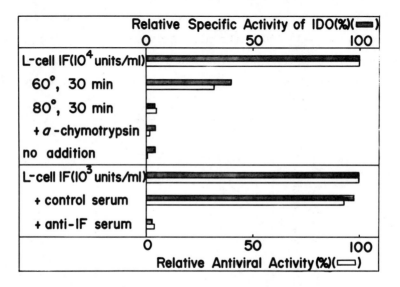

Fig. 11. Effect of heat, α-chymotrypsin or anti-interferon
treatment of interferon on pulmonary indoleamine 2,3-
dioxygenase activity and antiviral activity

After heat (60° or 80°, 30 min), α-chymotrypsin (for
6 h) or anti-interferon serum treatment of interferon,
the changes in the antiviral activity and in the
ability to increase the pulmonary indoleamine 2,3-
dioxygenase activity were examined 36 h later. The
actual activities of the enzyme induced by interferon
at the concentrations of 10^4 and 10^3 units/ml were
2.4 and 0.76 (nmole/h/mg protein) for a mean value in
duplicate, respectively. These values represent 100%
activity. ▭, antiviral activity; ■, specific activity of indoleamine 2,3-dioxygenase.

about 60%, while the interferon heated at 80° for 30 minutes
all but lost the activity. The observed degree of inactivation by heat treatment was essentially proportional to that
of antiviral activity of interferon. Fig. 11 also shows that
incubation with α-chymotrypsin for 6 h reduced the activity
of mouse L cell interferon (10^4 units/ml) by more than 99%
($<10^2$ units/ml), and that the ability to increase the enzyme
activity was all but lost by this treatment. However, the
effect of LPS or poly I·poly C on the enzyme activity was
retained intact by the treatment. Furthermore, when interferon was preincubated with anti-interferon serum, the antiviral activity of interferon (10^3 units/ml) was almost completely abolished (<5 units/ml) and the enzyme activity did
not increase significantly. These results suggest that the
increase in the indoleamine 2,3-dioxygenase activity is
caused by the biological activities of interferon and not by
possible contaminants, such as LPS or double stranded RNA.

Actinomycin D (2 μg/ml) and cycloheximide (12.5 μg/ml)
blocked the increase in the enzyme activity measured 6 to 24 h
after addition of interferon if given 2 h before the addition
of interferon (Fig. 12), suggesting that the increase in the
enzyme activity was probably due to net synthesis of the
enzyme protein.

Interferon is reported to induce several enzymes
such as 2',5'-oligoadenylate synthetase and protein kinase.
The antiviral activity of interferon has been assumed to be
mediated by these enzymes because oligoadenylate and phosphorylated proteins were found to be synthesized in extracts
of interferon-treated mouse L cells (15), chicken embryo
cells (16), and in the reticulocyte lysates (17) if supplemented with double stranded RNA and ATP. However, the

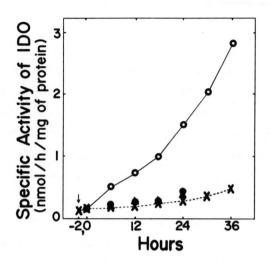

Fig. 12. Effects of actinomycin D and cycloheximide on the
increase in the pulmonary indoleamine 2,3-dioxygenase
activity induced by interferon treatment

Arrow indicates the time when these agents were added
to the culture medium; 2 h before interferon treatment (10^4 units/ml). Each point represents the mean
value in duplicate. o, mouse L cell interferon (10^4
units/ml) only; •, interferon plus actinomycin D (2
μg/ml); ▲, interferon plus cycloheximide (12.5 μg/ml)
; X, no addition.
IDO, indoleamine 2,3-dioxygenase.

interrelationships of these enzyme activities and relevance

to the mechanism by which interferon blocks replication are

unclear at present. On the other hand, no information on

the molecular mechanism of interferon actions other than

antiviral activity is available. Comparative studies on the

induction of these enzymes may be important in understanding

the mechanism of action of interferon.

CONCLUSION

It has been almost a quarter of a century since the

discovery of oxygenases, and extensive studies have been
carried out on the properties and functions of oxygenases.
Indoleamine 2,3-dioxygenase is a unique oxygenase, because
it requires the superoxide anion for the initiation of the
reaction and for maintenance of the catalytic cycle during
the steady state. This enzyme is widely distributed in
nature and plays a crucial role in the newly discovered
metabolic pathway of indoleamines.

A dramatic induction of the indoleamine 2,3-dioxygen-
ase was observed in the mouse lung during virus infection or
when bacterial endotoxin had been administered in vivo, or
lung slices were exposed to endotoxin or interferon in vitro.
Further studies are required to elucidate the biological
significance of this phenomenon and the biological events
in the lung inflammation and/or the mode of action of interferon.

REFERENCES

1. Hayaishi, O. (1962) in Oxygenases, (O. Hayaishi, ed.)
    pp. 1-29, Academic Press, New York.
2. Hayaishi, O., Katagiri, M., and Rothberg, S. (1955) J.
    Amer. Chem. Soc. 77, 5450-5451.
3. Mason, H. S., Fowlks, W. L., and Peterson, L. (1955) J.
    Amer. Chem. Soc. 77, 2914-2915.
4. Kotake, Y. and Ito, N. (1937) J. Biochem. 25, 71-77.
5. Higuchi, K., Kuno, S., and Hayaishi, O. (1967) Arch.
    Biochem. Biophys. 120, 397-403.
6. Shimizu, T., Nomiyama, S., Hirata, F., and Hayaishi, O.
    (1978) J. Biol. Chem. 253, 4700-4706.
7. Hirata, F. and Hayaishi, O. (1971) J. Biol. Chem. 246,
    7825-7826.
8. Hirata, F. and Hayaishi, O. (1975) J. Biol. Chem. 250,
    5960-5966.
9. Hayaishi, O., Hirata, F., Ohnishi, T., Henry, J. P.,
    Rosenthal, I., and Katoh, A. (1977) J. Biol. Chem. 252,
    3548-3550.
10. Hirata, F., Hayaishi, O., Tokuyama, T., and Senoh, S.

(1974) J. Biol. Chem. 249, 1311-1313.

11. Yoshida, R., Nukiwa, T., Watanabe, Y., Fujiwara, M., Hirata, F., and Hayaishi, O. (1980) Arch. Biochem. Biophys. 203, 343-351.

12. Feigelson, P. and Greengard, O. (1962) J. Biol. Chem. 237, 1903-1907.

13. Yoshida, R. and Hayaishi, O. (1978) Proc. Natl. Acad. Sci. U.S.A. 75, 3998-4000.

14. Fisher, H. K. (1977) in Biochemical Mechanisms of Paraquat Toxicity, (A. P. Author, ed.) pp. 57-65, Academic Press, New York.

15. Roberts, W. K., Hovanessian, A., Brown, R. E., Clemens, M. J., and Kerr, I. M. (1976) Nature 264, 477-480.

16. Ball, L. A. and White, C. N. (1978) Proc. Natl. Acad. Sci. U.S.A. 75, 1167-1171.

17. Hovanessian, A. and Kerr, I. M. (1978) Eur. J. Biochem. 84, 149-159.

THE TOPOLOGY OF ENZYMES IN MICROSOMES FROM LIVER

Henri Beaufay, Alain Amar-Costesec
and Christian de Duve

Laboratoire de Chimie Physiologique,
Université Catholique de Louvain
and
International Institute of Cellular and Molecular Pathology
B-1200 Brussels, Belgium

INTRODUCTION

Several excellent reviews have dealt recently with the structural and functional organization of enzymes in microsomal membranes (1-3). They illustrate the complexity of this multi-faceted topic, which, in its present stage of extensive but still largely incomplete development, can be covered adequately only if detailed critical attention is given to the materials, methods and research strategies employed by each individual investigator or group of investigators. This is clearly not possible within the restricted space of the present survey, and we must therefore limit ourselves to some sort of bird's eye view of the field, leaving the reader to fill in details with the help of the reviews quoted above.

C. P. Lee, G. Schatz, G. Dallner (eds.), Mitochondria and Microsomes
in honor of Lars Ernster   ISBN 0-201-04576-1

629

For understandable reasons, we will take this survey
from the vantage point of the results that have been obtained in our laboratory on the microsomal fraction from rat
liver (4-12, reviewed in 13 and 14).

COMPOSITION OF THE MICROSOMAL FRACTION

Quantitative analytical fractionation of total rat
liver microsomes by means of density equilibration in a
continuous sucrose gradient, using a high performance, low
hydrostatic pressure zonal rotor specially constructed for
this purpose (15), has disclosed that enzymes and other
biochemical constituents of the microsomal fraction display one of four distinct distribution patterns, termed a,
b, c and d (4, 5). Subsequently, the a group could be
divided further into three subgroups, on the basis of
sedimentation velocity (8), and especially of magnitude
of the "digitonin shift" (9), i.e. of the displacement
of the distribution pattern towards higher densities induced by digitonin treatment (presumably by digitonin
binding to accessible cholesterol molecules) : a_1 showing
no detectable digitonin shift, a_2 showing the largest
shift, of the order of 0.03 density units, and a_3 showing
an intermediate shift, of about 0.02 density units. The
groups b, c and d are either not shifted, or only very
slightly shifted by digitonin treatment.

A representative example of the six distribution
patterns and of their differential displacement by digitonin is shown in Fig. 1. The vertical arrow at density
value 1.20 provides a convenient marker whereby to
distinguish the various patterns from each other. In Table I

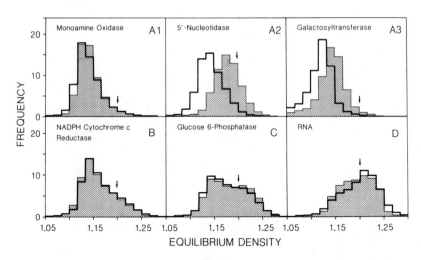

Fig. 1.Density distribution of some typical constituents of
the microsome fraction after isopycnic centrifugation in a linear gradient of sucrose.

Microsomes were prepared from rat liver according to
ref. (7). The density distributions obtained after
treatment of the microsomes with digitonin under the
conditions given previously (9) are shown by the
shaded histograms and compared to those obtained from
untreated microsomes (solid lines). Frequency histograms are divided into 15 fractions of identical density increment and averaged. They include results
reported previously (8, 9) and more recent, unpublished data. The number of determinations, given in
the order a_1, a_2, a_3, b, c and d, is 19, 20, 12, 9,
25 and 10 in experiments on untreated microsomes,
and 8, 11, 6, 2, 12 and 3 in experiments on digito-
nin-treated microsomes. Comparison of the median
density values reveals an average shift of 0.036 and
0.022 unit for 5'-nucleotidase and galactosyltransferase, respectively, significant at P < 0.001. A
shift of 0.006 unit, significant at 0.02 < P < 0.05,
is noted for the glucose-6-phosphatase activity.
Other differences are not significant at P ≤ 0.05.
Arrows indicate demarcation at density 1.2, allowing
easier comparison of distribution profiles.

Table I. Classification of rat liver microsomal enzymes according to distribution pattern and putative location.

Putative Location	Enzyme or Constituent	Group	References
Outer mitochondrial membrane	Monoamine oxidase	a_1	8,9,67
	Minor part of NADH cytochrome c reductase		68
	Minor part of cytochrome b_5		11,68
Plasma membrane	5'-Nucleotidase	a_2	8,9,69-72
	Alkaline phosphatase	a_2	8,9,70
	Alkaline phosphodiesterase I	a_2	8,9,20,70
	$Na^+ K^+$-ATPase	a_2	73,74,(a) *
	NAD-glycohydrolase	a_2	75,76
	Adenylate cyclase	a_2	77,(b) *
	γ-Glutamyltransferase	a_2	78,(b)
	Bulk of sialoglycoprotein	a_2	20,26,70
	Nucleoside diphosphatase (ADP)	a_2	79,(a)
	Bulk of cholesterol	a_2	8,9,70,72
Golgi complex membranes	Galactosyltransferase	a_3	8,9,80-82
	N-acetylglucosaminyltransferase	a_3	10,81-83

	Enzyme		References
	Sialyltransferase	a_3	10,82
	Some sialoglycoprotein		23,26
	Some cholesterol		(c)*
ER cytoplasmic surface	Most of NADH cytochrome c reductase	b	8,9,84–86
	Most of cytochrome b_5	b	8,9,12,84–86
	Stearoyl-CoA desaturase	n.i.++	87
	NADPH cytochrome P-450 (c) reductase	b	8,9,86,88
	Cytochrome P-450	b	8,9,86,89–91
	Hydroxymethylglutaryl-CoA reductase	b	22
	Cholesterol 7α hydroxylase	b	22
	Lysophosphatidic acid acyltransferase	n.i.	92,93
	Fatty acid CoA ligase	n.i.	92,94
	sn-Glycerol-3-phosphate acyltransferase	n.i.	92,94
	Phosphatidic acid phosphatase	n.i.	92
	Diacylglycerol acyltransferase	n.i.	92,94
	Diacylglycerol cholinephosphotransferase	n.i.	92,94,95
	Diacylglycerol ethanolaminephosphotransferase	n.i.	92,94,95
	Phosphatidylethanolamine SAM§ methyltransferase	n.i.	95
	Cholic acid CoA ligase	n.i.	96

Table I. (continued)

Putative Location	Enzyme or Constituent	Group	References
ER luminal surface or content	Haem oxygenase	n.i.	98
	Part of protein disulfide isomerase	n.i.	97
	Glucose-6-phosphatase	c	8,9,99-102
	Glucuronosyltransferase	c	8,9,(see 29)
	Esterase	c	8,9,103
	Nucleoside diphosphatase	c	8,9,104,105
	Signal peptidase	n.i.	49
	Part of protein disulfide isomerase	n.i.	97
Rough ER cytoplas-mic surface	Ribosomes	d	6
	Ribophorins	d(?)	32
	Acyl-CoA cholesterol acyltransferase	d	106-108

(*)Unpublished results of : (a) M. Prado Figueroa ; (b) F. Pecker and M. Prado Figueroa ; (c) M. Wibo, D. Thinès-Sempoux, A. Amar-Costesec, H. Beaufay and D. Godelaine.

(+)Not investigated by density equilibration in a sucrose gradient.

(§) S-Adenosylmethionine.

are listed the enzymes and other constituents so far found
in each group.

In this list, only the b, c and d groups are recognized as bona fide constituents of the endoplasmic reticulum (ER), whereas all three a groups belong to other
membrane systems : a_1 most likely to stripped off outer
mitochondrial membranes, a_2 to plasma membranes (PM), and
a_3 to Golgi membranes. If we assume that a group enzymes
have the same specific activity in the membrane fragments
present in the microsomal fraction as they have in purified
preparations of the corresponding membranes, the contribution of each type of membrane to the total microsomal protein is that shown in Table II. After correcting further
for the small contamination by larger granules, we find that
only about 77 % of the total microsomal protein belongs to
ER. According to Losa et al. (16), who have used freeze
fracture and cytochemical staining for glucose-6-phosphatase
as recognition means, the contribution of authentic ER
elements to the microsomal fraction could be even lower,
of the order of 62-63 %.

Such estimates, although surrounded by a certain
margin of uncertainty, make it abundantly clear that microsomes are a highly heterogeneous mixture of different kinds
of membranes, and that gross mistakes can result from
taking this fraction as consisting simply of fragmented ER.
In many cases, the contamination of the microsomes by non-
ER elements could be even higher than estimated above,
since most workers discard more microsomes with the large
granules than we do, and probably isolate fractions that
are poorer than ours in rapidly sedimenting rough-surfaced

Table II.   Estimated composition of rat liver microsomal
            fraction

Component	% of Total Protein
Plasma membranes	7-8
Golgi membranes	4-5
Outer mitochondrial membranes	3
Mitochondria	6
Lysosomes	1
Peroxisomes	1
Endoplasmic reticulum	∿77

Estimates are based on the assumption that marker enzymes
for each component have same specific activity in microsomal contaminant as in purified sample. Contribution of
ER is obtained by difference (from ref. 14).

vesicles. As to preparations of so-called "smooth microsomes", they are likely to be even more contaminated.

Let it be noted also that cytochemical findings such
as that of Widnell (17) indicating the association of some
5'-nucleotidase activity with authentic ER vesicles in no
way invalidate our conclusions. A simple comparison of the
distributions shown in Fig. 1 for 5'-nucleotidase and for
glucose-6-phosphatase makes it clear that the amount of
5'-nucleotidase associated with ER must be small in comparison with the total amount present in the microsomal
fraction. Also, other PM enzymes besides 5'-nucleotidase
show the same characteristic properties of the a_2 group
(Table I).

According to our results, the bulk of the microsomal
cholesterol belongs to non-ER components. Dallner and co-
workers (18, 19), on the other hand, have reported the

presence of substantial amounts of cholesterol in what they
consider authentic ER elements, including purified rough
microsomal subfractions. We have to point out that microsomal cholesterol behaves essentially as an a_2 constituent in
density gradient analysis. Its distribution pattern does
not extend significantly beyond the 1.20 density limit,
where almost half the rough microsomes are located (Fig. 1),
and it undergoes a substantial digitonin shift (0.03 density units). On the other hand, Glaumann and Dallner (19) are
correct in pointing out that microsomes contain more cholesterol than can be accounted for by their content in PM
fragments identified by the 5'-nucleotidase activity, a
point that is confirmed by the comparative measurements of
Touster et al. (20). There is, however, no clear evidence
indicating that cholesterol and 5'-nucleotidase are similarly distributed along the periphery of hepatocytes, or
between the various populations of liver cells. As a consequence, the amount of non-PM cholesterol in microsomal
fractions is not accurately known. Most likely, Golgi
membranes (which show a moderate digitonin shift), and
VLDL particles contained in Golgi vesicles, account for
the major part of it. Small amounts of cholesterol may
occur, in addition, along the ER proper, as product and
substrate of the enzymes of cholesterol metabolism located
in ER membranes (21, 22).

Also contrasting with findings of the Dallner group
(23-25) are our results on the distribution of sialoglyco-
protein, which we find to be absent from rough ER elements
and probably also from authentic smooth ER components (26).
Differences in assay technique are probably responsible for

this discrepancy. In agreement with our results, Rodriguez Boulan et al. (27) have found no sialic acid end groups on the glycoproteins of rough ER, and have obtained no evidence to support the proposal by the Swedish group of a retrograde transport of completed glycoproteins from the Golgi complex to the ER.

BIOCHEMICAL HETEROGENEITY OF THE ENDOPLASMIC RETICULUM

The nonrandom distribution of ribosomes on the surface of the ER provides visible evidence of the biochemical heterogeneity of this system. Whether the simple dualistic distinction between rough and smooth ER suffices to account for this heterogeneity, or whether each of these two parts itself displays some sort of polarity, is not yet clear. At least, upon isopycnic centrifugation in a sucrose gradient, such a polarity is generated. As shown by Wibo et al. (6), rough vesicles become ordered along the gradient according to the number of ribosomes they bear relative to their area. It has further been found that this kind of ordering is mainly responsible for the distinction between the b and c groups. When ribosomes are stripped off from the vesicles, enzymes of the two groups show closely similar distributions (9). A simple explanation for these observations (28) is that binding of ribosomes to the membrane crowds out enzymes of the b group more than it does those of the c group (see Fig. 1).

Experiments aiming at localizing enzymes in the transverse plane of the membrane (see 2 and 29 for reviews) have yielded results that tend to support the above

hypothesis (discussed in ref. 28). They have shown, in
effect, that practically every enzyme of the b group is
at least partly accessible to nonpenetrating reagents in
intact microsomal vesicles, whereas the converse is true
of the enzymes of the c group. This correlation would seem
to indicate that ER enzymes display the b or the c pattern
of distribution in our gradient system, depending on
whether their molecule does or does not face the cytoplasm.
Relating this rule to our hypothesis requires simply that
molecules facing the cytoplasm be displaced to a greater
extent by ribosome binding than are those facing the lumen,
or contained in it (28).

Although plausible, such an assumption cannot be
taken simply for granted. Ribosome binding is not a simple
surface phenomenon, but is associated with a profound
reorganization of the ER membrane that permits the vectorial discharge of the product of protein synthesis into the
ER lumen (reviewed in 30). Thus, both faces of the membrane
must be affected. In agreement with this requirement, the
main protein species believed to be concerned with ribosome binding (31, 32),or translocation of nascent protein
(49), have been characterized as proteins with a transmembrane insertion. This does not mean, however, that they
occupy the same surface area on the cytoplasmic and on the
luminal sides of the membrane. It is not unreasonable to
assume that more space needs to be cleared on the side of
the membrane where the ribosomes are attached than on the
other. Another uncertainty is that we do not know whether
unoccupied ribophorins occur in any significant amounts on
ER membranes, and if they do, how they may be distributed
within the gradient. Our hypothesis requires that the bulk of

ribophorins should show a d type distribution pattern. That
this may well be so is suggested by the fact that ribophorins are not found in significant amounts in smooth microsomal fractions (31).

In view of the various facts considered above, one
could propose as a sort of minimal hypothesis that ribosome clustering in certain parts of the ER system, represents the main constraint that prevents the constituents
of the ER membrane from achieving a completely uniform
distribution throughout the whole system. Viewing these
clusters as anchored in the membrane, we may visualize
the b and c constituents as moving freely around them, and
to some extent also behind them (luminally oriented c
constituents) by lateral diffusion. Such a model does not,
of course, exclude the possibility that some constituents
move in the form of multi-enzyme aggregates, the existence
of which is suggested by some experiments (18, 33, 34).

Although probably too simplistic, this model has
the virtue that it accounts satisfactorily for the remarkable degree of homogeneity that is observed within each
group of constituents. It is consistent also with our
present knowledge of the biosynthesis of ER membrane
proteins. Several such proteins have been found to be
made by bound ribosomes, and then to spread from the
rough-surfaced to the smooth-surfaced part of the ER.
The rapidity of the latter process, which is measured in
minutes (see, for instance, reference 35), is highly
suggestive of free lateral diffusion. (36, 37, reviewed
in 38).

Accepting this model as a basis for discussion,

we remain faced with two key questions : a) what causes
membrane constituents to belong to a given group ? and
b) what is responsible for the nonuniform distribution
of the ribosomes on the membrane surface ?

The first question has in recent years given rise
to a great deal of experimental work, as well as speculation, and is too complex to be discussed here. It has
recently been reviewed by Wickner (39) who points out that
two main factors must be considered. One is the structural
conformation of the protein, in particular the location
of the hydrophobic sequences in the polypeptide chain,
which determines the manner in which the molecule can be
inserted into a lipid bilayer. The other is its mode of
biosynthetic delivery, which may be important in bringing
the molecule to the appropriate side of the membrane. In
addition, co- and post-translational processing, by limited
proteolysis, glycosylation and other means, probably also
plays an important role.

The second question is to some extent related to
the first one, inasmuch as it depends on a proper posi-
tioning of the ribophorins in the membrane. More specifically, however, it refers to the mechanism whereby ribosomal binding sites are prevented from drifting into the
smooth-surfaced areas of the ER, as other membrane constituents seem able to do.

The simplest answer to this question is : by the
attached ribosomes, which are themselves kept clustered
by their connecting mRNA threads. It has indeed been found
by Ojakian et al. (40) that bound ribosomes acquire lateral mobility after cleavage of mRNA by RNase. On the other

hand, these ribosomes form large, tightly packed aggregates on the surface of microsomes treated with RNase,
and evidence from the same laboratory suggests that the
ribophorins may be associated with each other to form an
"intramembranous network which may interconnect the ribo-
some-binding sites" (31).

Closely related to the above question is the problem of the morphological disposition of the two parts
of the ER. It is known that rough ER forms large flat
cisternae, whereas smooth ER generally adopts a much more
convoluted and tubular configuration. One may well ask
whether ribosome binding and/or the presence of an intramembranous ribophorin network plays a role in stabilizing
the large expanses of membrane that characterize the
rough ER. Or is it the other way round, and does the
convoluted structure of the smooth ER act as a barrier
against further spreading of the polysomes through the
system ? Arguments in favor of the former explanation
have been put forward by Kreibich et al. (31).

Recent work from our laboratory (41) has shown
that stripped rough microsomal vesicles are induced to
fuse into large cisternae when exposed to GTP, a treatment
which also stimulates certain dolichol-phosphate dependent
glycosylation reactions (42-44). The nature and possible
relationship of the morphological and biochemical changes
caused by GTP raise intriguing questions, sofar unanswered.

THE ER-GOLGI JUNCTION

A number of recent publications (11, 45-47) have

reported the occurrence in Golgi preparations of typical
ER constituents, including glucose-6-phosphatase, NADH
cytochrome $\underline{b}_5$ ($\underline{c}$) reductase, cytochrome $\underline{b}_5$, NADPH cytochrome $\underline{c}$ reductase, cytochrome P-450, and "cytoplasmically
oriented" AMPase (48). In most cases, the obvious explanation of contamination by ER vesicles has been ruled out
by what appears to be adequate morphological evidence.

Yet, other experiments specifically designed to
test the possible association of ER enzymes with typical
Golgi markers such as terminal glycosyltransferases have
given negative results. In our laboratory, a clear dissociation could be demonstrated by means of the "digitonin
shift" between, on one hand, galactosyltransferase, $\underline{N}$
acetylglucosaminyltransferase and sialyltransferase, and,
on the other hand, glucose-6-phosphatase and NADH cytochrome $\underline{c}$ reductase : the three transferases exhibited a
distinct shift, whereas the ER markers did not (10,
and unpublished results of M. Wibo and D. Godelaine).
Recently, Borgese and Meldolesi (47) have reported somewhat different results. In their hands, ER markers were
shifted somewhat by digitonin, though less than galactosyltransferase. They attribute this difference to the
fact that our Golgi preparations, isolated according to
Morré $\underline{et}$ $\underline{al}$. (50), are not as rich in small VLDL-filled
vesicles, as are their fractions, which are separated
according to Ehrenreich $\underline{et}$ $\underline{al}$. (51). While this may be
so, the fact that digitonin treatment distinguished the
two sets of markers less sharply in the latter preparation in no way constitutes proof that both sets of en-
zymes occur together $\underline{in}$ $\underline{the}$ $\underline{same}$ $\underline{piece}$ $\underline{of}$ $\underline{membrane}$. In

actual fact, convincing evidence to the contrary has been
obtained by Ito and Palade (45), who have succeeded in
separating the two sets to a considerable extent from
a GF_{1+2} fraction prepared according to Ehrenreich et al.
(51), by selectively immunoadsorbing the vesicles bearing
the ER markers on polyacrylamide beads coated with anti-
NADPH cytochrome c reductase antibody. These vesicles
were examined in the electron microscope, and were "re-
liably identified as Golgi vesicles derived from either
the trans side of the Golgi complex or the dilated rims
of Golgi cisternae". Similarly, Hino et al. (52) concluded
that NADH and NADPH cytochrome c reductase activities are
localized in a restricted region of the Golgi apparatus,
for these activities behaved differently from galactosyltransferase when Golgi fragments were subjected to
partitioning in aqueous polymer two-phase systems. Contrasting with the dissociation of galactosyltransferase
from ER markers in various fractionation systems, significant difference in the distribution of the terminal
glycosyltransferases of the Golgi apparatus has not been
observed so far (10, 53).

The simplest way out of this dilemma is by assuming
that the Golgi membranes bearing the ER markers are in
direct continuity with the ER proper, receive their
markers by lateral diffusion, and have in fact the composition of ER membranes. What they should be called then
becomes a matter of semantics, depending on whether one
attaches more significance to morphological or to biochemical criteria.

Such an explanation is consistent with the numerous

observations indicating the existence of permanent connections between the smooth ER and the Golgi system (50,
54, 55), and also with the fact that ER markers are found
mainly on components of the GF_3 fraction, believed to
originate mostly from the cis face of the Golgi apparatus
(51). The identification by Ito and Palade (45) quoted
above is somewhat at variance with this interpretation,
but can hardly be considered decisive. As shown by Claude
(55), VLDL particles are assembled in the smooth ER, from
which they appear to be conveyed to the Golgi by tubular
connections. Thus it would not be surprising to find
VLDL particles accumulating already in the ER-connected
cis part of the Golgi, especially in a system artificially engorged by ethanol administration to the animals.

Another virtue of the proposed explanation is that
it makes more readily understandable the abrupt biochemical discontinuity that seems to exist between ER-type
and Golgi-type membranes. That such a discontinuity could
survive in spite of the permanent connections between
the two domains revealed by morphological observations
requires a rigid constraint on lateral diffusion for
which so far no structural support has been demonstrated.
But this difficulty is removed if the true boundary should
be located beyond the conventional ER-Golgi junction,
somewhere within the morphologically defined Golgi complex
itself, where it could be represented by a real anatomical
discontinuity. One could imagine that it is at this level
that the energy-dependent step in the transport of secretory proteins evidenced by Jamieson and Palade (56) takes
place, and that it involves the participation of clathrin-
coated vesicles shuttling across the gap between the two

domains (see, for instance, reference 57). With rapid
enough "fusion-fission" events, intermingling of membrane
constituents could be minimized, especially under the
constraint that may be imposed by the clathrin basket.
Many recent studies on membrane recycling between Golgi,
PM and lysosomes suggest that membrane patches may retain
their individuality through a large number of cycles of
this sort (58-62). Indications of the participation of
clathrin-coated vesicles in such events have been obtained
in a number of cases (63-66).

SUMMARY AND CONCLUSION

According to the schematic model proposed in this
paper, the endoplasmic reticulum of rat hepatocytes is
visualized as a continuous closed membranous domain,
extending from the deepest recesses of the rough-surfaced
cisternae, through highly convoluted tubular smooth-
surfaced connections, up to some undefined portion of the
cis or forming part of what is conventionally described
as the Golgi apparatus.

The main factor responsible for generating biochemical heterogeneity within this system is believed
to be the segregation to its more distal parts, of the
bound ribosomes and of their binding sites and associated
factors responsible for the vectorial discharge of nascent
polypeptides into the ER lumen. Interlocking of the ribosomes by mRNA threads, and of the binding sites themselves by intramembranous connections, could play a
role in this phenomenon. Together, each string of ribosomes

with their supporting base may be seen as a relatively
stable platform moving slowly on the surface of the rough-
surfaced cisternae, as detachment of ribosomes at the 3'
tail of the mRNA, and capture of new ribosomes at its 5'
head progressively displace its anchorage points.

All other membrane constituents not linked to these
platforms are pictured as diffusing freely in the lateral
plane of the membrane, either as single molecules or as
small functional complexes, thereby filling all areas not
occupied by the polysomal platforms with two apposed homogeneous sheets, sealed together by the lipid bilayer and
by their transmembrane constituents. The inner one of
these sheets (luminal face of the membrane, c constituents)
is assumed to contribute proportionately more to the rough-
surfaced parts of the ER than does the outer sheet (cytoplasmic face, b constituents), because the polysomal platforms are wider on their surface than on their base.

This system is believed to end somewhere within
the Golgi complex, perhaps by coalescence of its terminal
tubules into one or more fenestrated plates. Stacked
against these plates, but not continuous with them, would
be a distinct membrane system characterized by a different biochemical composition, including a set of glycosyltransferases which, unlike the ER transferases, operate
without the participation of dolichol phosphate. This is
the system which, at least biochemically, would be termed
Golgi. Transfer of contents from the ER to this system
would occur at the interface without accompanying transfer
of container material, perhaps by means of an energy-
dependent mechanism involving clathrin-coated vesicles.

Such, in brief, is the hypothetical model we propose as being the simplest one consistent with presently
known facts. Reality, no doubt, will prove considerably
more complex. In particular, it is difficult to imagine
how such a crude form of biochemical organization as that
postulated by our model could account for the high degree
of morphological differentiation of the ER system.

REFERENCES

1. DePierre, J.W. and Dallner, G. (1975) Biochim.
    Biophys. Acta 415, 411-472.
2. DePierre, J.W. and Ernster, L. (1977) Annu. Rev.
    Biochem. 46, 201-262.
3. Meldolesi, J., Borgese, N., De Camilli, P. and
    Ceccarelli, B. (1978) in Membrane Fusion, (G. Poste
    and G.L. Nicolson, eds.) pp. 509-627, Elsevier/
    North-Holland Biomedical Press.
4. Amar-Costesec, A., Beaufay, H., Feytmans, E.,
    Thinès-Sempoux, D. and Berthet, J. (1969) in Micro-
    somes and Drug Oxidations, (J.R. Gillette et al.,
    eds.) pp. 41-58, Academic Press Inc., New York.
5. Thinès-Sempoux, D., Amar-Costesec, A., Beaufay, H.
    and Berthet, J. (1969) J. Cell Biol. 43, 189-192.
6. Wibo, M., Amar-Costesec, A., Berthet, J. and
    Beaufay, H. (1971) J. Cell Biol. 51, 52-71.
7. Amar-Costesec, A., Beaufay, H., Wibo, M., Thinès-
    Sempoux, D., Feytmans, E., Robbi, M. and Berthet, J.
    (1974) J. Cell Biol. 61, 201-212.
8. Beaufay, H., Amar-Costesec, A., Thinès-Sempoux, D.,
    Wibo, M., Robbi, M. and Berthet, J. (1974) J. Cell
    Biol. 61, 213-231.
9. Amar-Costesec, A., Wibo, M., Thinès-Sempoux, D.,
    Beaufay, H. and Berthet, J. (1974) J. Cell Biol.
    62, 717-745.
10. Wibo, M., Godelaine, D., Amar-Costesec, A. and
    Beaufay, H. (1976) in The Liver : Quantitative
    Aspects of Structure and Function, (R. Preisig
    et al., eds.) pp. 70-83, Editio Cantor, Aulendorf
    Publ.

11. Fowler, S., Remacle, J., Trouet, A., Beaufay, H., Berthet, J., Wibo, M. and Hauser, P. (1976) J. Cell Biol. 71, 535-550.

12. Remacle, J., Fowler, S., Beaufay, H., Amar-Costesec, A. and Berthet, J. (1976) J. Cell Biol. 71, 551-564.

13. de Duve, C. (1971) J. Cell Biol. 50, 20D-55D.

14. de Duve, C. (1976) Pontificae Academiae Scientiarum Scripta Varia, 40, 47-64.

15. Beaufay, H. and Amar-Costesec, A. (1976) in Methods in Membrane Biology, (E.D. Korn, ed.) Vol. 6, pp. 1-100, Plenum Press, New York-London.

16. Losa, G.A., Weibel, E.R. and Bolender, R.P. (1978) J. Cell Biol. 78, 289-308.

17. Widnell, C.C. (1972) J. Cell Biol. 52, 542-558.

18. Dallman, P.R., Dallner, G., Bergstrand, A. and Ernster, L. (1969) J. Cell Biol. 41, 357-377.

19. Glaumann, H. and Dallner, G. (1970) J. Cell Biol. 47, 34-48.

20. Touster, O., Aronson, N.N., Dulaney, J.T. and Hendrickson, H. (1970) J. Cell Biol. 47, 604-618.

21. Balasubramaniam, S., Venkatesan, S., Mitropoulos, K.A. and Peters, T.J. (1978) Biochem. J. 174, 863-872.

22. Mitropoulos, K.A., Venkatesan, S., Balasubramaniam, S. and Peters, T.J. (1978) Eur. J. Biochem. 82, 419-429.

23. Autuori, F., Svensson, H. and Dallner, G. (1975) J. Cell Biol. 67, 687-699.

24. Autuori, F., Svensson, H. and Dallner, G. (1975) J. Cell Biol. 67, 700-714.

25. Elhammer, A., Svensson, H., Autuori, F. and Dallner, G. (1975) J. Cell Biol. 67, 715-724.

26. Amar-Costesec, A. (1981) J. Cell Biol. (in press).

27. Rodriguez Boulan, E., Kreibich, G. and Sabatini, D.D. (1978) J. Cell Biol. 78, 874-893.

28. Amar-Costesec, A. and Beaufay, H. (1981) J. Theoret. Biol. (in press).

29. Hallinan, T. and Ronhild de Brito, A.E. (1981) in Hormones and Cell Regulation, (G. Dumont and G. Nunez, eds.) Vol. 5, Elsevier/North Holland (in press).

30. Sabatini, D.D., Kreibich, G. (1976) in The Enzymes of Biological Membranes, (A. Martonosi, ed.) Vol. 2, pp. 531-579, Plenum Publishing Co., New York.

31. Kreibich, G., Ulrich, B.L. and Sabatini, D.D. (1978) J. Cell Biol. 77, 464-487.

32. Kreibich, G., Freienstein, C.M., Pereyra, B.N., Ulrich, B.L. and Sabatini, D.D. (1978) J. Cell Biol. 77, 488-506.

33. Franklin, M.R. and Estabrook, R.W. (1971) Arch. Biochem. Biophys. 143, 318-329.

34. Ito, A. (1974) J. Biochem. (Tokyo) 75, 787-793.

35. Fujii-Kuriyama, Y., Negishi, M., Mikawa, R. and Tashiro, Y. (1979) J. Cell Biol. 81, 510-519.

36. Poo, M.M. and Cone, R.A. (1974) Nature (London) 247, 438-441.

37. Edidin, M. and Wei, T. (1977) J. Cell Biol. 75, 475-482.

38. Shinitzky, M. and Henkart, P. (1979) Int. Rev. Cytol. 60, 121-147.

39. Wickner, W. (1980) Science 210, 861-868.

40. Ojakian, G.K., Kreibich, G. and Sabatini, D.D. (1977) J. Cell Biol. 72, 530-551.

41. Paiement, J., Beaufay, H. and Godelaine, D. (1980) J. Cell Biol. 86, 29-37.

42. Godelaine, D., Beaufay, H. and Wibo, M. (1977) Proc. Natl. Acad. Sci. USA 74, 1095-1099.

43. Godelaine, D., Beaufay, H., Wibo, M. and Amar-Costesec, A. (1979) Eur. J. Biochem. 96, 17-26.

44. Godelaine, D., Beaufay, H. and Wibo, M. (1979) Eur. J. Biochem. 96, 27-34.

45. Ito, A. and Palade, G.E. (1978) J. Cell Biol. 79, 590-597.

46. Jarasch, E.D., Kartenbeck, J., Bruder, G., Fink, A., Morré, D.J. and Franke, W.W. (1979) J. Cell Biol. 80, 37-52.

47. Borgese, N. and Meldolesi, J. (1980) J. Cell Biol. 85, 501-515.

48. Farquhar, M.G., Bergeron, J.J.M. and Palade, G.E. (1974) J. Cell Biol. 60, 8-25.

49. Walter, P., Jackson, R.C., Marcus, M.M., Lingappa, V.R. and Blobel, G. (1979) Proc. Natl. Acad. Sci. USA 76, 1795-1799.

50. Morré, D.J., Hamilton, R.L., Mollenhauer, H.H., Mahley, R.W., Cunningham, W.P., Cheetham, R.D. and Lequire, V.S. (1970) J. Cell Biol. 44, 484-491.

51. Ehrenreich, J.H., Bergeron, J.J.M., Siekevitz, P. and Palade, G.E. (1973) J. Cell Biol. 59, 45-72.

52. Hino, Y., Asano, A. and Sato, R. (1978) J. Biochem. (Tokyo) 83, 935-942.
53. Bretz, R., Bretz, H. and Palade, G.E. (1980) J. Cell Biol. 84, 87-101.
54. Fleischer, B., Fleischer, S. and Ozawa, H. (1969) J. Cell Biol. 43, 59-79.
55. Claude, A. (1970) J. Cell Biol. 47, 745-766.
56. Jamieson, J.D. and Palade, G.E. (1968) J. Cell Biol. 39, 589-603.
57. Rothman, J.E., Bursztyn-Pettegrew, H. and Fine, R.E. (1980) J. Cell Biol. 86, 162-171.
58. Schneider, Y.-J., Tulkens, P., de Duve, C. and Trouet, A. (1979) J. Cell Biol. 82, 466-474.
59. Muller, W.A., Steinman, R.M. and Cohn, Z.A. (1980) J. Cell Biol. 86, 304-314.
60. Stahl, P., Schlesinger, P.H., Sigardson, E., Rodman, J.S. and Lee, Y.C. (1980) Cell 19, 207-215.
61. Ottosen, P.D., Courtoy, P.J. and Farquhar, M.G. (1980) J. Exp. Med. 152, 1-19.
62. Herzog, V. and Reggio, H. (1980) Eur. J. Cell Biol. 21, 141-150.
63. Pearse, B.M.F. (1975) J. Molec. Biol. 97, 93-98.
64. Pearse, B.M.F. (1976) Proc. Natl. Acad. Sci. USA 73, 1255-1259.
65. Goldstein, J.L., Anderson, R.G.W. and Brown, M.S. (1979) Nature 279, 679-685.
66. Bretscher, M.S., Thomson, J.N. and Pearse, B.M.F. (1980) Proc. Natl. Acad. Sci. USA 77, 4156-4159.
67. Schnaitman, C., Erwin, V.G. and Greenawalt, J.W. (1967) J. Cell Biol. 32, 719-735.
68. Sottocasa, G.L., Kuylenstierna, B., Ernster, L. and Bergstrand, A. (1967) J. Cell Biol. 32, 415-438.
69. Novikoff, A.B. and Essner, E. (1962) Fed. Proc. 21, 1130-1142.
70. Emmelot, P., Bos, C.J., Benedetti, E.L. and Rümke, Ph. (1964) Biochim. Biophys. Acta 90, 126-145.
71. El-Aaser, A.A., Reid, E., Klucis, E., Alexander, P. Lett, J.T. and Smith, J. (1966) Natl. Cancer Inst. Monographs 21, 323-331.
72. Coleman, R. and Finean, J.B. (1966) Biochim. Biophys. Acta 125, 197-206.
73. Essner, E., Novikoff, A.B. and Masek, B. (1958) J. Biophys. Biochem. Cytol. 4, 711-716.

74. Emmelot, P. and Bos, C.J. (1962) Biochim.
     Biophys. Acta 58, 374-375.
75. Bock, K.W., Siekevitz, P. and Palade, G.E.
     (1971) J. Biol. Chem. 246, 188-195.
76. Amar-Costesec, A. and Beaufay, H. (1977) Arch.
     Int. Physiol. Biochim. 85, 949-950.
77. Marinetti, G.V., Ray, T.K. and Tomasi, V. (1969)
     Biochem. Biophys. Res. Commun. 36, 185-193.
78. Smith, G.D. and Peters, T.J. (1980) Eur. J.
     Biochem. 104, 305-311.
79. Wattiaux-De Coninck, S. and Wattiaux, R. (1969)
     Biochim. Biophys. Acta 183, 118-128.
80. Fleischer, B. and Fleischer, S. (1970) Biochim.
     Biophys. Acta 219, 301-319.
81. Morré, D.J., Merlin, L.M. and Keenan, T.W.
     (1969) Biochem. Biophys. Res. Commun. 37, 813-
     819.
82. Schachter, H., Jabbal, I., Hudgin, R.L., Pinteric,
     L., McGuire, E.J. and Roseman, S. (1970) J. Biol.
     Chem. 245, 1090-1100.
83. Wagner, R.R. and Cynkin, M.A. (1969) Biochem.
     Biophys. Res. Commun. 35, 139-143.
84. Strittmatter, P., Rogers, M.J. and Spatz, L.
     (1972) J. Biol. Chem. 247, 7188-7194.
85. Nilsson, O.S. and Dallner, G. (1977) J. Cell
     Biol. 72, 568-583.
86. Nilsson, O.S., DePierre, J. and Dallner, G.
     (1978) Biochim. Biophys. Acta 511, 93-104.
87. Enoch, H.G., Catala, A. and Strittmatter, P.
     (1976) J. Biol. Chem. 251, 5095-5103.
88. Morimoto, T., Matsuura, S., Sasaki, S., Tashiro,
     Y. and Omura, T. (1976) J. Cell Biol. 68, 189-
     201.
89. Welton, A.F. and Aust, S.D. (1974) Biochim.
     Biophys. Acta 373, 197-210.
90. Matsuura, S., Fujii-Kuriyama, Y. and Tashiro,
     Y. (1978) J. Cell Biol. 78, 503-519.
91. Cooper, M.B., Craft, J.A., Estall, M.R. and
     Rabin, B.R. (1980) Biochem. J. 190, 737-746.
92. Coleman, R. and Bell, R.M. (1978) J. Cell Biol.
     76, 245-253.
93. Coleman, R.A. and Bell, R.M. (1980) Biochim.
     Biophys. Acta 595, 184-188.
94. Ballas, L.M. and Bell, R.M. (1980) Biochim.
     Biophys. Acta 602, 578-590.
95. Vance, D.E., Choy, P.C., Farren, S.B., Lim, P.H.
     and Schneider, W.J. (1977) Nature 270, 268-269.

96. Polokoff, M.A., Coleman, R.A. and Bell, R.M. (1979) J. Lipid Res. 20, 17-21.
97. Ohba, H., Harano, T. and Omura, T. (1977) Biochem. Biophys. Res. Commun. 77, 830-836.
98. Hino, Y.,Asagami, H. and Minakami, S. (1979) Biochem. J. 178, 331-337.
99. Leskes, A., Siekevitz, P. and Palade, G.E. (1971) J. Cell Biol. 49, 264-287.
100. Arion, W.J., Wallin, B.K., Lange, A.J. and Ballas, L.M. (1975) Mol. Cell. Biochem. 6, 75-83.
101. Nilsson, O.S., Arion, W.J.,DePierre, J.W., Dallner, G. and Ernster, L. (1978) Eur. J. Biochem. 82, 627-634.
102. Zoccoli, M.A. and Karnovsky, M.L. (1980) J. Biol. Chem. 255, 1113-1119.
103. Akao, T. and Omura, T. (1972) J. Biochem. (Tokyo) 72, 1245-1256.
104. Ernster, L. and Jones, L.C. (1962) J. Cell Biol. 15, 563-578.
105. Kuriyama, Y. (1972) J. Biol. Chem. 247, 2979-2988.
106. Venkatesan, S., Mitropoulos, K.A., Balasubramaniam, S. and Peters, T.J. (1980) Eur. J. Cell Biol. 21, 167-174.
107. Lichtenstein, A.H. and Brecher, P. (1980) J. Biol. Chem. 255, 9098-9104.
108. Hashimoto, S. and Fogelman, A.M. (1980) J. Biol. Chem. 255, 8678-8684.

LIPID CARRIERS IN MICROSOMAL MEMBRANES

Gustav Dallner and Frank W. Hemming

Department of Biochemistry, University of Stockholm,
Department of Pathology, Karolinska Institutet,
Stockholm, Sweden
and
Department of Biochemistry, Medical School, University
of Nottingham, Nottingham, England

Glycoproteins are present in all organs and membranes and, consequently, their biosynthesis is of considerable interest. A large amount of experimental data demonstrates the participation of lipid intermediates in the synthesis of the oligosaccharide chain, but at present very limited information exists concerning lipid intermediate-mediated glycosylation of individual and characterized protein species. Most intracellular membranes contain chiefly or exclusively proteins with oligosaccharide chains involving a N-glycosidic linkage between asparagine and N-acetylglucosamine (GlcNAc) and such binding is also present in most secretory proteins. The two most common types of oligosaccharide structures are shown in Fig. 1.

C. P. Lee, G. Schatz, G. Dallner (eds.), Mitochondria and Microsomes
in honor of Lars Ernster   ISBN 0-201-04576-1

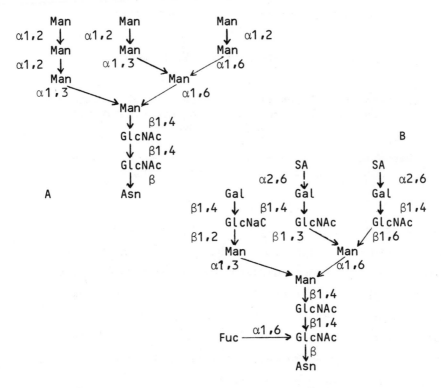

Fig. 1. Structure of two common oligosaccharides
         A) High mannose type (ovary cell membrane)
         B) Complex type (thyroglobulin B)

BIOSYNTHESIS OF THE CORE PART

The biosynthetic pathway for the complex oligosaccharides is now well established, at least for certain secretory proteins and proteins transported to the cell surface
(1-4). Initiation of the core takes place by the transfer
of GlcNAc-1-P from the nucleotide-activated form to dolichol-
P and the accumulation of subsequent sugars occurs on the
pyrophosphate molecule (Fig. 2). The following addition of
1 GlcNAc, 1 mannose in β-linkage and 4 mannose in α-linkage
occurs directly from the nucleotide-activated form. Only

P-Dolichol
$\downarrow\!\leftarrow\!$UDP-GlcNAc
GlcNAc-PP-Dolichol
$\downarrow\!\leftarrow\!$UDP-GlcNAc
$(GlcNAc)_2$-PP-Dolichol
$\downarrow\!\leftarrow\!(GDP\text{-}Man)_5$
Man_5-$(GlcNAc)_2$-PP-Dolichol
$\downarrow\!\leftarrow\!(Man\text{-}P\text{-}Dolichol)_4$
Man_9-$(GlcNAc)_2$-PP-Dolichol
$\downarrow\!\leftarrow\!(Glc\text{-}P\text{-}Dolichol)_3$
Glc_3-Man_9-$(GlcNAc)_2$-PP-Dolichol

Fig. 2. Assembly of core oligosaccharide on dolichol-PP

the last 4 mannose residues and the terminal 3 glucose residues are transferred from phosphoryl dolichol.

PROCESSING AND COMPLETION

The core of the oligosaccharide chain is transferred
to the protein, where it is bound in the case of N-glycosidic binding, to asparagine in the Asn-X-Thr(or Ser) tripeptide sequence (Fig. 3). It is generally accepted that most
if not all glycosylation occurs while the polypeptide is
still attached to the large subunit (5). The fact that in
hen oviduct lipid-linked glycosyltransferases are enriched
in rough microsomes does not exclude the possibility that
some glycosylation may also be initiated in the smooth ER in
other organs (6). Hen oviduct smooth microsomes are probably composed to a large extent of Golgi and plasma membrane
elements, while in liver smooth microsomes are highly enriched in components of the endoplasmic reticulum. Therefore, it is possible that the transfer of the oligosaccharide core from diphosphoryl dolichol may occur in smooth
microsomes in those cases where this fraction consists predominantly of elements of the endoplasmic reticulum (7).

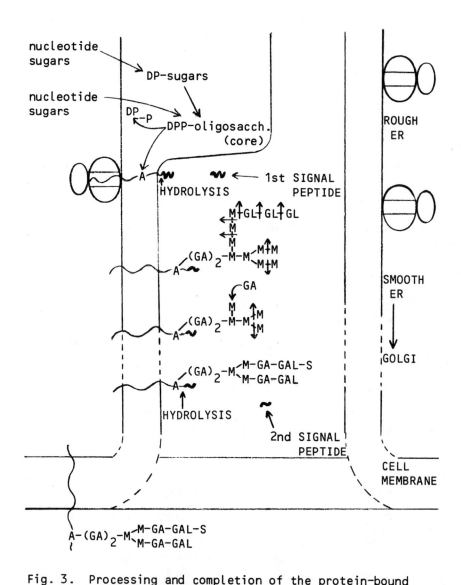

Fig. 3. <u>Processing and completion of the protein-bound</u>
<u>oligosaccharide in the ER-Golgi system</u>

DP = dolichol-P; DPP = dolichol-PP; A = asparagine;
GA = N-acetylglucosamine; M = mannose; GL = glucose;
GAL = galactose; S = sialic acid.

Those proteins which are produced at the bound ribosomal level and transported to the cell membrane and out
from the cell are processed and completed along the line
given in Fig. 3. This illustration depicts a specific sequence of events, e.g., an organized, consecutive interaction of transferases and glycosidases is required. It is
well established that both rough and smooth microsomes are
heterogeneous and individual enzymes are enriched in separate subfractions (8). Transferases and glycosidases are
very probably enriched in certain regions of the ER and that
may be one of the mechanisms behind the ordered processing and completion of the oligosaccharide chain and the resulting transport process. However, to establish this
principle is a difficult task. For all individual sugars
there are several transferases, depending on the type of the
binding to be established and the nature of the acceptor
(9). To test the lateral distribution of individual glyco-
syltransferase it will be absolutely necessary to use well
specified - probably endogenous - acceptors and to characterize the product chemically.

The situation is even more complicated with the Golgi-
membranes. This fraction possesses high activities of a
number of glycosidases and represents the exclusive localization of some terminal glycosylation reactions (10). One
would expect very high specialization and heterogeneity in
this fraction. Investigations along these lines have so
far established an increasing capacity for glycosylation
from the cis to the trans direction (from the ER to the
plasma membrane). However, a heterogeneous distribution of
the individual transferases was not observed (11,12). Here
again one should try to prepare specific acceptors which
may reveal a heterogeneous localization of enzymes in the

lateral plane. Such investigations meet with additional
problems in the case of Golgi, e.g., extensive lipid peroxidation (13), the uncertainty of dolichol involvement in
some type of reactions and the unsolved problem as to how
large and highly charged nucleotide-sugar molecules penetrate the Golgi membranes.

MEMBRANE PROTEINS

   All our knowledge on protein glycosylation at the
cellular level is based on studies of secretory proteins or
proteins which are in a way secretory for the ER-Golgi system, like those of the plasma membrane or viral proteins.
A real possibility is that the membrane proteins of the ER
system are glycosylated in a different manner than the
secretory proteins. Protein-bound sugars in limited amounts
(1-2 % of the dry weight) are present in endoplasmic membranes (14). Only few glycoproteins in the ER system have
been functionally identified to date, but a number of reports indicate that several microsomal enzymes both on the
cytoplasmic and luminal surfaces belong to this group
(15-20).

   In contrast to secretory proteins membrane-bound proteins possess probably only one short oligosaccharide chain
per molecule protein, e.g., in the case of cytochrome P-450
one glucosamine and two mannoses (18). The lack of the di-
acetylchitobiose residue on cytochrome P-450 indicates that
for this enzyme, and probably for most of the other membrane-bound enzymes, the established dolichol-mediated
pathway is not involved. Also, the fact that lysosomal enzymes may contain glucose is inconsistent with a biosynthetic pathway which is common with that of the secretory pro-

teins. The limited amount and low turnover of membrane enzymes make it difficult to study the mechanism of glycosylation of these proteins.

One explanation for the differences between the oligosaccharide chains of secretory and membrane proteins may be
that sugars in the non-secretory proteins function as signals governing their transport to the final destination.
Lysosomal enzymes appear to receive, already in the rough
ER, glucosamine-1-phosphate on a terminal mannose, followed
by the hydrolysis of the amino sugar to leave a terminal
mannose-6-phosphate on the molecule (21,22). This sugar
interacts with receptors at the luminal side of the membrane
channels and the enzyme is directed to the lysosomes, where
dephosphorylation and, possibly, completion are carried out.
Terminal sugar or sugars may play the role of a signal
function, like the peptide signals, and it is possible
that all microsomal glycoprotein enzymes contain this type
of recognition mechanism.

GLYCOSYLATION IN OTHER INTRACELLULAR MEMBRANES

Dolichol and dolichol-mediated glycosylation is obviously present not only in the endoplasmic reticulum-Golgi-
system but in a number of other or in all cellular membranes.

Mitochondria are the site of dolichol synthesis (23,24)
and both phosphoryl dolichol and glycosyl transferases are
found here (25,26). The outer membrane is probably much
more active in glycosylation than the inner membrane and
obviously most of the glycosyl transfer reactions take
place with the involvement of phosphoryl dolichol. The mitochondrial fraction is always to some extent contaminated

with microsomes and, therefore, one may question the validity of the mitochondrial localization. However, the specific activity of some reactions of the glycosylating pathway is as high here as in microsomes, which excludes the
possibility that the only source of mitochondrial transferases is microsomal contamination (27). The high capacity
of the outer mitochondrial membrane for dolichol-mediated
glycosylation may be necessary to initiate and probably
also to complete the glycosylation of some proteins which
enter the mitochondria after synthesis on free ribosomes.
The transport of the polypeptide through a defined point on
the outer membrane (temporary channel) may require a streth-
ching of the peptide, thereby eliminating the steric hind-
er for N-glycosylation of the suitable asparagine residue.

Another membrane which has also been investigated in
some detail is the plasma membrane (28,29). Beside their
role in cell-cell contact, glycosyl transferases, some of
which are dolichol associated, may perform a number of reactions for completion of the oligosaccharide chain. Repair of the sialidated surface glycoproteins occurs easily
and this function may also involve other sugars. Sugar
substrates are present in the blood, which also raises the
possibility that the glycosyl transferases of the blood
participate, under certain conditions, in completion of
hydrolytically damaged sugar sequences.

Dolichol appears to be a component of several cellular
membranes and it is very possible that a number of glycosyl
transferases also have a broad distribution. Difficulties
in isolation, low recovery and relatively high contamination may, however, cause a sizeable problem in the evaluation of the results. To establish that a certain type of
glycosylation occurs in an organelle which is isolated in

small quantities requires evidence in addition to the measurement of sugar transfer to an acceptor with broad specificity.

TRANSVERSE DISTRIBUTION OF GLYCOPROTEINS

The distribution of dolichols within membranes is not
known. It is difficult to devise an approach to study this
problem, since the size of the dolichol (around 100 carbon
atoms in fully extended form) would allow this substance to
span the membrane. Consequently, it is important to know
the three dimensional arrangement of dolichol in the membrane. In spite of the lack of experimental evidence concerning the position of dolichol the data summarized below
indicate that phosphoryl dolichol-sugar interaction has a
broad distribution and occurs on both sides of the microsomal membrane.

The final products of the glycosylation, the protein-
bound oligosaccharide chains, are present in rough and
smooth microsomes and also in Golgi membranes, both at the
cytoplasmic and the luminal surface (14,30). Available
data indicate clearly that most carbohydrates of the plasma
membrane are found on the outer surface (31). In contrast
to the carbohydrate residues of the other membranes, the
sugar moieties at this location can be related to functional properties such as antigenicity or receptor capacity.
Experimental evidence also demonstrates that a considerable
portion of the protein-bound carbohydrates of the outer
mitochondrial membrane (32), lysosomes (33) and chromaffin
granula (34) are also on the outer, cytoplasmic surface.

Concerning the endoplasmic reticulum we can differentiate three types of glycoproteins: 1) the membrane type
which remains at the site or close to the site of synthe –

sis; 2) the membrane type which is transported to other organelles and which is a sort of secretory protein; 3) secretory proteins which are synthesized in the ER but shortly
after transported out of the cell. It appears that secretory proteins in many cases possess long, branching oligosaccharide chains and, consequently, they are carbohydrate
rich; while membrane proteins often contain only a few sugar residues and, consequently, are carbohydrate poor.
Furthermore, secretory proteins are synthesized in large
amounts and have a short half life in the ER-system (15-30
min), in contrast to membrane proteins of the ER, which are
present in small amounts and have a half life of 3-4 days.
On the basis of these facts, one would expect that a majority of the glycosylation reactions which take place in an
experimental system involve secretory proteins. Surprisingly, this is not the case. Immunoprecipitation of secretory proteins after in vitro glycosylation reveals that
only 20 % of the sugar is transferred to secretory proteins.
Usually in an in vitro system no protein synthesis takes
place and probably in the case of secretory proteins the
amount of acceptor available is extremely limited. It is
possible that N-glycosylation of membrane proteins occurs
not only in the early phase of nascent protein discharge
from ribosomes, but also later in both rough and smooth ER
membranes.

TRANSVERSE DISTRIBUTION OF GLYCOSYL TRANSFERASES

The different types of proteins glycosylated both at the
inner and outer surface of the membrane raise the possibility that different glycosylating systems exist in various
compartments of the membrane and the individual systems are

devoted to glycosylation of different types of proteins.
The nucleotide-activated sugars are present in the cytoplasm and they cannot penetrate the microsomal membranes.
Some of the glycosyl transferases, at least those involved
in the initial phase of glycosylation, must be at the cytoplasmic surface in order to interact with the substrate.
However, the complex nature of the reactions taking place
in the microsomes is suggested by a number of experiments
(35-39). In the case of rough microsomes: a) Inhibitors of
the synthesis of the phosphoryl dolichol-sugar intermediate
(tunicamycin, amphomycin) inhibited only about half of the
protein glycosylation. b) Surface probes (proteolysis,
diazobenzene sulfonate) inhibit a part of the sugar transfer to phosphoryl dolichol and also to protein. c) The
glycosylation remaining after treatment with surface probes
is partially inhibited by tunicamycin and amphomycin.
d) Surface probes also inhibit protein glycosylation when
phosphoryl dolichol sugars are used as donors. These types
of experiments indicate that the majority of, but not all
glycosylation in rough microsomes involves phosphoryl dolichol, that glycosyl transferases are present both on the
outer and the inner surfaces of the microsomal membranes,
and that several compartments of dolichol may exist.

The situation with smooth microsomes is different.
a) Inhibitors of phosphoryl dolichol-sugar interaction do
not inhibit protein glycosylation. b) Surface probes inhibit protein glycosylation only to a small extent.
c) Phosphoryl dolichol-sugars are substrates to some extent
for protein glycosylation and this reaction is partially
inhibited by surface probes. Judging from such results,
the main glycosylating reactions in smooth microsomes do not
seem to involve dolichol phosphate and the enzymes trans-

ferring sugar residues to the protein are not associated
with that part of the cytoplasmic surface of the membrane
which is affected by surface probes. Dolichol phosphate is
present in smooth microsomes and can also participate in
protein glycosylation, but the exact nature of this reaction remains to be elucidated.

CHEMISTRY, ASSAY AND DISTRIBUTION OF DOLICHOLS

Dolichols (Dolikos, Gr: long-alcohols) are members of
the group of polycis-isoprenoid alcohols (Fig. 4). Detailed

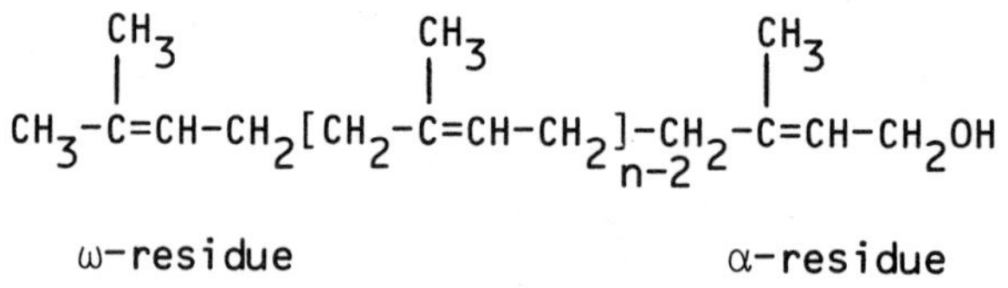

$$CH_3-\underset{\underset{CH_3}{|}}{C}=CH-CH_2[CH_2-\underset{\underset{CH_3}{|}}{C}=CH-CH_2]_{n-2}-CH_2-\underset{\underset{CH_3}{|}}{C}=CH-CH_2OH$$

ω-residue α-residue

Fig. 4. General structure of polyisoprenoid alcohols

reviews (40,41) conclude that dolichols differ from other
polyprenols in having a saturated (2,3-dihydro)α-residue
and high value for n (14-22). In animal dolichols all but
the three isoprene residue at the ω-end of the chain and
its α-residue have the cis configuration (CH_3 relative to
H across the double bound). From any one source the range
in value of n is small giving rise to a family of isoprenologues spanning a difference in size of four or five residues. The composition of the family of dolichols varies
from one species to another but within the species variation among tissues is relatively minor and not always consistent.

The most sensitive and straightforward assay for un-
esterified polyprenols is by high performance liquid chrom-

Table I. Distribution of dolichol in vertebrate tissue

Tissue	Concentr.[a] µg/g	Main[b] components	Ref.
Adrenal, human	1273 (20)	ND	46
Blood, total, rat	<2	ND	50
" plasma, chicken	<0.5	ND	44
" erythrocytes, "	<0.5	ND	44
Bone marrow, rat	<2	ND	50
Brain, rat	17	18,19,17	43,50
Heart, human	262	ND	46
" rat	8	18,19,17	43
Kidney, human	13 (<5)	ND	46
" rat	240 (12)	18,19,17	43,50
Liver, chicken	182	ND	44
" human	1226 (11)	20,21,19	46,49
" pig	69-129 (52-63)	19,20,18[c]	23,44
" rabbit	10-40	18,19,17,20	d
" rat	23	18,19,17	43
Lung, human	82	ND	46
" rat	14	18,19,17	43,50
Muscle, rat	4	18,19,17	43
Oviduct, chicken	164-316	ND	44
Pancreas, human	943 (20)	ND	46
" rat	26	ND	50
Pituitary, bovine	130-170 (25)	20,19,18[e]	47,48
" human	1400 (10)	ND	47
Prostate, human	268	ND	46
Skin, rat	<2	ND	50
Small intestine, rat	14	ND	50
Spleen, human	161	ND	46
" rat	100 (25)	18,19,17	43,50
Testis, human	3226 (3)	ND	46
" rat	11	18,20,19	43
Thyroid, bovine	200 (5)	ND	45
" human	1145 (5)	ND	46

a: values in brackets give dolichol esterified to fatty
   acid (% of total)
b: value of n in Figure 4, main component underlined
c: also contains small quantities of dolichol-11 (51)
d: Pawson, S. and Hemming, F.W. (1979) unpublished results
e: also contains 25 µg/g decaprenol (48)

atography (HPLC) using a UV detector set at 210 nm (42).
The distribution of dolichol in tissues of vertebrates is
summarized in Table I. Assays reported in reference 43
used the method described above but all others used somewhat less sensitive procedures, especially those appearing
in the older reports. It can be seen that considerable
variation in tissue content occurs between species and that
human tissues are particularly rich in dolichol.

Dolichols have been found in all eucaryotic cells so
far investigated. These include invertebrates, plants,
fungi and yeast. In marine invertebrates the dolichols-19
and -20 are probably the major comparents of the family
present (52). Dolichols of Saccharomyces cerevisiae are
composed mainly of dolichols-14,15 and -16 with dolichol-15
predominating (42). The main isoprenologue in higher
plants is probably larger than this, n being close to 20
(53). In some filamentous fungi the value of n ranges from
21 to 23 and the ω- and ω-1-isoprene residues may be further
substituted (40,41).

PHOSPHORYLATED DERIVATIVES OF DOLICHOL

It was shown earlier that phosphoryl dolichol functions
as a coenzyme in protein N-glycosylation. During the process mono-glycosylated derivatives of monophosphoryl dolichol are formed and oligoglycosyl diphosphoryl derivatives
of dolichol are built up as intermediates in the phosphoryl dolichol pathway (Fig. 2). Diphosphoryl dolichol is an
intermediate both in the phosphoryl dolichol pathway and
also probably in the biosynthesis of dolichol and of monophosphoryl dolichol. All of the glycosylated derivatives
contain a sugar-1-phosphate link which hydrolyses upon mild

treatment with acid to liberate the mono- and diphosphoryl derivatives of dolichol. Slightly stronger treatment with acid converts diphosphoryl to monophosphoryl dolichol. It follows that treatment with acid of lipid extracts of tissues will convert all phosphorylated derivatives of dolichol to monophosphoryl dolichol. The amount of monophosphoryl dolichol so released in tissue is very small. For example in pig liver the concentration is approximately 0.1 µM (140 µg/kg tissue). Currently there is no reliable sensitive assay for monophosphoryl dolichol. Useful HPLC systems have not yet been devised. Enzymic assays in which monophosphoryl dolichol is used as an acceptor for glucosyl or mannosyl transferases have been used but are rarely reliable when used to assay tissue extracts - probably because other components of the extracts effect a change in enzymatic activity and also complicate solubilization of the acceptor. However, using this approach in studies on subcellular distribution, the concentration of monophosphoryl dolichol has been shown to be highest in nuclear, Golgi and rough endoplasmic membranes (25).

The widespread occurrence of phosphoryl dolichol in different cell membranes and the range of isoprenologues concerned raised the possibility of the specificity of different isoprenologues for different cell fractions or even for different glycosyl transferases. At the moment there is no firm evidence for cell fractions or enzymes being capable of distinguishing one endogenous dolichol from another. However, it has been demonstrated (54) that glycosyl transferases of different cell fractions of rat liver are able to differentiate between phosphoryl dolichol-11 and phosphoryl dolichol-17 when presented exogenously.

THE FORMATION OF DOLICHOL AND PHOSPHORYL DOLICHOL

Several features of the biosynthetic pathway of phosphoryl dolichol are summarized in Fig. 5. It has been

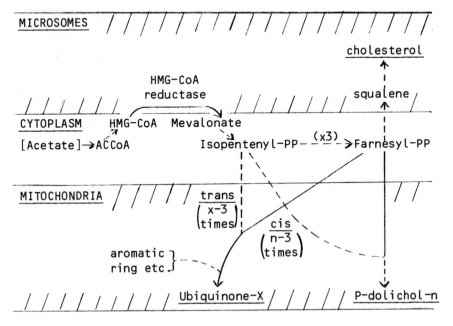

Fig. 5. Diagrammatic representation of the relationships between the biosynthetic pathways of phosphoryldolichol, cholesterol and ubiquinone.

shown that in rat liver dolichols can be formed from mevalonate by a process that involves cis-addition of isoprene residues to trans, trans-farnesyl pyrophosphate (55). Acetate can also be incorporated into dolichol and into phosphoryl dolichol (43). It appears that the early steps in the biosynthesis of cholesterol are also common to the biosynthesis of farnesyl pyrophosphate that is to act as a precursor of both dolichol and ubiquinone. However, whether or not separate pools of these intermediates exist, one for each pathway, is not yet clear. It is reported

that in liver the post-farnesyl pyrophosphate stages in the
synthesis of both ubiquinone (56) and phosphoryl dolichol
(24) are concentrated in the mitochondria; the former being
formed in the inner membrane and the latter in the outer
membrane. Although the specific activity of the phosphoryl
dolichol synthetase in microsomal fractions was lower than
in mitochondria the total synthetase activity in microsomes was still significant (24) making it uncertain whether
or not a mechanism for transport of phosphoryl dolichol
from mitochondria to microsomes needs to be considered. A
similar subcellular distribution for phosphoryl dolichol
synthetase has been reported in the pea (57) and in the
alga Prototheca zopfii (53).

   Fig. 6 describes the interrelationships of compounds
that readily give rise to monophosphoryl dolichol. Most

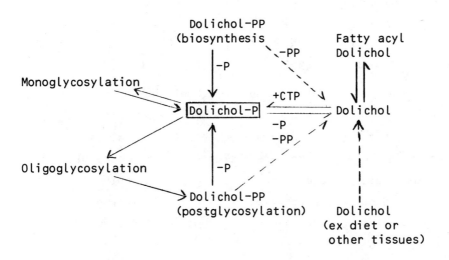

Fig. 6. Pathways leading to the formation of monophos-
         phoryl dolichol.

probably this is formed from the biosynthetic route by the
direct removal of one phosphate from diphosphoryl dolichol.

An enzymic activity catalyzing this step has been described in both lymphocytes (58) and nervous tissue (59) but its specificity is not yet established. The same groups also reported on phosphatase activity capable of liberating dolichol from monophosphoryl dolichol.

That dolichol kinase is CTP specific was first described by Allen (60). Later Waechter reported (61) that incubation of microsomal vesicles with CTP in the absence of detergent led to the phosphorylation of endogenous dolichol, suggesting that both the dolichol and kinase were at the cytoplasmic face of the vesicles. The monophosphoryl dolichol so formed is capable of being glycosylated (61) and is presumably closely associated with the enzymes of the dolichyl phosphate (phosphoryl dolichol) cycle. Hepatocytes have been shown to take up a short chain dolichol from liposomes and catalyze its phosphorylation and glycosylation (62).

FURTHER REACTIONS OF DOLICHOL AND RELATED COMPOUNDS

The essential role of phosphoryl dolichol in protein N-glycosylation has been discussed. The pathway (Fig. 2) illustrates the specificity of phosphoryl dolichol as an acceptor of sugars. The direct transfer of mannose, glucose and N-acetylglucosamine phosphate are the only well-established natural steps that utilize this coenzyme although there is evidence for the formation of its xylose derivative (63) in oviduct and disputed reports of the transfer of galactose (64). The further transfer of sugars from these glycosylated intermediates is also highly specific. The en bloc transfer of the oligosaccharide specifically only to an asparagine residue in the sequence Asn-X-

Thr(Ser) of a β-loop of a peptide chain has been described
although not all asparagines in this environment are glycosylated (65). The glucosylated coenzyme appears to utilize
only the oligosaccharide intermediate as an acceptor while
the mannosylated coenzyme may donate its sugar to a shorter
chain oligosaccharide derivative or, in yeast and fungi,
to a serine or threonine residue of a peptide chain.

In fact the specificity of the enzymes transferring
mannose to and from phosphoryl dolichol is not sufficient
to exclude transfer of 2-deoxyglucose from GDP-2-deoxyglu-
cose (66). It is this fact that allows 2-deoxyglucose to
act as an inhibitor of protein glycosylation. The glycosyl
transferases utilizing phosphoryl dolichol as a lipid
acceptor also show some specificity to the nature of the
lipid. Early studies with membrane preparations of pig
liver showed that a fairly wide range of polyprenols could
substitute for dolichol. More recent work has established
that these transferases function best if the polyprenol is
reasonably long and has a saturated α-residue, there also
being a slight preference for a poly-cis-isoprenoid chain.
These same specificities were particularly important in the
further transfer of its sugar (or oligosaccharides) from
the glycosylated coenzyme to other acceptors (67-68). The
possibility has been raised that the glycosylation of phosphoryl retinol (and of a metabolite of retinoic acid) is a
consequence of the broad specificity of glycosyl transferases for phosphoryl dolichol. However, the formation in liver
of the mannosyl derivatives of the liver lipids has been
reported to be catalyzed by different enzymes (69). The
role of mannosyl phosphoryl retinol remains to be established.

Table I shows that in some tissues much of the dolichol

is esterified with fatty acid. Early work showed that in
pig liver and in Aspergillus fumigatus the ester is more
widely distributed among cell organelles than is unesterifi-
ed dolichol, but is concentrated in a crude nuclear and
cell debris fraction (23,70). These observations need extending using current methods. The Aspergillus fatty acyl
dolichols differed from fatty acyl esters of ergosterol
both in subcellular distribution and in containing a higher
proportion of saturated fatty acids (70). In mammalian
cells the fatty acyl residues are transferred directly to
dolichol from phospholipids especially lecithins (71). The
transacylase concerned showed a preference for short-chain
polyprenols and for α-saturated over unsaturated or multip-
ly saturated versions (72). The enzyme can be found in
liver, spleen, kidney, pancreas, brain and intestinal muco-
sa. The activity is highest in liver where it is present in
all membrane fractions, a microsomal fraction being the
richest source. No function other than a storage form of
dolichol has been suggested for these esters. Esterases
active upon fatty acyl dolichols have not been described
although clearly regulation of their activity as well as
that of the transacylase activity could be important in
controlling the relative concentrations of dolichol and
fatty acyl derivatives. The literature is also devoid of
reports of studies on the enzymic degradation of dolichols.

REGULATION OF THE CONCENTRATION OF PHOSPHORYL DOLICHOL

Because of its key role in protein N-glycosylation and
its low concentration interest has quickened in the regulation of the concentration of phosphoryl dolichol and its
role in the control of protein glycosylation. HMGCoA re-

ductase (see Fig. 5) is known to be reduced in cells treated with 25-hydroxycholesterol. This compound when administered to cultured smooth muscle cells was shown to cause a
reduced biosynthesis of phosphoryl dolichol with a consequential reduction in protein N-glycosylation (73). It has
since been shown that in L-cells sterol biosynthesis may be
inhibited by as much as 75 % before the pathway leading to
phosphoryl dolichol is affected suggesting a control point
in these cells after farnesyl pyrophosphate (74).

HMGCoA reductase is also susceptible to competitive
inhibition by compactin. The presence of this compound in
the growth medium of developing embryos of the sea urchin
caused a fall in the concentration of phosphoryl dolichol
and in the extent of protein N-glycosylation (75). Concurrently, a block at a specific gastrulation stage in the
development of the embryos occurred. Development and glycosylation returned to normal when dolichol or phosphoryl
dolichol, but not ubiquinone or cholesterol, were added to
the medium.

Hypercholesterolaemia in rats causes a reduced activity
both in hepatic HMGCoA reductase and in a step between farnesyl pyrophosphate and cholesterol, probably squalene
synthetase (76). Since normally in liver the amount of
farnesyl pyrophosphate utilized for synthesis of phosphoryl
dolichol is less than 1 % of that converted to cholesterol,
partial inhibition of these two enzymes results in more
farnesyl pyrophosphate being available for phosphoryl dolichol synthesis than normal. It is thus not surprising to
find that hypercholesterolaemia in rats leads to increased
synthesis and concentration of dolichol and of phosphoryl
dolichol with a consequential increase in protein N-glycosylation (43). A similar phenomenon has been reported in

rabbits (77). In other tissues the concentration of dolichol fell during this treatment suggesting a different
control mechanism in liver from that in other tissues (76).

Whether or not the dolichol in a cell is derived from
the diet, by de novo biosynthesis or by transport from another biosynthetic site, control of the concentration of
phosphoryl dolichol is clearly going to depend on the relative activities of dolichol kinase and phosphoryl dolichol
phosphatase. Results of studies on the possible regulation
of these enzymes are awaited with interest.

THE FUNCTIONAL SIGNIFICANCE OF PHOSPHORYL DOLICHOL

During the membrane catalyzed, phosphoryl dolichol-
mediated protein N-glycosylation at least some of the sugar
residues are transferred across the membrane. There is no
evidence in favour of the glycosylated phosphoryl dolichol
flipping from one side to the other. Indeed this would be
expected to be slow if the glycosylated lipid was freely
diffusible in the hydrophobic phase. If, as seems likely,
phosphoryl dolichols are intimately linked to glycosyl
transferases arranged in a tightly bound and highly organized complex the possibilities of movement of the intermediate across the membrane will be very limited. Studies
with synthetic membranes show that although phosphoryl dolichol readily moves in the plane of the membrane it doesn't
aggregate and movement across the membrane is very slow
(78). The appearance of the sugar residues as part of a
polymer on the opposite side of the membrane from their
nucleotide donors is most probably the result of the spatial
arrangement of the enzymes within the complex and of steric
aspects of the glycosylation mechanism. It is also relevant that in several other membrane-catalyzed glycosylations,

for example Golgi-mediated glycosylation of secreted glycoproteins, polymers on one side of a membrane receive sugars
from donors on the other side of the membrane without the
involvement of phosphorylated dolichol.

Evidence is appearing that the rate limiting steps in
protein N-glycosylation are immediately prior to the transfer en bloc of oligosaccharide to peptide. This would mean
that a pool of almost complete oligosaccharide is available
in the membrane for rapid transfer to peptide. This may be
particularly important if the oligosaccharide is to be modified further as part of a zip code system for addressing
proteins to different parts of the cell (79). A hydrophobic carrier would be an essential feature of the available
pool of oligosaccharide in the membrane. Some of the glycosyl transferases would be deeply embedded in the membrane
leading to the need for hydrophobic donors of mannose and
glucose. Possibly this sort of rationalisation will lead
to a clearer understanding of the function of phosphoryl
dolichol.

Not only will the fourteen to sixteen enzymes involved
in the N-glycosylation cycle (Figs. 2 and 3) be expected to
function as a complex but probably enzymes concerned with
the formation and hydrolysis of phosphoryl dolichol (the
kinase and phosphatase) will be closely associated. The
concept of a domain of hydrophobic proteins emerges. The
relationship of this to the hydrophobic channels envisaged
(80) in endoplasmic reticulum whereby hydrophobic signals
of nascent peptides, destined to become secreted glycoproteins, bind ribosomes to the membrane remains to be ex-
plored.

REFERENCES

1. Kornfeld, S., Li, E. and Tabas, I. (1978) J. Biol. Chem. 253, 7771-7778.
2. Hubbard, S.C. and Robbins, P.W. (1979) J. Biol. Chem. 254, 4568-4576.
3. Spiro, M.J., Spiro, R.G. and Bhoyroo, V.D. (1979). J. Biol. Chem. 254, 7668-7674.
4. Chapman, A., Fujimota, K. and Kornfeld, S. (1980) J. Biol. Chem. 255, 4441-4446.
5. Kronquist, K.E. and Lennarz, W.J. (1978) J. Supramol. Struct. 8, 51-65.
6. Czichi, M. and Lennarz, W.J. (1977) J. Biol. Chem. 252, 7901-7904.
7. Vargas, V.I. and Carminatti, H. (1977) Mol. Cell. Biochem. 16, 171-176.
8. DePierre, J.W. and Dallner, G. (1975) Biochim. Biophys. Acta 415, 411-472.
9. Beyer, T.A., Rearick, J.I., Paulson, J.C., Prieels, J.P., Sadler, J.E. and Hill, R.L. (1979) J. Biol. Chem. 254, 12531-12541.
10. Schachter, H., Jabbal, I., Hudgin, R.L., Pinteric, L., McGuire, E.J. and Roseman, S. (1970) J. Biol. Chem. 245, 1090-1100.
11. Hino, Y., Asano, A. and Sato, R. (1978) J. Biochem. 83, 935-942.
12. Bretz, R., Bretz, H. and Palade, G.E. (1980) J. Cell Biol. 84, 87-101.
13. Howell, K.E., Ito, A. and Palade, G.E. (1978) J. Cell Biol. 79, 581-589.
14. Bergman, A. and Dallner, G. (1976) Biochim. Biophys. Acta 433, 496-508.
15. Ozols, J. (1972) J. Biol. Chem. 247, 2242-2254.
16. Evans, W.H. and Gurd, J.W. (1973) Biochem. J. 133, 189-199.
17. Bischoff, E., Tran-thi, T. and Decker, K.F.A. (1975) Eur. J. Biochem. 51, 353-361.
18. Haugen, D. and Coon, M. (1976) J. Biol. Chem. 251, 7929-7939.
19. Elhammer, A., Dallner, G. and Omura, T. (1978) Biochem. Biophys. Res. Commun. 84, 572-580.
20. Imai, Y., Hashimoto-Yutsudo, C., Satake, M., Girardin, A. and Sato, R. (1980) J. Biochem. 88, 489-503.
21. Tabas, I. and Kornfeld, S. (1980) J. Biol. Chem. 255, 6633-6639.
22. Varki, A. and Kornfeld, S. (1980) J. Biol. Chem. 255, 8398-8401.

23. Butterworth, P.H.W. and Hemming, F.W. (1968) Arch. Bio-
    chem. Biophys. 128, 503-508.
24. Daleo, G.R., Hopp, E., Romero, P.A. and Pont Lezica, R.
    (1977) FEBS Letters 81, 411-414.
25. Dallner, G., Behrens, N., Parodi, A.J. and Leloir, L.F.
    (1972) FEBS Letters 24, 315-317.
26. Gateau, O., Morelis, R. and Louisot, P. (1978) Eur. J.
    Biochem. 88, 613-622.
27. Coolbear, T., Mookerjea, S. and Hemming, F.W. (1979)
    Biochem. J. 184, 391-397.
28. Arnold, D., Hommel, E. and Risse, H.J. (1976) Mol. Cell.
    Biochem. 11, 137-145.
29. Cacan, R., Hoflack, B. and Verbert, A. (1980) Eur. J.
    Biochem. 106, 473-479.
30. Appelkvist, E.L. and Dallner, G. (1978) Biochim. Biophys.
    Acta 513, 173-178.
31. Cook, G.M.W. and Stoddart, R.W. (1973) Surface Carbo-
    hydrates of the Eukaryotic Cell. Acad. Press, London.
32. Lindsay, J.G. and D'Souza, M.P. (1979) Biochem. Soc.
    Trans. 7, 210-212.
33. Milsom, D.W. and Aston, S.M. (1979) In Glycoconjugates
    (R. Schauer, ed.), pp. 601-602, Thieme, Stuttgart.
34. Meyer, D.I. and Burger, M.M. (1976) Biochim. Biophys.
    Acta 443, 428-436.
35. Nilsson, O.S., DeTomás, M.E., Peterson, E., Bergman, A.,
    Dallner, G. and Hemming, F.W. (1978) Eur. J. Biochem.
    89, 619-628.
36. Bergman, A. and Dallner, G. (1978) Biochim. Biophys.
    Acta 512, 123-135.
37. Hanover, J.A. and Lennarz,W.J.(1980) J. Biol. Chem. 255,
    3600-3604.
38. Snider, M.D., Sultzman, L.A. and Robbins, P.W. (1980)
    Cell 21, 385-392.
39. Eggens, I. and Dallner, G. (1980) FEBS Letters 122,
    247-250.
40. Hemming, F.W. (1974) Lipids in Glycan Biosynthesis.
    In Biochemistry, Series One, Vol. 4 (T.W. Goodman, ed.),
    pp. 39-97, Butterworth, London.
41. Hemming, F.W. (1981) The Biosynthesis of Dolichols and
    Related Compounds. In Polyisoprenoid Biosynthesis
    (J.W. Porter, ed.) Acad. Press, in press.
42. Tavares, I.A., Johnson, M.J. and Hemming, F.W. (1977)
    Biochem. Trans. 5, 1771-1773.
43. Tavares, I.A., Coolbear, T. and Hemming, F.W. (1981)
    Arch. Biochem. Biophys. in press.
44. Keller, R.K. and Adair, W.L. (1977) Biochim. Biophys.
    Acta 489, 330-336.

45. Van Dessel, G., Lagrou, A., Hilderson, J. Dommisse, R., Esmans, E. and Dierich, W. (1979) Biochim. Biophys. Acta 573, 296-300.
46. Rupar, C.A. and Carroll, K.K. (1978) Lipids 13, 291-293.
47. Carroll, K.K., Vilim, A. and Woods, M.C. (1973) Lipids 8, 246-248.
48. Radominska-Pyrek, A., Chojnacki, T. and Pyrek, S. (1979) Biochem. Biophys. Res. Commun. 86, 395-401.
49. Ekström, T., Chojnacki, T. and Dallner, G. (1981) in preparation.
50. Butterworth, P.H.W. (1964) Ph.D. Thesis. University of Liverpool.
51. Mankowski, T.W., Jankowski, T., Chojnacki, T. and Franke, P. (1976) Biochemistry 15, 2125-2130.
52. Walton, M.J. and Pennoch, J.F. (1972) Biochem. J. 127, 471-479.
53. Hopp, H.E., Daleo, G.R., Romero, P.A. and Pont Lezica, R. (1978) Plant Physiol. 61, 248-251.
54. Bergman, A., Mankowski, T., Chojnacki, T., DeLuca, L.M., Peterson, E. and Dallner, G. (1978) Biochem. J. 172, 123-127.
55. Gough, D.P. and Hemming, F.W. (1970) Biochem. J. 118, 163-166.
56. Momose, K. and Rudney, H. (1972) J. Biol. Chem. 247, 3930-3940.
57. Daleo, G.R. and Pont Lezica, R. (1977) FEBS Letters 74, 247-250.
58. Wedgwood, J.F. and Strominger, J.L. (1980) J. Biol. Chem. 255, 1120-1123.
59. Idogooga-Vangas, V., Belocopitow, E., Mentaberry, A. and Carminatti, H. (1980) FEBS Letters 112, 63-66.
60. Allen, C.M., Kalin, J.R. and Sach, J. (1978) Biochemistry 17, 5020-5026.
61. Burton, W.A., Scher, M.G. and Waechter, C.J. (1979) J. Biol. Chem. 254, 7129-7136.
62. Chojnacki, T., Ekström, T. and Dallner, G. (1980) FEBS Letters, 113, 218-220.
63. Waechter, C.J., Lucas, J.J. and Lennarz, W.J. (1974) Biochem. Biophys. Res. Commun. 56, 343-350.
64. Parodi, A.J. and Leloir, L.F. (1979) Biochim. Biophys. Acta 559, 1-37.
65. Marshall, R.D. (1979) Biochem. Soc. Trans. 7, 800-805.
66. Schwarz, R.T., Schmidt, M.F.G. and Datema, R. (1979) Biochem. Soc. Trans. 7, 322-326.
67. Mankowski, T., Sasak, W., Janczura, E. and Chojnacki, T. (1977) Arch. Biochem. Biophys. 181, 393-401.

68. Palamarczyk, G., Lehle, L., Mankowski, T., Chojnacki, T. and Tanner, W. (1980) Eur. J. Biochem. 105, 517-523.
69. Rosso, G., De Luca, L.M., Warren, C.D. and Wolf, G. (1975) J. Lipid Res. 16, 235-243.
70. Stone, K.J. and Hemming, F.W. (1968) Biochem. J. 109, 877-882.
71. Keenan, R. and Kruczek, H. (1976) Biochemistry 15, 1586-1591.
72. Radominska-Pyrek, A., Chojnacki, T. and Zulczyk, W. (1979) Acta Biochimica Polonica 26, 125-134.
73. Mills, J.T. and Adamany, A.M. (1978) J. Biol. Chem. 253, 5270-5273.
74. James, M.J. and Kandutsch, A.A. (1979) J. Biol. Chem. 254, 8442-8446.
75. Carson, D.D. and Lennarz, W.J. (1979) Proc. Nat. Acad. Sci. 76, 5709-5713.
76. Gould, R.G. and Swyrud, E.A. (1966) J. Lipid Res. 7, 698-707.
77. White, D.A., Middleton, B., Pawson, S., Bradshow, P., Clegg, R.J., Hemming, F.W. and Bell, G.D. (1981) Arch. Biochem. Biophys., in press.
78. McCloskey, M.A. and Iroy, J.A. (1980) Biochemistry 19, 2056-2060.
79. Sly, W.S. (1980) In Structure and Function of Ganglio-sides (L. Svennerholm, P. Mandel, H. Dreyfus and P.F. Urban, eds.), pp. 433-451, Plenum, New York.
80. Blobel, G. and Dobberstein, B. (1975) J. Cell Biol. 67, 835-851.

CYTOCHROMES AND FLAVOPROTEINS OF THE MICROSOMAL MEMBRANE:

PAST HYPOTHESES AND CURRENT CONCEPTS REGARDING

STRUCTURE — FUNCTION RELATIONSHIPS[1]

R. W. Estabrook, B. S. S. Masters, J. Capdevila,

R. A. Prough[2], J. Werringloer, and J. A. Peterson

Department of Biochemistry
Southwestern Medical School
The University of Texas Health Science Center at Dallas
Dallas, Texas 75235

PRELUDE

Lars Ernster's thirty years of contributions to biochemistry, cell biology and biophysics have centered largely on the chemistry, composition, and functional role of membranes and membrane-bound enzymes. His work, and that of his students and colleagues, on mitochondrial and microsomal electron transport reactions have increased

[1]Supported in part by Research grants from the USPHS (NIGMS 16488, NIGMS 19036, NHLBI 19654, NHLBI 13619), The American Cancer Society (BC-336), and The Robert A. Welch Foundation (I-616, I-405, I-453).

[2]Recipient of a USPHS Research Career Development Award (1-K04-HL00255)

C. P. Lee, G. Schatz, G. Dallner (eds.), Mitochondria and Microsomes
in honor of Lars Ernster   ISBN 0-201-04576-1

significantly   our   present   understanding   of   these
experimentally   and   intellectually   challenging   reaction
systems. The purpose of this presentation is to focus on
our current understanding of two facets of microsomal
electron transport reactions which are manifest during the
oxidative metabolism of certain organic compounds. The
first is the role of the membrane-bound flavoprotein,
NADPH-cytochrome P-450 reductase, as it participates in
electron transport and the second is the possible
relationship of lipid peroxidation and lipid peroxides to
the catalytic cycle of cytochrome P-450. We know that both
of these important problems are of keen interest to Lars
and he continues to provide new and fundamental results
related to these topics.   It is a privilege to dedicate
this offering to Lasse on the occasion of his sixtieth
year.

- - HAPPY BIRTHDAY, LASSE! - -

PRESENTATION

The presence of an amorphous "ground substance" in
cells, in addition to organelles visible by the light
microscope, led to the hypothesis that most cells contained
a   complex   network   of   internal   structure (1). This
hypothesis was convincingly proven by the pioneer work of
Albert   Claude   with   his   discovery (2) of a
membrane-containing subcellular fraction that he termed
"microsomes" (3,4). Today we know that the microsomal
fraction consists primarily of a mixture of membrane
vesicles - the vast majority of which arise from a

pinching-off of the endoplasmic reticulum of the cell during homogenization. Thus, the microsomal fraction is recognized as a mixture dominated by "smooth" and "rough" membrane fragments representing the continuum of cellular constituents present in vivo as the endoplasmic reticulum.

The 1950's saw a surge of activities in an attempt to understand the properties and functions of membrane-bound enzymes. Studies of the enzymes of liver microsomes proceeded on two fronts: a) an examination of the role of bound ribosomes as they participate in protein synthesis, and b) the action of NADPH- and oxygen-dependent enzymes in the biotransformation of drugs and steroids. During the Spring of 1958, Lars Ernster performed collaborative experiments at the Rockefeller Institute with P. Siekevitz and G. Palade. It was at this time that his interest in the properties of microsomes was kindled. One need only review the first publication resulting from these studies (5) to gain an appreciation of his perspicacity and intuitive thinking. While examining the effect of the detergent, deoxycholate, on the properties of various enzymes of liver microsomes, Ernster et al. (5) concluded "our experiments, finally, demonstrate that certain microsomal enzymes, namely NADH-cytochrome c reductase, NADH diaphorase, Mg^{++}-activated ATPase and glucose-6-phosphatase, either are part of these membranes or are tightly bound to them. Other enzymes, e.g., NADPH-cytochrome c reductase, DT diaphorase, and cytochrome b_5 do not seem to be part of the membranes or appear to be loosely attached thereon". Thus, a clear perception of enzyme distribution and topography was presented and this has   served   as   one   of   the   fundamental   criteria   for evaluating   enzyme   polarity   and   accessibility   for membrane-bound   proteins.   Today it is widely accepted that those enzymes which Ernster et al. (5) recognized as loosely   associated   with   microsomes,   such   as NADPH-cytochrome c reductase or cytochrome b_5, are tenuously attached to the cytosolic face of the endoplasmic reticulum   by   a   hydrophobic   "anchor"   composed   of approximately 40 to 80 amino acid residues (6-11). The amphipathic properties of these electron transfer proteins influence   their   ability   to   function   during reduction-oxidation reactions. A number of speculations concerning the organization of proteins associated with the microsomal membrane has been presented during the years since Ernster et al. (5) initially posed the question of protein topology, but even today we still rely on mental manipulation when constructing models on which to develop hypotheses.   One might consider, as diagrammatically designated in Figure 1, that some of these membrane-bound enzymes sail on a sea of lipid dominated by partially submerged proteins (icebergs of cytochrome P-450) with only a few free-floating companions (buoys of cytochrome b_5) similarly adrift without apparent guidance. This animated analogy conjures up visions of the chaos that would result from a stormy sea, e.g., an increase in entropy as the temperature of the reaction system is increased - what some consider critical for the elaboration of "fluidity" in a membrane-bound enzyme system.   One must agree with the perceptive statement contained in the Ernster et al. paper (5) which concludes "--- that the activity of an enzyme

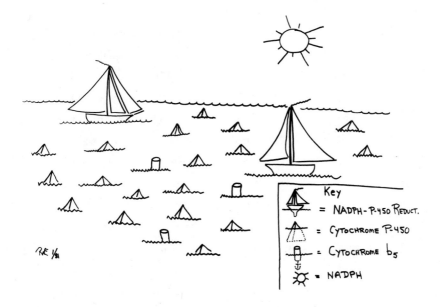

Fig. 1. A cartoon of the microsomal membrane depicting the flavoprotein, NADPH-cyto-chrome P-450 reductase, as a sailboat and cytochrome P-450 as an iceberg.

The hemeprotein, cytochrome b_5, is represented as a partially submerged buoy while NADPH is represented by the sun shining on this tranquil sea of lipid.

could be influenced and controlled by agents acting not upon the enzyme itself but upon its position within the cellular structure to which it belongs".

We have spent considerable time and effort during the last decade in an attempt to better define the constraints which the membrane may impose on the function of microsomal electron transfer proteins, and it is one aspect of these studies that we briefly summarize in the following section.

THE MEMBRANE-BOUND MICROSOMAL ELECTRON TRANSPORT SYSTEM. The microsomal fraction prepared from livers of animals exposed to repeated insults with various xenobiotics is rich in the hemeprotein, cytochrome P-450. The ability of cytochrome P-450 to function in a variety of oxygenation or oxidation reactions is critically dependent on the ability of the flavoprotein, NADPH-cytochrome P-450-reductase, to transfer reducing equivalents (electrons) from the reduced pyridine nucleotide to the hemeprotein. This flavoprotein is not a simple enzyme - it possesses two prosthetic groups (FAD and FMN), as well as essential sulfhydryl groups (12-14). Its ability to function in electron transfer reactions to cytochrome P-450 is dependent on retention of its amphipathic characteristics (15). Recently, a number of investigators have carried out elegant experiments on the redox properties of this flavoprotein. It is well established (16,17) that one can chemically (with a reductant such as sodium dithionite) or enzymatically (with NADPH as reductant) reduce the flavoprotein, thereby establishing discrete redox steps. Full reduction of the flavoprotein (i.e., a four-electron reduced state) can only be attained by chemical reduction and the probability that this state participates in physiological reactions is highly unlikely. Further, it is well established that the FMN prosthetic group of the flavoprotein can be removed readily and that a concomitant loss of ability to reduce cytochrome P-450 occurs (18,19). However, electron transfer from NADPH to an oxidizing agent, such as potassium ferricyanide, is retained to a great extent in the remaining FAD-containing protein. This type of experiment suggests (but does not prove) that a sequence of

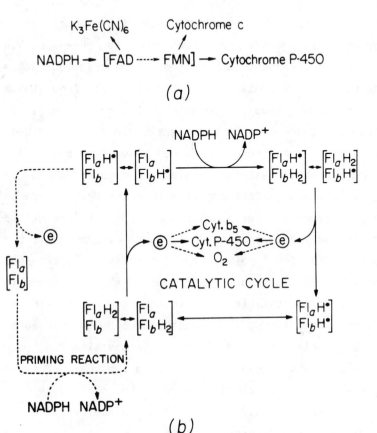

Fig. 2. Proposed schemes for the function of microsomal NADPH-cytochrome P-450 reductase.

(a) The sequence of electron transfer via
the two flavin prosthetic groups from NADPH
to various electron acceptors. (b) The
various intermediates formed during the
priming reaction or catalytic cycle of
flavoprotein reduction and oxidation.

electron transfer, such as shown in Figure 2A, may occur. Another way of illustrating the interaction of the different prosthetic groups of this flavoprotein as it carries out its transfer of electrons is presented in Figure 2B. This scheme is based on the observation that a priming reaction is needed which is distinct from those reductive steps which participate in the catalytic cycle. The presence of an electron paramagnetic resonance signal at about g=2.0 in freshly prepared liver microsomes (13) may represent the air-stable, 1-electron reduced state of this flavoprotein. This suggests that the primed form of the enzyme occurs physiologically and that it pre-exists within the cell and persists during the preparation of the microsomal fraction. Since the catalytic cycle of cytochrome P-450 encompasses two reducing steps, which have different redox potentials, and since the physiologically significant oxidation of the reductase can be envisioned to include 2 one-electron oxidations, it is tempting to pair these complementary redox processes on the basis of their potentials. Since one cannot assign with confidence the discrete steps participating in these unique reactions of microsomal electron transport, it is not possible to assess the comparative efficiencies of the various oxidation-reduction states of the flavoprotein although one would intuitively expect marked differences to exist. Clearly, advances in understanding the participation of the flavoprotein in electron transfer from NADPH to the cytochrome will depend on our ability to develop new techniques for monitoring the various states of reduction of this pigment.

A number of facts are known about the reaction of the membrane-bound flavoprotein which may provide clues as to its mode of action. Some of these are as follows:

a) The rate of electron transfer from the reduced flavoprotein to various hemeproteins and other acceptors differs markedly. The rate of reduction of cytochrome b_5 (or cytochrome c) is at least 10 times faster than the rate of reduction of cytochrome P-450 (20-22);

b) The rate of enzymatic reduction of cytochrome P-450, when measured under an atmosphere of carbon monoxide, is multiphasic (23,24) and consists of at least two clearly distinguishable first-order reactions (22);

c) The concentration of flavoprotein relative to cytochrome P-450 is significantly less than stoichiometric and ratios as small as 1 molecule of flavoprotein to 30 molecules of hemeprotein may exist. Assuming a microsomal vesicle of 1000 Å diameter and a membrane thickness of about 100 Å one can estimate that only 8 to 10 molecules of flavoprotein and 250 molecules of cytochrome P-450 are associated with each microsomal particle (25,26);

d) Foci of these redox proteins, in which the flavoprotein appears to play a dominant role, may exist amongst the membrane-bound electron transfer proteins of the microsomes (27). Recent results have shown that the steady-state concentration of oxycytochrome P-450, formed during the NADPH-dependent metabolism of many substrates, is readily perturbed by $NADP^+$ (28). Likewise, the kinetics of oxidation of cytochrome b_5, reduced by addition of limiting concentrations of NADPH, is very sensitive to the NADPH to $NADP^+$ ratio. These types of studies, as well as others, showing an apparent redox equilibrium between

cytochrome b_5 and cytochrome P-450, point to
the presence of an intimate biochemical
communication between proteins that function
in concert during metabolism.

These and many other observations have led to
speculation about the organization of these proteins on the
surface of the membrane vesicle. Clearly, one cannot
exclude a role for membrane fluidity in altering the rates
of interactions of the requisite proteins and one cannot
exclude the need for protein aggregation and/or association
for facilitating electron transfer between flavoprotein and
the hemeproteins. The question arises of how one can
visualize the types of association which may exist.
Recently, we have been interested in determining the
effects of ions on the reactivity of these proteins. It
was recognized many years ago (29) that the ability of the
purified flavoprotein (then called TPNH-cytochrome c
reductase) to transfer electrons to purified cytochrome b_5
was dependent on the presence of a high concentration of
ions in the reaction medium. It is known (30) that salt
concentration can markedly influence the rate of electron
transfer via the flavoprotein and iron sulfur protein
participating in the reduction of adrenal cortical
mitochondrial cytochromes P-450 or the cytochrome P-450
isolated and purified from the bacterium, Pseudomonas
putida (31). As shown in Figure 3, the kinetics of
reduction of liver microsomal cytochrome P-450 are greatly
enhanced by increasing the buffer concentration of the
reaction medium. This effect of ions not only influences
the initial rate of the reaction, but also the extent of
biphasicity of the reaction kinetics, i.e., the amount of
cytochrome P-450 reduced during the "fast phase" of

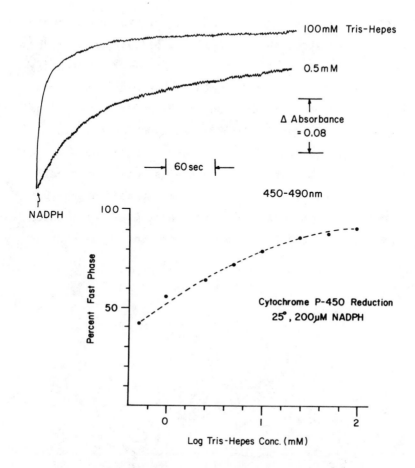

Fig. 3. The kinetics of reduction of liver microsomal cytochrome P-450.

Upper curve: The effect of Tris-HEPES buffer concentration on the rate of NADPH reduction of liver microsomal cytochrome P-450. Experiments were carried out at 25° using an Aminco-Morrow stopped-flow apparatus as described elsewhere (32). Lower curve: The extent of the rapid phase of cytochrome P-450 reduction as a function of Tris-HEPES buffer concentration. The kinetic curves of cytochrome P-450 reduction were resolved by computer program as described by Matsubara et al (22).

kinetics is significantly decreased at low ionic strength (Figure 3). Likewise, the overall rate of cytochrome P-450 catalyzed metabolism of many substrates is markedly influenced by the ionic strength of the reaction medium. This effect of salt concentration on microsomal reactions has been known for many years (33), although an explanation for the observed changes has not been established. One interpretation of these results is diagrammatically proposed in Figure 4. In this Figure we suggest a possible role for surface charges on the interacting proteins which can dictate the rates of association and dissociation

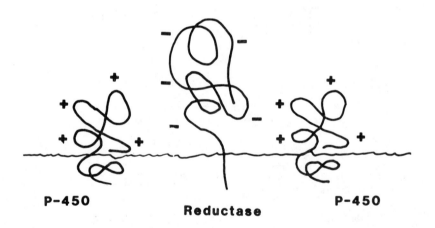

P-450   **Reductase**   **P-450**

Fig. 4.Proposed surface charges on the exposed portion of microsomal proteins.

The flavoprotein, NADPH-cytochrome P-450 reductase, is shown to have a net negative charge at neutral pH. Cytochrome P-450 is shown to have an excess of positive charges.

during the time course of the reaction. Increasing salt concentrations may modify these interactions and thereby increase the probability of forming productive complexes for electron transfer. One can now extend the prediction of Ernster et al. (5) to include the need to consider not only those agents which influence the position of an enzyme within a cellular structure, but also those factors, such as surface charge, conformation, etc., which influence the efficiency and effectiveness of interactions between two proteins bound to the same membrane.

LIPID PEROXIDATION AND MICROSOMAL CYTOCHROME P-450. Lars Ernster developed a keen interest in lipid peroxidation from his pioneer work with Paul Hochstein. Using liver microsomes, they (34) demonstrated the key role played by iron chelates in the propagation of lipid peroxidation initiated during the concomitant oxidation of NADPH. At about this time (1963 to 1964), Ernster and his colleagues were concerned with the role of cytochrome P-450 in microsomal protein-catalyzed oxygenation reactions (35-38). It is prophetic that one of their papers published in 1964 (36) starts by stating, "The demonstration of a TPNH-linked peroxidation of lipids in microsomes has raised the question as to the possible relationship of this process to other microsomal TPNH-dependent reactions". Today many assume that there might be a direct as well as an indirect relationship between the generation of lipid peroxides and one of the actions of cytochrome P-450 in microsomal preparations.

The membrane-bound cytochrome P-450 of liver microsomes presumably can function as an oxygenase, oxidase or peroxygenase (39). In the presence of NADPH and oxygen, cytochrome P-450 can catalyze the "activation of oxygen" for the monooxygenation of a broad spectrum of substrates. Alternatively, it can reduce atmospheric oxygen for the generation of hydrogen peroxide - a reaction which probably involves superoxide as an intermediate (40). In recent years, considerable interest has centered on the peroxidase-like action of cytochrome P-450. In the presence of organic hydroperoxides, such as cumene hydroperoxide and tertiary butyl hydroperoxide, or hydrogen peroxide, liver microsomal cytochrome P-450 can catalyze the oxidative metabolism of a number of substrates (41-44). Our current hypothesis regarding these various reaction systems is diagrammatically shown in Figure 5. This scheme depicts the concomitant operation of cytochrome P-450 in these three types of oxidative reactions. It is proposed that during the course of NADPH oxidation by oxygen, as catalyzed by the electron transport proteins present in liver microsomes, the various types of oxygen-dependent reactions are all occurring to various degrees.

When cytochrome P-450 is reduced enzymatically by the flavoprotein, NADPH-cytochrome P-450 reductase, it reacts rapidly with oxygen to form oxycytochrome P-450 which represents a branch point in the metabolic sequence of reactions involved in the function of this hemeprotein. This intermediate may decay to generate superoxide and restore ferricytochrome P-450 or it can undergo further reduction to the functional intermediate needed for substrate oxidation reactions. Indeed some substrates,

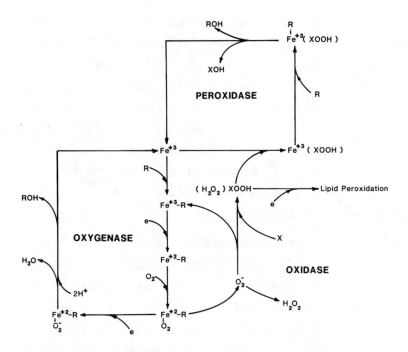

Fig. 5.A scheme showing the relationship of the oxygenase, oxidase and peroxidase-like activities of liver microsomal cytochrome P-450.

The formation of lipid peroxides may serve as a key to link the diverse roles of cytochromes P-450 in substrate metabolism to various products.

such as perfluorohexane or hexobarbital, can serve as uncouplers (45-47) thereby facilitating the breakdown of oxycytochrome P-450 with the concomitant generation of hydrogen peroxide - presumably via the superoxide anion. In our scheme (Figure 5), it is proposed that the superoxide anion may react with unsaturated fatty acids thereby generating lipid peroxides. Alternatively, the superoxide anion may dismutate to form hydrogen peroxide. We recognize that lipid peroxides may be formed by reactions other than the one depicted. When one carries out a series of experiments in which the balance of the amounts of NADPH oxidized and oxygen utilized are related to the amount of products formed, it is noted that a significant extent of "endogenous substrate metabolism" occurs (48). We now suggest that these "endogenous substrates" may serve as precursors (X) of the enzymatically formed lipid peroxides.

Incubation of liver microsomal cytochrome P-450 with organic hydroperoxides results in the formation of oxidized metabolites of many substrates (41-44). Recently, we have directed our attention (49,50) to the metabolism of benzo(a)pyrene - a polycyclic hydrocarbon of great interest to those concerned with chemical carcinogenesis. When benzo(a)pyrene serves as a substrate for the NADPH- and oxygen-dependent function of rat liver microsomal cytochrome P-450, a large variety of metabolites are formed, including dihydrodiols, phenols, and quinones (51-53). A comparable reaction system, in which cumene hydroperoxide replaces NADPH, results principally in the formation of quinones (49). Studies in the presence of the heavy isotope of oxygen ($^{18}O_2$) revealed the expected incorporation of molecular oxygen into the dihydrodiols and phenols formed during the NADPH-supported reaction (Table I). However, the observation that only a part of the oxygen incorporated into the quinone metabolites was derived from molecular oxygen interested and puzzled us. Additional experiments provided a clue to explain these results (54). Termination of the reaction at various times

Table I. Oxygen-18 incorporation into benzo(a)pyrene
          metabolites

(Percent Incorporation)

Product	Zero ^{18}O	One ^{18}O	Two ^{18}O
3-Phenol	12	88	-
4,5-diol	10	90	0
9,10-diol	14	86	0
1,6-quinone	16	42	42
3,6-quinone	14	42	44

Liver microsomes (1 mg/ml) from phenobarbital treated rats were incubated with 1 mM NADPH and 80 µM benzo(a)pyrene for 5 minutes at $37^{o}C$ with an atmosphere of 20% oxygen which was 95% enriched with $^{18}O_2$. Products were separated by high performance liquid chromatography and identified by mass spectrometry as described elsewhere (55).

after initiation and subsequent examination of the quinone metabolites showed that the amount of ^{18}O incorporated was very dependent on the time of incubation of the reaction system (Figure 6). This difference presumably reflects the concomitant operation of two pathways of metabolism and the progressive dominance of the "classic" monooxygenase reaction as time of incubation increases. The question was asked: "Why is there only partial incorporation of molecular oxygen during short reaction times"? A further clue (55) was provided when the peroxide-dependent

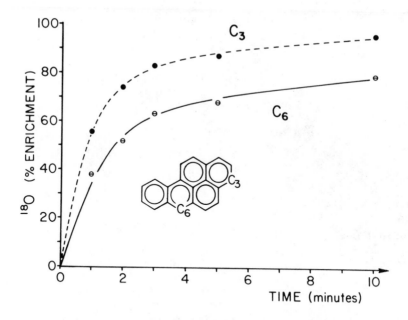

Fig. 6.The effect of time of incubation on the extent of incorporation of molecular oxygen into benzo(a)pyrene quinones formed during the NADPH-supported microsomal catalyzed reaction.

A suspenson of liver microsomes (1mg protein per ml) was incubated with 80 µM benzo(a)pyrene and 1mM NADPH at 37°C in an atmosphere of 20% $^{18}O_2$ and 80% N_2. At the indicated time points, the 3,6-quinone was isolated by h.p.l.c. and analyzed by mass spectrometry.

metabolism of benzo(a)pyrene was examined. It has previously been assumed that the peroxide-dependent metabolism of substrates, which involves the hemeprotein cytochrome P-450, would occur to the same degree independent of the presence of atmospheric oxygen. As shown in Table 2, this is not true for the metabolism of benzo(a)pyrene. The presence of oxygen markedly enhances

Table II. The   peroxygenatic   metabolism   of
          benzo(a)pyrene by cumene hydroperoxide

a) Rates of metabolism

    Aerobic = 2.0 nmoles/min/mg

    Anaerobic = 0.8 nmoles/min/mg

b) Extent of atmospheric oxygen incorporation
    into 3,6-quinone of benzo(a)pyrene

                    Quinone   0.66

                    C-3 position   0.14

                    C-6 position   0.52

Liver microsomes (0.5 mg/ml) from phenobarbital-treated
rats were incubated with 80 µM benzo(a)pyrene and 100 µM
cumene hydroperoxide for 1 minute at 37°C.   The system
was made anaerobic by glucose oxidase and glucose.
Oxygen incorporation was determined using   nitrogen
containing 20% oxygen which was 95% enriched with $^{18}O_2$.

the rate of quinone formation from benzo(a)pyrene when
cumene hydroperoxide is used to support the cytochrome
P-450-dependent reaction. Indeed, a significant amount of
oxygen in the quinone product is derived from atmospheric
oxygen. This suggests, that during the peroxide-driven
reaction, highly reactive intermediates of benzo(a)pyrene
are formed and these reactive compounds can interact with
molecular   oxygen.   One candidate for these reactive
intermediates may   be   free   radicals   formed   during
metabolism.   Studies of the type illustrated in Figure 6
suggest that similar highly reactive intermediates (free
radicals?)   are also generated, albeit in much lower

quantity, in the presence of NADPH and oxygen suggesting a catalytic role for lipid peroxides as proposed in the scheme shown in Figure 5. Obviously, many more experiments will be required before the marriage of lipid peroxidation and cytochrome P-450 dependent metabolism can be established. However, these preliminary results are encouraging and again point to the perceptive and penetrating incisiveness of L. Ernster when he early recognized the potential existence of this relationship.

POSTSCRIPT

Lars Ernster is known as an innovative thinker. He enjoys the speculation and free thinking that goes with the generation of new approaches and new concepts. The above description of two facets of the microsomal electron transport system represents some highly speculative ideas. These are presented as topics to ruminate in hopes that some element of truth will emerge that can further extend our base of knowledge of this system. The odds are that Lars Ernster will be at the center of these additional studies since his active and fertile mind will rapidly grasp the implications of these sketchy and preliminary experiments.

REFERENCES

1. Weibel, E.R., Staubli, W., Gnagi, H.R., and Hess, F.A. (1969) J. Cell Biol. 42, 68-112.
2. Claude, A. (1938) Proc. Soc. Exp. Biol. Med. 39, 398-403.
3. Claude, A. (1943) Science 97, 451-456.
4. Claude, A. (1969) in Microsomes and Drug Oxidations (J.R. Gillette et al., eds.) pp. 3-39, Academic Press.
5. Ernster, L., Siekevitz, P. and Palade, G.E. (1962) J. Cell. Biol. 15, 541-558.
6. Spatz, L. and Strittmatter, P. (1971) Proc. Natl. Acad. Sci. U.S.A. 68, 1042-1046.
7. Ito, A. and Sato, R. (1968) J. Biol. Chem. 243, 4922-4923.
8. Vermilion, J.L. and Coon, M.J. (1974) Biochem. Biophys. Res. Commun. 60, 1315-1322.
9. Dignam, J.D. and Strobel, H.W. (1975) Biochem. Biophys. Res. Commun. 63, 845-852.
10. Yasukochi, Y. and Masters, B.S.S. (1976) J. Biol. Chem. 251, 5337-5344.
11. Satake, H. and Sato, R. (1973) Seikagaku 45, 582-585.
12. Williams, C.H., Jr. and Kamin, H. (1962) J. Biol. Chem. 237, 587-595.
13. Iyanagi, T. and Mason, H.S. (1973) Biochemistry 12, 2297-2308.
14. Masters, B.S.S., Prough, R.A. and Kamin, H. (1975) Biochemistry 14, 607-613.
15. Lu, A.Y.H., Junk, K.W., and Coon, M.J. (1969) J. Biol. Chem. 244 3714-3721.
16. Yasukochi, Y., Peterson, J.A. and Masters, B.S.S. (1979) J. Biol. Chem. 254, 7097-7104.
17. Iyanagi, T., Makino, N., and Mason, H.S. (1974) Biochemistry 13, 1701-1710.
18. Vermilion, J.L. and Coon, M.J. (1978) J. Biol. Chem. 253, 2694-2704.
19. Alexander, L.M., Hersh, L.B. and Masters, B.S.S. (1980) in Microsomes, Drug Oxidations, and Chemical Carcinogenesis (M.J. Coon et al., eds.) Vol. 1, pp. 285-288, Academic Press.
20. Masters, B.S.S., Kamin, H., Gibson, Q.H., and Williams, C.H. Jr. (1965) J. Biol. Chem. 240, 921-931.

21. Strittmatter, P. (1968) in Biological Oxidations (T.P. Singer, ed.) pp. 171-191, Interscience.

22. Matsubara, T., Baron, J., Peterson, L.L., and Peterson, J.A. (1976) Arch. Biochem. Biophys. 172, 463-469.

23. Diehl, H., Schadelin, J., and Ullrich, V. (1970) Hoppe-Seyler Z. Physiol. Chem. 351, 1359-1371.

24. Gigon, P.L., Gram, T.E., and Gillette, J.R. (1968) Biochem. Biophys. Res. Commun. 31, 558-562.

25. Estabrook, R.W., Franklin, M.R., Cohen, B., Shigamatzu, A., and Hildebrandt, A.G. (1971) Metabolism 20 187-199.

26. Peterson, J.A., Ebel, R.E., O'Keeffe, D.H., Matsubara, T., and Estabrook, R.W. (1976) J. Biol. Chem. 251, 4010-4016.

27. Werringloer, J. and Kawano, S. (1980) in Microsomes, Drug Oxidations, and Chemical Carcinogenesis (M.J. Coon et al., eds.) pp. 469-478, Academic Press.

28. Werringloer, J., Archakov, A.J., Kawano, S., and Estabrook, R.W. Unpublished results.

29. Bilimoria, M.H. and Kamin, H. (1973) Ann. N.Y. Acad. Sci. 212, 428-448.

30. Lambeth, J.D., Seybert, D.W., and Kamin, H. (1979) J. Biol. Chem. 254, 7255-7264.

31. Hintz, M.J. and Peterson, J.A. (1981) J. Biol. Chem. In Press.

32. Peterson, J.A. and Mock, D.M. (1975) Anal. Biochem. 68, 545-553.

33. Peters, M.A. and Fouts, J.R. (1970) Biochem. Pharm. 19, 533-544.

34. Hochstein, P. and Ernster, L. (1963) Biochem. Biophys. Res. Commun. 12, 388-394.

35. Orrenius, S., Ericsson, J.L.E., and Ernster, L. (1965) J. Cell Biol. 25, 627-634.

36. Orrenius, S., Dallner, G. and Ernster, L. (1964) Biochem. Biophys. Res. Commun. 14, 329-334.

37. Ernster, L. and Orrenius, S. (1965) Fedn. Proc. 24 1190-1199.

38. DePierre, J.W. and Ernster, L. (1978) Biochim. Biophys. Acta 473, 149-186.

39. Estabrook, R.W., Baron, J., Peterson, J. and Ishimura, Y. (1972) in Biological Hydroxylation Mechanisms (G. S. Boyd and R. M. S. Smellie, eds.) pp. 159-186, Academic Press.

40. Kuthan, H., Tsuji, H., Graf, H., Ullrich, V., Werringloer, J., and Estabrook, R.W. (1978) FEBS Letters, 91, 343-345.

41. Kadlubar, F.F., Morton, K.C., and Ziegler, D.M. (1973) Biochem. Biophys. Res. Commun. 54, 1255-1261.

42. O'Brien, P.J. and Rahimtula, A.D. (1980) in Microsomes, Drug Oxidations, and Chemical Carcinogenesis (M. J. Coon et al., eds.) pp. 263-271, Academic Press.

43. Hrycay, E.G. and O'Brien, P.J. (1972) Arch. Biochem. Biophys. 153, 480-494.

44. Groves, J.T., Akinbote, O.F., and Avaria, G.E. (1980) in Microsomes, Drug Oxidations, and Chemical Carcinogenesis (M. J. Coon et al., eds.) pp. 243-252, Academic Press.

45. Ullrich, V. and Diehl, H. (1971) Eur. J. Biochem. 20, 509-512.

46. Hildebrandt, A.G., Speck, M., and Roots, I. (1974) Naunyn-Schniedeberg's Arch. Pharmacol. 281, 371-376.

47. Werringloer, J. (1977) in Microsomes and Drug Oxidations (V. Ullrich et al., eds.) pp. 261-268, Pergamon Press.

48. Werringloer, J., Hildebrandt, A.G., and Estabrook, R.W. Unpublished results.

49. Capdevila, J., Estabrook, R.W., and Prough, R.A. (1980) Arch. Biochem. Biophys. 200, 186-195.

50. Renneberg, R., Capdevila, J., Chacos, N., Estabrook, R.W., and Prough, R.A. (1981) Biochem. Pharmacol. 29, In Press.

51. Selkirk, J.K., Croy, R.G., and Gelboin, H.V. (1974) Science, 184, 169-171.

52. Sims, P. and Grover, P.L. (1975) Adv. Cancer Res. 20, 165-274.

53. Capdevila, J. and Orrenius, S. (1980) FEBS Letters, 119, 33-37.

54. Capdevila, J., Saeki, Y., Prough, R.A., and Estabrook, R.W. (1980) in Carcinogenesis: Fundamental Mechanisms and Environmental Effects (B. Pullman, P.O.P. Ts'o and H.V. Gelboin, eds.) Vol. 13, pp. 113-124, D. Reidel Pub. Co.

55. Estabrook, R.W., Saeki, Y., Chacos, N., Capdevila, J., and Prough, R.A. (1981) in Advances in Enzyme Regulation (G. Weber, ed.) In press, Pergamon.

STRUCTURE AND FUNCTION OF NADPH-CYTOCHROME P-450 REDUCTASE AND OF ISOZYMES OF CYTOCHROME P-450[1,2]

M. J. Coon, S. D. Black, D. R. Koop, E. T. Morgan, and
A. V. Persson

Department of Biological Chemistry, Medical School,
The University of Michigan, Ann Arbor, Michigan 48109

INTRODUCTION

Solubilization and resolution of the liver microsomal
polysubstrate monooxygenase system into its components,
cytochrome P-450, NADPH-cytochrome P-450 reductase, and
phosphatidylcholine (1-3), was an essential first step in
the determination of whether one or more catalysts are involved in the hydroxylation of a variety of substrates.
These include fatty acids, steroids, and prostaglandins
as well as numerous foreign compounds such as drugs, pesti-

---

[1]Dedicated to Professor Lars Ernster on the occasion of
his sixtieth birthday.

[2]Supported by NSF grant PCM76-14947 and NIH grant AM-10339.
S.D.B. was the recipient of a Merck Predoctoral Fellowship, D.R.K. of a Chaim Weizmann Postdoctoral Fellowship
for Scientific Research, and A.V.P. of a Postdoctoral Fellowship from the Damon Runyon-Walter Winchell Cancer Research Fund.

C. P. Lee, G. Schatz, G. Dallner (eds.), Mitochondria and Microsomes
in honor of Lars Ernster   ISBN 0-201-04576-1

cides, anesthetics, petroleum products, and carcinogens.
The second necessary step, enzyme purification, was hampered by the limited solubility of these hydrophobic proteins and their tendency to form mixed aggregates. However, with the use of nonionic detergents and of glycerol
as a protective agent, ion exchange column chromatography
and other techniques eventually led to isolation of cytochrome P-450 (4,5) and the reductase (6) from phenobarbital-
induced rabbits in electrophoretically homogeneous form.
Shortly thereafter, evidence was obtained by further enzyme
fractionation for the occurrence of multiple forms of P-450
in rabbit liver microsomes (7). These isozymes, which differ in their inducibility and chemical and physical properties, were found to have slightly different but overlapping
activities toward a variety of substrates.

In the present paper the recent contributions of several
laboratories to our knowledge of the function and properties
of the isozymes of microsomal P-450 are reviewed. We shall
limit our discussion primarily to the enzymes isolated
from the rabbit, which are designated according to their
relative mobilities when submitted to SDS-polyacrylamide
gel electrophoresis. Beginning with the microsomal band
of greatest electrophoretic mobility, the bands are numbered
according to decreasing mobility and increasing molecular
weight. Thus, for example, the phenobarbital- and 5,6-
benzoflavone-inducible isozymes are referred to as $P-450_{LM_2}$
and $P-450_{LM_4}$, respectively, or simply as forms or isozymes
2 and 4. This method, which is in accord with general re-
commendations of the Commission on Biochemical Nomenclature,
has proved particularly useful for this group of enzymes
having a variety of induction patterns and acting on a multitude of substrates, but an eventual nomenclature based on

function is to be desired. We are grateful to colleagues
in a number of laboratories who have been willing to exchange rabbit liver and lung microsomal isozymes of P-450.
Much needless confusion has thereby been avoided. The reader is referred to an article by Lu and West (8) for a more
general review of multiple forms of P-450, including those
isolated from rat liver microsomes.

RECENT RESULTS

Isozymes of cytochrome P-450. Five forms of liver microsomal cytochrome P-450 have been purified to electrophoretic
homogeneity and characterized sufficiently so that they
can be called distinct isozymes. Of these, the two major
inducible forms, $P-450_{LM_2}$ and LM_4, were the first to be
studied. $P-450_{LM_2}$, which is barely detectable in liver
microsomes from untreated rabbits, was isolated after the
subcutaneous or oral administration of phenobarbital to
the animals for several days (4,5,7,9-11). In contrast,
$P-450_{LM_4}$ is present at significant levels in liver microsomes from untreated rabbits and about doubled in amount
upon induction by various agents. $P-450_{LM_4}$ was first isolated in this laboratory from 5,6-benzoflavone-induced animals
(7,9,10); the resulting enzyme preparation appeared to be
identical in its general properties to that obtained from
animals treated with phenobarbital (7,9) or untreated animal-
s (9,10). The same form was isolated from 3-methylcholan-
threne-induced rabbits by Kawalek et al. (12), Hashimoto and
Imai (13), and Imai et al. (11) from 2,3,7,8-tetrachlorodi-
benzo-p-dioxin (TCDD)- treated rabbits by Johnson and

Muller-Eberhard (14,15), and from isosafrole-treated rabbits in this laboratory (16). As an indication of the usefulness of the electrophoretic method of nomenclature, purified form 4 from 3-methylcholanthrene-treated animals, kindly furnished by Dr. Lu and his associates, or from TCDD-treated animals, kindly furnished by Drs. Johnson and Muller-Eberhard, was found to be electrophoretically indistinguishable from 5,6-benzoflavone-induced $P-450_{LM_4}$. Thus, it could be assumed, at least tentively, that our three laboratories were working with the same protein, thereby avoiding confusion by the use of separate names based on the inducer used or the substrates tested by a particular research group.

That forms 2 and 4 are distinct proteins was indicated by their different spectral properties, amino acid compositions, and C-terminal amino acid residues (9) as well as by differences in their N-terminal amino acid residues (17) and immunological properties (18). Rabbits and goats were inoculated with these purified proteins. Antibodies against $P-450_{LM_2}$ but not against LM_4 were produced by rabbits, whereas antibodies against both cytochromes were produced by goats. In Ouchterlony double diffusion studies, all antisera yielded only one precipitin band with both purified and crude preparations of the corresponding antigen, thereby providing additional evidence for the homogeneity of the cytochromes. No cross-reactions observable by precipitin band formation were detected between anti-LM_2 sera and $P-450_{LM_4}$, or between anti-LM_4 sera and $P-450_{LM_2}$. Competitive binding studies with radiolabeled cytochromes confirmed that anti-LM_2 does not cross-react with $P-450_{LM_4}$; however, slight but significant cross-reactions were detected by this technique between goat anti-LM_4 and $P-450_{LM_2}$. These

results indicate that the two cytochromes have significant
structural differences. Johnson and colleagues (19,20) made
the important finding that treatment with TCDD increased a
single cytochrome species in microsomes of neonatal rabbit
liver as demonstrated by SDS-polyacrylamide gel electrophoresis. This species had the same mobility as form 6 isolated
from liver microsomes of TCDD-treated adult rabbits, and its
identity was confirmed by comparison of its peptide finger-
print with that of purified adult form 6. This assignment
was further supported by immunological studies which showed
that monospecific antibody developed against adult form 6
reacted with microsomes from TCDD-treated newborns but not
with microsomes from control newborns. As shown by these
investigators, two forms of cytochrome P-450 are induced by
TCDD in the adult rabbit: form 4, which is the major species
induced by TCDD, and form 6. This difference in the response of adult and neonatal rabbits to TCDD suggests that
one or more age-dependent factors control the induction
of form 4.

We have recently turned our attention to the forms of
P-450 which are constitutive — that is, which occur in untreated animals and for which no inducer has been found.
At least two of these occur in the electrophoretic region
between forms 2 and 4 and are therefore designated as form
3 with suitable subscripts. A fraction tentatively labeled
$P-450_{LM_{3b}}$ was partially purified earlier as a by-product
of $P-450_{LM_2}$ isolation (9,10,21) and shown to be active in
the hydroxylation of benzo[a]pyrene and benzo[a]pyrene-
7,8-dihydrodiol (22) as well as warfarin (23). We have
recently purified a form designated $P-450_{LM_{3b}}$ from control
rabbits to electrophoretic homogeneity and a specific content of over 19 nmol per mg of protein by chromatographic

procedures carried out in the presence of detergents (24)
while P-450$_{LM_{3c}}$ has also been obtained in pure form as a
by-product of P-450$_{LM}$ isolation from microsomes of phenobarbital-induced rabbits (25). The isozymes we are now
designating LM$_{3c}$ corresponds to the main component of the
partially purified fraction which was tentatively termed
LM$_{3b}$ (9). Several other research groups have also studied
cytochromes from rabbit liver microsomes with electrophoretic migration in the 3 region. Ingelman-Sundberg et al.
(26,27) have described the purification of "P-450$_{LM_3}$" from
phenobarbital-induced animals by an extension of our procedure for P-450$_{LM_2}$ and have reported that their preparation
catalyzes the hydroxylation of androstenedione at a higher
rate in the presence of reconstituted phospholipid vesicles.
Yamano and colleagues (28-30) have reported that a form
of cytochrome P-450 with a high affinity for cytochrome
$\underline{b}_5$ was isolated from liver microsomes of untreated animals
with the aid of chromatography on a column of immobilized
cytochrome $\underline{b}_5$; the form appears to correspond to a P-450$_{LM_3}$
in its electrophoretic mobility. Furthermore, Johnson (31,
32) has reported the electrophoretic properties and immuno-
reactivity of purified "form 3" from control (uninduced)
animals; forms 2, 3, 4, and 6 were distinguished by the
monospecificity of their antisera and by other methods.
In addition, Hansson and Wikvall (33) have isolated a form
3 active toward both ethylmorphine and 5β-cholestane-3α,-
7α-diol.

The important question therefore arises as to whether
the form 3's obtained by other investigators correspond
to P-450$_{LM_{3b}}$ or P-450$_{LM_{3c}}$ isolated in Ann Arbor. The ex-

change of small amounts of these proteins has allowed us to conclude by electrophoretic comparison that the enzyme kindly furnished by Dr. Johnson corresponds to our P-450$_{LM3b}$, and those furnished by Dr. Ingelman-Sundberg and Dr. Yamano correspond to our P-450$_{LM3c}$.

Philpot and colleagues (34) have obtained highly interesting results on the possible similarity of the forms of P-450 in rabbit liver and lung. Two forms of P-450 were partially purified from liver microsomes (35) and from lung microsomes (36). The lung cytochromes, designated forms I and II, were then obtained in an electrophoretically homogeneous state and shown to be immunochemically distinct, whereas form I, unlike form II, was indistinguishable from hepatic P-450$_{LM2}$ (37,38). Various lines of evidence showed that P-450$_{LM2}$ and lung form I are identical, even though the latter protein is not inducible by phenobarbital administration (39).

The data in Table I summarize some of the properties of the isozymes of microsomal P-450 from rabbit liver and lung. A striking characteristic of P-450$_{LM4}$ is that, as isolated in the ferric form, it is in the high spin state whereas the other proteins are all in the low spin state. The absorption maximum at 392 nm, which has been correlated with EPR evidence (9), indicates this difference. Although the minimal molecular weights of P-450$_{LM2}$ and lung form I appear to be slightly different, when we exchanged liver cytochrome preparations with Dr. Philpot the enzymes were found to have identical electrophoretic behavior. Accordingly, we conclude that this variation is within the range of experimental error because of the slightly different conditions employed.

Table I. Properties of Purified Isozymes of Cytochrome P-450 from Rabbit Tissues

Isozyme	Inducer	Minimal Molecular Weight	Absorption Maxima Ferric (nm)	Absorption Maxima Ferrous-Carbonyl (nm)
Liver 2	Phenobarbital	49,000	418	451
" 3b	(Constitutive)	52,000	418	450
" 3c	(Constitutive)	53,000	418	449
" 4	5,6-Benzoflavone[a]	55,000	392	447
" 6	TCDD	57,000	418	448
Lung I	(Constitutive)	51,000	418	451
" II	(Constitutive)	53,000	418	449

[a]Other compounds which serve as inducers for liver isozyme 4 are 3-methycholanthrene, TCDD, and isosafrole.

The results presented in Table II show a comparison of
the activities of the various purified cytochromes toward
selected substrates. All of the liver isozymes are active
with all of the compounds tested, but exhibit different turnover numbers. The data indicate that, under these experimental conditions with the cytochrome present in limiting
amount and other components at apparently saturating levels,
form 2 is most effective with benzphetamine, 3b with amino-
pyrene and testosterone, 4 with acetanilide, and 6 with benzo-
[a]pyrene. As might be expected, lung form I is similar
to liver form 2, but lung form II is active with benzo[a]-
pyrene and not benzphetamine.

NADPH-Cytochrome P-450 Reductase. In contrast to cytochrome
P-450, the flavoprotein in this enzyme system occurs as a
single protein species. After the liver microsomal hydroxylation system was reconstituted from detergent-solubilized
P-450 and NADPH-cytochrome P-450 reductase in the presence
of phospholipid (1,2), we found that trypsin-solubilized
"NADPH-cytochrome c reductase" (40) could not replace the
latter enzyme (41,42). More recently, the detergent-solubilized reductase has been obtained in highly purified form
from rat (6,43-45) and rabbit liver microsomes (46-47), thereby permitting a structural comparison with the corresponding
protease-treated preparations. As shown in Table III, the
rat and rabbit reductases are highly similar in their properties, whether one compares the intact or trypsin-treated
forms. Upon cleavage by this protease the enzyme exhibits
a significant increase in activity toward cytochrome c, and,
as already stated, loses its ability to donate electrons
to cytochrome P-450.

Table II. Substrate Specificity of Isozymes of Cytochrome P-450 from Rabbit Tissues

| Substrate | Turnover Number with Various Isozymes (nmol/min/nmol P-450) | | | | | | |
| | Liver | | | | | Lung | |
	2	3b	3c	4	6	I	II
Benzphetamine	63	12.5	6.4	4.9	4.0	75	0
Aminopyrine	8.0	22	1.2	11.6	2.4		
Acetanilide	0.06	0.56	1.0	5.01	1.4		
Testosterone, 6β	0.08	1.8	0.37	0.60			
" 7α	0.05	0.04	0.04	0.03			
" 16α	0.18	2.1	0.05	0.03			
Benzo(a)pyrene (total metabolites)	0.71	0.29	0.02	0.16	4.0	1.0	0.8

Table III. Properties of Reductase Purified from Rat and Rabbit Liver Microsomes

Enzyme	Minimal Molecular Weight	Absorption Maxima (nm)	Turnover number	
			P-450 as acceptor	Cytochrome c as acceptor
Rat reductase, detergent-solubilized	76,000	384,456,485(s)	235	4,010
Rat reductase, trypsin-treated	69,000	384,456,485(s)	<1	5,000
Rabbit reductase, detergent-solubilized	78,000	382,456,485(s)	93	4,030
Rabbit reductase, trypsin-treated	71,000	382,456,484(s)	<1	5,950

Cytochrome c reduction was determined at 30° under aerobic conditions in phosphate buffer, pH 7.7. P-450$_{LM_2}$ reduction was estimated at 30° from the rate of NADPH oxidation in the complete reconstituted enzyme system containing benzphetamine.

We have recently described the isolation and properties
of a small hydrophobic peptide derived from the amino-terminal region of the native reductase obtained from rabbit
liver microsomes upon treatment with trypsin (48). The
large, soluble peptide obtained in this process has previously been purified and characterized in other laboratories
(40,49). The small peptide appears to be essential in the
binding of the native reductase to cytochrome P-450 and
to the microsomal membrane, and exhibits considerable hydrophobic character (48). Gum and Strobel (50) have recently
concluded that a membrane-binding domain is present in the
rat liver reductase, but the amino acid composition of the
domain, calculated by the difference between the detergent-
and steapsin-solubilized reductase preparations, was not
especially hydrophobic.

Detergent-solubilized NADPH-cytochrome P-450 reductase
contains one molecule each of FMN and FAD (6), as shown
earlier by Iyanagi and Mason (49) for the trypsin-solubilized
preparation. Anaerobic spectral titrations established
that the higher molecular weight, detergent-solubilized preparation contains no additional electron acceptors (45).
Selective removal of FMN by treatment with KBr proved to
be a particularly useful way of determining the properties
of the individual flavins (51,52). Especially striking
was the loss in ability to catalyze electron transfer to
the phenobarbital-inducible form of liver microsomal cytochrome P-450 in the reconstituted hydroxylation system as
well as to cytochrome c and some other artificial acceptors,
but activity was retained toward ferricyanide and 3-acetyl-
pyridine adenine dinucleotide phosphate. All catalytic
activities were restored when the depleted enzyme was incu-

bated with FMN. From fluorescence measurements a value
of 1.3 x 10^{-8} M was determined for the FMN dissociation
constant under the conditions used. Riboflavin and FAD
were also bound by the FMN-depleted enzyme, but less effectively than FMN.

A series of spectrophotometric experiments were carried
out to determine whether the properties of the FMN-depleted
enzyme correspond to those of the high or low potential
flavin of the native enzyme, with E' values of -0.190 and
-0.328 V, respectively (52). Addition of NADP to the fully
reduced, FMN-depleted reductase resulted in significant
oxidation of flavin, indicating a midpoint potential for
FAD near, rather than above, that of the pyridine nucleotide couple. The semiquinone form of the FMN-depleted reductase, which was produced during air oxidation of NADPH-
reduced enzyme or during stepwise photochemical reduction
of oxidized enzyme under anaerobic conditions, had spectral characteristics similar to those of the semiquinone
of the low potential flavin of the native enzyme and was
readily oxidized under aerobic conditions. Addition of
oxidized FMN to 1-electron-reduced, FMN-depleted reductase
under anaerobic conditions produced an enzyme species with
properties similar to those of the 1-electron-reduced form
of the native enzyme, thereby indicating that electron
transfer from FAD to FMN is thermodynamically favorable.
These observations establish that the low and high potential flavins of the reductase are FAD and FMN, respectively. Thus, the midpoint potentials determined by Iyanagi
et al. (53) could be assigned to specific flavins, as
follows:

$$FMN \rightleftharpoons FMNH \cdot \quad (E'_o = -0.110 \text{ V})$$
$$FMNH \cdot \rightleftharpoons FMNH_2 \quad (E'_o = -0.270 \text{ V})$$
$$FAD \rightleftharpoons FADH \cdot \quad (E'_o = -0.290 \text{ V})$$
$$FADH \cdot \rightleftharpoons FADH_2 \quad (E'_o = -0.365 \text{ V})$$

Since the FMN-depleted enzyme is capable of accepting electrons from NADPH (as judged by catalysis of ferricyanide reduction) but incapable of transferring electrons to $P\text{-}450_{LM_2}$, the pattern of electron transfer was concluded to be: NADPH $\longrightarrow$ FAD $\longrightarrow$ FMN $\longrightarrow$ P-450.

Rat liver microsomal NADPH-cytochrome P-450 reductase was recently prepared free of detectable amounts of FMN by a new procedure based on the exchange of this flavin into apoflavodoxin (54). The resulting FMN-free reductase binds NADP in the oxidized state with the same affinity (K_d = 6.4 μM) as does the native enzyme (K_d = 4.4 μM), and in both cases the stoichiometry is 1 mol of NADP bound per mol of enzyme. Both the native and FMN-free reductase catalyze rapid reduction of ferricyanide, but the ability to reduce $P\text{-}450_{LM_4}$ is lost upon removal of FMN, as shown earlier with the FMN-depleted enzyme.

The FMN-free enzyme was reconstituted with artificial flavins which in the free state have oxidation-reduction potentials ranging from -152 to -290 mV, including several FMN analogs with a halogen or sulfur substituent on the dimethylbenzene portion of the flavin ring system, and 5-carba-5-deaza-FMN. Enzyme reconstituted with 5-carba-5-deaza-FMN has catalytic properties which are not significantly different from those of the FMN-free reductase, and is unable to reduce $P\text{-}450_{LM_4}$. On the other hand, the ability to reduce to $P\text{-}450_{LM_4}$ and the other FMN-dependent activities of the native reductase are restored by substitution of several other analogs for FMN. The reduction of

$P-450_{LM_4}$ by enzyme containing functional analogs was studied under anaerobic conditions by stopped-flow spectrophotometry, and it was found that the kinetics of this reaction are significantly altered when compared to the native enzyme.

The spectral and oxidation-reduction properties of enzyme reconstituted with 7-nor-7-Br-FMN and 8-nor-8-mercapto-FMN were examined in some detail. In the former case, the oxidation-reduction behavior is substantially different from that of the native enzyme, and there is less thermodynamic stabilization of the semiquinone of this flavin analog. The oxidation-reduction properties of enzyme containing 8-nor-8-mercapto-FMN are similar to those of the native enzyme, but there is a substantial difference in the spectral properties of this artificial enzyme. It was shown in a stopped-flow experiment that reduction of this FMN analog precedes reduction of $P-450_{LM_4}$ when a complex of the flavoprotein and $P-450_{LM_4}$ is allowed to react with NADPH. Our experiments support the sequence of electron transfer in this system discussed above, NADPH $\rightarrow$ FAD $\rightarrow$ FMN $\rightarrow$ P-450.

Even before the physiological role of the reductase was fully realized, studies on the reaction with cytochrome c and other artificial acceptors were initiated by Masters et al. (55,56) as a result of general interest in the mechanism of flavin-catalyzed electron transfer. Since it contains no known oxidation-reduction groups other than flavin, the enzyme falls into the category of a simple flavoprotein of the dehydrogenase-electron transferase class (57). Another well studied example of an enzyme of this type with nonequivalent flavins is a component of the bacterial sulfite reductase. This protein also contains FMN

and FAD and has a similar electron transfer function, but
is a large multi-enzyme complex (58) containing other oxidation-reduction centers. Although there are several differences between this flavoprotein and P-450 reductase in both
the catalytic and oxidation-reduction properties, the mechanism proposed by Siegel et al. (59,60), which involves separate roles for FMN and FAD, may also apply to the reaction catalyzed by P-450 reductase. This mechanism dictates
that the enzyme cycle between a 1e- and 3e-reduced state,
with reducing equivalents entering the enzyme in the reaction of NADPH with FAD. The earlier studies on the reaction of the protease-treated P-450 reductase (55,56) have
already provided evidence that in the reduction of cytochrome c the enzyme is never fully oxidized during turnover,
but cycles between two different reduced states; it is now
understood that the more oxidized state is the 1e-reduced
species, FMNH·-FAD, referred to as the air-stable semiquinone, and that this form of the enzyme will not reduce
P-450$_{LM}$ until it first reacts with NADPH (61). Figure 1A
is a representation of a turnover event for a two-flavin
enzyme in which the role of FAD is to accept reducing equivalents from NADPH, with subsequent transfer of electrons
to an oxidized acceptor in a reaction carried out by FMN.
This mechanism is essentially that proposed by Siegel et
al. (59,60) and, as these authors point out, provides a
means by which reducing equivalents from NADPH may eventually be transferred in two equipotential 1e-transfer steps.
Figure 1B shows an alternate scheme which accomplishes the
same overall process, but the role of FMN in this case is
indirect; transfer of electrons to oxidized acceptors is
carried out by FADH·, which is produced in the intramolecular disproportionation of FADH$_2$ by oxidized FMN. Electron

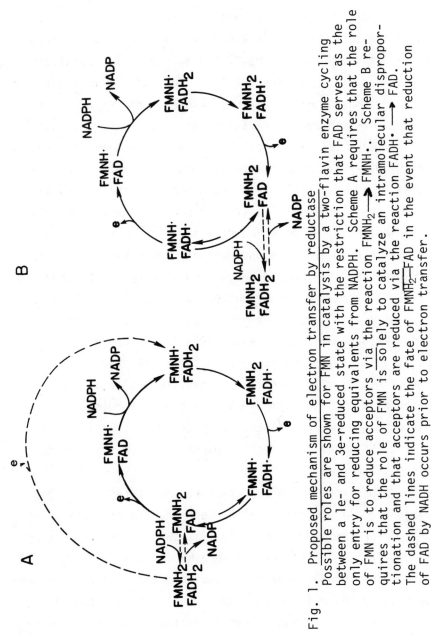

Fig. 1. Proposed mechanism of electron transfer by reductase
Possible roles are shown for FMN in catalysis by a two-flavin enzyme cycling
between a 1e- and 3e-reduced state with the restriction that FAD serves as the
only entry for reducing equivalents from NADPH. Scheme A requires that the role
of FMN is to reduce acceptors via the reaction FMNH$_2$ → FMNH·. Scheme B requires that the role of FMN is solely to catalyze an intramolecular disproportionation and that acceptors are reduced via the reaction FADH· → FAD.
The dashed lines indicate the fate of FMNH$_2$—FAD in the event that reduction
of FAD by NADH occurs prior to electron transfer.

transfer reactions catalyzed assuming this mechanism would show an FMN dependence even though it is the FAD which interacts directly with acceptors. As a mechanism for catalysis by P-450 reductase, Scheme B falls into disfavor in light of the following considerations. There is no reason to expect that, once formed, the intermediate 2e-reduced species which contains FAD in the oxidized state is not readily reduced by NADPH to yield the fully reduced enzyme. The fate of this 2e-reduced species is indicated in both schemes by dashed lines. Under the restrictions of Scheme B, the fully reduced enzyme would be unreactive in electron transfer and represent a dead end. In the mechanism outlined in Scheme A, however, the 4e-reduced enzyme should be perfectly capable of reducing acceptors and re-entering the cycle as shown. Oprian et al. (61) have shown that the reaction of the P-450$_{LM_4}$·reductase complex with NADPH in the concentration range from 2 μM to 1 mM does not alter either the extent or the kinetics of the reaction; thus, reduction is not inhibited at high levels of NADPH. This behavior is incompatible with Scheme B, making it very probable that it is indeed the FMN which is directly responsible for electron transfer to cytochromes.

Our experiments with reductase reconstituted with FMN analogs are all consistent with the mechanism of electron transfer for the reductase which is shown in Fig. 1A. We have provided evidence that the enzyme contains a single binding site for pyridine nucleotide and that the rate of reduction by NADPH is only minimally affected by removal of FMN. Preparation and characterization of FAD-free reductase will be required to fully rule out a possible role for FMN in reactions with NADPH. In addition, our preliminary experiments with the 8-mercapto-FMN-reductase suggest that study of this artificial enzyme, which is as efficient in the reduction of P-450$_{LM_4}$ as the native enzyme, may be one of the few experimental approaches which allow the reactions of each flavin to be monitored independently.

SUMMARY

Recent studies in this and other laboratories have shown that rabbit liver microsomes contain multiple forms of cyto-chrome P-450 whose physical and chemical properties are sufficiently different to justify their being termed isozymes. These proteins vary in their inducibility and have somewhat different but overlapping substrate specificities.

NADPH-cytochrome P-450 reductase has been obtained free of FMN and reconstituted with artificial flavins. Various experiments support the following electron transfer sequence in the microsomal hydroxylation system: NADPH $\longrightarrow$ FAD $\longrightarrow$ FMN $\longrightarrow$ P-450 $\longrightarrow$ O_2. A scheme is presented to indicate the possible redox states occurring when FMNH$_2$ reacts with P-450 to produce the flavin semiquinone, FMNH·, and the ferrous cytochrome.

REFERENCES

1. Lu, A. Y. H., and Coon, M. J. (1968) J. Biol. Chem. 243, 1331-1332.
2. Coon, M. J., and Lu, A. Y. H. (1969) in Microsomes and Drug Oxidations (J. R. Gillette et al., eds.) pp. 151-166, Academic Press, New York.
3. Strobel, H. W., Lu, A. Y. H., Heidema, J., and Coon, M. J. (1970) J. Biol. Chem. 245, 4851-4854.
4. van der Hoeven, T. A., Haugen, D. A., and Coon, M. J. (1974) Biochem. Biophys. Res. Commun. 60, 569-575.
5. Imai, Y., and Sato, R. (1974) Biochem. Biophys. Res. Commun. 60, 8-14.

6. Vermilion, J. L., and Coon, M. J. (1974) Biochem. Biophys. Res. Commun. 60, 1315-1322.
7. Haugen, D. A., van der Hoeven, T. A., and Coon, M. J. (1975) J. Biol. Chem. 250, 3567-3570.
8. Lu, A. Y. H., and West, S. B. (1980) Pharmacol. Rev. 31, 277-295.
9. Haugen, D. A., and Coon, M. J. (1976) J. Biol. Chem. 251, 7929-7939.
10. Coon, M. J., van der Hoeven, T. A., Dahl, S. B., and Haugen, D. A. (1978) Methods Enzymol. 52, 109-117.
11. Imai, Y., Hashimoto-Yutsudo, C., Sataki, H., Girardin, A., and Sato, R. (1980) J. Biochem. (Tokyo) 88, 489-503.
12. Kawalek, J. C., Levin, W., Ryan, D., Thomas, P. E., and Lu, A. Y. H. (1975) Mol. Pharmacol. 11, 874-878.
13. Hashimoto, C., and Imai, Y. (1976) Biochem. Biophys. Res. Commun. 68, 821-827.
14. Johnson, E. F., and Muller-Eberhard, U. (1977) Biochem. Biophys. Res. Commun. 76, 652-658.
15. Johnson, E. F., and Muller-Eberhard, U. (1977) J. Biol. Chem. 252, 2839-2845.
16. Oprian, D. D., and Coon, M. J. (1980) in Biochemistry, Biophysics and Regulation of Cytochrome P-450, J.-A. Gustafsson et al., eds.) pp. 323-330, Elsevier/North Holland Biomedical Press, Amsterdam.
17. Haugen, D. A., Armes, L. G., Yasunobu, K. T., and Coon, M. J. (1977) Biochem. Biophys. Res. Commun. 77, 967-973.
18. Dean, W. L., and Coon, M. J. (1977) J. Biol. Chem. 252, 3255-3261.
19. Norman, R. L., Johnson, E. F., and Muller-Eberhard, U. (1978) J. Biol. Chem. 253, 8640-8647.
20. Johnson, E. F. (1980) in Biochemistry, Biophysics and Regulation of Cytochrome P-450 (J.-A. Gustafsson et al., eds.) pp. 65-72, Elsevier/North Holland Biomedical Press, Amsterdam.
21. Coon, M. J., Vermilion, J. L., Vatsis, K. P., French, J. S., Dean, W. L., and Haugen, D. A. (1977) in Drug Metabolism Concepts (D. M. Jerina, ed.) American Chemical Society Symposium Series, No. 44, pp. 46-71.
22. Deutsch, J., Leutz, J. C., Yang, S. K., Gelboin, H. V., Chiang, Y. L., Vatsis, K. P., and Coon, M. J. (1978) Proc. Natl. Acad. Sci. USA 75, 3123-3127.
23. Fasco, M. J., Vatsis, K. P., Kaminsky, L. S., and Coon, M. J. (1978) J. Biol. Chem. 253, 7813-7820.
24. Koop, D. R., and Coon, M. J. (1979) Biochem. Biophys. Res. Commun. 91, 1075-1081.

25. Coon, M. J., Koop, D. R., Persson, A. V., and Morgan, E. T. (1980) in Biochemistry, Biophysics and Regulation of Cytochrome P-450 (J.-A. Gustafsson et al., eds.) pp. 7-15, Elsevier/North Holland Biomedical Press, Amsterdam.

26. Ingelman-Sundberg, M., Johansson, I., and Hansson, A. (1979) Acta Biol. Med. Ger., 38, 379-388.

27. Ingelman-Sundberg, M., Glaumann, H., and Johansson, I. (1980) in Microsomes, Drug Oxidations, and Chemical Carcinogenesis (M. J. Coon et al., eds.) pp. 513-516, Academic Press, New York.

28. Miki, N., Sugiyama, T., and Yamano, T. (1980) in Microsomes, Drug Oxidations, and Chemical Carcinogenesis (M. J. Coon et al., eds.) pp. 27-36, Academic Press, New York.

29. Sugiyama, T., Miki, N., and Yamano, T. (1980) J. Biochem. (Tokyo) 87, 1457-1467.

30. Miki, N., Sugiyama, T., and Yamano, T. (1980) J. Biochem. (Tokyo) 88, 307-316.

31. Johnson, E. F. (1980) in Microsomes, Drug Oxidations, and Chemical Carcinogenesis (M. J. Coon et al., eds.) pp. 143-146, Academic Press, New York.

32. Johnson, E. F. (1980) J. Biol. Chem. 255, 304-309.

33. Hansson, R., and Wikvall, K. (1980) J. Biol. Chem. 255, 1643-1649.

34. Slaughter, S. R., Wolf, C. R., Serabjit-Singh, C. J., Marciniszyn, J. P., and Philpot, R. M. (1980) in Biochemistry, Biophysics and Regulation of Cytochrome P-450 (J.-A. Gustafsson et al., eds.) pp. 41-48, Elsevier/North Holland Biomedical Press, Amsterdam.

35. Philpot, R. M., and Arinc, E. (1976) Mol. Pharmacol. 12, 483-493.

36. Wolf, C. R., Szutowski, M. M., Ball, L. M., and Philpot, R. M. (1978) Chem.-Biol. Interactions 21, 29-43.

37. Wolf, C. R., Smith, B. R., Ball, L. M., Serabjit-Singh, C. J., Bend, J. R., and Philpot, R. M. (1979) J. Biol. Chem. 254, 3658-3663.

38. Serabjit-Singh, C. J., Wolf, C. R., and Philpot, R. M. (1979) J. Biol. Chem. 254, 9901-9907.

39. Wolf, C. R., Slaughter, S. R., Marciniszyn, J. P., and Philpot, R. M. (1980) Biochim. Biophys. Acta 624, 409-419.

40. Masters, B. S. S., Williams, C. H., Jr., and Kamin, H. (1967) Methods Enzymol. 10, 565-573.

41. Lu, A. Y. H., Junk, K. W., and Coon, M. J. (1969) J. Biol. Chem. 244, 3714-3721.

42. Coon, M. J., Strobel, H. W., and Boyer, R. F. (1973) Drug Metab. Disp. 1, 92-97.
43. Yasukochi, Y., and Masters, B. S. S. (1976) J. Biol. Chem. 251, 5337-5344.
44. Dignam, J. D., and Strobel, H. W. (1977) Biochemistry 16, 1116-1123.
45. Vermilion, J. L., and Coon, M. J. (1978) J. Biol. Chem. 253, 2694-2704.
46. Iyanagi, T., Anan, F. K., Imai, Y., and Mason, H. S. (1978) Biochemistry 17, 2224-2230.
47. French, J. S., and Coon, M. J. (1979) Arch. Biochem. Biophys. 195, 565-577.
48. Black, S. D., French, J. S., Williams, C. H., Jr., and Coon, M. J. (1979) Biochem. Biophys. Res. Commun. 91, 1528-1535.
49. Iyanagi, T., and Mason, H. S. (1973) Biochemistry 12, 2297-2308.
50. Gum, J. R., and Strobel, H. W. (1979) J. Biol. Chem. 254, 4177-4185.
51. Vermilion, J. L., and Coon, M. J. (1976) in Flavins and Flavoproteins (T. P. Singer, ed.) pp. 674-678, Elsevier Scientific Publishing Co., Amsterdam.
52. Vermilion, J. L., and Coon, M. J. (1978) J. Biol. Chem. 253, 8812-8819.
53. Iyanagi, T., Makino, N., and Mason, H. S. (1974) Biochemistry 13, 1701-1710.
54. Vermilion, J. L., Ballou, D. P., Massey, V., and Coon, M. J. (1981) J. Biol. Chem. 256, 266-277.
55. Masters, B. S. S., Kamin, H., Gibson, Q. H., and Williams, C. H., Jr. (1965) J. Biol. Chem. 240, 921-931.
56. Masters, B. S. S., Bilimoria, M. H., Kamin, H., and Gibson, Q. H. (1965) J. Biol. Chem. 240, 4081-4088.
57. Hemmerich, P., Massey, V., and Fenner, H. (1977) FEBS Lett. 84, 5-21.
58. Faeder, E. J., Davis, P. S., and Siegel, L. M. (1974) J. Biol. Chem. 249, 1599-1609.
59. Siegel, L. M., Faeder, E. G., and Kamin, H. (1972) Z. Naturforsch. Teil. B. 276, 1087-1089.
60. Siegel, L. M., Kamin, H., Rueger, D. C., Presswood, R. P., and Gibson, Q. H. (1971) in Flavins and Flavoproteins (H. Kamin, ed.) pp. 523-554, University Park Press, Baltimore.
61. Oprian, D. D., Vatsis, K. P., and Coon, M. J. (1979) J. Biol. Chem. 254, 8895-8902.

MEMBRANE-ASSOCIATED REACTIONS INVOLVING GLUTATHIONE[1]

Bengt Mannervik

Department of Biochemistry, Arrhenius Laboratory,
University of Stockholm, S-106 91 Stockholm, Sweden.

INTRODUCTION

An overview of the multitude of chemical reactions which
are associated with biological membranes is difficult to
achieve within reasonable limits of space. Likewise, to review the reactions dependent on the tripeptide glutathione
(γ-L-glutamyl-L-cysteinylglycine) is a formidable task. Hence, a selection of topics among the membrane-associated reactions involving glutathione must, of necessity, have a fla-
vor of personal bias. The work in this area of research can
be expected to grow at an accelerating rate in the future
owing to the cross-fertilization achieved by combining knowledge and experimental methods from both fields of investigation. At present, our perception of membrane-associated reactions involving glutathione is only fragmentary as evidenced

[1] Work from the author's laboratory was supported by the Swedish Natural Science Research Council, the Swedish Cancer
Society, and the Swedish Council for Planning and Coordination of Research.

C. P. Lee, G. Schatz, G. Dallner (eds.), Mitochondria and Microsomes
in honor of Lars Ernster   ISBN 0-201-04576-1

by searching for this topic in the various recent symposium volumes and review articles on glutathione (1-8).

OCCURRENCE OF GLUTATHIONE

Glutathione is the most abundant low-molecular-weight thiol in all kinds of organisms and cells (9). Nevertheless, the total concentration of glutathione, including reduced glutathione and glutathione disulfide, ranges from 0.5 to 10 mM, and differs from organ to organ in the same organism. In view of the various protective functions proposed for glutathione, such differences may to some extent explain why certain tissues or parts of them are more susceptible than others to necrotic or carcinogenic chemicals. For example, the concentration of cytosolic glutathione in rat lung was determined as 2 mM, which is only about 20% of the average concentration in liver (10). Similarly, the glandular stomach of rodents, which like liver is highly resistant to carcinogenesis by polycyclic aromatic hydrocarbons, has significantly higher glutathione concentration than more susceptible parts of the gastro-intestinal tract (11). Moreover, the concentration of glutathione in the rat liver lobule is significantly lower (<50%) in hepatocytes located within 100 μm of the central vein than in cells located elsewhere in the same tissue (12). Intracellular gradients of glutathione concentration should also be considered as possible, but the measured values for cytosol (8.2 mM) and mitochondrial matrix (11.0 mM) in rat liver (13) do not indicate any major differences. Possibly, local depletion by glutathione--consuming reactions might cause transient spatial concentration gradients within a cell.

The strongly polar and zwitter-ionic nature of the glutathione molecule makes it highly water-soluble. The molecule has low solubility in a hydrophobic milieu and any glutathione associated with membranes would be expected to be bound to glutathione-dependent proteins or to be linked covalently as mixed disulfides with sulfhydryl-containing constituents of the membranes. A large proportion of the cytosolic glutathione has actually been found to exist as mixed disulfides with proteins (see Ref. 14, for a review), but to what extent this applies to membranes appears to be unknown.

The majority of functions established for glutathione involves the reduced form. The maintenance of a high level of reduced glutathione (GSH) is secured by glutathione reductase, which catalyzes the reduction of glutathione disulfide (GSSG):

$$GSSG + NADPH + H^+ \rightarrow 2\ GSH + NADP^+$$

Glutathione reductase in rat liver has been found to be present in cytosol (65.8%), mitochondria (32.4%), and in small amounts in peroxisomes and microsomes; no significant activity was demonstrated in the nucleus or lysosome fractions (15). Subfractionation of mitochondria showed that the enzyme was partitioned between the matrix and the intermembrane space, no significant activity being associated with the outer or inner mitochondrial membranes (15). The ratio [GSSG]/[GSH] appears to be near 0.05 in various cells (see Ref. 16, for a review). This value corresponds to much more glutathione disulfide than calculated for the thermodynamic equilibrium of the reaction catalyzed by glutathione reductase. An intriguing question is therefore how this ratio can be maintained at the normal poise of the NADPH/NADP$^+$ redox couple (17,18).

PROTECTIVE FUNCTIONS OF GLUTATHIONE

Three classes of reactions which may be deleterious to cells and their membrane systems are counteracted by glutathione: reactions involving (i) free radicals, (ii) oxidative chemical species, and (iii) electrophilic (e.g. alkylating and arylating) components.

RADICAL SCAVENGING. Clear-cut examples of generation of free radicals include exposure of cells to ionizing radiation. It is well-known that free sulfhydryl groups are highly sensitive to radicals. A hydrogen atom of the thiol group (-SH) is easily removed by the radical (R·) to yield a thiyl radical (-S·) as a product:

R· + -SH → RH + -S·

Sensitivity of cells to irradiation appears to be related to the presence of susceptible sulfhydryl groups, and parameters such as membrane permeability (as a measure of damage) are correlated to oxidation of membrane thiol groups (see Refs. 19 and 20 for reviews). It seems obvious that intracellular glutathione may afford protection by scavenging free radicals which may otherwise damage cells. A particularly interesting proposal is that the chemical radioprotector cysteamine exerts its effect by liberating endogenous glutathione which is bound in the form of mixed disulfides with protein sulfhydryl groups (21). Even if the role of glutathione as an endogenous radioprotector in the cell needs further clarification, the capacity of this abundant thiol to react with radicals produced by irradiation as well as by cellular redox reactions is well established. Some examples of the latter category of reactions are the generation of radicals from chloroform and other halogen-substituted hydrocarbons as well as from aromatic compounds (see Refs. 22 and 23 for reviews). Many of these radical-generating reactions are catalyzed by the microsome fraction of the cell.

PROTECTION AGAINST OXIDATIVE PROCESSES

Lipid peroxidation. In this volume it may be considered appropriate to discuss the function of glutathione in preventing cellular damage by lipid peroxidation. Hochstein and Ernster (24,25) discovered lipid peroxidation as a microsome-associated process stimulated by NADPH and iron chelates (Fe^{2+}--ADP). Such peroxidation of polyunsaturated fatty acids of lipids takes place also in mitochondria and lysosomes (26).

Recent investigations in many laboratories indicate that lipid peroxidation may be a fundamental mechanism of toxicity which can be initiated by singlet oxygen as well as by free radicals produced in the cell by various xenobiotics (26). For example, the hepatotoxic effect of carbon tetrachloride has been ascribed to initiation of lipid peroxidation by the trichloromethyl radical formed from the parent compound by the action of the microsomal monooxygenase system. The endoplasmic reticulum and the mitochondria, which are rich in polyunsaturated fatty acids, are particularly prone to peroxidative degradation. Glutathione provides several ways of protecting the lipids against peroxidation: (i) it can act as a scavenger of free radicals; (ii) it can reduce hydrogen peroxide and lipid hydroperoxides; and, in some cases, (iii) it can deactivate compounds by conjugation, thereby facilitating their excretion. A protective function of glutathione in vivo is indicated by studies on whole ani-

mals (27) and isolated hepatocytes (28-30) that have been depleted of glutathione. In both systems extensive decreases
of the glutathione concentration gave rise to lipid peroxidation and concomitant cell injury.

Attempts to demonstrate the protective role of glutathione in cells have been made by introduction of diazene reagents aiming at specific oxidation of glutathione to its disulfide form (31). Even if the effect of the diazene reagents
is not limited to oxidation of glutathione (32), it appears
clear that the use of the reagents has demonstrated the crucial role of glutathione in maintaining the integrity of biological membranes that are challenged by the oxidative stress
of an aerobic milieu (31,33).

Role of glutathione peroxidases. Oxygen gives rise to hydrogen peroxide and superoxide anion in the cell. The latter
product reacts further to give hydrogen peroxide as well as
organic hydroperoxides from unsaturated hydrocarbon chains
in cellular membranes. Both mitochondria and microsomes produce substantial amounts of hydrogen peroxide, and peroxisomes provide another source of this toxic oxidant (34).
Well-known inducers of drug-metabolizing enzymes have been
found to increase the production of hydrogen peroxide in microsomes without changing the steady-state levels in perfused
rat liver (see Ref. 35). Thus it appears essential to have
mechanisms for protection against this oxygen metabolite. In
addition to the reaction catalyzed by catalase, inactivation
of hydrogen peroxide can be effected by glutathione. The latter reaction is catalyzed by glutathione peroxidase:

$$2 \text{ GSH} + \text{H}_2\text{O}_2 \rightarrow \text{GSSG} + 2 \text{ H}_2\text{O}$$

This enzyme has been found in the cytosol (73.3%) and the

mitochondria (25.9%) of rat liver, but only in negligible
amounts in the peroxisomes (15). In fact, glutathione peroxidase has been identified with one of the contraction factors
of mitochondria (36). A role for the enzyme in the modulation
of mitochondrial oxidations in liver has also been proposed
(37). It appears that catalase and glutathione peroxidase
act complementarily in the protection of the cell. Glutathione peroxidase that is active with hydrogen peroxide contains
selenium in the form of selenocysteine in its active site
(38). The functional role of selenium would seem to explain
the importance of this element in protection against oxidative damage of the cell.

The selenium-containing glutathione peroxidase is specific for glutathione as the thiol substrate, but unspecific
for the hydroperoxide substrate (39). Organic hydroperoxides
(ROOH) of lipids and nucleic acid components as well as simpler aliphatic compounds can be reduced to their corresponding
alcohols (ROH):

$$2 \; GSH + ROOH \rightarrow GSSG + ROH + H_2O$$

Thus, various cellular components which may form hydroperoxides may be salvaged by the action of glutathione peroxidase.
Rat liver mitochondria suspensions were found to be capable
of reducing various organic hydroperoxides added (40). Glutathione peroxidase appeared to be involved in the reduction.
The entry of the hydroperoxides into the mitochondrial matrix was an energy-requiring process as judged by the effects
of uncouplers and inhibitors of the respiratory chain. More
recently, it was discovered that glutathione transferases,
known to catalyze conjugations of electrophiles with glutathione, also have glutathione peroxidase activity (41,42).
In rat liver, glutathione transferases B and AA were found to

have particularly high specific activities, whereas transferases A and C were only slightly active (43-45). The different types of glutathione transferase identified in human
tissues have also been found to have significant differences
in specific activity with organic hydroperoxides (45,46). An
important difference between the selenium-containing enzyme
and the glutathione transferases, also called non-selenium-
-dependent glutathione peroxidases, is that the latter can
not reduce hydrogen peroxide. Some species, such as the guinea pig, appear to lack the selenium-dependent enzyme and
have to rely entirely on the activity of the glutathione
transferases (47). The available evidence indicates that, in
general, the selenium- and the non-selenium-dependent activities are complementary.

Recently an apparently new glutathione peroxidase activity was detected in liver and cardiac mitochondria of mice
(48). It was reported to be located in the inter-membrane
space of the mitochondrion. The activity was present in selenium-deficient mice, demonstrating its distinction from selenium-dependent glutathione peroxidase. The finding that hydrogen peroxide was a substrate for the enzyme (48) indicates
that it is distinct from the glutathione transferases, provided that these enzymes in mice, like those in rats and humans (38,45,46), are inactive with this hydroperoxide.

Glutathione oxidase activity. In addition to hydroperoxides,
oxygen itself can be reduced by an enzyme-catalyzed reaction:

$$2 \text{ GSH} + O_2 \rightarrow \text{GSSG} + H_2O_2$$

The highest activity was found in kidneys (49,50) and recent
studies have shown that the oxidase activity is associated
with the plasma membranes (51,52). Purified preparations of

γ-glutamyltranspeptidase have been found to exhibit glutathione oxidase activity as well (53), but chromatographic
separation of the oxidase and transpeptidase activities gives strong support to the conclusion that two separate enzymes exist (51). The enzyme is inhibited by chelators and
activity is restored by addition of Cu^{2+}. The oxidase can
use other thiols than glutathione as substrates (51,54). Its
biological function is unknown.

Efflux of glutathione disulfide. Active transport of glutathione disulfide as a response to oxidative stress was first
demonstrated in erythrocytes (55). The transport system was
characterized kinetically by use of inside-out vesicles from
erythrocytes (56). This efflux of glutathione disulfide appears to be a more general mechanism for elimination of "oxidizing equivalents" from cells. It has been studied in perfused rat livers exposed to t-butyl hydroperoxide and compounds giving rise to hydrogen peroxide (57,58). Liver cells
release both reduced glutathione and glutathione disulfide
under normal conditions, but the increased efflux caused by
oxidative stress is due entirely to glutathione disulfide
(59). Rats treated with the inducer phenobarbital demonstrate an increased efflux rate of glutathione disulfide during
drug oxidation catalyzed by cytochrome P-450 (35). This finding is consistent with the view that reduced glutathione is
involved as a substrate in the reduction of oxidation products, such as hydroperoxides, of the monooxygenase-catalyzed
reactions. Even though the hepatic glutathione reductase level
is increased by inducers of drug-metabolizing enzymes (I.
Carlberg, J.W. DePierre, and B. Mannervik, unpublished results), the enhanced efflux of glutathione disulfide may be

a necessary complement to reduction in order to prevent exhaustion of reduced pyridine nucleotides (primarily NADPH)
in the cell. Hence, active export of glutathione disulfide
should be regarded as part of the protective mechanism against
oxidative stress on cells.

BIOTRANSFORMATION OF XENOBIOTICS AND ENDOGENOUS SUBSTRATES

Conjugation with glutathione. Electrophilic carbon, sulfur,
nitrogen, or oxygen atoms in organic molecules may react with
the nucleophilic sulfur atom of glutathione. The group of
glutathione transferases catalyze such reactions with a very
large number of xenobiotics as well as with some endogenous
substances (60,61). The latter category of substrates include
prostaglandins (62) and leukotrienes (63). Most of the glutathione transferase activity is localized in the cytosol (61),
but activity has also been found in microsomes (64,65) and
in mitochondria (13,66). The mitochondrial matrix space has
about 7% of the activity in the cytosol (13), and evidence
for three different forms of glutathione transferase in mitochondria has been reported (66). Likewise, the microsomal
glutathione transferase activity has been resolved into at
least three forms (64). The microsomal glutathione transferases were found to be very similar to the cytoplasmic enzymes, and it was concluded that the microsomal activity results from a firm association of the cytoplasmic glutathione
transferases with the microsomal membranes (64). The hydrophobic nature of the cytosolic glutathione transferases would
be expected to promote binding to biological membranes.

An interesting and, as yet, unexplained finding was the
up to 8-fold activation of microsomal glutathione transferase

from rat liver elicited by treatment of microsomes with
N-ethylmaleimide (65). The activation is only obtained with
enzyme bound to freshly prepared microsomes and not with
enzyme solubilized before treatment or with the cytosolic
activities. The possible implications of activation of microsomal glutathione transferase by products with reactive
groups are evident. Electrophilic compounds, such as epoxides produced by the microsomal cytochrome P-450 system, might
react with thiol groups and thereby activate the membrane-
-bound glutathione transferase activity and increase the capacity for conjugation with glutathione. It is noteworthy
that in contrast to glutathione transferases A, B, and C in
rat liver which are all inducible (67), the microsomal glutathione transferase activity is not increased by treatment
of rats with inducers of drug-metabolizing enzymes (64,65).

γ-Glutamyl transpeptidase and the formation of mercapturic
acids. Glutathione conjugates formed by the action of glutathione transferases are exported by liver cells and may be
found in bile as well as in the blood. Conjugates in the
bile may be eliminated via the alimentary canal or be metabolized in the intestine, absorbed, and recirculated via the
blood. In the kidney S-substituted glutathione derivatives,
formed in situ or transported to the kidney, are degraded
by γ-glutamyl transpeptidase in the plasma membrane of renal
cells (68). The enzyme has been shown to be an extrinsic
brush border membrane protein, which catalyzes the extracellular, intraluminal removal of the γ-glutamyl group of glutathione and glutathione conjugates (69). Various additional
biological functions of γ-glutamyl transpeptidase, including
amino acid transport, have been reviewed elsewhere (5,68).

S-Substituted cysteinylglycine derivatives produced by the action of γ-glutamyl transpeptidase are hydrolyzed by dipeptidases (68). The resulting S-substituted cysteine derivatives may be N-acetylated by acetylcoenzyme A to a corresponding mercapturic acid (S-substituted N-acetylcysteine) and excreted in the urine. Both dipeptidase and N-acetyltransferase activities have been identified in the microsomal fraction (70,71).

More recently an alternative metabolic pathway was discovered, which leads from a glutathione conjugate, via the corresponding S-substituted cysteine derivative, to a methylthio-containing metabolite (see ref. 72 for a review). The S-substituent of the cysteine derivative is cleaved off by a C-S lyase present predominantly in the cytosol fraction of liver and kidney (73-75). The thiol formed is then S-methylated by S-adenosylmethionine; a reaction catalyzed by a microsomal thiol S-methyltransferase (76,77). By this biotransformation the xenobiotic, although initially conjugated with glutathione, retains only the sulfur atom of the glutathione molecule. Thus, the methylthio-metabolite formation branches off from the classical mercapturic acid pathway (78) at the level of S-substituted cysteine.

THIOL-DISULFIDE INTERCHANGE

Many components in the cell contain thiol or disulfide groups that are important for their structure and biological function. Reversible thiol-disulfide interchange may modulate biochemical functions in a manner similar to phosphorylation - dephosphorylation of proteins (79). Enzymes catalyzing thiol-disulfide interchange have been named thioltrans-

ferases (80), but the inadequate name of transhydrogenases
is also in common use. At least two types of thioltransferases have been identified. A low-molecular-weight thioltransferase (M_r 11 000) has been isolated from rat liver cytosol
(81) and a larger protein (M_r 60 000) displaying the same
activity was obtained in highly purified form from bovine
liver microsomes (82). The latter enzyme differs in many re-
spects from the cytosolic thioltransferase although both are
active with glutathione as a substrate.

PROTEIN BIOSYNTHESIS. Proteins containing disulfide bridges
have to be formed from polypeptide chains containing cysteine
residues. In classical experiments it was found that the correct pairing of the half-cystine moieties of disulfide bonds
in bovine pancreatic ribonuclease could be facilitated by
means of an isomerase identified in microsomes (83,84). The
isomerization reaction can be initiated by a thiol such as
glutathione. However, in the de novo biosynthesis of disulfide-containing proteins it is likely that an oxidant reacts
enzymatically with the thiol groups of the cysteine residues
of the polypeptide chain. Disulfide bonds have to be formed
after the peptide bonds because the genetic code has codons
for cysteine but not for cystine. Even though the ultimate
oxidant is oxygen, it has been suggested that the disulfide
cystamine, which is formed by microsomal oxidation of the thiol
cysteamine, is the reactant which interacts directly with the
thiol groups of the nascent polypeptide (85). The cytosolic
thioltransferase from rat liver (81) catalyzes such reactions
of disulfides with protein thiol groups (86,87). The thioltransferase from microsomes has not been investigated in this

capacity. Further work is necessary to establish the proposed role of cystamine in the biosynthesis of disulfide containing proteins.

HORMONES AND RECEPTORS. Evidence for thiol-disulfide reactions involved in the expression of hormonal signals is accumulating. Some hormones are disulfide-containing polypeptides. Well--known examples are insulin, oxytocin and vasopressin. Reduction of the disulfide bonds of these molecules cause inactivation of the hormones. Both the microsomal (88) and the cytosolic (81) thioltransferases catalyze reduction of these hormones by use of glutathione as a reductant. Glutathione disulfide formed as a product can be reduced by NADPH in the reaction catalyzed by glutathione reductase.

Membrane-bound receptors of hormones may also be dependent on disulfide bonds. This is the case of the insulin receptor, which changes its affinity for insulin and its structural organization upon reduction of its disulfide bonds (89-91). It has even been suggested that the disulfide groups of the hormone react with the protein receptor (92). The role, if any, of glutathione at the level of receptors is uncertain, but it appears relevant that the thioltransferase identified in microsomes has been found in the plasma membrane (93), where also many receptors are located.

It may be added that the feeding response of Hydra is elicited by glutathione (94). The levels of cyclic GMP are elevated in the animal upon exposure to glutathione (95). Evidence for a specific receptor of glutathione has been obtained (96). Thus, glutathione itself is sensed by a chemical receptor in a biological membrane.

CONCLUSION

Glutathione is involved in numerous kinds of biochemical processes. Most of those that have been studied so far have been localized to the cytosol fraction of the cell. The present survey indicates many new fields of investigation which involve glutathione and biological membranes. These areas include protection against oxidative stress, mutagens, and carcinogens as well as biochemical control mechanisms at the level of membranes. Some of the topics are focal points in toxicological and biological research.

DEDICATION

This contribution is dedicated to Lars Ernster not only as a token of my deep friendship to a great scientist with encyclopedic knowledge of biochemistry, but, above all, to acknowledge the lessons in human relations I have learnt by his example.

REFERENCES

1. Flohé, L., Benöhr, H.C., Sies, H., Waller, H.D., and Wendel, A. (eds.) (1974) Glutathione, Georg Thieme Publishers, Stuttgart.
2. Arias, I.M. and Jakoby, W.B. (eds.) (1976) Glutathione: Metabolism and Function, Raven Press, New York.
3. Sies, H. and Wendel, A. (eds.) (1978) Functions of Glutathione in Liver and Kidney, Springer-Verlag, Berlin.
4. Meister, A. (1975) in Metabolic Pathways (D.M. Greenberg, ed.) 3rd edn., Vol. 7, pp. 101-188, Academic Press, New York.
5. Meister, A. and Tate, S.S. (1976) Annu. Rev. Biochem. 45, 559-604.

6. Kosower, N.S. and Kosower, E.M. (1976) in Free Radicals in Biology (W.A. Pryor, ed.) Vol. 2, pp. 55-84, Academic Press, New York.

7. Benöhr, H.C. and Waller, H.D. (1975) Klin. Wochenschr. 53, 789-802.

8. Kosower, N.S. and Kosower, E.M. (1978) Int. Rev. Cytol. 54, 109-160.

9. Jocelyn, P.C. (1972) Biochemistry of the SH Group, pp. 10-11, Academic Press, London,

10. Moron, M.S., DePierre, J.W., and Mannervik, B. (1979) Biochim. Biophys. Acta 582, 67-78.

11. Boyd, S.C., Sasame, H.A., and Boyd, M.R. (1979) Science 205, 1010-1012.

12. Smith, M.T., Loveridge, N., Wills, E.D., and Chayer, J. (1979) Biochem. J. 182, 103-108.

13. Wahlländer, A., Soboll, S., and Sies, H. (1979) FEBS Lett. 97, 138-140.

14. Mannervik, B. and Eriksson, S.A. (1974) in Glutathione (L. Flohé, H.C. Benöhr, H. Sies, H.D. Waller, and A. Wendel, eds.) pp. 120-131, Georg Thieme Publishers, Stuttgart,

15. Flohé, L. and Schlegel, W. (1971) Hoppe-Seyler's Z. Physiol. Chem. 352, 1401-1410.

16. Sies, H., Wahlländer, A., Waydhas, C., Soboll, S., and Häberle, D. (1980) Adv. Enzyme Regul. 18, 303-320.

17. Sies, H., Akerboom, T.P.M., and Tager, J.M. (1977) Eur. J. Biochem. 72, 301-307.

18. Viña, J., Hems, R., and Krebs, H.A. (1978) Biochem. J. 170, 627-630.

19. Hope, D.B. (1959) in Glutathione (E.M. Crook, ed.) pp. 93-114, Cambridge University Press, Cambridge.

20. Rink, H. (1974) in Glutathione (L. Flohé, H.C. Benöhr, H. Sies, H.D. Waller, and A. Wendel, eds.) pp. 206-215, Georg Thieme Publishers, Stuttgart.

21. Modig, H.G., Edgren, M., and Révész, L. (1971) Int. J. Radiat. Biol. 22, 257-268.

22. Pohl, L.R. (1979) in Reviews in Biochemical Toxicology (E. Hodgson, J.R. Bend, and R.M. Philpot, eds.) Vol. 1, pp. 79-107, Elsevier/North-Holland Biomedical Press. Amsterdam.

23. Mason, R.P. (1979) in Reviews in Biochemical Toxicology (E. Hodgson, J.R. Bend, and R.M. Philpot, eds.) Vol. 1, pp. 151-200, Elsevier/North-Holland Biomedical Press, Amsterdam.

24. Hochstein, P. and Ernster, L. (1964) in Ciba Foundation Symposium on Cellular Injury (A.V.S. de Reuck and J. Knight, eds.) pp. 123-134, Churchill Ltd, London.

25. Hochstein, P., Nordenbrand, K., and Ernster, L. (1964) Biochem. Biophys. Res. Commun. 14, 323-328.
26. Bus, J.S. and Gibson, J.E. (1979) in Reviews in Biochemical Toxicology (E. Hodgson, J.R. Bend, and R.M. Philpot, eds.) Vol.1, pp. 125-149, Elsevier/North-Holland Biomedical Press, Amsterdam.
27. Wendel, A., Feuerstein, S., and Konz, K.-H. (1978) in Functions of Glutathione in Liver and Kidney (H. Sies and A. Wendel, eds.) pp. 183-188, Springer-Verlag, Berlin.
28. Högberg, J., Orrenius, S., and O'Brien, P.J. (1975) Eur. J. Biochem. 59, 449-455.
29. Högberg, J. and Kristoferson, A. (1977) Eur. J. Biochem. 74, 77-82.
30. Anundi, I., Högberg, J., and Stead, A.H. (1979) Acta Pharmacol. Toxicol. 45, 45-51.
31. Kosower, E.M. and Kosower, N.S. (1969) Nature 224, 117-120.
32. Harris, J.W. and Biaglow, J.E. (1972) Biochem. Biophys. Res. Commun. 46, 1743-1749.
33. Kosower, N.S., Song, K.-R., and Kosower, E.M. (1969) Biochim. Biophys. Acta 192, 23-28.
34. Chance, B., Sies, H., and Boveris, A. (1979) Physiol. Rev. 59, 527-605.
35. Sies, H., Bartoli, G.M., Burk, R.F., and Waydhas, C. (1978) Eur. J. Biochem. 89, 113-118.
36. Neubert, D., Wojtczak, A.B., and Lehninger, A.L. (1962) Proc. Natl. Acad. Sci. USA 48, 1651-1658.
37. Sies, H. and Moss, K.M. (1978) Eur. J. Biochem. 84, 377-383.
38. Wendel, A. (1980) in Enzymatic Basis of Detoxication (W.B. Jakoby, ed.) Vol. 1, pp. 333-353, Academic Press, New York.
39. Flohé, L. and Günzler, W.A. (1974) in Glutathione (L. Flohé, H.C. Benöhr, H. Sies, H.D. Waller, and A. Wendel, eds.) pp. 132-143, Georg Thieme Publishers, Stuttgart.
40. Jocelyn, P.C. and Dickson, J. (1980) Biochim. Biophys. Acta 590, 1-12.
41. Lawrence, R.A. and Burk, R.F. (1976) Biochem. Biophys. Res. Commun. 71, 952-958.
42. Prohaska, J.R. and Ganther, H.E. (1977) Biochem. Biophys. Res. Commun. 76, 437-445.
43. Lawrence, R.A., Parkhill, L.K., and Burk, R.F. (1978) J. Nutr. 108, 981-987.
44. Prohaska, J.R. (1980) Biochim. Biophys. Acta 611, 87-98.
45. Mannervik, B., Guthenberg, C., and Åkerfeldt, K. (1980) in Microsomes, Drug Oxidations, and Chemical Carcinoge-

nesis (M.J. Coon, A.H. Conney, R.W. Estabrook, H.V. Gelboin, J.R. Gillette, and P.J. O'Brien, eds.) Vol. 2,
pp. 663-666, Academic Press, New York.

46. Warholm, M., Guthenberg, C., Mannervik, B., and von Bahr,
C. (1981) Biochem. Biophys. Res. Commun. 98, 512-519.

47. Lawrence, R.A. and Burk, R.F. (1978) J. Nutr. 108,
211-215.

48. Katki, A.G. and Myers, C.E. (1980) Biochem. Biophys. Res.
Commun. 96, 85-91.

49. Ames, S.R. and Elvehjem, C.A. (1945) J. Biol. Chem. 159,
549-562.

50. Ziegenhagen, A.J., Ames, S.R., and Elvehjem, C.A. (1947)
J. Biol. Chem. 167, 129-133.

51. Ashkar, S., Binkley, F., and Jones, D.P. (1981) FEBS Lett.
124, 166-168.

52. Jones, D.P., Moldéus, P., Stead, A.H., Ormstad, K., Jörnvall, H., and Orrenius, S. (1979) J. Biol. Chem. 254,
2782-2792.

53. Griffith, O.W. and Tate, S.S. (1980) J. Biol. Chem. 255,
5011-5014.

54. Ormstad, K. (1980) Doctoral thesis, Karolinska Institutet,
Stockholm.

55. Srivastan, S.K. and Beutler, E. (1969) J. Biol. Chem.
244, 9-16.

56. Kondo, T., Dale, G.L., and Beutler, E. (1980) Proc. Natl.
Acad. Sci. USA 77, 6359-6362.

57. Sies, H., Gerstenecker, C., Menzel, H., and Flohé, L.
(1972) FEBS Lett. 27, 171-175.

58. Oshino, N. and Chance, B. (1977) Biochem. J. 162,
509-525.

59. Bartoli, G.M. and Sies, H. (1978) FEBS Lett. 86, 89-91.

60. Chasseaud, L.F. (1979) Adv. Cancer Res. 29, 175-274.

61. Jakoby, W.B. and Habig, W.H. (1980) in Enzymatic Basis
of Detoxication (W.B. Jakoby, eds.) Vol. 2, pp. 63-94,
Academic Press, New York.

62. Christ-Hazelhof, E. and Nugteren, D.H. (1978) in Functions
of Glutathione in Liver and Kidney (H. Sies and A. Wendel,
eds.) pp. 201-206, Springer-Verlag, Berlin.

63. Rådmark, O., Malmsten, C., and Samuelsson, B. (1980)
Biochem. Biophys. Res. Commun. 96, 1679-1687.

64. Friedberg, T., Bentley, P., Stasiecki, P., Glatt, H.R.,
Raphael, D., and Oesch, F. (1979) J. Biol. Chem. 254,
12028-12033.

65. Morgenstern, R., Meijer, J., DePierre, J.W., and Ernster,
L. (1980) Eur. J. Biochem. 104, 167-174.

66. Kraus, P. (1980) Hoppe-Seyler's Z. Physiol. Chem. 361,
9-15.

67. Guthenberg, C., Morgenstern, R., DePierre, J.W., and Mannervik, B. (1980) Biochim. Biophys. Acta 631, 1-10.
68. Tate, S.S. (1980) in Enzymatic Basis of Detoxication (W.B. Jakoby, ed.) Vol. 2, pp. 95-120, Academic Press, New York.
69. Silbernagl, S., Pfaller, W., Heinle, H., and Wendel, A. (1978) in Functions of Glutathione in Liver and Kidney (H. Sies and A. Wendel, eds.) pp. 60-69, Springer-Verlag, Berlin.
70. Hughey, R.P., Rankin, B.B., Elce, J.S., and Curthoys, N. P. (1978) Arch. Biochem. Biophys. 186, 211-217.
71. Green, R.M. and Elce, J.S. (1975) Biochem. J. 147, 283-289.
72. Mannervik, B. (1981) in Metabolic Basis of Detoxication (W.B. Jakoby, J.R. Bend, and J. Caldwell, eds.) Academic Press, New York, in the press.
73. Binkley, F. (1950) J. Biol. Chem. 186, 287-296.
74. Anderson, P.M. and Schultze, M.O. (1965) Arch. Biochem. Biophys. 111, 593-602.
75. Tateishi, M., Suzuki, S., and Shimizu, H. (1978) J. Biol. Chem. 253, 8854-8859.
76. Bremer, J. and Greenberg, D.M. (1961) Biochim. Biophys. Acta 46, 217-224.
77. Weisiger, R.A. and Jakoby, W.B. (1979) Arch. Biochem. Biophys. 196, 631-637.
78. Boyland, E. and Chasseaud, L.F. (1969) Adv. Enzymol. 32, 173-219.
79. Krebs, E.G. and Beavo, J.A. (1979) Annu. Rev. Biochem. 48, 923-959.
80. Askelöf, P., Axelsson, K., Eriksson, S., and Mannervik, B. (1974) FEBS Lett. 38, 263-267.
81. Axelsson, K., Eriksson, S., and Mannervik, B. (1978) Biochemistry 17, 2978-2984.
82. Carmichael, D.F., Morin, J.E., and Dixon, J.E. (1977) J. Biol. Chem. 252, 7163-7167.
83. Goldberger, R.F., Epstein, C.J., and Anfinsen, C.B. (1963) J. Biol. Chem. 238, 628-635.
84. Venetianer, P. and Straub, F.B. (1963) Biochim. Biophys. Acta 67, 166-168.
85. Ziegler, D.M. and Poulsen, L.L. (1977) Trends Biochem. Sci. 2, 79-81.
86. Axelsson, K. and Mannervik, B. (1980) Biochim. Biophys. Acta 613, 324-336.
87. Mannervik, B. and Axelsson, K. (1980) Biochem. J. 190, 125-130.
88. Morin, J.E., Carmichael, D.F., and Dixon, J.E. (1978) Arch. Biochem. Biophys. 189, 354-363.

89. Pilch, P.F. and Czech, M.P. (1980) J. Biol. Chem. 255, 1722-1731.
90. Schweitzer, J.B., Smith, R.M., and Jarett, L. (1980) Proc. Natl. Acad. Sci. USA 77, 4692-4696.
91. Jacobs, S. and Cuatrecasas, P. (1980) J. Clin. Invest. 66, 1424-1427.
92. Czech, M.P. (1977) Annu. Rev. Biochem. 46, 359-384.
93. Varandani, P.T., Raveed, D., and Nafz, M.A. (1978) Biochim. Biophys. Acta 538, 343-353.
94. Loomis, W.F. (1955) Ann. N.Y. Acad. Sci. 66, 209-228.
95. Cobb, M.H., Heagy, W., Danner, J., Lenhoff, H.M., Marshall, G.R. (1980) Comp. Biochem. Physiol. 65C, 111-115.
96. Koizumi, O. and Kijima, H. (1980) Biochim. Biophys. Acta 629, 338-348.

REACTIONS RELATED TO MONOOXYGENASE FUNCTIONS IN ISOLATED

MAMMALIAN CELLS

Dean P. Jones and Sten Orrenius

Department of Biochemistry, Emory University, Atlanta, GA 30322, USA and Department of Forensic Medicine, Karolinska Institutet, S-104 01 Stockholm, Sweden.

INTRODUCTION

It is now about twentyfive years since the role of the hepatic endoplasmic reticulum in the oxidative biotransformation of lipid-soluble drugs and other foreign compounds was first established. The pioneering studies by Elizabeth and James Miller and their associates (1) and by B.B. Brodie and his colleagues (2) revealed the presence of drug metabolizing enzymes in the liver microsomal fraction and clarified several characteristics of these reactions including the requirement for NADPH and molecular oxygen. The existence of a common oxygen activating component involved in microsomal drug oxidations was postulated early (2). The identity of this component remained however unknown until the discovery of the microsomal hemoprotein first known as the CO-binding pigment (3,4) and subsequently termed cytochrome P-450 (5,6).

C. P. Lee, G. Schatz, G. Dallner (eds.), Mitochondria and Microsomes
in honor of Lars Ernster   ISBN 0-201-04576-1

The discovery and early characterization of cytochrome P-450 was soon followed by experimental support for its role as terminal oxidase in microsomal hydroxylation reactions. Evidence for this was based originally on the spectral characteristics of the light reversibility of CO-inhibited steroid hydroxylation in adrenal microsomes (7). Subsequently, a similar experimental approach was used to show the involvement of cytochrome P-450 in the oxidation of both endogenous and exogenous compounds in liver and kidney microsomes (8-10). In addition, the observation of characteristic substrate binding spectra with cytochrome P-450 (11), and the demonstration of enhanced drug metabolism associated with increased cytochrome P-450 levels as result of various induction treatments (12), provided indirect evidence for the function of cytochrome P-450 in the hydroxylation of a variety of substances by microsomes from liver and other tissues.

Much of this development occurred during the early 1960's and was based on experiments with microsomal fractions isolated from various organs. More recently, studies with solubilized and reconstituted enzyme systems have provided further detailed information on cytochrome P-450-linked monooxygenases. However, there is a number of important questions related to monooxygenase functions which can only be studied with experimental models that retain a high degree of organization like the intact cell or tissue. These include problems related to cellular uptake and distribution of drugs, competition between exogenous and endogenous substrates, availability of cofactors including oxygen, interrelationship between various reactions involved in drug biotransformation, disposal of metabolites, etc. For this purpose several experimental models have been employed including isola-

ted, perfused tissues, tissue slices, freshly isolated mammalian cells and cells in culture. In particular, the use of
isolated mammalian cells in studies of monooxygenase functions has recently led to a better understanding of several
aspects of drug metabolism and toxicity. Many factors have
contributed to this progress including improved methodology
for rapidly obtaining sufficient quantities of metabolically
active cells, detailed information on monooxygenase function
in subcellular systems, and development of experimental conditions to approach diverse questions. In this discussion,
we shall briefly review these methodological approaches and
the recent contributions that have been made by the study of
monooxygenase functions in isolated mammalian cells.

PREPARATION AND CHARACTERISTICS OF ISOLATED CELLS. The most
extensively used cell preparation is that of rat hepatocytes,
originally developed in the late 1960's (13,14), and subsequently modified by several research groups. The currently
used procedure was designed for drug metabolism studies (15)
and involves perfusion of the liver through the portal vein
with a calcium-free Hanks' balanced salts medium containing
EGTA and albumin, a second perfusion with Hanks' medium containing collagenase, dispersion of liver, and washing cells
two or three times with Krebs-Henseleit buffer. Modifications
of this procedure have been used to obtain isolated cell suspensions for drug metabolism studies from small intestine
(16,17) and kidney (18,19); Table I compares some properties
of cell suspensions from different rat tissues. These methods
have also been successfully applied for cell isolation from
mice and guinea pig tissues.

Table I. Characteristics of Cell Suspensions Freshly Isolated from Rat Liver, Small Intestine and Kidney[a]

	Liver	Small Intestine	Kidney
Yield, cells/g tissue	$\sim 5 \times 10^7$	$\sim 6 \times 10^6$	$\sim 10^7$
Trypan blue exclusion frequency, per cent	>95	>90	>90
NADH exclusion frequency, per cent	>90	>85	>85
Cytochrome P-450, nmol/10^6 cells	~ 0.25	~ 0.03	~ 0.07
GSH, nmol/10^6 cells	~ 50	~ 10	~ 30
Endogenous respiration, nmol $O_2/10^6$ cells/min	~ 15	~ 10	~ 25

a/ Data compiled from refs 15-19.

The isolated cells maintain metabolic integrity for several hours of incubation, and can be prepared under sterile conditions for establishment of primary cultures. Cytochrome P-450 content in isolated liver cells is constant for at least 4 h, but declines during longer incubation (20). Glutathione level in hepatocytes is stable provided that either cysteine or methionine is present in the medium (21). Addition of other amino acids or carbohydrates to the medium is typically not necessary with cells from fed animals for experiments lasting less than 2 h, unless a particular metabolic stress is created. Cells from starved animals are more dependent upon extracellular metabolite sources, and have been used for study of supply of reducing equivalents and metabolic intermediates (22-24). In liver cell incubations, ATP, ADP, AMP, NADH and NADPH levels and rates of glucose

synthesis, albumin synthesis and endogenous respiration have
been found to be constant up to 4 h, and to be similar to
values obtained for intact liver or isolated perfused liver
(25).

Pretreatment of animals with various agents has been
used to obtain hepatocytes with altered metabolic activities.
Phenobarbital, 3-methylcholanthrene, β-naphthoflavone and
ethanol have been used to increase the concentrations of various forms of cytochrome P-450 (15,20). Clofibrate has been
administered to increase the content of peroxisomes (20),
selenium-deficient diet and aminotriazole pretreatment have
been used to decrease the levels of glutathione peroxidase
and catalase, respectively (26) and diethylmaleate pretreatment has been employed to lower GSH content in liver and
kidney cells (15,18,27). Incubations of isolated cells in
different media, such as sulfate-free Krebs-Henseleit buffer
(28) and Krebs-Henseleit buffer with varying concentrations
of calcium (29), have been used to study specific aspects of
drug metabolism. In general, with proper selection of conditions, these manipulations can be made such that there is no
significant effect on the viability of the cells.

Improved subfractionation procedures for isolated cells
have also contributed significantly to the utility of these
cells for study of intracellular compartmentation. Isolated
cells are difficult to homogenize efficiently in conventional homogenizers so that various subcellular fractions can
be isolated and examined. However, selective disruption of
the plasma membrane with digitonin has allowed isolation of
mitochondria from liver cells, and has been used to study
the supply of reducing equivalents for drug metabolism (23).
Another method that allows similar fractionation procedures

involves rapid disruption of cells by sonication (30). This procedure similarly permits subfractionation of cells so that metabolites in specific compartments can be examined.

HEMOPROTEIN QUANTITATION. The access to suspensions of isolated mammalian cells has allowed direct application of spectrophotometric techniques to quantitate various hemoproteins and to assess changes in their levels due to different treatments. Table II exemplifies this by showing concentrations of several hemoproteins in hepatocytes isolated from rats subjected to different pretreatments.

Methods were developed (20) to determine the content of cytochrome P-450 based upon the original procedure of Omura and Sato (5) or by differential reduction techniques (31,32). In cells from animals which had been treated with phenobarbital or 3-methylcholanthrene to increase the cytochrome P-450 content, application of the former method (CO-difference spectrum of dithionite-reduced forms) results in only a 2-3% underestimate of the cytochrome P-450 content due to interference by cytochrome c oxidase. However, in control cells, the relatively large negative absorbance at 445 nm of the CO-difference spectrum of reduced cytochrome c oxidase results in at least 10% underestimation of cytochrome P-450. Consequently, accurate quantitation of this hemoprotein in control cells requires application of a differential reduction technique.

Cytochrome b_5 can be quantitated from the t-butylhydroperoxide difference spectrum of hepatocytes in the presence of a mitochondrial uncoupler such as carbonyl cyanide p-trifluoromethoxyphenylhydrazone (FCCP)(20). Under these condi-

Table II. Hemoprotein Concentrations in Hepatocytes Isolated from Rats Pretreated with Various Agents[a]

Component	Control	Pheno-barbital	3-Me-thyl-cholan-threne	Etha-nol	Amino-tria-zole	Clo-fib-rate
Catalase	0.22	0.20	0.21	0.17	0.02	0.27
Cytochromes $a+a_3$	0.18	0.20	0.16	0.09	0.17	0.20
Cytochromes $b_{561} + b_{566}$	0.14	0.17	0.16	0.13	0.17	
Cytochromes $c+c_1$	0.17	0.20	0.17	0.19	0.19	
Cytochrome P-450	0.26	0.90	0.84	0.40	0.16	
Cytochrome b_5	0.09	0.29	0.24	0.09	0.10	

a/ Data (expressed as $nmol/10^6$ cells) were compiled from refs 20,26.

tions, the mitochondrial cytochromes are oxidized and therefore do not contribute to the absorbance change upon addition of t-butylhydroperoxide. The principal absorbance change is due to the oxidation of cytochrome b_5 by t-butylhydroperoxide. The facility of this oxidation appears to depend upon electron flow through cytochrome P-450 and this approach seems to be reliable only for hepatocytes in which cytochrome P-450 content has been increased by phenobarbital or 3-methylcholanthrene treatment. With control hepatocytes, or intestinal and kidney cells, this procedure always gives an underestimate of cytochrome b_5 content (17,19,20). Under these conditions, an alternative approach is to use methylviologen ($+O_2$) as the oxidizing agent. The methylviologen-induced difference spectrum of renal cells in the presence of FCCP gives a characteristic oxidized minus reduced difference spectrum of cytochrome b_5 and allows a more accurate calculation of its content in these cells.

Spectral studies of cytochrome P-450 in isolated cells is complicated by the appearance of this hemoprotein in the mitochondrial compartment as well as in the endoplasmic reticulum. Subfractionation studies of isolated intestinal cells have allowed identification of at least three forms of cytochrome P-450 (17). One form is found in the mitochondrial fraction, is not inducible by 3-methylcholanthrene and does not metabolize benzo(a)pyrene. Two forms are present in the microsomal fraction, one which is present in control animals and metabolizes benzo(a)pyrene at a slow rate and one that is inducible by 3-methylcholanthrene and metabolizes benzo(a)pyrene at a rapid rate.

Altered levels of cytochrome P-450 due to various agents have been reported in isolated liver cells for both in vivo and in vitro treatments. The most thoroughly studied inducer has been phenobarbital which results in a 4 to 5-fold increase in cytochrome P-450 content in hepatocytes after repeated administration to rats in vivo. This increase has long been recognized to be due to a proliferation of the endoplasmic reticulum within the cells as well as to an increase in the specific, membranal content of cytochrome P-450 (12). Pretreatment of rats with 3-methylcholanthrene causes a similar increase in hepatocyte cytochrome P-450 level, whereas chronic ethanol treatment results in only about 50% increase on a cellular basis (20).

A decrease in cellular cytochrome P-450 concentration during establishment of primary cultures of hepatocytes has been reported (33) and also occurs in liver cells maintained in suspension for more than 4 - 5 h (20). Loss of other cytochromes parallelled the loss of cell viability measured by

trypan blue exclusion or NADH penetration, but the decrease
in the levels of cytochromes P-450 and $\underline{b}_5$ was more rapid.
This loss was partially blocked by cycloheximide, and was
enhanced by inclusion of cobalt in the medium, as reported
for the primary cultures. However, there was an overriding
loss of cell viability ($\sim$30% loss) over the period from 6 -
10 h incubation that made suspensions of isolated hepatocytes a poor model for study of these changes in different hemoprotein concentrations.

DRUG UPTAKE. Among the first questions approached in studies using isolated liver cells was drug uptake. Moldéus and
associates (34) used the formation of the Type I spectral
change of cytochrome P-450 as a measure of uptake of hexobarbital, lidocaine and nortriptyline by isolated hepatocytes. They found that drug uptake occurred rapidly in control
cells, but was even faster in cells from phenobarbital-treated rats. Additional studies (35) showed that the rate of
hexobarbital uptake increased with temperature in a manner
compatible with that expected from passive diffusion. No
effect on drug uptake was seen due to inhibition of ATP synthesis with rotenone, and uptake did not show saturation kinetics. The results led to the conclusion that hexobarbital
is taken up by a non-energy-requiring diffusion process.
This hypothesis is further supported by the observation that
the apparent K_m-values for both alprenolol and hexobarbital
metabolism by liver cells are similar to those observed with
liver microsomes.

  The relationship of hepatic drug uptake to the binding
of drug to cytochrome P-450 was further studied with alpre-

nolol (36). Displacement of ^{3}H-alprenolol from liver cells was observed when imipramine or metyrapone was added to the medium suggesting that the rapid uptake and high-affinity binding to cytochrome P-450 may be important in the first-pass elimination of this drug.

Other drugs have been used to study carrier-mediated transport systems in liver cells. Bromosulfophthalein, an anionic dye often used to test liver function, is cleared by the liver in a multistep process that involves uptake by the tissue, binding to ligandin, conjugation to glutathione and excretion into the bile. Studies by Schwenk and associates (37,38) and van Bezooijen et al (39) showed energy requirement for uptake of this drug by hepatocytes. Schwarz et al (40) have subsequently compared the rates of conjugation and release of the conjugate to the rate of drug uptake. They found that the initial rate of uptake of the drug is two to three times as fast as the rate of conjugation, and over ten times as fast as the rate of efflux of the conjugate. This results in a transient accumulation of both the drug and the conjugate within the hepatocyte. Laperche et al (41) examined the effect of rifampicin on bromosulfophthalein uptake and found that rifampicin inhibits the uptake of bromosulfophthalein competitively, whereas bromosufophthalein does not inhibit the hepatic uptake of rifampicin. Since rifampicin is also taken up by a carrier-mediated process, they concluded that several carriers may be involved in the uptake of anionic drugs by rat liver.

DRUG INTERACTIONS WITH CYTOCHROME P-450. Direct spectrophotometry of isolated cells in suspension has provided a use-

ful means for study of drug interactions with cytochrome
P-450 in the intact cell. Type I spectral changes have been
observed with several compounds incubated with isolated hepatocytes (34,35,42). Studies with other cell types have
been less successful in this regard (Jones and Orrenius,
unpublished observations) due primarily to the lower concentrations of cytochrome P-450 in extrahepatic tissues.

Studies of the time dependence of absorbance changes
during the first few minutes after addition of drugs to isolated hepatocytes led to the recognition of absorbance maxima at 437 nm and 446 nm in addition to the typical Type I
spectral change elicited by the drug (34). The nature of the
species responsible for these absorbance maxima is not known
but may represent intermediates in the reaction cycle.

Addition of carbon monoxide to hepatocytes results in
only a small fraction of the total cytochrome P-450 being
converted to the species absorbing at 450 nm. This indicates
that the hemoprotein is present largely in the substrate-
free, oxidized form in the intact cell. Addition of hexobarbital to the medium increases the proportion of the hemoprotein converted to the CO-bound form. However, since the
binding of carbon monoxide changes the half-reduction potential of the hemoprotein it is not possible to readily obtain
the fraction of the cytochrome present in the reduced form
by this approach.

Isolated hepatocytes have also provided a suitable model to examine various cytochrome P-450 product complexes
including those formed with amphetamine derivatives (43).
These complexes, which are characterized by absorption maxima in the region of 450 nm, are relatively stable and are
catalytically inactive. If formed in vivo, they may inhibit

cytochrome P-450 by converting it to an inactive form. Formation of cytochrome P-450 product complexes in isolated hepatocytes is however slower and less extensive than in liver microsomes (43). A comparison of complex formation in liver microsomes and hepatocytes after addition of amphetamine derivatives suggests that conjugation with glutathione may protect the intact cell against the conversion of cytochrome P-450 to the inactive complex, and that inactivation of cytochrome P-450 in vivo by this mechanism may occur only under special conditions, such as involving the depletion of GSH.

Drug-induced spectral changes offer a useful tool for study of drug uptake and interaction with cytochrome P-450 in intact cells. However, it is necessary to distinguish between spectral changes due to interaction with cytochrome P-450 and those due to interaction of drugs with other hemoproteins in the cells. Drugs that act as inhibitors of electron transport, or uncouple oxidative phosphorylation, may cause spectral changes due to effects on the oxidation-reduction state of the mitochondrial cytochromes. Similarly, agents such as t-butylhydroperoxide and methylviologen result in oxidation of cytochrome b_5. Other changes can be observed due to variations in the steady-state concentration of the primary hydrogen peroxide - catalase complex (Compound I), which occur upon addition of suitable hydrogen donors such as ethanol, methanol or formic acid to hepatocytes (27). Further, drugs that result in formation of formaldehyde can also cause these spectral changes due to generation of formic acid which binds to hemoproteins. Finally, conversion of catalase to the secondary H_2O_2 complex (Compound II) has also been detected in suspensions of isolated hepatocy-

tes following addition of α-methyldopa (44). The formation
of this complex, which is also catalytically inactive, is
dependent on the rate of intracellular H_2O_2 production as
well as on the α-methyldopa concentration.

SUPPLY OF COSUBSTRATES. Isolated cells offer a useful model
also for study of supply of cosubstrates for drug oxidations.
Perfused organ systems are cumbersome for such studies because of the difficulty of obtaining repeated tissue samples
and because subfractions of the tissue cannot be obtained
rapidly. In contrast, a suspension of cells from a single
liver can be divided into several flasks and can therefore
be examined under a wide variety of experimental manipulations. In addition, the time course of metabolite changes
can be readily studied and the cells can be rapidly separated into several subcellular fractions. For studies of oxygen dependence, tissue oxygen gradients are eliminated which
allows direct observation of oxygen delivery on a cellular
basis.

NADPH generation from endogenous substrates was found
to be sufficient to support optimal alprenolol metabolism in
liver cells from control rats, whereas the addition of lactate or glucose to the medium stimulated drug metabolism in liver cells from starved, phenobarbital-treated rats (15). These results suggest that the level of cytosolic NADPH may become rate-limiting for drug metabolism linked to cytochrome
P-450 under certain conditions. Further, inhibitors of mitochondrial respiration such as rotenone, antimycin A, KCN,
and the uncoupler FCCP, inhibit drug metabolism in isolated
liver cells suggesting that regeneration of reducing equiva-

lents in the cytosol is dependent upon mitochondrial supply. The details of this energy dependence have not yet been elucidated. More detailed studies of NADPH supply for drug oxidations seems especially warranted also in light of the contributions of Ernster and his colleagues on the energy dependent transhydrogenase (45,46). Isolated cells may be a suitable preparation to probe the function of this system. Extensive studies are already available concerning supply of reducing equivalents for gluconeogenesis, ureogenesis and lactate synthesis from pyruvate (24,47-49), the relationship between energy supply and oxidation-reduction reactions (50), the disposition of reducing equivalents during ethanol metabolism (51,52), and the supply of NADPH by the hexose monophosphate pathway in liver cells (53).

Oxygen dependence of monooxygenase reactions in hepatocytes was found to be similar to that for microsomal metabolism (25). Data indicated that although there appears to be a marked gradient of oxygen into the mitochondrial membrane under hypoxic conditions, there is at most a minimal gradient to the endoplasmic reticulum. Metabolism of hexobarbital and alprenolol was oxygen dependent below 20-35 μM. Since normal tissue oxygen concentrations are in this range (54), the results suggest that drug metabolism may be impaired during hypoxia (55).

DRUG METABOLISM AND DISPOSITION OF PRODUCTS. The availability of isolated cells from various tissues has allowed comparative studies of drug biotransformation in different cell types. The metabolism of a large number of drugs has been examined (cf 56). Important contributions to our understanding

of overall drug biotransformation have come from studies of
interrelationships of various drug metabolizing systems, the
disposition of products formed by initial oxidation reactions (formaldehyde metabolism, conjugation reactions) and
reconstitution of entire drug biotransformation pathways by
incubations of combinations of cell types.

Formation of glucuronide, sulfate and glutathione conjugates have been studied in liver, kidney and intestinal
cells. Rates of formation of the various conjugates are dependent on substrate concentration and relate to the affinity of the drug for the particular enzyme systems (57). The
relative proportions of metabolites can be varied by selective inhibition of individual pathways, such as use of metyrapone to inhibit cytochrome P-450, galactosamine or ethanol
to inhibit glucuronidation, sulfate-free medium to inhibit
sulfation, or depletion of glutathione to inhibit formation
of glutathione conjugates (28,57).

Disposition of formaldehyde produced during N-demethy-
lation of ethylmorphine and related drugs was shown to be
glutathione-dependent in isolated liver cells (27), presumably due to involvement of the cytosolic formaldehyde dehydrogenase system (58). The formic acid that is produced by
the oxidation of formaldehyde is largely metabolized by the
tetrahydrofolate-dependent pathway (59). Under some circumstances, methionine may become rate-limiting for this pathway in isolated cells and must be added exogenously. Catalase may also contribute to the oxidation of formate, but this
contribution appears to be relatively small in hepatocytes
(27).

Reconstitution of the entire biotransformation pathway
for conversion of a drug to a mercapturic acid derivative

has been possible by sequential incubations of different
cell types (60-62). In liver cells, acetaminophen is oxidized by the cytochrome P-450-linked monooxygenase system to
an electrophilic product which reacts with glutathione to
form a conjugate. The glutathione conjugate, which is normally released from the liver into the bile, is recovered in
the medium of liver cell incubations. Subsequent incubation
with either isolated intestinal or renal cells results in
rapid conversion of the glutathione conjugate to the cysteine derivative which is finally acetylated in kidney cells to
form the mercapturic acid derivative. Fig. 1 illustrates the
tissue localization of the major metabolic steps involved in
the conversion of acetaminophen to its mercapturic acid derivative.

FORMATION AND METABOLISM OF HYDROGEN PEROXIDE IN HEPATOCY-
TES.   Incubation of liver microsomes with NADPH has long
been known to result in formation of H_2O_2 (63). More recently, microsomal H_2O_2 production has been shown to be affected
by cytochrome P-450 inducers and by the presence of various
drug substrates in the microsomal incubation (64), and available evidence indicates that it is due to autooxidation of
cytochrome P-450 and mediated by release of superoxide from
the oxy-complex of the hemoprotein (65).

Studies of ethylmorphine metabolism in hepatocytes suggested that drug oxidation can be associated with stimulated
H_2O_2 generation also in the intact cell (27). Thus, incubation of hepatocytes from phenobarbital-treated rat with
ethylmorphine resulted in oxidation of intracellular GSH and
release of glutathione disulfide into the medium. When cells

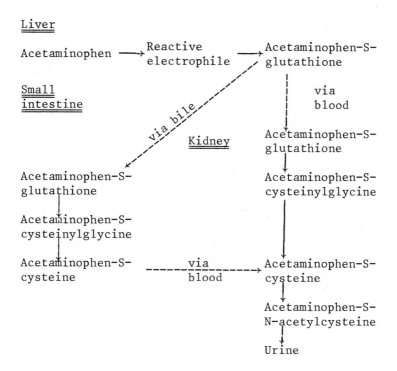

Fig. 1. Tissue localization of major reactions involved in conversion of acetaminophen to the corresponding mercapturic acid derivative.

depleted of glutathione by pretreatment of the animal with diethylmaleate (15) were used for incubation, titration with methanol revealed that a relatively larger fraction of catalase was present as the primary H_2O_2 complex, Compound I, when ethylmorphine or related drugs were present in the medium (27).

These early observations have recently been extended into a more detailed study of H_2O_2 generation and catabolism in hepatocytes (26). In this investigation, ethylmorphine and glycolate have been used to stimulate H_2O_2 generation in different cellular compartments, and the contribution of catalase and glutathione peroxidase to the metabolism of endogenously generated H_2O_2 has been compared (Fig. 2). Evidence for compartmentation of H_2O_2 catabolism has been obtained (26). The mechanism underlying the release of glutathione disulfide to the medium as a result of stimulation of the glutathione peroxidase reaction in hepatocytes is presently under detailed investigation.

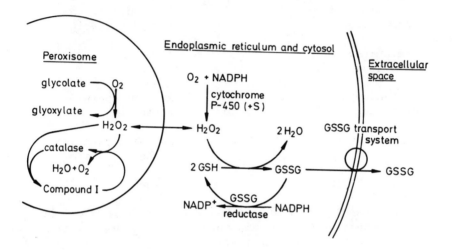

Fig. 2.   Formation and catabolism of hydrogen peroxide in isolated hepatocytes.

ISOLATED CELLS FOR STUDIES OF DRUG TOXICITY. During recent years isolated hepatocytes have been used frequently in studies of drug toxicity. In fact, they seem to be an almost ideal experimental model for such studies, since they are capable of metabolic activation and/or inactivation of a variety of hepatotoxic agents and, at the same time, serve as a sensitive indicator system for various toxic effects.

Changes in plasma membrane permeability are easy to monitor
(15) and normally used as a measure of cytotoxicity in suspensions of hepatocytes. However, other toxic effects have
also been registered in this system including lipid peroxidation (66), alkylation of protein (67) and nucleic acid
(68) and inhibition of specific enzymes (67).

Isolated liver cells have been used for detailed investigations of cytochrome P-450-mediated metabolic activation
of hepatotoxic drugs including bromobenzene (69,70) and acetaminophen (57). Series of toxic responses have been registered and data suggesting that the plasma membrane may be a
primary target in bromobenzene toxicity have been reported
(67). In another study (68), reactions affecting the formation of DNA-binding metabolites of benzo(a)pyrene in hepatocytes have been characterized. Recently, the metabolism of
benzo(a)pyrene-3,6-quinone has been compared in liver microsomes and hepatocytes (71).

The studies of drug toxicity in isolated hepatocytes
have emphasized the importance of glutathione conjugation
as a major defence mechanism against accumulation and toxic
effects of reactive drug metabolites. Often, glutathione depletion precedes other signs of drug toxicity which, in turn,
can be prevented by facilitated glutathione biosynthesis
(69,70). These observations have led to more detailed investigations of glutathione turnover and utilization during
drug biotransformation in hepatocytes (21,27,72) which have
illustrated the utility of isolated cells for studies of
this kind.

CONCLUSION

From this presentation it is clear that suspensions of freshly isolated mammalian cells have become an experimental model frequently used in biochemical and toxicological studies. This is particularly true for reactions related to monooxygenase function which can now be approached with improved methodology. Thus, development of refined spectrophotometric procedures has allowed direct quantitation of various hemoproteins and detailed analysis of interactions between drugs and cytochrome P-450 in the intact cell. In addition, improved chromatographic methods now permit quantitation of various classes of drug metabolites and of certain cosubstrates required for drug biotransformation at very high sensitivities. The generation of cosubstrates for drug biotransformation appears to be one of the questions which can be favorably approached in future studies with isolated cells. This is also true for the elucidation of the interrelationship of reactions involved in drug biotransformation and for further studies of cytochrome P-450-mediated drug toxicity. For this purpose, in particular, isolated mammalian cells seem to be an ideal experimental model.

REFERENCES

1. Conney, A.H., Miller, E.C. and Miller, J.A. (1957) J.
    Biol. Chem. 228, 753-766.
2. Brodie, B.B., Gillette, J.R. and LaDu, B.N. (1958) Ann.
    Rev. Biochem. 27, 427-454.
3. Klingenberg, M. (1958) Arch. Biochem. Biophys. 75,
    376-386.
4. Garfinkel, D. (1958) Arch. Biochem. Biophys. 77, 493-
    509.
5. Omura, T. and Sato, R. (1964) J. Biol. Chem. 239,
    2370-2378.
6. Omura, T. and Sato, R. (1964) J. Biol. Chem. 239,
    2379-2385.
7. Estabrook, R.W., Cooper, D.Y. and Rosenthal, O. (1963)
    Biochem. Z. 338, 741-755.
8. Cooper, D.Y., Levin, S., Narasimhulu, S., Rosenthal,
    O. and Estabrook, R.W. (1965) Science, 147, 400-402.
9. Diehl, H., Capalna, S. and Ullrich, V. (1969) FEBS
    Letters, 4, 99-102.
10. Ellin, Å., Jakobsson, S.W., Schenkman, J.B. and Orrenius,
    S. (1972) Arch. Biochem. Biophys. 150, 64-71.
11. Schenkman, J.B., Remmer, H. and Estabrook, R.W. (1967)
    Mol. Pharmacol. 3, 113-123.
12. Ernster, L. and Orrenius, S. (1965) Fed. Proc. 24,
    1190-1199.
13. Howard, R.B. and Pesch, L.A. (1968) J. Biol. Chem. 243,
    3105-3109.
14. Berry, M.N. and Friend, D.S. (1969) J. Cell Biol. 43,
    506-520.
15. Moldéus, P., Högberg, J. and Orrenius, S. (1978) Meth.
    Enzymol. 51, 60-70.
16. Stohs, S.J., Grafström, R.C., Burke, M.D. and Orrenius,
    S. (1977) Arch. Biochem. Biophys. 179, 71-80.
17. Jones, D.P., Grafström, R. and Orrenius, S. (1980) J.
    Biol. Chem. 255, 2383-2390.
18. Jones, D.P., Sundby, G.-B., Ormstad, K. and Orrenius,
    S. (1979) Biochem. Pharmacol. 28, 929-935.
19. Ormstad, K., Orrenius, S. and Jones, D.P. (1981) Meth.
    Enzymol. in press.
20. Jones, D.P., Orrenius, S. and Mason, H.S. (1979) Bio-
    chim. Biophys. Acta, 576, 17-29.
21. Thor, H., Moldéus, P. and Orrenius, S. (1979) Arch.
    Biochem. Biophys. 192, 405-413.

22. Moldéus, P., Grundin, R., Vadi, H. and Orrenius, S. (1974) Eur. J. Biochem. 46, 351-360.

23. Weigl, K. and Sies, H. (1977) Eur. J. Biochem. 77, 401-408.

24. Meijer, A.J., Gimpel, J.A., Deleeuw, G., Tischler, M. E., Tager, J.M. and Williamson, J.R. (1978) J. Biol. Chem. 253, 2308-2320.

25. Jones, D.P. and Mason, H.S. (1978) J. Biol. Chem. 253, 4874-4880.

26. Jones, D.P., Eklöw, L., Thor, H. and Orrenius, S. (1981) Arch. Biochem. Biophys. in press.

27. Jones, D.P., Thor, H., Andersson, B. and Orrenius, S. (1978) J. Biol. Chem. 253, 6031-6037.

28. Moldéus, P., Andersson, B. and Gergely, V. (1979) Drug Metab. Disp. 7, 416-419.

29. Andersson, B., Jones, D.P. and Orrenius, S. (1979) Biochem. J. 184, 709-711.

30. Gellerfors, P. and Nelson, B.D. (1979) Anal. Biochem. 93, 200-203.

31. Orrenius, S., Ellin, Å., Jakobsson, S.W., Thor, H., Cinti, D.L., Schenkman, J.B. and Estabrook, R.W. (1973) Drug Metab. Disp. 1, 350-357.

32. Ghazarian, J.G., Jefcoate, C.R., Knutson, J.C., Orme-Johnson, W.H. and DeLuca, H.F. (1974) J. Biol. Chem. 249, 3026-3033.

33. Guzelian, P.S. and Bissell, D.M. (1976) J. Biol. Chem. 251, 4421-4427.

34. Moldéus, P., Grundin, R., von Bahr, C. and Orrenius, S. (1973) Biochem. Biophys. Res. Commun. 55, 937-944.

35. von Bahr, C., Vadi, H., Grundin, R., Moldéus, P. and Orrenius, S. (1974) Biochem. Biophys. Res. Commun. 59, 334-339.

36. Grundin, R., Moldéus, P., Orrenius, S., Borg, K.O., Skånberg, I. and von Bahr, C. (1974) Acta Pharmacol. Toxicol. 35, 242-260.

37. Schwenk, M., Burr, R., Schwarz, L. and Pfaff, E. (1976) Eur. J. Biochem. 64, 189-197.

38. Schwenk, M., Burr, R. and Pfaff, E. (1976) Naunyn-Schmiedeberg's Arch. Pharmacol. 295, 99-102.

39. van Bezooijen, C.F.A., Grell, T. and Knook, D.L. (1976) Biochem. Biophys. Res. Commun. 69, 354-361.

40. Schwarz, L.R., Summer, K.-H. and Schwenk, M. (1979) Eur. J. Biochem. 94, 617-622.

41. Laperche, Y., Graillot, C., Arondel, J. and Berthelot, P. (1979) Biochem. Pharmacol. 28, 2065-2069.

42. Jones, D.P. (1976) Ph.D. Dissertation, Univ. of Oregon

Health Sciences Center, Portland, Oregon, USA.

43. Hirata, M., Högberg, J., Thor, H. and Orrenius, S. (1977) Acta Pharmacol. Toxicol. 41, 177-189.

44. Jones, D.P., Meyer, D.B., Andersson, B. and Orrenius, S. (1981) Mol. Pharmacol. in press.

45. Rydström, J., Teixeira da Cruz, A. and Ernster, L. (1970) Eur. J. Biochem. 17, 56-62.

46. Rydström, J., Teixeira da Cruz, A. and Ernster, L. (1971) Eur. J. Biochem. 23, 212-219.

47. Sies, H., Akerboom, T.P.M. and Tager, J.M. (1977) Eur. J. Biochem. 72, 301-307.

48. Berry, M.N. (1971) Biochem. Biophys. Res. Commun. 44, 1449-1456.

49. Meijer, A.J. and Williamson, J.R. (1974) Biochim. Biophys. Acta, 333, 1-11

50. Wilson, D.F., Stubbs, M., Veech, R.L., Erecinska, M. and Krebs, H.A. (1974) Biochem. J. 140, 57-64.

51. Meijer, A.J., van Woerkom, G.M., Williamson, J.R. and Tager, J.M. (1975) Biochem. J. 150, 205-209.

52. Cederbaum, A.I., Dicker, E. and Rubin, E. (1977) Arch. Biochem. Biophys. 183, 638-646.

53. Junge, O. and Brand, K. (1975) Arch. Biochem. Biophys. 171, 398-406.

54. Kessler, M. (1974) Microvasc. Res. 8, 283-290.

55. Jones, D.P. (1981) Biochem. Pharmacol. in press.

56. Thurman, R.G. and Kauffman, F.C. (1980) Pharmacol. Rev. 31, 229-251.

57. Moldéus, P. (1978) Biochem. Pharmacol. 27, 2859-2863.

58. Uotila, L. and Koivusalo, M. (1974) J. Biol. Chem. 249, 7653-7663.

59. Waydhas, C., Weigl, K. and Sies, H. (1978) Eur. J. Biochem. 89, 143-150.

60. Moldéus, P., Jones, D.P., Ormstad, K. and Orrenius, S. (1978) Biochem. Biophys. Res. Commun. 83, 195-200.

61. Jones, D.P., Moldéus, P., Stead, A.H., Ormstad, K., Jörnvall, H. and Orrenius, S. (1979) J. Biol. Chem. 254, 2787-2792.

62. Grafström, R., Ormstad, K., Moldéus, P. and Orrenius, S. (1979) Biochem. Pharmacol. 28, 3573-3579.

63. Gillette, J.R., Brodie, B.B. and LaDu, B.N. (1957) J. Pharm. Exp. Therap. 119, 532-540.

64. Hildebrandt, A.G., Speck, M. and Roots, I. (1973) Biochem. Biophys. Res. Commun. 54, 968-975.

65. Kuthan, H., Tsuji, H., Graf, H., Ullrich, V., Werringloer, J. and Estabrook, R.W. (1978) FEBS Letters, 91, 343-345.

66. Högberg, J., Orrenius, S. and Larson, R.E. (1975) Eur.
    J. Biochem. 50, 595-602.
67. Thor, H. and Orrenius, S. (1980) Arch. Toxicol. 44,
    31-43.
68. Burke, M.D., Vadi, H., Jernström, B. and Orrenius, S.
    (1977) J. Biol. Chem. 252. 6424-6431.
69. Thor, H., Moldéus, P., Kristoferson, A., Högberg, J.,
    Reed, D.J. and Orrenius, S. (1978) Arch. Biochem. Bio-
    phys. 188, 114-121.
70. Thor, H., Moldéus, P., Hermanson, R., Högberg, J.,
    Reed, D.J. and Orrenius, S. (1978) Arch. Biochem. Bio-
    phys. 188, 122-129.
71. Capdevila, J. and Orrenius, S. (1980) FEBS Letters,
    119, 33-37.
72. Reed, D.J. and Orrenius, S. (1977) Biochem. Biophys.
    Res. Commun. 77, 1257-1264.

Numbers set in *italics* designate pages on which complete literature citations are given.

A set of inclusive page numbers immediately following a name represents the first and last pages of the contribution associated with the author.

Thiery, J.P., 81(67), *90*
Thinès-Sempoux, D., 552(200), *561*
  630(4, 5, 7-9), 631(7-9), 632-
  634(8, 9), 638(9), *648*
Thomas, D.Y., 69(14), *87*
Thomas, P.E., 709(12), *726*
Thomson, J.N., 646(66), *651*
Thor, H., 752(21), 753(27-29),
  759(43), 760(27, 43), 763(28),
  764(27), 765(26, 27), 766(26),
  767(21-27, 67, 69, 70), *769-772*
Thore, A., 376(8), *402*, 532-
  533(57), *539*
Thorgeirsson, S., 549(147, 149),
  559
Thorn, G.D., 163(33, 34), 168(54),
  186-187
Thurman, R.G., 762(56), *771*
Tice, W., 542(14), *553*
Timkovich, R., 521(23), 525(23),
  527(23), *537*
Tiozzo, R., 358(13), 359(17), 361-
  362(29), *372-373*
Tischler, M.E., 144(88), *152*,
  752(24), 762(24), *770*
Tisdale, H.D., 157-158(13), 160(13),
  186
Titheradge, M.A., 239(116, 117),
  247
Tokuyama, T., 614(10), *627*
Tomasi, V., 632(77), *652*
Tomita, M., 571(80), *582*
Tomkins, G.M., 545(87), *556*
Toneguzzo, F., 572(87), *582*
Topali, V.P., 378(12), *402*
Topitsch, P., 305(50), *316*
Torndal, U.-B., 9(17), *12*
Touster, O., 632(20), 637, *649*
Tran-thi, T., 660(17), *678*
Trentham, D.R., 418(27), *426*,
  447(114), *457*
Trissl, H.W., 387(41, 42), *404*
Trouet, A., 630(11), 632(11),
  646(58), *649, 651*
Trumpower, B.L., 200(33), 212(33),
  215,219(10, 11), 225(44), 226(10,
  11, 58), 232, 235(97, 100), 237(10,
  58, 100, 109, 115), 238(10, 100),
  242-244, 246
Tsai, D.K., 551(179), *560*
Tsibakowa, E.T., 26, *41*
Tsofina, L.M., 377(9), 378(12), *402*
Tsou, C.L., 24, *40*, 191(5), 192(5),
  213
Tsuboi, M., 430-431(37), 438(37),
  454
Tsuji, H., 696(40), *705*, 764(65),
  771
Tu, C., 83(78), *91*

Tu, Y.L., 132(38), *150*
Tulkens, P., 646(58), *651*
Tustanoff, E.R., 155(2), *186*
Tuttle, J.H., 436(69), *455*
Tweedle, M.F., 252(26), *266*
Tyler, D.D., 126(21), *149*
Tzagoloff, A., 32(94), *42*, 46(5, 7),
  63, 69(19), 79(59), 81(59, 60),
  81(69), 82(69), 84-86(82), 85(87),
  88, 90, 91, 229(79), *245*, 338(17),
  354, 520(12), *537*

Ui, N., 428-431(12), 435-438(12),
  453
Ullrich, V., 691(23), 696(40),
  697(45), *704-705*, 750(9),
  764(65), *769, 771*
Ulrich, B.L., 576(99, 100, 105-106),
  577(99), *583*, 634(32), 639(31, 32),
  640(31), 642(31), *650*
Ulrich, J.T., 162(31), *187*
Underbrink-Lyon, K., 85(89), *91*
Underwood, C., 439(92), *456*
Uotila, L., 763(58), *771*
Upchurch, R.G., 173(91), *189*
Urbanski, G.J., 566(34), *579*
Uribe, E., 30, *41*, 489, *514*
Uzzell, T., 519(11), 525(11), *537*

Vadi, H., 752(22), 757(35),
  767(68), *770-771*
Vänngård, T., 252(25), 257(73),
  266, 268
Vagelos, P.R., 551(186), *561*
Vallejos, R.H., 439(91), 449(91),
  456
Vallin, I., 38(117), *43*, 124(11, 12),
  134(41), 135(11), *149*, 151
van Bezooijen, C.F.A., 758, *770*
Van Bruggen, E.F.J., 34(107), *42*,
  82(74), *91*
Van Buuren, K.J.H., 257(71), *268*
Vance, D.E., 551(180), *560*,
  633(95), *652*
Van Dam, K., 136(51), 137(56),
  140(56), *151*, 343(34), *354*
Van Deenen, L.L.M., 546(101),
  551(176, 181), *557, 560*
Van der Drift, C., 382(26), *403*
van der Hoeven, T.A., 548(130,
  138), *558*, 708(4, 7), 709(4, 7,
  10), 711(10), *725, 726*
Van der Kraan, I., 438(87), *456*
Van Dessel, G., 667(45), *680*
Van de Stadt, R.J., 136(51), *151*,
  343(34), *354*